WAVES IN LAYERED MEDIA

APPLIED MATHEMATICS
AND MECHANICS

A Series of Monographs Prepared Under the Auspices of
the Applied Physics Laboratory, The Johns Hopkins University

Volume 1
K. OSWATITSCH: GAS DYNAMICS
ENGLISH VERSION BY G. KUERTI

Volume 2
G. BIRKHOFF and E. H. ZARANTONELLO: JETS, WAKES, AND CAVITIES

Volume 3
R. VON MISES: THEORY OF COMPRESSIBLE FLUID FLOW
REVISED AND COMPLETED BY HILDA GEIRINGER AND G. S. S. LUDFORD

Volume 4
F. L. ALT: ELECTRONIC DIGITAL COMPUTERS—THEIR USE IN SCIENCE AND ENGINEERING

Volume 5
W. D. HAYES and R. F. PROBSTEIN: HYPERSONIC FLOW THEORY

Volume 6
L. M. BREKHOVSKIKH: WAVES IN LAYERED MEDIA
TRANSLATED FROM THE RUSSIAN BY D. LIEBERMAN

ACADEMIC PRESS • PUBLISHERS • NEW YORK • LONDON

WAVES IN LAYERED MEDIA

BY

LEONID M. BREKHOVSKIKH

Director, Acoustics Institute, Academy of Sciences, USSR

TRANSLATED FROM THE RUSSIAN BY

DAVID LIEBERMAN

*Under the direction of the American Institute of Physics and with
the support of the National Science Foundation*

TRANSLATION EDITED BY

ROBERT T. BEYER

Department of Physics, Brown University, Providence, Rhode Island

1960

ACADEMIC PRESS • PUBLISHERS • NEW YORK • LONDON

PRINTED IN GREAT BRITAIN BY JOHN WRIGHT & SONS LTD.,
THE STONEBRIDGE PRESS, BATH ROAD, BRISTOL 4

PREFACE

A SYSTEMATIC exposition of the theory of the propagation of elastic and electromagnetic waves in layered media is given in this monograph. A considerable part of the material originated with the author, and has appeared earlier in a number of journal articles. I have endeavored to present the results of other authors in the spirit of my own, as far as possible, in order to avoid methodological "disharmony". Furthermore as one of my primary tasks, I attempted to give the reader a clear physical picture of the phenomena under investigation. As regards the mathematical rigor, it is possible that it was not attained to a sufficient degree everywhere, partly due to the fear of making the presentation too cumbersome.

The simultaneous presentation of the theory of propagation of elastic and of electromagnetic waves, followed in the book, is quite advantageous, since the same mathematical methods may be applied in both cases. Also, as a result of the common presentation, each region is enriched by the methods applied in the other. Thus, for example, the impedance method developed in acoustics and radio-engineering may be quite successfully applied in calculations of multilayer reflection reduction of optical systems and interference filters.

The bibliography given at the end of the book is comparatively complete, but is far from exhaustive.

I would like to express my deep gratitude to V. A. Polianskii and I. F. Treshchetenkovii, who were of great help in checking the equations and in preparing the manuscript for publication.

February 1, 1956   L. BREKHOVSKIKH

TRANSLATION EDITOR'S PREFACE

The appearance of this work by Professor Brekhovskikh climaxed a great deal of research by him on various aspects of wave propagation through laminated media. The book presents a thorough survey of such wave propagation—both acoustic and electromagnetic. While it is by no means restricted to Russian work in the field, it does present a complete picture of Soviet researches on wave propagation through layered media.

Because of the significance of the subject matter, and because of the insight the text provides to Russian wave propagation studies, the American Institute of Physics undertook the translation, operating with the aid of a grant from the National Science Foundation.

The Editor is joined by the translator in expressing gratitude to Professor Brekhovskikh for his cooperation in the translation, and especially for supplying a list of corrections. They are also grateful to Professor Arnold Schoch, now of C.E.R.N., Geneva, Switzerland, for furnishing original photographs of Figs. 32, 33 and 35.

ROBERT T. BEYER

March 10, 1960.

CONTENTS

Chapter I

PLANE WAVES IN LAYERS

THE theory of wave reflection from interfaces and from layers will be developed in this chapter. Principal attention will be given to plane harmonic waves. The behavior of beams bounded in space and pulses bounded in time are not considered before § 8. In all cases, the media in which the waves propagate are assumed to be homogeneous and to be bounded by parallel planes.

For completeness of presentation, the first sections of the chapter are devoted to relatively simple questions such as plane waves in homogeneous media, the reflection and refraction of waves at an interface, etc. However, even in these sections, the reader will find some comparatively new problems, such as the theory of inhomogeneous plane waves and their refraction and reflection, an analysis of Leontovich's approximate boundary conditions (which are satisfied in cases of so-called "locally reacting" surfaces), and others.

Acoustic and electromagnetic waves will be considered simultaneously.

§ 1. PLANE WAVES IN HOMOGENEOUS UNBOUNDED MEDIA

1. *Fundamental concepts and definitions*

The plane wave is the simplest form of wave motion. The most general analytic expression for a plane wave is the function

$$F\left(\frac{n_x x + n_y y + n_z z}{c} - t\right), \tag{1.1}$$

where n_x, n_y, and n_z are three numbers which satisfy the condition

$$n_x^2 + n_y^2 + n_z^2 = 1,$$

and are the projections on the coordinate axes of the unit vector normal to the wave front, i.e. normal to planes of constant phase.

The function (1.1) is a solution of the wave equation

$$\frac{\partial^2 F}{\partial x^2} + \frac{\partial^2 F}{\partial y^2} + \frac{\partial^2 F}{\partial z^2} = \frac{1}{c^2}\frac{\partial^2 F}{\partial t^2}. \tag{1.2}$$

It describes a disturbance which is propagated through the medium with the velocity c. The form of the wave, which is determined by the form of the function F, remains unchanged as the wave propagates.

We shall use the so-called spectral method for the study of wave and vibrational phenomena, which is quite widely used in physics and engineering. When the principle of superposition holds, the analysis of the behavior of waves of any form can be reduced to the analysis of the behavior of the simplest "harmonic" waves by using the spectral method.

We let

$$\xi = \frac{n_x x + n_y y + n_z z}{c} - t$$

in Eq. 1.1, and represent the function $F(\xi)$ as the real part of a Fourier integral

$$F(\xi) = \mathrm{Re} \int_0^\infty \Phi(\omega)\, e^{i\omega\xi}\, d\omega \tag{1.3}$$

where the symbol Re denotes the real part.

Since the real part of any complex number a can be written in the form $\mathrm{Re}\, a = \frac{1}{2}(a + a^*)$, the last expression can also be written in the form

$$F(\xi) = \frac{1}{2}\int_0^\infty \Phi(\omega)\, e^{i\omega\xi}\, d\omega + \frac{1}{2}\int_0^\infty \Phi^*(\omega)\, e^{-i\omega\xi}\, d\omega. \tag{1.3a}$$

We multiply this expression by $e^{-i\omega'\xi}\, d\xi$ and integrate over ξ from $-\infty$ to $+\infty$. We then easily obtain

$$\Phi(\omega) = \frac{1}{\pi}\int_{-\infty}^{+\infty} F(\xi)\, e^{-i\omega\xi}\, d\xi \tag{1.4}$$

for the spectral density function.†

The integrand in Eq. 1.3, corresponding to a definite value of ω

$$f(\omega, x, y, z, t) = \Phi(\omega) \exp{(i\omega\xi)}$$

$$= \Phi(\omega) \exp{\left[i\omega\left(\frac{n_x x + n_y y + n_z z}{c} - t\right)\right]}, \tag{1.5}$$

represents a plane harmonic wave.

---

† The derivation of Eq. 1.4 becomes especially simple if we use the Dirac function

$$2\pi\delta(x) = \int_{-\infty}^{+\infty} e^{ix\xi}\, d\xi$$

and take account of its fundamental property

$$\int_{-\infty}^{+\infty} \Phi(x)\, \delta(x)\, dx = \Phi(0),$$

where $\Phi(x)$ is a continuous function at $x = 0$.

The Fourier integral and the expression for an individual harmonic wave are written in a complex form. As has already been mentioned, only the real parts of these expressions have physical meaning. Therefore, in the end, a harmonic plane wave must be written as the real part of Eq. 1.5, i.e.

$$A(\omega)\cos\left[\omega\,\frac{n_x x + n_y y + n_z z}{c} - \omega t + \phi(\omega)\right],\qquad(1.6)$$

where, in going from Eq. 1.5 to Eq. 1.6, we represent the (generally complex) function $\Phi(\omega)$ in the form

$$\Phi(\omega) = A(\omega)\,e^{i\phi(\omega)}.$$

We use the usual notation

$$\frac{\omega}{c} = k = \frac{2\pi}{\lambda},\quad kn_x = k_x,\quad kn_y = k_y,\quad kn_z = k_z,$$

where k, k_x, k_y, k_z are the modulus of the propagation vector and its components along the coordinate axes, respectively, and λ is the wavelength. Then Eq. 1.5 can be written in the form

$$f = \Phi(\omega)\exp\left[i(k_x x + k_y y + k_z z - \omega t)\right] = \Phi(\omega)\exp\left[i(\mathbf{k}\cdot\mathbf{r} - \omega t)\right].$$
$$(1.7)$$

Since the time differentiation of a function of this form reduces simply to multiplication by $-i\omega$, the wave equation for f can be written in the form

$$\frac{\partial^2 f}{\partial x^2} + \frac{\partial^2 f}{\partial y^2} + \frac{\partial^2 f}{\partial z^2} + k^2 f = 0.\qquad(1.8)$$

The spectral approach to wave phenomena has attained widespread application as a result of the following features:

1. The comparatively simple analysis of the behavior of each of the harmonic waves.

2. The possibility of the expansion of any wave process into harmonic waves, when the principle of superposition holds.

3. The extremely high monochromaticity of many of the radiators used in practice. As a result, the radiated waves are close to harmonic.

The expansion of a complex wave process into harmonic waves and the reduction of the problem to Eq. 1.8, where the frequency is considered as given, is the most convenient method of analysis when dispersion is present. In this case, even Eq. 1.2 has no meaning because of the vagueness of the meaning of the quantity c.

In what follows we shall consider harmonic waves (Eq. 1.7) almost exclusively; when necessary, we shall use them to construct wave disturbances of more complex forms.

2. *Inhomogeneous plane waves*

The expression for a plane harmonic wave (1.7) admits of an interesting generalization, which will be of importance in the future. It was indicated above that k_x, k_y, k_z are the components of the propagation vector along the coordinate axes, and it was assumed that these quantities could be any triplet of *real* numbers, satisfying the relation

$$k_x^2 + k_y^2 + k_z^2 = k^2. \tag{1.9}$$

We now abandon the graphical description in the treatment of these numbers, and assume that the set k_x, k_y, k_z is a triplet of complex numbers

$$k_x = k_x' + i k_x'', \quad k_y = k_y' + i k_y'', \quad k_z = k_z' + i k_z''. \tag{1.10}$$

We again require that Eq. 1.9 be satisfied for a real value of $k = \omega/c$. Then, as previously, Eq. 1.7 satisfies the wave equation (1.8).

Let us see what is represented by a wave described by Eq. 1.7 with complex values of k_x, k_y, k_z. Substituting Eq. 1.10 into Eq. 1.7, we obtain

$$f = \Phi(\omega) \exp\left[i(k_x' x + k_y' y + k_z' z - \omega t) - (k_x'' x + k_y'' y + k_z'' z)\right]. \tag{1.11}$$

This expression describes a wave with varying amplitude. As is easily seen, planes of constant amplitude of this wave are given by the equation

$$k_x'' x + k_y'' y + k_z'' z = c_1, \tag{1.12}$$

and planes of constant phase by the equation

$$k_x' x + k_y' y + k_z' z = c_2, \tag{1.13}$$

where c_1 and c_2 are constants. It can be shown that the planes of equal phase are orthogonal to the planes of equal amplitude. In fact, substituting Eq. 1.10 into Eq. 1.9, and equating the imaginary parts on both sides of the equation, we obtain.

$$k_x' k_x'' + k_y' k_y'' + k_z' k_z'' = 0.$$

This equation expresses the condition of orthogonality of the families of planes (1.12) and (1.13).

A wave of the form (1.11) is usually called an *inhomogeneous plane wave*. This wave propagates in the direction given by the vector $\mathbf{k}'$ (k_x', k_y', k_z'), and its amplitude falls off in the perpendicular direction.

The coordinate system can always be chosen in such a way that $k_y' = k_y'' = 0$. Then, just as with ordinary plane waves, we can introduce the angle ϑ and set

$$k_x = k \sin \vartheta, \quad k_z = k \cos \vartheta. \tag{1.14}$$

However, since, according to (1.10), k_x and k_z are complex, ϑ will be a complex angle.

Let us consider, for example, the case $\vartheta = (\pi/2) - i\alpha$, where α is real. From (1.14) we obtain $k_x = k \cosh \alpha$, $k_z = ik \sinh \alpha$, and expression (1.7) for a plane wave is written

$$f = \Phi(\omega) \exp(ik \cosh \alpha \cdot x - k \sinh \alpha \cdot z - i\omega t). \tag{1.15}$$

Thus, we obtain a wave which is propagated in the x-direction and has an amplitude which falls off exponentially in the z-direction. As follows from (1.15), the velocity of propagation is

$$\frac{\omega}{k \cosh \alpha} = \frac{c}{\cosh \alpha},$$

i.e. it is always less than the velocity of propagation of the ordinary plane wave c. The corresponding wavelength is $(2\pi/k \cosh \alpha) = \lambda/\cosh \alpha$, i.e. it is less than the wavelength of the ordinary wave at the same frequency. The greater α, the smaller the wavelength and the greater the damping coefficient of the wave in the z-direction.

All that has been said here refers to the case in which there is no absorption in the medium (real k). The introduction of the concept of inhomogeneous waves in the presence of absorption involves no fundamental difficulties. Here, the planes of equal phase and of equal amplitude will naturally no longer be perpendicular to one another.

Running somewhat ahead of ourselves, we shall show that upon refraction of plane waves at an interface, inhomogeneous plane waves can be transformed into ordinary homogeneous waves, and vice versa. This is immediately evident from the law of refraction

$$n \sin \vartheta_1 = \sin \vartheta, \tag{1.16}$$

where n is the index of refraction, ϑ is the angle of incidence and ϑ_1 is the angle of refraction.

If $n < 1$ and $\sin \vartheta > n$, then it follows from (1.16) that $\sin \vartheta_1 > 1$, i.e. ϑ_1 is complex, and the refracted wave is inhomogeneous. This is a well known occurrence and is realized in the case of the total internal reflection of waves.

If, on the contrary, $\sin \vartheta > 1$, i.e. the incident wave is inhomogeneous, but $\sin \vartheta < n$ (in this case, of course, $n > 1$), then we obtain $\sin \vartheta_1 < 1$, i.e. the refracted wave will be of the ordinary homogeneous kind. We shall meet both cases below in § 23, where the refraction of spherical waves is considered. A spherical wave can be expanded into a sum of plane waves, including inhomogeneous plane waves; hence the question of the refraction of the latter arises.

§ 2. THE REFLECTION AND REFRACTION OF ELECTROMAGNETIC WAVES

1. *General relations*

In the investigation of the reflection and refraction of electromagnetic waves, we start from Maxwell's equations for a homogeneous isotropic medium, in Gaussian units:

$$\operatorname{curl} \mathbf{H} = \frac{4\pi\sigma}{c}\mathbf{E} + \frac{\epsilon}{c}\cdot\frac{\partial \mathbf{E}}{\partial t}, \quad \operatorname{div} \mathbf{E} = 0,$$

$$\operatorname{curl} \mathbf{E} = -\frac{\mu}{c}\cdot\frac{\partial \mathbf{H}}{\partial t}, \qquad \operatorname{div} \mathbf{H} = 0. \tag{2.1}$$

In the case of harmonic time dependence of the vectors $\mathbf{E}$ and $\mathbf{H}$, differentiation with respect to time is reduced to multiplication by $-i\omega$. Therefore, Eq. 2.1 can be written in the form

$$\operatorname{curl} \mathbf{H} = -\frac{i\omega}{c}\epsilon'\mathbf{E},$$

$$\operatorname{curl} \mathbf{E} = \frac{i\omega}{c}\mu\mathbf{H}, \tag{2.2}$$

where

$$\epsilon' = \epsilon + \frac{4\pi\sigma}{\omega}i \tag{2.3}$$

denotes the complex dielectric permeability of the medium. Taking the curl of the second of Eqs. 2.2, eliminating curl $\mathbf{H}$ between these equations, and using the relation curl curl $\mathbf{E} = -\nabla^2\mathbf{E}$ (where we have used div $\mathbf{E} = 0$), we obtain the wave equation

$$\nabla^2\mathbf{E} + k^2\mathbf{E} = 0, \tag{2.4}$$

where

$$k = \frac{\omega}{c}\sqrt{(\epsilon'\mu)}. \tag{2.5}$$

The simplest solution of this equation is a plane harmonic wave

$$\mathbf{E} = \mathbf{E}_0 \exp\left[i(\mathbf{k}\cdot\mathbf{r} - \omega t)\right], \tag{2.6}$$

where $\mathbf{E}_0$ is a constant vector.

The corresponding expression for $\mathbf{H}$ is found from the second of Eqs. 2.2

$$\mathbf{H} = -\frac{ic}{\mu\omega}\,\text{curl}\,\mathbf{E},$$

which after substitution of Eq. 2.6 gives

$$\mathbf{H} = \frac{c}{\mu\omega}\,\mathbf{k}\times\mathbf{E}. \tag{2.7}$$

2. *Reflection and refraction at an interface*

We choose a rectangular system of coordinates such that the xy-plane coincides with the boundary separating the media, and the xz-plane

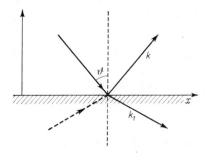

Fig. 1. The reflection and refraction of an electromagnetic wave at a plane boundary.

coincides with the plane of incidence of the wave (here, $k_y = 0$). As is well known, any plane electromagnetic wave can be represented in the form of a superposition of two waves, in one of which the vector $\mathbf{E}$ is perpendicular to the plane of incidence, while in the other it lies in this plane.

We consider first the reflection and refraction of a wave of the first kind. The complex dielectric and magnetic permeabilities in the upper and lower media will be denoted by ϵ', μ and ϵ'_1, μ_1, respectively. The angle of incidence of the wave will be denoted by ϑ, and the angle of refraction by ϑ_1 (Fig. 1).

The electric field in the upper medium can be written in the form

$$E_y = E_0 \exp\left[i(k_x x - k_z z)\right] + V_\perp E_0 \exp\left[i(k_x x + k_z z)\right], \tag{2.8}$$

where the first term represents the incident wave, and the second the reflected wave. The amplitude of the incident wave is denoted by E_0, and the coefficient of reflection by $V_\perp$. The components of the propagation vector $\mathbf{k}$ are denoted by k_x and k_z and are connected with the angle of incidence ϑ by Eqs. 1.14. Substituting (2.8) into (2.7), we obtain

$$H_x = \frac{ck_z E_0}{\mu\omega}\{\exp[i(k_x x - k_z z)] - V_\perp \exp[i(k_x x + k_z z)]\} \qquad (2.9)$$

for H_x in the upper medium. The expression for H_z can be obtained similarly.

In the lower medium, the electric field can be written in the form

$$E_{1y} = W_\perp E_0 \exp[i(k_{1x} x - k_{1z} z)], \qquad (2.10)$$

where $W_\perp$ is the transmission coefficient of the boundary, and

$$k_{1x} = k_1 \sin\vartheta_1, \quad k_{1z} = k_1 \cos\vartheta_1, \quad k_1 = k\sqrt{\left(\frac{\epsilon'_1 \mu_1}{\epsilon' \mu}\right)}. \qquad (2.11)$$

Moreover, using Eq. 2.7, we obtain

$$H_{1x} = \frac{cW_\perp k_{1z}}{\mu_1 \omega} E_0 \exp[i(k_{1x} x - k_{1z} z)]. \qquad (2.12)$$

The constants $V_\perp$, $W_\perp$ (and the angle ϑ also) are determined by the boundary conditions, which, as is well known, require that the tangential components of **E** and **H** be continuous across the boundary, i.e., in this case,

$$E_y = E_{1y}, \quad H_x = H_{1x}. \qquad (2.13)$$

Substituting Eqs. 2.8, 2.9, 2.10, and 2.12 into these equations, we obtain the well known expression for the law of refraction

$$k_{1x} = k_x \quad \text{or} \quad n\sin\vartheta_1 = \sin\vartheta, \qquad (2.14)$$

where

$$n = \frac{k_1}{k} = \sqrt{\left(\frac{\epsilon'_1 \mu_1}{\epsilon' \mu}\right)} \qquad (2.15)$$

is the index of refraction of the boundary. We also obtain expressions for the reflection and transmission coefficients:

$$V_\perp = \frac{(\mu_1/\mu)\cos\vartheta - n\cos\vartheta_1}{(\mu_1/\mu)\cos\vartheta + n\cos\vartheta_1} = \frac{(\mu_1/\mu)\cos\vartheta + \sqrt{(n^2 - \sin^2\vartheta)}}{(\mu_1/\mu)\cos\vartheta - \sqrt{(n^2 - \sin^2\vartheta)}}, \qquad (2.16)$$

$$W_\perp = 1 + V_\perp. \qquad (2.17)$$

In the case of waves polarized in the plane of incidence, the x-component of the electric field in the upper and lower media can be expressed by equations similar to (2.8) and (2.10). We denote the reflection and transmission coefficients in this case by $V_\parallel$ and $W_\parallel$, which indicates that the electric vector is parallel to the plane of incidence.

Proceeding as above, we now obtain

$$V_{\parallel} = \frac{n\cos\vartheta_1 - (\epsilon_1'/\epsilon')\cos\vartheta}{n\cos\vartheta_1 + (\epsilon_1'/\epsilon')\cos\vartheta} = \frac{\sqrt{(n^2-\sin^2\vartheta)} - (\mu/\mu_1)\,n^2\cos\vartheta}{\sqrt{(n^2-\sin^2\vartheta)} + (\mu/\mu_1)\,n^2\cos\vartheta}, \quad (2.18)$$

$$W_{\parallel} = 1 + V_{\parallel}. \quad (2.19)$$

Relation (2.14) remains valid in this case.† Had we defined the reflection coefficient as the ratio of the H_y (rather than the E_x) components in the reflected and incident waves, we would have obtained Eq. 2.18 with the opposite sign.

Let us note one detail, usually omitted in the derivation of the fundamental formulas for the reflected and refracted waves. For definiteness, we again consider the case of polarization perpendicular to the plane of incidence.

We assume a field of the form (2.10) in the lower medium. At the same time, we note that the conditions at the boundary could have been satisfied had we assumed a field of the form

$$E_{1y} = W_{\perp}E_0\exp\left[i(k_{1x}x - k_{1z}z)\right] + W_{\perp}'E_0\exp\left[i(k_{1x}x + k_{1z}z)\right].$$

Here, the boundary conditions would be satisfied by using an infinite sum of the possible values of the coefficients $W_{\perp}$ and $W_{\perp}'$. However, the latter term, containing the coefficient $W_{\perp}'$, cannot be added because of the requirements of a radiation principle of which we have not yet spoken. In the present case, this principle requires that for $z \to -\infty$, the field in the lower medium must correspond to waves leaving the boundary, whereas the additional term corresponds to a wave arriving from infinity (its direction of propagation is shown in Fig. 1 by the dotted arrow).‡

---

† The square root $\sqrt{(n^2-\sin^2\vartheta)}$ enters into the expressions (2.16) and (2.18) for the reflection coefficient. The sign must be chosen such that with complex n, the condition $\mathrm{Im}\sqrt{(n^2-\sin^2\vartheta)} > 0$ is satisfied. In fact, since

$$\sqrt{(n^2-\sin^2\vartheta)} = n\cos\vartheta_1 = (k_1/k)\cos\vartheta_1 = k_{1z}/k,$$

the contrary assumption would give $\mathrm{Im}\,k_{1z} < 0$, and this, according to Eq. 2.10, would lead to an infinite increase of the amplitude of the field as $z \to -\infty$ (in the lower medium).

‡ In the light of these arguments, an interesting situation arises in regard to the refraction of waves when the group and phase velocities in the medium have different signs (the existence of waves in a crystal lattice, possessing this property, was pointed out by L. I. Mandel'shtam). In this case, in the lower medium, of the two possible waves indicated above, only the one providing energy flow away from the boundary will be present. But the phase velocity, on the contrary, will be directed towards the boundary. This will be exactly like the wave shown in Fig. 1 by the dotted arrow, which we have until now left out.

The index of refraction $n = \sqrt{(\epsilon'_1 \mu_1 / \epsilon' \mu)}$ is generally complex. Therefore, the coefficients $V_\perp$, $V_\parallel$, $W_\perp$, $W_\parallel$ are generally complex. This means that not only the amplitude of the wave, but also its phase, changes upon reflection and refraction. It is convenient to use the notation

$$V_\perp = \rho_\perp \exp(i\phi_\perp), \quad V_\parallel = \rho_\parallel \exp(i\phi_\parallel), \tag{2.20}$$

where ρ and ϕ, with appropriate indices, denote the modulus and the phase of the reflection coefficients.

The modulus of the reflection coefficient as a function of the angle of incidence at an air–sea water boundary is shown in Fig. 2. In this case

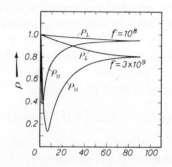

Fig. 2. The modulus of the reflection coefficient of an electromagnetic wave at an air–sea water boundary ($\epsilon = 81$, $\sigma = 3 \times 10^{10}$ cgs electrostatic units) for two frequencies and the various states of polarization. The abscissa represents the grazing angle $\chi = (\pi/2) - \vartheta$.

the index of refraction is complex, and can be determined from the relation

$$n = \sqrt{\left(\epsilon + \frac{4\pi\sigma}{\omega}i\right)}, \tag{2.21}$$

with $\epsilon = 81$ and $\sigma = 3 \times 10^{10}$ cgs (electrostatic).

It is clear from Fig. 2 that in this case the reflection coefficient also depends on the frequency. Two values of the frequency were used in Fig. 2; $f = 10^8$ and 3×10^9 cps. Note that $\rho_\perp$ and $\rho_\parallel$ coincide at $\vartheta = 0$ and $\vartheta = \pi/2$. Between these two values of ϑ, $\rho_\perp$ varies monotonically, whereas $\rho_\parallel$ goes through a minimum at some particular angle. For real n, this minimum degenerates into a cusp lying on the horizontal axis, on which $\rho_\parallel = 0$. The angle at which this minimum occurs is known as Brewster's angle.

When $\vartheta \to \pi/2$, Eqs. 2.16 and 2.18 give $V_\perp \to -1$, and $V_\parallel \to +1$.

It is sometimes convenient to represent the reflection coefficient V in the complex plane, which makes the values of its modulus and phase immediately evident. We separate V into its real and imaginary parts

$$V = A + iB,$$

and let the horizontal and vertical axes represent the real and the imaginary part, respectively.

The reflection coefficient $V_\parallel$ for the case $n > 1$, $\mu = \mu_1 = 1$ is represented in the complex plane by the heavy line in Fig. 3. According to

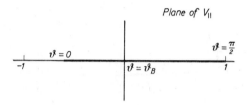

Fig. 3. The reflection coefficient for the case $n > 1$.

Eq. 2.18, $V_\parallel$ is real for all ϑ. It is therefore shown as a segment of the real axis. Its value varies from some finite value $V_\parallel = (1-n)/(1+n) < 1$ at $\vartheta = 0$ to $V_\parallel = 0$ at Brewster's angle $\vartheta = \vartheta_b = \arctan n$, and approaches $V_\parallel = 1$ as the angle of incidence is increased further toward $\vartheta = \pi/2$.

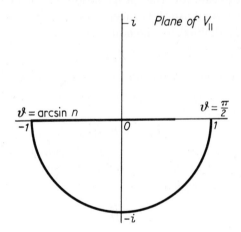

Fig. 4. The reflection coefficient for the case $n < 1$.

The case $n < 1$ is shown in Fig. 4. Starting at $\vartheta = 0$, the reflection coefficient decreases and becomes equal to -1 at $\sin \vartheta = n$ (see Eq. 2.18, again). As ϑ is increased further, the modulus remains equal to 1, and only the phase changes. This is the case of total internal reflection.

As $\vartheta \to \pi/2$, we again have $V_\parallel \to 1$. As a result, all the values lie on the segment of a straight line and on the semicircle.

Finally, $V_\parallel$ and $V_\perp$ are shown in Fig. 5 for the real case of the reflection of an electromagnetic wave of frequency $f = 3 \times 10^9$ cps from a sea water surface. The numbers on the curves indicate the values of the angle of incidence in degrees.

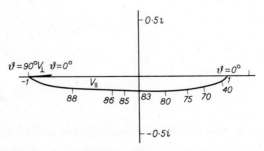

Fig. 5. The reflection coefficients $V_\parallel$ and $V_\perp$ for both polarizations of an electromagnetic wave of frequency $f = 3 \times 10^9$ cps undergoing reflection at a sea water surface. The numbers on the curves indicate the angle of incidence in degrees.

3. *The concept of impedance*

In the investigation of the propagation of plane waves in unbounded space, we shall use the concept of the characteristic impedance of the medium Z_0, defined by the equation

$$E = Z_0 H,$$

where E and H are the amplitudes of the electric and magnetic field vectors in a plane wave. Using Eq. 2.5 and the orthogonality of the vectors $\mathbf{E}$ and $\mathbf{k}$, Eq. 2.7 gives

$$Z_0 = \sqrt{(\mu/\epsilon')}.$$

In the problem of wave reflection from plane boundaries, the normal impedance Z, defined as the ratio of the tangential components of the electric and magnetic fields, i.e.

$$Z = \frac{E_t}{H_t},$$

is extremely useful.

Let us consider a plane wave, impinging on the boundary at an angle ϑ. When the vector $\mathbf{E}$ is parallel to the boundary (horizontal polarization), we have $E_t = E$, $H_t = H \cos \vartheta$, and consequently,

$$Z = \frac{E}{H \cos \vartheta} = \frac{Z_0}{\cos \vartheta}. \tag{2.22}$$

For the reflected wave, the ratio E_t/H_t will be equal to $-Z$. In the case of vertical polarization, we obtain, similarly,

$$Z = Z_0 \cos \vartheta. \tag{2.23}$$

In the case of normal incidence ($\cos \vartheta = 1$) the normal impedance coincides in absolute value with the characteristic impedance of the medium. The use of impedance simplifies the solution of the problem of the reflection of a plane electromagnetic wave at an interface. As an example, we consider the case in which $\mathbf{E}$ is perpendicular to the plane of incidence, and use the notation:

$Z = (1/\cos \vartheta) \sqrt{(\mu/\epsilon')}$ is the normal impedance of the medium through which the incident wave travels (the "upper" medium),

$Z_1 = (1/\cos \vartheta_1) \sqrt{(\mu_1/\epsilon_1')}$ is the normal impedance of the medium from which the wave is reflected,

E_t^0 is the amplitude of the electric field in the incident wave.

The total electric field (directed parallel to the boundary) in the upper medium is

$$E_t = E_t^0[\exp(-ikz \cos \vartheta) + V_\perp \exp(ikz \cos \vartheta)] \exp(ikx \sin \vartheta),$$

where $V_\perp$ is the reflection coefficient.

Setting $z = 0$, we have

$$E_t = E_t^0(1 + V_\perp) \exp(ikx \sin \vartheta)$$

at the boundary. We obtain

$$H_t = E_t^0 \frac{1}{Z} (1 - V_\perp) \exp(ikx \sin \vartheta),$$

for the tangential component of the magnetic field at the boundary.

Hence, for the ratio of the quantities E_t and H_t at the boundary, we obtain

$$\frac{E_t}{H_t} = Z \frac{1 + V_\perp}{1 - V_\perp}.$$

However, since the tangential components of the field, i.e. E_t and H_t, are continuous across the boundary, the last expression must be equal to Z_1, the ratio of these components in the lower medium, i.e.

$$Z \frac{1 + V_\perp}{1 - V_\perp} = Z_1,$$

whence

$$V_\perp = \frac{Z_1 - Z}{Z_1 + Z}. \tag{2.24}$$

Substituting for Z and Z_1 in this equation, we find an expression coinciding with Eq. 2.16. We can also obtain the reflection coefficient $V_{\parallel}$ from Eq. 2.24. However, the value of the impedance will be different (see Eq. 2.23). Thus, we can write the important general relation for the reflection coefficient

$$V = \frac{Z_1 - Z}{Z_1 + Z}. \tag{2.25}$$

The transmission coefficient, according to Eqs. 2.17 and 2.19, can be written

$$W = \frac{2Z}{Z + Z_1}. \tag{2.26}$$

4. The Leontovich approximate boundary conditions

As we have already seen, the reflection coefficient can be determined from the single condition

$$\frac{E_t}{H_t} = Z_1, \tag{2.27}$$

where E_t and H_t are the tangential components of the field in the upper medium, and Z_1 is the normal impedance in the lower medium.

This boundary condition is unusual in that only quantities characterizing the field in the upper medium enter into it. It would seem that Eq. 2.27 would have meaning only for plane waves, since ϑ_1, the angle of refraction for plane waves, enters in the definition of Z_1. Indeed, an arbitrary wave field can be represented as a superposition of plane waves with various angles of incidence, so that a Z_1 referring to a definite ϑ_1 would seem to lose all meaning.

However, this is not completely true. Let us look a little more closely at the quantity ϑ_1 for any of the plane waves. From Eq. 2.14, we have

$$\cos \vartheta_1 = \sqrt{\left(1 - \frac{\sin^2 \vartheta}{n^2}\right)}. \tag{2.28}$$

In a number of important cases, in particular, in the propagation of radio waves over well conducting ground, the modulus of n^2 is quite large, so that the term $\sin^2 \vartheta / n^2$ can be neglected with respect to unity. Then $\cos \vartheta \approx 1$, independently of the angle of incidence of the wave, and we have

$$Z_1 = \sqrt{\frac{\mu_1}{\epsilon_1'}},$$

for Z_1, while the boundary condition can be written

$$\frac{E_t}{H_t} = \sqrt{\frac{\mu_1}{\epsilon_1'}}. \tag{2.29}$$

In this form the condition contains nothing which refers specifically to plane waves and can be applied to the general case. This boundary condition, which was first suggested by M. A. Leontovich, has been used successfully for the solution of a number of problems in the theory of radio wave propagation (see Refs. 57 and 58). In comparison with the usual pair of exact boundary conditions, expressing the continuity of the tangential components of E_t and H_t, this boundary condition has the advantage that it contains only quantities characterizing the field in the upper medium. As we see, the criterion for the validity of boundary condition (2.29) is the absence of waves not satisfying the condition $(\sin^2 \vartheta / n^2) \ll 1$ in the expansion of the field into plane waves.

§ 3. The Reflection of a Plane Sound Wave at an Interface Separating Liquid and Gaseous Media

1. *The reflection and transmission coefficients*

We shall characterize the sound field by the acoustic potential ψ. The particle velocity and acoustic pressure in a harmonic wave (time

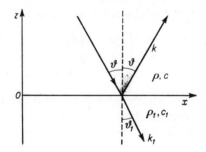

Fig. 6. The reflection and refraction of a sound wave.

dependence given by the factor $e^{-i\omega t}$) will be expressed through ψ by the equations[9, 25]

$$\mathbf{v} = -\operatorname{grad} \psi, \quad p = -i\omega\rho\psi. \tag{3.1}$$

We consider the problem of the reflection of a plane sound wave at a plane boundary separating two media. The density of the medium from which the wave is incident (in what follows, this medium will be called the upper medium) will be denoted by ρ, and the acoustic velocity by c. The corresponding quantities in the lower medium will be denoted by ρ_1 and c_1. As in § 2, the angle of incidence will be denoted by ϑ, and the angle of refraction by ϑ_1. We assume that the normal to the wave front lies in the plane of the diagram (Fig. 6).

With these assumptions, and omitting the factor $e^{-i\omega t}$, the incident wave can be written in the form

$$\psi_{inc} = A \exp\left[ik(x \sin\vartheta - z \cos\vartheta)\right], \tag{3.2}$$

where A is the amplitude of the wave. The reflected wave can be written in the form

$$\psi_{ref} = VA \exp\left[ik(x \sin\vartheta + z \cos\vartheta)\right], \tag{3.3}$$

where V is the reflection coefficient. The total potential in the upper medium will be

$$\psi = \psi_{inc} + \psi_{ref} = A\left[\exp\left(-ikz\cos\vartheta\right) + V\exp\left(ikz\cos\vartheta\right)\right]\exp\left(ikx\sin\vartheta\right). \tag{3.4}$$

The refracted wave can be written in the form†

$$\psi_1 = WA \exp\left[ik_1(x \sin\vartheta_1 - z\cos\vartheta_1)\right], \tag{3.5}$$

where W is the transmission coefficient, and $k_1 = \omega/c_1$ is the wave number in the lower medium.

The acoustic pressure and the normal component of the particle velocity must be continuous across the boundary of separation $z = 0$. Taking account of Eq. 3.1, these conditions can be written in the form

$$z = 0 \begin{cases} \rho\psi = \rho_1\psi_1, & (3.6) \\[2mm] \dfrac{\partial\psi}{\partial z} = \dfrac{\partial\psi_1}{\partial z}, & (3.7) \end{cases}$$

from which the two unknown coefficients V and W, as well as the angle of refraction ϑ, can be determined.

Substituting Eqs. 3.4 and 3.5 into 3.6, we obtain

$$\frac{\rho}{\rho_1}(1 + V) = W \exp\left[i(k_1\sin\vartheta_1 - k\sin\vartheta)x\right]. \tag{3.8}$$

Since the left hand side is independent of x, the right hand side must also be independent of x, whence we obtain the well known refraction law

$$k \sin\vartheta = k_1 \sin\vartheta_1. \tag{3.9}$$

This relation expresses the equality of the phase velocities of waves propagating along the interface in the lower and the upper media. It can also be written in the form

$$\frac{\sin\vartheta}{\sin\vartheta_1} = n, \quad \text{where } n = \frac{k_1}{k} = \frac{c}{c_1}. \tag{3.10}$$

---

† In prescribing such a refracted wave, we have implicitly used the radiation principle. In this regard, see § 2, sec. 2.

Then Eq. 3.8 takes the form

$$W = \frac{1}{m}(1 + V), \quad m = \frac{\rho_1}{\rho}. \tag{3.11}$$

Moreover, the substitution of Eqs. 3.4 and 3.5 into 3.7 gives

$$\cos \vartheta (1 - V) = n \cos \vartheta_1 W. \tag{3.12}$$

From the last two relations, we find

$$V = \frac{m \cos \vartheta - n \cos \vartheta_1}{m \cos \vartheta + n \cos \vartheta_1}, \tag{3.13}$$

or, taking account of Eq. 3.10,

$$V = \frac{m \cos \vartheta - \sqrt{(n^2 - \sin^2 \vartheta)}}{m \cos \vartheta + \sqrt{(n^2 - \sin^2 \vartheta)}}. \tag{3.14}$$

Let us analyze the expressions we have obtained for the reflection and transmission coefficients. At normal incidence of the wave ($\vartheta = \vartheta_1 = 0$), Eq. 3.13 gives

$$V = \frac{m - n}{m + n} = \frac{\rho_1 c_1 - \rho c}{\rho_1 c_1 + \rho c}. \tag{3.15}$$

The quantity $z = \rho c$ is called the wave resistance or the characteristic impedance of the medium. (The impedance is considered in more detail in §3, sec. 4.) Using impedance, the reflection coefficient can also be written in the form

$$V = \frac{Z_1 - Z}{Z_1 + Z}. \tag{3.16}$$

This formula remains valid for oblique incidence, if we use

$$Z = \frac{\rho c}{\cos \vartheta}, \quad Z_1 = \frac{\rho_1 c_1}{\cos \vartheta_1}, \tag{3.17}$$

for the impedances Z and Z_1. This is easily verified by starting with Eq. 3.13 and taking account of the definitions of m and n. As $\vartheta \to \pi/2$ we have $Z \to \infty$, $V \to -1$, $W \to 0$, i.e. the reflection coefficient approaches -1, and the transmission coefficient approaches zero. In essence, this result indicates the impossibility of the existence of plane waves propagating along the boundary, since in such a case the incident and reflected waves would completely cancel one another. It will be shown in §4 that the situation will be different at a boundary between a liquid and a *solid medium*. As follows from Eq. 3.14, the reflection coefficient will become zero at an angle ϑ satisfying the equation

$$m \cos \vartheta - \sqrt{(n^2 - \sin^2 \vartheta)} = 0.$$

In this case, there is no reflected wave, and the boundary will be completely transparent. From the last equation, we find $\vartheta = \vartheta_b$, where

$$\sin \vartheta_b = \sqrt{\left(\frac{m^2 - n^2}{m^2 - 1}\right)}. \tag{3.18}$$

This angle is analogous to Brewster's angle for electromagnetic waves. It should be kept in mind that ϑ_b will not necessarily be a real angle, and complete transmission will not necessarily be observed for any relation between the parameters of the media, which are characterized by the quantities m and n. Rather, as is seen from Eq. 3.18, the condition

$$0 < \frac{m^2 - n^2}{m^2 - 1} < 1$$

must be satisfied. Hence, when $m > 1$, we must have $1 < n < m$, and when $m < 1$, we must have $1 > n > m$.

2. *Total internal reflection*

When $n < 1$ $(c_1 > c)$, and the angle of incidence satisfies the condition $\sin \vartheta > n$, total internal reflection will occur. In this case, Eq. 3.14 can be written

$$V = \frac{m \cos \vartheta - i \sqrt{(\sin^2 \vartheta - n^2)}}{m \cos \vartheta + i \sqrt{(\sin^2 \vartheta - n^2)}}, \tag{3.19}$$

or, if we represent the reflection coefficient in the form

$$V = \rho \, e^{i\phi},$$

where ρ is its modulus (not to be confused with the density ρ) and ϕ is its phase, then

$$\rho = 1, \quad \phi = -2 \arctan \frac{\sqrt{(\sin^2 \vartheta - n^2)}}{m \cos \vartheta}, \tag{3.20}$$

and the square root is assumed to be positive.

Under conditions of total internal reflection, as we shall see below (§§ 8 and 9), the relation between the phase of the reflection coefficient and the angle of incidence causes extremely interesting effects in the reflection of bounded pencils of rays, and also pulses bounded in time.

Borrowing the terminology used in transmission line theory, we can say that at total internal reflection, the boundary presents a purely reactive (inductive) impedance to the incident wave, while in the usual case, the impedance is purely resistive.

It is convenient to draw the reflection coefficient V in the complex plane, just as was done in § 2 for electromagnetic waves. Letting the abscissa represent the real part and the ordinate represent the imaginary part of the reflection coefficient, we obtain curves such as those shown in Fig. 7 for various relations among the parameters of the two media (cases a–d). It is assumed here that attenuation in the medium can be neglected.

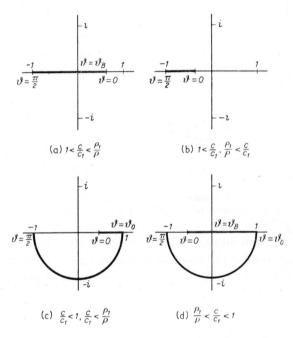

Fig. 7. The reflection coefficient (in the complex plane) of a sound wave for various relations among the parameters of the media.

In cases a and b $(n > 1)$ the reflection coefficient is real. Therefore, for various values of ϑ, it is confined to a segment of a straight line lying on the real axis. In case a, the reflection coefficient is zero for some value $\vartheta = \vartheta_b$ of the angle of incidence. In case b, it is negative for all values $0 \leqslant \vartheta \leqslant \pi/2$, and is never zero.

Cases c and d $(n < 1)$ correspond to total internal reflection. Here, when $\vartheta > \vartheta_0 = \arcsin n$, the points corresponding to the complex values of the reflection coefficient lie on a semicircle of unit radius. This shows graphically that the modulus of the reflection coefficient is unity, and that as ϑ varies, only its phase changes.

The relation between the modulus and phase of the reflection coefficient and the angle of incidence is shown in Fig. 8 for the case $m = \rho_1/\rho = 2.7$, $n = c/c_1 = 0.83$ (solid curves), which corresponds, for example, to the reflection of a wave incident from water on to a packed sandy ocean bottom.[13]

At angles of incidence greater than 56°, we have $\rho = 1$; this is the region of total internal reflection. As the angle of incidence decreases, beginning at 56°, the modulus of the reflection coefficient decreases sharply. Here, the phase is equal to zero.

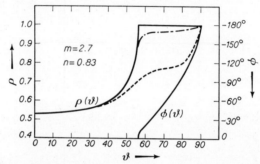

Fig. 8. The modulus and phase of the reflection coefficient of a sound wave at a water–packed sea sand interface, with different wave attenuations in the sand.

Total internal reflection will not occur at any angle of incidence, if we take into account the absorption of sound in the medium from which reflection takes place. Indeed, when absorption is present, n will be complex:

$$n = n_0(1 + i\alpha). \tag{3.21}$$

We consider the case $\alpha \ll 1$, so that

$$n^2 = n_0^2(1 + 2i\alpha),$$

and use the notation

$$\sin^2 \vartheta - n_0^2 = A, \quad 2n_0^2 \alpha = B. \tag{3.22}$$

Then, taking into account that

$$\surd(A - iB) = M_1 + iM_2, \tag{3.23}$$

where

$$M_1 = \frac{1}{\sqrt{2}} \surd[\surd(A^2 + B^2) + A], \quad M_2 = -\frac{1}{\sqrt{2}} \surd[\surd(A^2 + B^2) - A],$$

Eq. 3.14 gives

$$V = \frac{m\cos\vartheta + M_2 - iM_1}{m\cos\vartheta - M_2 + iM_1}, \tag{3.24}$$

whence

$$\rho^2 = |V|^2 = \frac{(m\cos\vartheta + M_2)^2 + M_1^2}{(m\cos\vartheta - M_2)^2 + M_1^2}. \tag{3.25}$$

The relation between ρ and the angle ϑ is shown in Fig. 8 for $n = 0.83$ and $m = 2.7$, with $\alpha = 0.1$ (dashed curve) and $\alpha = 0.01$ (dot-dash curve). It is clear from the Figure that ρ decreases rather rapidly as the angle ϑ departs from $90°$, and also that ρ is less than unity for all angles $\vartheta \neq 90°$.

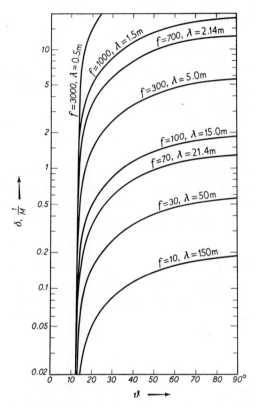

Fig. 9. Nomogram for the determination of the penetration depth of a sound wave into the water under conditions of total internal reflection at an air–water interface.

Let us now consider the refracted wave in the case of total internal reflection. From Eq. 3.9 we have

$$k_1 \cos \vartheta_1 = k \sqrt{(n^2 - \sin^2 \vartheta)} = ik \sqrt{(\sin^2 \vartheta - n^2)}.$$

Making use of Eqs. 3.5 and 3.11, we obtain for the potential in the lower medium

$$\psi_1 = \frac{1}{m} A(1 + V) \exp \left[ikx \sin \vartheta + k \sqrt{(\sin^2 \vartheta - n^2} z) \right]. \qquad (3.26)$$

Thus, we have an inhomogeneous wave (see §1), the amplitude of which decreases exponentially with distance from the boundary of separation (in the negative z direction).

Let us consider, for example, a sound wave incident from air on a surface of water.

We shall use $\rho = 1.3 \times 10^{-3}$ gm/cm³, $\rho_1 = 1$ gm/cm³, $c = 333$ m/sec, $c_1 = 1500$ m/sec. Then $m = \rho_1/\rho \approx 880$, $n = 0.22$.

Total internal reflection will occur at angles of incidence $\vartheta > \arcsin 0.22 = 13°42'$. Since m is quite large, Eq. 3.17 yields, approximately, $V \approx 1$ for any angle ϑ. Therefore, according to Eq. 3.26, we can write

$$|\psi_1| = \frac{2}{m} A \, e^{-\delta D}, \qquad (3.27)$$

for the wave amplitude $|\psi_1|$ in the lower medium, where we have used $\delta = k\sqrt{(\sin^2 \vartheta - n^2)}$, and $D \equiv -z$.

The "attenuation exponent" δ as a function of the angle of incidence for various frequencies is given as a nomogram in Fig. 9. For example, at a frequency of 100 cps and an angle of incidence $\vartheta = 45°$, we obtain $\delta = 1.3$ m⁻¹, i.e. the wave is attenuated by a factor of e at a depth of 77 cm of water.

3. *Energy relations*

According to what has been presented above, the amplitude of the acoustic potential in the transmitted wave will exceed that of the incident wave by a factor $(1/m)(1 + V)$ (see Eq. 3.11). Since the potential is multiplied by the density ρ to obtain the acoustic pressure (see Eq. 3.1), the acoustic pressure in the transmitted wave will be greater than that of the incident wave by a factor $(1 + V)$. Thus, in the example considered above of the reflection of a wave incident from air on a surface of water, we have $V \approx 1$, and the amplitude of the acoustic pressure in the water will be twice as great as the acoustic pressure in the incident wave.

On the other hand, if a wave is incident from the water on its boundary with air, then, since the density of air is considerably less than that of water, i.e. $m = \rho_1/\rho \ll 1$, we have $V \approx -1$, $W = 1 + V \approx 0$, and therefore the acoustic pressure in the wave transmitted into the air will be much smaller than the acoustic pressure in the incident wave. Thus, there is no symmetry with respect to the values of the acoustic pressure in the transmission of a sound wave from one medium to another and inversely.

However, as regards the energy flux, symmetry must appear in one form or another. We shall now go on to the consideration of this question.

Let us consider the intensity in the incident, reflected and refracted waves. As is well known, the intensity, given by the absolute value of the Umov vector, is

$$I = \frac{|p|^2}{2\rho c} = \frac{(\omega\rho)^2}{2} \cdot \frac{|\psi|^2}{\rho c}, \qquad (3.28)$$

where ρ and c are the density and acoustic velocity in the medium in which the energy flux is being calculated, and $|\psi|$ is the amplitude of the acoustic potential. Taking into account that the amplitudes of the incident, reflected and refracted (transmitted) waves are

$$|\psi_{\text{inc}}| = A, \quad |\psi_{\text{refl}}| = |V|A, \quad |\psi_{\text{tr}}| = \frac{\rho}{\rho_1}|1+V|A,$$

we obtain

$$I_{\text{inc}} = \frac{(\omega\rho)^2 A^2}{2\rho c}, \quad I_{\text{refl}} = \frac{(\omega\rho)^2 |V|^2 A^2}{2\rho c}, \quad I_{\text{tr}} = \frac{(\omega\rho)^2 |1+V|^2 A^2}{2\rho_1 c_1},$$

for the intensity in the directions of the incident, reflected and refracted waves, respectively.

The ratio of the intensity in the transmitted and the incident waves is then

$$\frac{I_{\text{tr}}}{I_{\text{inc}}} = \frac{\rho c}{\rho_1 c_1}|1+V|^2. \qquad (3.29)$$

We shall begin with the case of normal incidence. Using Eq. 3.15 for the reflection coefficient, the last equation gives

$$\frac{I_{\text{tr}}}{I_{\text{inc}}} = \frac{4nm}{(m+n)^2}.$$

It is thus quite apparent that there is complete symmetry as regards the passage of energy from one medium to another and the reverse. Indeed, if we interchange the medium numbers, i.e. we assume that the wave is now incident from the medium into which it was previously transmitted, we must make the replacements $\rho \to \rho_1$, $c \to c_1$, and the reverse, i.e. $m \to 1/m$ and $n \to 1/n$. Then the last equation gives

$$\frac{I_{\text{tr}}}{I_{\text{inc}}} \to \frac{4(1/nm)}{[(1/m)+(1/n)]^2} = \frac{4mn}{(m+n)^2},$$

i.e. the ratio of the intensities in the incident and the transmitted waves remains unchanged.

The picture is somewhat more complicated in the case of oblique incidence. In this case the intensities in the two media are both different, for the following two reasons:

(a) the reflection of a portion of the acoustic energy at the boundary;

(b) the change of area of the ray tube upon refraction.

Referring to Fig. 10, we see that even if all of the acoustic energy contained in the ray bundle $ABA'B'$ were transmitted into the lower medium without reflection, the intensity would decrease upon transmission from the upper to the lower medium because in the lower medium the energy would be distributed over the ray bundle $BCB'C'$, which in this case has a greater transverse cross sectional area than the ray bundle in the upper medium.

According to Fig. 10, the increase of the cross sectional area of the ray bundle upon transition from the upper to the lower medium is described by the factor

$$\frac{B'D}{BD'} = \frac{BB'\cos\vartheta_1}{BB'\cos\vartheta} = \frac{\cos\vartheta_1}{\cos\vartheta}.$$

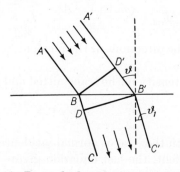

Fig. 10. Ray tube broadening at refraction.

The intensity in the lower medium is obtained by dividing by this quantity, or by multiplying by its inverse, $\cos\vartheta/\cos\vartheta_1$.

One would expect that if the factor $\cos\vartheta/\cos\vartheta_1$, which arises because of the change of the transverse cross section of the ray tube, were separated out of the expression for the ratio of the intensities I_{tr}/I_{inc}, the remaining part, characterizing the penetrability of the boundary to sound waves, would remain unchanged if the order of the media were interchanged.

We shall prove that this is actually the case. Substituting Eq. 3.13 for the reflection coefficient at an arbitrary angle of incidence into Eq. 3.29, we obtain†

$$\frac{I_{tr}}{I_{inc}} = \frac{4nm\cos^2\vartheta}{(m\cos\vartheta + n\cos\vartheta_1)^2} = \frac{\cos\vartheta}{\cos\vartheta_1}\frac{4nm\cos\vartheta\cos\vartheta_1}{(m\cos\vartheta + n\cos\vartheta_1)^2}.$$

† Here, we exclude the case of total internal reflection (complex ϑ), in which the energy flux in the lower medium will be only in a direction parallel to the boundary.

Clearly, the second factor in the last expression remains unchanged if we reverse the numbers of the media and the direction of the rays, i.e. we make the replacements $m \to 1/m$, $n \to 1/n$, $\vartheta \gtrless \vartheta_1$, which completes the proof.

It is clear that, aside from symmetry considerations, a definite connection between the reflection and transmission coefficients must also be a consequence of the law of conservation of energy. We may formulate the law of conservation of energy as follows: *The energy carried to the boundary by the incident wave must be equal to the energy carried away from the boundary by the reflected and refracted waves.* Of the two components of the energy flux (normal and tangential to the boundary), only the normal component will enter into the law of conservation of energy.

Taking into account the expression (Eq. 3.28) for the intensity along the ray, and also the amplitudes of the incident, reflected and refracted rays A, VA, and WA, we obtain a mathematical expression for the law of conservation of energy:

$$\frac{\cos \vartheta}{\rho c} = \frac{V^2}{\rho c} \cos \vartheta + \frac{W^2 m^2}{\rho_1 c_1} \cos \vartheta_1.$$

Here, the multiplication by $\cos \vartheta$ and $\cos \vartheta_1$ corresponds to the projection of the intensity on to the normal to the boundary. From the last relation, we obtain

$$m \cdot W^2 = (1 - V^2) \frac{\cos \vartheta}{n \cos \vartheta_1}.$$

This equation is consistent with the values for W and V obtained earlier. In particular, if we use the connection between W and V given by Eq. 3.11, the latter equation immediately gives the expression for the reflection coefficient (Eq. 3.14).

4. *Locally reacting surfaces. Impedance*

It was indicated above that the reflection coefficient can be expressed in a very simple way through the impedances of the media by using Eq. 3.16.

The physical meaning of impedance is quite simple. We will show that the quantity Z is nothing else than the ratio of the acoustic pressure to the normal component of the particle velocity in a plane wave propagating in the direction of the positive z-axis. Indeed, the particle velocity in a plane wave is obtained from Eq. 3.1:

$$\mathbf{v} = -\frac{i}{\omega \rho} \operatorname{grad} p = \frac{\mathbf{k} p}{\omega \rho}. \tag{3.30}$$

Hence, remembering that $k = \omega/c$, we obtain

$$v_n = \frac{p}{\rho c}\cos\vartheta, \quad \text{or} \quad \frac{p}{v_n} = \frac{\rho c}{\cos\vartheta},$$

which completes the proof. If the direction of propagation is reversed, the ratio p/v_n will be equal to $-Z$.

In a number of cases, the impedance of the medium turns out to be independent of the angle of incidence. From the law of refraction (3.10) we have $\cos\vartheta_1 = \sqrt{[1-(\sin^2\vartheta/n^2)]}$. When $n \gg 1$, i.e. when the acoustic velocity in the lower medium is considerably less than in the upper, we obtain $\cos\vartheta_1 \approx 1$, and consequently,

$$Z_1 = \frac{\rho_1 c_1}{\cos\vartheta_1} = \rho_1 c_1.$$

The use of this simplified impedance simplifies the boundary conditions and makes it much easier to obtain solutions of all kinds of boundary value problems, including the reflection of plane waves, considered above.

Indeed, in this case, the two boundary conditions (3.6) and (3.7) can be replaced by the single condition

$$\frac{p}{v_n} = -Z_1. \tag{3.31}$$

This equation is a consequence of the fact that we have $p_1/v_{1n} = -Z_1$ near the boundary in the lower medium; the minus sign occurs because in the lower medium the refracted wave propagates in the direction of negative z. Furthermore, since p and v_n remain unchanged as we cross the boundary, the quantity $-Z$ must also be equal to the ratio p/v_n.

The boundary condition (3.31) is remarkable in that only the quantities p and v_n characterizing the field of the sound wave in the upper medium are involved.

The solution of the problem of the reflection of a plane sound wave, using condition (3.31), is carried out in the following simple manner.

In accordance with Eqs. 3.2 and 3.3, the total field of the direct and reflected waves is written

$$\psi = A[\exp(-ikz\cos\vartheta) + V\exp(ikz\cos\vartheta)]\exp(ikx\sin\vartheta),$$

whence

$$v_n = v_z = -\frac{\partial\psi}{\partial z}$$

$$= ikA\cos\vartheta\,[\exp(-ikz\cos\vartheta) - V\exp(ikz\cos\vartheta)]\exp(ikx\sin\vartheta),$$

$$p = -i\omega\rho\psi = -i\omega\rho A[\exp(-ikz\cos\vartheta) + V\exp(ikz\cos\vartheta)]\exp(ikx\sin\vartheta).$$

Substituting p and v_n into Eq. 3.31, we obtain a single equation for the reflection coefficient V, the solution of which is

$$V = \frac{\rho_1 c_1 - (\rho c/\cos \vartheta)}{\rho_1 c_1 + (\rho c/\cos \vartheta)},$$

which coincides with Eq. 3.16, if we take account of the values of Z_1 and Z.

There is, in acoustics, a class of anisotropic media for which the boundary condition of the form (3.31), with an impedance Z_1 independent of angle, is not approximate but exact. As an example, we consider a wave incident from air on to a medium with open vertical pores (Fig. 11), the number of which is great compared to the wavelength. In this case, the normal component of the particle velocity, equal to the particle velocity in the pores, will evidently depend only on the acoustic pressure

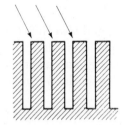

Fig. 11. Model of a surface at which the reflection is characterized by an impedance independent of angle.

at the particular point of the surface, and will not depend on the angle of incidence of the wave. This is valid for all types of media in which the acoustic disturbance is not transmitted along the boundary of the reflecting plane, and consequently the normal component of the particle velocity at each point of the surface will be determined exclusively by the local value of the acoustic pressure at that point.

Surfaces of separation between media, satisfying this condition, are called locally reacting surfaces. They are met quite frequently in architectural acoustics.

It is now clear why the boundary could be considered locally reacting and the simplified expression for the impedance could be used in the investigation of wave reflection from a boundary when the condition $\sin^2 \vartheta/n^2 \ll 1$ is satisfied. The refracted wave travels practically along the normal to the boundary, and does not propagate along the boundary.

The reader will find a more complete analysis of the applicability of the concept of an impedance independent of the angle of incidence in Ref. 34.

§ 4. THE REFLECTION OF WAVES AT A LIQUID–SOLID INTERFACE. SURFACE WAVES

In this section we generalize the problem considered above by assuming that one of the two mutually bounding media is a solid. Later (see § 6), we shall also consider the penetration of waves through a solid plate and through an arbitrary system of solid layers.

The system of reflection and transmission coefficients for a boundary between two solid media is considered in §§ 24 and 32. This question is also considered in seismological literature, for example, Refs. 17, 143, 162.

1. *Fundamental equations and boundary conditions*

The particle velocity at any point of a solid medium can be expressed through a scalar and a vector potential, using the equation

$$\mathbf{v} = \operatorname{grad} \phi + \operatorname{curl} \psi \tag{4.1}$$

(see Ref. 24, Chap. XII, § 1, and also Ref. 17).

In the special case of a plane problem, assuming that all quantities depend only on the coordinates x and z, and that the particle trajectories also lie in the xz-plane, the potential ψ can be chosen such that only its y-component, which we will denote by ψ, differs from zero. Then, according to Eq. 4.1, $\mathbf{v}$ will be a vector with the components

$$v_x = \frac{\partial \phi}{\partial x} - \frac{\partial \psi}{\partial z}, \quad v_y = 0, \quad v_z = \frac{\partial \phi}{\partial z} + \frac{\partial \psi}{\partial x}; \tag{4.2}$$

ϕ and ψ can be called the potentials of longitudinal and transverse (or shear) waves. It can be shown that these potentials will satisfy the wave equations

$$\nabla^2 \phi = \frac{1}{c^2} \frac{\partial^2 \phi}{\partial t^2}, \quad \nabla^2 \psi = \frac{1}{b^2} \frac{\partial^2 \psi}{\partial t^2}, \tag{4.3}$$

where c and b are the velocities of the longitudinal and the transverse waves, respectively. They may be expressed in terms of the Lamé parameters λ and μ and the density ρ of the solid:

$$c = \sqrt{\left(\frac{\lambda + 2\mu}{\rho}\right)}, \quad b = \sqrt{\frac{\mu}{\rho}}. \tag{4.4}$$

The normal components of the stress and displacement must be continuous across the boundary between the solid and the liquid. The tangential components of the stress tensor must also be continuous, but since the tangential stresses in the liquid vanish, this condition reduces simply to the requirement that the tangential components of the stress tensor be zero at the boundary of the solid medium.

In the case of a plane problem we have the following expressions for the components of the stress tensor of interest to us (Refs. 17, 24)

$$Z_z = \lambda \left(\frac{\partial u_x}{\partial x} + \frac{\partial u_z}{\partial z} \right) + 2\mu \frac{\partial u_z}{\partial z},$$

$$Z_x = \mu \left(\frac{\partial u_x}{\partial z} + \frac{\partial u_z}{\partial x} \right), \quad Z_y = 0, \tag{4.5}$$

where u_x and u_z are the displacements along the x- and z-axes, respectively. We assume, as previously, that the z-axis is normal to the boundary. It is useful to express the displacements and stresses in terms of the potentials ϕ and ψ. For this purpose we must use Eqs. 4.2, and take into account that the displacement components u_x and u_z are obtained from the velocity components v_x and v_z by dividing by $-i\omega$.

Quantities referring to the solid medium will be denoted by the subscript 1, and quantities referring to the liquid will be left without subscripts. The elasticity of the liquid will be characterized by the Lamé constant λ, connected with the acoustic velocity c and the density ρ by the first of Eqs. 4.4, in which we must set $\mu = 0$.

We characterize the sound field in the liquid by the potential ϕ. Clearly, all relations obtained for the solid medium can be extended to the liquid medium by setting $\psi = 0$, $\mu = 0$. In particular, according to Eq. 4.1, the connection between $\mathbf{v}$ and ϕ will be of the form

$$\mathbf{v} = \operatorname{grad} \phi. \tag{4.6}$$

Then for the acoustic pressure, taking account of Eq. 3.1, we obtain†

$$p = i\omega\rho\phi. \tag{4.7}$$

The boundary conditions at $z = 0$ are written:
continuity of Z_z:

$$\lambda \nabla^2 \phi = \lambda_1 \nabla^2 \phi_1 + 2\mu_1 \left(\frac{\partial^2 \phi_1}{\partial z^2} + \frac{\partial^2 \psi_1}{\partial x \partial z} \right); \tag{4.8}$$

Z_x equal to zero: $\quad 2\dfrac{\partial^2 \phi_1}{\partial x \partial z} + \dfrac{\partial^2 \psi_1}{\partial x^2} - \dfrac{\partial^2 \psi_1}{\partial z^2} = 0; \tag{4.9}$

continuity of u_z: $\quad \dfrac{\partial \phi}{\partial z} = \dfrac{\partial \phi_1}{\partial z} + \dfrac{\partial \psi_1}{\partial x}. \tag{4.10}$

We now go on to the consideration of concrete problems involving the influence of a liquid–solid boundary on wave propagation.

---

† The quantity ϕ introduced here differs in sign from the quantity ψ used in the preceding section.

2. *The reflection of a sound wave incident on a solid from a liquid*

Let a plane sound wave be incident from a liquid on a liquid–solid interface, and let the sound wave be prescribed by the potential

$$\phi_{\text{inc}} = A \exp\left[ik(x \sin \vartheta + z \cos \vartheta)\right], \tag{4.11}$$

where ϑ is the angle of incidence and A is the amplitude of the wave.

The reflected wave may be written in the form

$$\phi_{\text{refl}} = A V \exp\left[ik(x \sin \vartheta + z \cos \vartheta)\right]. \tag{4.12}$$

Thus, the total sound field in the liquid will be

$$\phi = A\left[\exp\left(-ikz \cos \vartheta\right) + V \exp\left(ikz \cos \vartheta\right)\right] \exp\left(ikx \sin \vartheta\right). \tag{4.13}$$

A longitudinal and a transverse wave will be present in the solid. These waves can be written in the form

$$\phi_1 = A W \exp\left[ik_1(x \sin \vartheta_1 + z \cos \vartheta_1)\right], \tag{4.14}$$

$$\psi_1 = A P \exp\left[i\kappa_1(x \sin \gamma_1 + z \cos \gamma_1)\right]. \tag{4.15}$$

Here, the coefficients V, W and P must be determined; k, k_1 and κ_1 are wave numbers:

$$k = \frac{\omega}{c}, \quad k_1 = \frac{\omega}{c_1}, \quad \kappa_1 = \frac{\omega}{b_1} \tag{4.16}$$

and, finally, ϑ_1 and γ_1 are the angles between the z-axis and the normals to the wave fronts of the longitudinal and transverse waves in the solid.

Substituting Eqs. 4.13, 4.14 and 4.15 into Eqs. 4.8, 4.9 and 4.10, and setting $z = 0$, we obtain three equations from which the angles ϑ_1 and γ_1 and the coefficients V, W and P can be found. Thus, for example, Eq. 4.10 gives

$$k \cos \vartheta \,(V - 1) = -k_1 \cos \vartheta_1 \, W \exp\left[i(k_1 \sin \vartheta_1 - k \sin \vartheta)\,x\right]$$
$$+ \kappa_1 \sin \gamma_1 \, P \exp\left[i(\kappa_1 \sin \gamma_1 - k \sin \vartheta)\,x\right]. \tag{4.17}$$

Since the left hand side of this equation is independent of x, the right hand side must also be independent of x. This is possible only if the equation

$$k \sin \vartheta = k_1 \sin \vartheta_1 = \kappa_1 \sin \gamma_1 \tag{4.18}$$

is satisfied, whence the directions of the waves in the solid are determined.

Now, Eq. 4.17 can be written

$$k \cos \vartheta \,(V - 1) = -k_1 \cos \vartheta_1 \, W + \kappa_1 \sin \gamma_1 \, P. \tag{4.19}$$

Similarly, we obtain

$$k_1^2 \, W \sin 2\vartheta_1 + \kappa_1^2 \, P \cos 2\gamma_1 = 0 \tag{4.20}$$

from Eq. 4.9. We add and subtract $2\mu_1(\partial^2 \phi_1/\partial x^2)$ on the right hand side of Eq. 4.8, and take into account that

$$\nabla^2 \phi_1 \equiv \frac{\partial^2 \phi_1}{\partial x^2} + \frac{\partial^2 \phi_1}{\partial z^2}.$$

Then this equation can be rewritten in the form

$$\lambda \nabla^2 \phi = (\lambda_1 + 2\mu_1) \nabla^2 \phi_1 + 2\mu_1\left(\frac{\partial^2 \psi_1}{\partial x \partial z} - \frac{\partial^2 \phi_1}{\partial x^2}\right), \quad z = 0.$$

Remembering that

$$\lambda = \rho c^2 = \rho \frac{\omega^2}{k^2}, \quad \lambda_1 + 2\mu_1 = \rho_1 \frac{\omega^2}{k_1^2}, \quad \mu_1 = \rho_1 \frac{\omega^2}{\kappa_1^2}, \tag{4.21}$$

and also that, according to the wave equations,

$$\nabla^2 \phi = -k^2 \phi, \quad \nabla^2 \phi_1 = -k_l^2 \phi_1,$$

the equation under consideration can be written in the simpler form

$$\frac{1}{m} \phi = \phi_1 - \frac{2}{\kappa_1^2}\left(\frac{\partial^2 \psi_1}{\partial x \partial z} - \frac{\partial^2 \phi_1}{\partial x^2}\right), \quad z = 0, \tag{4.22}$$

where

$$m = \frac{\rho_1}{\rho}. \tag{4.23}$$

Now, substituting the values of ϕ, ϕ_1 and ψ_1 into Eq. 4.22, we obtain the third equation for the determination of the coefficients V, W and P

$$\frac{1}{m}(1 + V) = \left(1 - 2\frac{k_1^2}{\kappa_1^2}\sin^2 \vartheta_1\right)W - \sin 2\gamma_1 P. \tag{4.24}$$

Solving the system of equations (4.19), (4.20) and (4.24), and making some transformations in which Eq. 4.18 is used, we find

$$V = \frac{Z_1 \cos^2 2\gamma_1 + Z_t \sin^2 2\gamma_1 - Z}{Z_1 \cos^2 2\gamma_1 + Z_t \sin^2 2\gamma_1 + Z}, \tag{4.25}$$

where Z, Z_1 and Z_t denote respectively the impedances of sound waves in the liquid, and longitudinal and transverse waves in the solid:

$$Z = \frac{\rho c}{\cos \vartheta}, \quad Z_1 = \frac{\rho_1 c_1}{\cos \vartheta_1}, \quad Z_t = \frac{\rho_1 b_1}{\cos \gamma_1} \tag{4.26}$$

and furthermore,

$$W = \frac{\rho}{\rho_1} \frac{2Z_1 \cos 2\gamma_1}{Z_1 \cos^2 2\gamma_1 + Z_t \sin^2 2\gamma_1 + Z}, \tag{4.27}$$

$$P = -\frac{\rho}{\rho_1} \frac{2Z_t \sin 2\gamma_1}{Z_1 \cos^2 2\gamma_1 + Z_t \sin^2 2\gamma_1 + Z}. \tag{4.28}$$

3. *Analysis of the reflection coefficient*

At normal incidence of the sound wave on the boundary

$$(\vartheta = \vartheta_1 = \gamma_1 = 0),$$

we obtain $\qquad V = \dfrac{Z_1 - Z}{Z_1 + Z}, \qquad W = \dfrac{\rho}{\rho_1}\dfrac{2Z_1}{Z_1 + Z}, \qquad P = 0.$

In this case, shear waves are not excited, and we obtain the formulas of § 3 for wave reflection from a boundary separating two liquids. On the contrary, when

$$\vartheta = \arcsin\frac{c}{\sqrt{(2)\,b_1}},$$

in which case, according to Eq. 4.18, $\gamma_1 = 45°$, we have

$$V = \frac{Z_t - Z}{Z_t + Z}, \qquad W = 0, \qquad P = -\frac{\rho}{\rho_1}\frac{2Z_t}{Z_t + Z},$$

i.e. shear waves but no longitudinal waves will be excited in the solid.

In the majority of practical cases, the velocity of sound c in the liquid is less than the velocity of longitudinal waves c_1 in the solid. It may also be less than the velocity of transverse waves b_1. We shall start by considering the case $b_1 < c < c_1$. From Eq. 4.18, we have

$$\sin\vartheta_1 = \frac{c_1}{c}\sin\vartheta, \qquad \sin\gamma_1 = \frac{b_1}{c}\sin\vartheta. \qquad (4.29)$$

Hence, it is clear that for $\sin\vartheta > (c/c_1)$, the angle ϑ_1 will be complex. However, the angle γ_1 will be real for all ϑ. Thus, the longitudinal wave in the solid will be an inhomogeneous wave, "gliding" along the boundary, while the transverse wave will be an ordinary plane wave. Since $\sin\vartheta_1 > 1$ here, $\cos\vartheta_1$ and $Z_1 = (\rho_1 c_1/\cos\vartheta_1)$ are purely imaginary. By requiring that Eq. 4.14 be bounded as $z \to -\infty$, we find that $\cos\vartheta_1$ must be positive imaginary and, consequently, Z_1 must be negative imaginary: $\cos\vartheta_1 = i\,|\cos\vartheta_1|,\ Z_1 = -i\,|Z_1|.$

In this case, the reflection coefficient, Eq. 4.25, is written

$$V = \frac{Z_t \sin^2 2\gamma_1 - Z - i\,|Z_1|\cos^2 2\gamma_1}{Z_t \sin^2 2\gamma_1 + Z - i\,|Z_1|\cos^2 2\gamma_1} \qquad (4.30)$$

and will evidently be complex. The square of its modulus is

$$|V|^2 = \frac{(Z_t \sin^2 2\gamma_1 - Z)^2 + |Z_1|^2 \cos^4 2\gamma_1}{(Z_t \sin^2 2\gamma_1 + Z)^2 + |Z_1|^2 \cos^4 2\gamma_1}. \qquad (4.31)$$

The modulus of the reflection coefficient is less than unity, which could have been expected beforehand, since part of the energy is carried from the boundary by a transverse wave.

Using the terminology of transmission line theory, we can say that in this case the boundary will present a complex impedance to the incident wave; its reactive part is due to the longitudinal wave, and its resistive part to the transverse wave.

Let us now consider the case $c < b_1 < c_1$. When $0 < \sin \vartheta < c/c_1$, it is clear from Eq. 4.29 that ϑ_1 and γ_1 will be real angles, i.e. we shall have the usual case of reflection from a boundary, with the reflection coefficient V real and less than unity. The impedance of the boundary will be purely resistive.

When $c/c_1 < \sin \vartheta < c/b_1$, the angle γ_1 will be real, and ϑ_1 will be complex, i.e. we obtain the case considered above.

When $c/c_1 < \sin \vartheta$, both ϑ_1 and γ_1 will be complex, which signifies that the longitudinal and transverse waves in the solid are inhomogeneous waves, propagating along the boundary. Z_1 and Z_t will be purely imaginary, i.e. the boundary will present a purely reactive impedance to the incident wave. The coefficient of reflection can be written

$$V = \frac{i(|Z_1| \cos^2 2\gamma_1 + |Z_t| \sin^2 2\gamma_1) + Z}{i(|Z_1| \cos^2 2\gamma_1 + |Z_t| \sin^2 2\gamma_1) - Z}. \tag{4.32}$$

The modulus of the reflection coefficient is unity, i.e. total internal reflection occurs.

The modulus and phase of the reflection coefficient of a sound wave ($V = \rho e^{i\phi}$) at a water–aluminum boundary is shown in Fig. 12 as a function of the angle of incidence ϑ.

Irregularities are evident at $\vartheta = 13°$ and $\vartheta = 29°$. The first angle coincides with $\arcsin(c/c_1)$, and the second with $\arcsin(c/b_1)$.

Let us note one more circumstance that is not without interest.[73] Equation 4.25 for the reflection coefficient may be written in the form

$$V = \frac{Z_{\text{tot}} - Z}{Z_{\text{tot}} + Z}, \tag{4.33}$$

where $$Z_{\text{tot}} = Z_1 \cos^2 2\gamma_1 + Z_t \sin^2 2\gamma_1 \tag{4.34}$$

is the total impedance of the boundary, due to the presence of longitudinal and transverse waves.

We can find the ratio

$$\frac{Z_{\text{tot}}}{Z_1} = \cos^2 2\gamma_1 + \frac{Z_t}{Z_1} \sin^2 2\gamma_1,$$

which is easily transformed into

$$\frac{Z_{\text{tot}}}{Z_1} = 1 - \left(1 - \frac{\tan \gamma_1}{\tan \vartheta_1}\right) \sin^2 2\gamma_1. \tag{4.35}$$

2

Since b_1 is always greater than c_1, and consequently, in view of Eq. 4.29, $\gamma_1 < \vartheta_1$, we have $Z_{\text{tot}}/Z_1 < 1$.

Thus, the total impedance of the solid boundary is less than the impedance of the equivalent liquid with the same ρ_1 and c_1, i.e. taking account of the excitation of transverse waves during reflection corresponds to a "softening" of the boundary.

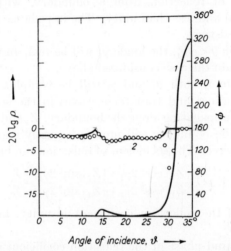

Fig. 12. The modulus and phase of the reflection coefficient of a sound wave at a water–aluminum interface. The circles indicate experimental data on the reflection coefficient at a frequency of 3.35×10^6 cps, for a beam with an angle of divergence of about $3°$. 1. Phase. 2. Reflection coefficient.

It is also shown in Ref. 73 that, as the angle of incidence changes, Z_{tot} changes less than the impedance of the equivalent liquid $Z_1 = \rho_1 c_1/\cos \vartheta_1$, so that in certain cases the reflection from the solid may be considered approximately as the reflection from a medium characterized by an impedance independent of angle (see § 3).

4. Other cases of reflection

We now consider the case in which a longitudinal wave, whose potential is of amplitude A_l, is incident from a solid onto a solid–liquid interface. This wave will excite the following system of three waves at the boundary:

(1) a reflected longitudinal wave (we denote its amplitude by B_l),

(2) a reflected transverse wave (of amplitude B_t),

(3) a sound wave in the liquid (of amplitude D).

The entire system of waves may be written in the form

$$\phi = D \exp \left[ik(x \sin \vartheta + z \cos \vartheta) \right]$$

for the sound wave in the liquid;

$$\phi_1 = A_l \exp\left[ik_1(x\sin\vartheta_1 + z\cos\vartheta_1)\right] + B_l \exp\left[ik_1(x\sin\vartheta_1 - z\cos\vartheta_1)\right]$$

for the incident and reflected longitudinal waves; and

$$\psi_1 = B_t \exp\left[i\kappa_1(x\sin\gamma_1 - z\cos\gamma_1)\right]$$

for the reflected transverse wave.

Using the boundary conditions (4.8) − (4.10) as previously, we find

$$V' \equiv \frac{B_l}{A_l} = \frac{Z + Z_t \sin^2 2\gamma_1 - Z_1 \cos^2 2\gamma_1}{Z + Z_t \sin^2 2\gamma_1 + Z_1 \cos^2 2\gamma_1}, \tag{4.36}$$

$$\frac{D}{A_l} = \frac{c\cos\vartheta_1}{c_1\cos\vartheta\cos^2 2\gamma_1}\left(1 - \frac{B_l}{A_l}\right), \tag{4.37}$$

$$\frac{B_t}{A_l} = \left(\frac{b_1}{c_1}\right)^2 \frac{\sin 2\vartheta_1}{\cos 2\gamma_1}\left(1 - \frac{B_l}{A_l}\right). \tag{4.38}$$

Finally, when a transverse wave (of amplitude A_t) in which the particle motion is in the xz-plane is incident from the solid onto the boundary, the system of waves can be written in the form

$$\phi = D \exp\left[ik(x\sin\vartheta + z\cos\vartheta)\right]$$

for the sound wave in the liquid;

$$\phi_1 = B_l \exp\left[ik_1(x\sin\vartheta_1 - z\cos\vartheta_1)\right]$$

for the reflected longitudinal wave in the solid; and

$$\psi_1 = A_t \exp\left[i\kappa_1(x\sin\gamma_1 + z\cos\gamma_1)\right] + B_t \exp\left[ix_1(x\sin\gamma_1 - z\cos\gamma_1)\right]$$

for the incident and reflected transverse waves.

Again, using the boundary conditions (4.8) − (4.10), we find

$$V'' = \frac{B_t}{A_t} = -\frac{Z + Z_1\cos^2 2\gamma_1 - Z_t\sin^2 2\gamma_1}{Z + Z_1\cos^2 2\gamma_1 + Z_t\sin^2 2\gamma_1}, \tag{4.39}$$

$$\frac{D}{A_t} = \frac{\tan\vartheta}{2\sin^2\gamma_1}\cdot\left(1 + \frac{B_t}{A_t}\right), \tag{4.40}$$

$$\frac{B_l}{A_t} = -\left(\frac{c_1}{b_1}\right)^2 \frac{\cos 2\gamma_1}{\sin 2\vartheta_1}\left(1 + \frac{B_t}{A_t}\right). \tag{4.41}$$

It is interesting to note that, except for a sign change, the reflection coefficients expressed by Eqs. 4.36 and 4.39 may be obtained from Eq. 4.25 for V by the cyclic replacement of the impedances $Z, Z_1 \cos^2 2\gamma_1$ and $Z_t \sin^2 2\gamma_1$.

It now remains to consider a transverse wave in which the particle velocity is parallel to the boundary of separation, incident onto the solid–liquid boundary.

A wave of this type creates neither normal displacements nor normal stresses at the boundary. Hence, it will excite neither a sound wave in the liquid nor a longitudinal wave in the solid. The single boundary condition that the component Z_y of the stress tensor be equal to zero turns out to be satisfied under the assumption that the direction of the particle velocity in the reflected wave is the same as in the incident wave (parallel to the boundary), and the amplitude of this wave is equal to the amplitude of the incident wave.

Thus, we have considered waves of any polarization, incident from a solid onto its boundary. Since, in an isotropic solid, a plane wave of arbitrary character can be represented in the form of a superposition of waves of various polarizations, we can consider our general investigation of the problem completed.

5. *The reflection of waves at a free boundary of a solid*

We will now make a more detailed analysis of the formulas obtained above for the case in which the solid body has a free boundary, i.e. is bounded by vacuum. With regard to wave reflection, a solid bounded by a gas is quite close to the idealized case.

We shall begin by considering the case in which a longitudinal wave is incident on the boundary of a solid from within the solid. Setting the "liquid" density, and, consequently, its impedance $Z = \rho c / \cos \vartheta$, equal to zero in Eqs. 4.36 and 4.38, we obtain

$$\frac{B_l}{A_l} = \frac{b_1 \cos \vartheta_1 \tan^2 2\gamma_1 - c_1 \cos \gamma_1}{b_1 \cos \vartheta_1 \tan^2 2\gamma_1 + c_1 \cos \gamma_1}, \qquad (4.42)$$

$$\frac{B_t}{A_l} = \frac{b_1 \sin 2\vartheta_1}{c_1 \cos 2\gamma_1} \frac{2b_1 \cos \gamma_1}{b_1 \cos \vartheta_1 \tan^2 2\gamma_1 + c_1 \cos \gamma_1}. \qquad (4.43)$$

These expressions give the ratio of the amplitudes of the *potentials* of the reflected longitudinal and transverse waves to the amplitude of the *potential* of the incident longitudinal wave. In practice, the particle velocity or displacement, rather than the potential, is usually measured directly. It follows from Eq. 4.1 that the amplitudes of the particle velocities in the longitudinal and the transverse waves will be equal to the amplitudes of the potentials multiplied by the corresponding wave numbers k_1 and κ_1. Hence, Eq. 4.42 also represents the reflection coefficient for the particle velocity (or displacement) in the longitudinal wave. On the other hand, to obtain the ratio of the amplitude of the particle velocity (or displacement) in the reflected transverse wave to the amplitude of the particle velocity in the incident longitudinal wave, Eq. 4.43 must be multiplied by $\kappa_1/k_1 = c_1/b_1$.

The reflection coefficient of longitudinal waves is shown graphically[190] in Fig. 13 as a function of the angle of incidence ϑ_1, for various values of the Poisson ratio σ. As is well known, σ is connected with the Lamé constants λ and μ and with the ratio of the velocities of longitudinal and transverse waves c_1/b_1 by the formula

$$\left(\frac{b_1}{c_1}\right)^2 = \frac{\mu}{\lambda+2\mu} = \frac{1-2\sigma}{2(1-\sigma)}. \tag{4.44}$$

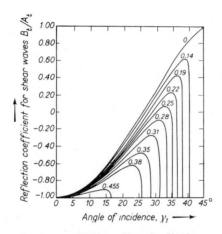

Fig. 13. The reflection coefficient of longitudinal waves at a free boundary of a solid, as a function of the angle of incidence and for various values of the Poisson ratio (indicated by the number next to each curve).

As $\sigma \to \frac{1}{2}$, we make the transition from a solid to a liquid medium, for which the reflection coefficient from a free surface is equal to -1 at all angles of incidence (see § 3), as is evident in Fig. 13.

Now let a transverse wave of amplitude A_t be incident on the boundary of a solid from within the solid. In this case, setting $Z = 0$ in Eqs. 4.39 and 4.41, and taking account of the values of Z_t and Z_1, we obtain the following expressions for the ratios of the amplitudes of the reflected transverse and longitudinal waves B_t and B_l to the amplitude of the incident wave A_t:

$$\frac{B_t}{A_t} = \frac{b_1 \cos \vartheta_1 \tan^2 2\gamma_1 - c_1 \cos \gamma_1}{b_1 \cos \vartheta_1 \tan^2 2\gamma_1 + c_1 \cos \gamma_1}, \tag{4.45}$$

$$\frac{B_l}{A_t} = -\frac{2c_1 \cos \gamma_1 \tan 2\gamma_1}{b_1 \cos \vartheta_1 \tan^2 2\gamma_1 + c_1 \cos \gamma_1}. \tag{4.46}$$

Thus, the reflection coefficient of transverse waves B_t/A_t coincides with the reflection coefficient of longitudinal waves found above. This coefficient is shown graphically in Fig. 14 as a function of the angle of incidence γ_1 for various values of the Poisson ratio σ.

When $\sin \gamma_1 > b_1/c_1$, we have the case of total "internal" reflection of a transverse wave. In this case, the angle ϑ_1, determined by the relation $\sin \vartheta_1 = (c_1/b_1) \sin \gamma_1$, is complex, and the longitudinal wave is an inhomogeneous wave, the amplitude of which decreases exponentially with distance from the boundary. Since the condition $b_1/c_1 < 1/\sqrt{2}$ is always fulfilled, the limiting angle $\gamma_1^{\lim} = \arc \sin b_1/c_1$ is always less than $45°$.

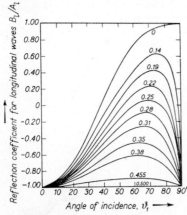

Fig. 14. The reflection coefficient of shear waves at a free boundary of a solid, as a function of the angle of incidence and for various values of the Poisson ratio.

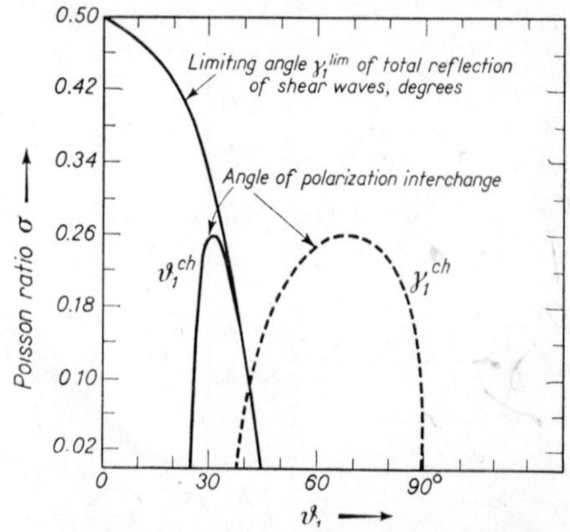

Fig. 15. Characteristic angles for shear wave reflection at a free surface of a solid.

When $\gamma_1 > \gamma_1^{\lim}$, the reflection coefficient B_t/A_t will be complex, with modulus unity and phase depending on γ_1. The angle $\gamma_1^{\lim}$ as a function of the Poisson ratio σ is shown in Fig. 15.

It is clear from Eqs. 4.42 and 4.45 that when the condition

$$b_1 \cos \vartheta_1 \tan^2 2\gamma_1 - c_1 \cos \gamma_1 = 0 \qquad (4.47)$$

is fulfilled, an incident longitudinal wave will be reflected as a transverse wave only, and vice versa, i.e. a change of polarization occurs at reflection. The angle γ_1^{ch} at which this occurs, and the corresponding angle ϑ_1^{ch}, determined from the condition $\sin \vartheta_1^{ch} = (c_1/b_1) \sin \gamma_1^{ch}$, is also shown in Fig. 15 as a function of σ.

6. Surface waves

Surface waves, also known as Rayleigh waves, on the boundary of a solid have been studied in great detail in seismology. Here, we shall approach this problem from a somewhat different point of view, considering Rayleigh waves as a degenerate case of the reflection of plane waves. Moreover, we shall investigate the surface waves for the case in which the solid is bounded by liquid.

The complex angle ϑ, satisfying the equation

$$Z_1 \cos^2 2\gamma_1 + Z_t \sin^2 2\gamma_1 + Z = 0, \qquad (4.48)$$

can be found purely formally. At this angle, according to Eq. 4.25, we shall have $V \to \infty$ for the reflection coefficient of a sound wave reflected from the solid. According to Eqs. 4.19, 4.20 and 4.24, the quantities W and P, proportional to the amplitudes of the longitudinal and transverse waves in the solid, will be of the same order as V. As a result, as is evident from Eqs. 4.11–4.15, we must make the amplitude of the incident wave approach zero in order that the amplitudes of the reflected wave and both waves in the solid remain finite. In this case, we shall have a wave process propagating along the boundary without an incident wave, i.e. we will have a surface wave.

In the same way, we could have convinced ourselves of the existence of a surface wave by starting from expressions (4.36) and (4.39) for the reflection coefficients of longitudinal and transverse waves in the solid. Investigation of all these cases shows that when condition (4.48) is fulfilled, a system of waves will exist, consisting of a longitudinal and a transverse wave in the solid and a longitudinal (sound) wave in the liquid. This system, as a whole, satisfies the boundary conditions and the wave equations for the longitudinal and the transverse waves.

Rayleigh[183] first studied the surface wave for the case in which the solid is bounded by a sufficiently rarefied gaseous medium ($Z = 0$). We shall examine this case in detail.

Here, Eq. 4.48 is written

$$Z_1 \cos^2 2\gamma_1 + Z_t \sin^2 2\gamma_1 = 0. \qquad (4.49)$$

It can also be written in another form.† For this purpose, we take into account the fact that the phase velocity v_R of waves along the boundary will be

$$v_R = \frac{b_1}{\sin \gamma_1}.$$

We also introduce the notation

$$q = (b_1/c_1)^2, \quad s = \sin^2 \gamma_1 = (b_1/v_R)^2. \qquad (4.50)$$

Now, taking account of Eqs. (4.18) and (4.26), Eq. 4.49 yields an equation for s

$$4s \sqrt{(1-s)} \cdot \sqrt{(q-s)} = -(1-2s)^2. \qquad (4.51)$$

---

† The following exposition of the problem of surface waves is based partially on Ref. 190.

Since the energy flux in a surface wave is only along the boundary, the wave will be undamped (we are neglecting absorption in the solid). Therefore, we are interested in the real roots of the last equation.

Any root of Eq. 4.51 is at the same time a root of the equation of the third degree obtained by squaring Eq. 4.51:

$$q(4s)^2(1-s) = 1 - 8s + 24s^2 - 16s^3.$$

The dependence of q on s, given by this equation, is shown in Fig. 16.[190] A more exact investigation indicates that the roots of the original Eq. 4.51 are the points of intersection of straight lines $q = $ const and the portion of the curve shown in Fig. 16 as a solid line. The corresponding values of q are between 0 and $\frac{1}{2}$. It is also clear from the Figure that the

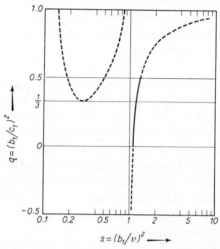

Fig. 16. Graphical representation of the equation for the Rayleigh wave velocity.

corresponding roots $s = s_R$ are somewhat greater than unity and depend only slightly on the value of q. The velocity of Rayleigh surface waves,

$$v_R = \frac{b_1}{\sin \gamma_1} = \frac{b_1}{\sqrt{s_R}},$$

is thus somewhat less than the velocity of transverse waves. The numerical value of the factor $1/\sqrt{s_R}$ is given in Table 1 for three values of q. From the results obtained above, it immediately follows that $\sin \gamma_1 = (b_1/v_R) > 1$, i.e. as could be expected for surface waves, γ_1 has the form $\gamma_1 = (\pi/2) - i\alpha$, where α is real. Here, $\cos \gamma_1 = i\sqrt{(s_R - 1)}$, $\cos \vartheta_1 = i\sqrt{[(s_R - q)/q]}$ are imaginary. Hence, according to Eqs. 4.14 and 4.15, as could be expected, the amplitudes of the longitudinal and transverse waves composing the surface wave decrease exponentially with distance from the boundary. The longitudinal wave behaves as $\exp[(-2\pi/\lambda_t)\sqrt{(s_R-q)}|z|]$ and the transverse as $\exp[(-2\pi/\lambda_t)\sqrt{(s_R-1)}|z|]$ (λ_t is the wavelength of a transverse wave in an unbounded solid).

Thus, the energy of the wave is concentrated near the boundary, in a layer of several wavelengths thickness.

It is not difficult to show that the trajectories of the particles taking part in the transfer of a surface wave are ellipses, the principal axes of which are parallel and perpendicular to the boundary.

TABLE 1

Values of the Ratio of the Rayleigh and the Shear Wave Velocities

$q = \left(\dfrac{b_1}{c_1}\right)^2$	σ	$\dfrac{1}{\sqrt{s_R}} = \dfrac{v_R}{b_1}$
0	0.5	0.9554
$\frac{1}{3}$	0.25	0.9194
$\frac{1}{2}$	0	0.8741

On the boundary itself ($z = 0$), the ratio of the displacements in the x and z directions is

$$\frac{u_x}{u_z} = -i\sqrt{\left(1-\frac{1}{s_R}\right)}\frac{2s_R(1+q)-1}{\sqrt{[s_R(s_R-1)(s_R-q)]}+2s_R-1}. \tag{4.52}$$

The factor i indicates the retardation of the displacement in the z direction with respect to the displacement in the x direction by a quarter of a period, which could be expected with an elliptical particle trajectory. Equation 4.52 could be obtained, for example, as follows: taking account of Eqs. 4.27 and 4.28, we substitute Eqs. 4.14 and 4.15 for the potentials in the solid in Eq. 4.2 for the velocity components, and then replace the angles ϑ_1 and γ_1 by their values for a surface wave, determined from the relations

$$\sin\gamma_1 = \sqrt{s_R}, \quad \sin\vartheta_1 = (c_1/b_1)\sin\gamma_1 = \sqrt{(s_R/q)},$$

after which the ratio $(u_x/u_z) = (v_x/v_z)$ at $z = 0$ is easily found.

It is interesting to note that the Rayleigh wave is not the most general form of a surface wave on the free boundary of a solid. In the Rayleigh wave the particle motion is confined to the xz-plane and the problem of wave propagation is two dimensional.

Uller[204] has studied the most general form of a surface wave without this limitation.

We shall now go on to the study of surface waves on the boundary between a solid and a liquid, which is limited to a two dimensional problem.

In this case, the velocity of surface waves will be determined by Eq. 4.48, which, if we take account of the values of Z_1, Z_t and Z (see Eq. 4.26), can be transformed into a form similar to Eq. 4.51:

$$4s \sqrt{(1-s)} \sqrt{(q-s)} + (1-2s)^2 = -\frac{\rho}{\rho_1} \sqrt{\left(\frac{q-s}{r-s}\right)} \tag{4.53}$$

where we have used the notation

$$r = (b_1/c)^2. \tag{4.50a}$$

The right hand side of Eq. 4.53 characterizes the influence of the liquid.

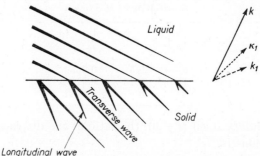

Fig. 17. Schematic illustration of the attenuating surface wave at a boundary between a liquid and a solid. The wave fronts are shown in the left half of the Figure, and the wave vectors are shown on the right.

We start by assuming that the solid is bounded by a very light medium, for example by a gas, so that $\rho/\rho_1 \ll 1$. Clearly, the velocity of the surface wave in this case will be close to the velocity found above for the Rayleigh wave at a free boundary, that is, $s \approx s_R$. Furthermore, realizing that the condition $1/c > 1$ will be fulfilled in all practical cases, the right hand side of Eq. 4.53 will be imaginary. Since, for a real s close to s_R, the left hand side of Eq. 4.53 will be purely real, it is clear that the equation can be satisfied only with complex s, and consequently only with complex values of the velocity of propagation.

From a physical point of view, this may be explained quite naturally: the surface wave will be attenuated because part of its energy is continuously carried away by a sound wave, excited in the medium adjacent to the solid. The wave vectors $\mathbf{k}_1$ and $\mathbf{\varkappa}_1$ in the solid will no longer be directed parallel to the boundary, but will be inclined to it, which indicates the presence of a continuous flow of energy from the solid medium to its boundary.

The wave fronts in the solid and in the adjacent medium are shown in Fig. 17, in which the line thickness indicates schematically the amplitude of the wave. For visual clarity, the degree of attenuation of the wave as it travels along the boundary has been strongly exaggerated in the Figure. The wave vectors of all three waves, which are naturally directed along the normals to the wave fronts, are shown on the right in Fig. 17. It is interesting to note that a system of waves in which the wave vectors are directed oppositely to those shown in Fig. 17 is also a solution of the problem. In this case, energy will flow from the liquid into the solid. In determining the mutual directions of the real and imaginary parts of the wave vectors in these cases, we must take account of a condition which is in a certain sense equivalent to the radiation condition, namely, we must require that the system of waves be attenuated along the boundary in the direction in which it is propagating.

Apart from the attenuating system of waves considered above, which approach Rayleigh waves as the liquid density approaches zero, an undamped surface wave having a somewhat different nature can propagate along the boundary between a solid and a liquid. Its velocity is less than the velocity c, and therefore, in the liquid, it will have the form of the usual inhomogeneous wave, decreasing in amplitude with distance from the boundary. The intensity in both media will be directed parallel to the boundary of separation.

In fact, when $\rho/\rho_1 \ll 1$, $r = (b_1/c)^2 \gg 1$ we have the approximate solution of Eq. 4.53

$$s_2 \approx r\left[1 + \frac{1}{4}\left(\frac{\rho c^2}{\rho_1(b_1^2 - c_1^2)}\right)^2\right], \qquad (4.54)$$

which is easily found by representing the right hand side of Eq. 4.53 in the form

$$4s^2\left[-\sqrt{\left(1 - \frac{1}{s}\right)}\sqrt{\left(1 - \frac{q}{s}\right)} + \left(1 - \frac{1}{2s}\right)^2\right]$$

and expanding the square brackets in powers of $1/s$, up to terms of the order 1 and $1/s$.

The wave velocity along the boundary will now be

$$v = \frac{b_1}{\sqrt{s_2}} = c\left[1 - \frac{1}{8}\left(\frac{\rho c^2}{\rho_1(b_1^2 - c_1^2)}\right)^2\right], \qquad (4.55)$$

i.e. somewhat less than the velocity of sound in the liquid.

The amplitude of the wave in the liquid will decrease with distance from the boundary according to the law

$$\exp\left[-\frac{\pi}{\lambda}\frac{\rho c^2}{\rho_1(c_1^2 - b_1^2)}|z|\right],$$

where λ is the wavelength of the sound wave in the liquid. In the solid, the decrease of the amplitude of the longitudinal and transverse waves will take place according to a law, an approximate expression of which has the form $\exp\left[-(2\pi/\lambda)\,|\,z\,|\,\right]$.

Thus, since $\rho c^2/\rho_1 b_1^2$ is small, the amplitude in the liquid decreases very slowly with distance from the boundary, while in the solid, the entire wave process is concentrated in a layer of thickness of the order of λ.

This type of surface wave is nothing other than a sound wave, incident on the solid boundary at "grazing" incidence. It was shown in §3 that when a plane sound wave strikes a boundary of separation between two liquids at grazing incidence, the reflected wave completely cancels the incident wave, as a result of which the total field is zero, i.e. no wave can exist. In the case under consideration, we see that a sound wave gliding along the boundary can exist, but it must then be "slightly" inhomogeneous, i.e. its amplitude must decrease, at least slowly, with distance from the boundary. Because of the compliance of the solid, the velocity of the wave will be somewhat lower than that of a sound wave in the unbounded liquid.

The presence of two types of surface waves on the boundary between a solid and a liquid is similar to the presence of two types of oscillation in two coupled systems, each of which has one degree of freedom. With weak coupling (which applies to the case considered above as a result of the assumption that $\rho c^2/\rho_1 b_1^2$ is small) each type of oscillation is the oscillation of an almost free individual system, and the vibrational energy is concentrated mainly in one or the other oscillating system. In the present case, the energy is concentrated principally in the solid for a wave of the first type, and principally in the liquid for a wave of the second type.

The reader will find an investigation of surface waves on the boundary between two solids in Refs. 147 and 197, and a detailed study of surface waves in a layer in Refs. 53, 188 and 193.

The problem of Rayleigh waves on the boundary between a liquid and a solid was also investigated in Ref. 48.

§5. REFLECTION FROM A PLANE LAYER AND FROM A SYSTEM OF PLANE LAYERS

We shall now proceed to the problem of the reflection of plane waves from a layer separating two homogeneous media, and from a system of plane layers. This problem has numerous applications in acoustics and in optics (for example, in the passage of waves through a plate,

the theory of acoustical and optical reflection-reducing layers, etc.).
The theory developed in this section will be equally applicable to acoustic
and to electromagnetic waves.

1. *The reflection and transmission coefficients for a single layer*

We shall suppose that a plane acoustic or electromagnetic wave is
incident on a plane layer of thickness d (Fig. 18) at an arbitrary angle
of incidence. The numbers 3, 2 and 1 will be used to denote, respectively,
the medium through which the incident wave travels, the layer, and
the medium into which the wave is transmitted. The angles between the
propagation directions in each of the media and the normal to the
boundaries of the layer will be denoted by ϑ_3, ϑ_2 and ϑ_1, respectively.
The xz-plane will again be considered the plane of incidence.

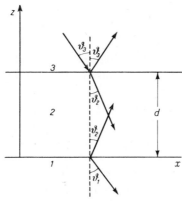

Fig. 18. The problem of the reflection of a plane wave at a layer.

The amplitudes of the reflected and transmitted waves can be determined most easily through the use of the normal impedance. The normal
impedances of the three media will be denoted by Z_3, Z_2 and Z_1,
respectively.

We recall that in acoustics, the normal impedance is understood to
be the ratio of the acoustic pressure to the projection of the particle
velocity on the normal to the boundaries of the layer.

Here, the impedance is expressed by the formula (see Eq. 3.17)

$$Z_j = \frac{\rho_j c_j}{\cos \vartheta_j}, \quad j = 1, 2, 3. \tag{5.1}$$

In electrodynamics, the impedance is understood to be the ratio
of the tangential components of the electric and magnetic fields, where
(see Eqs. 2.22 and 2.23)

$$Z_j = \frac{1}{\cos \vartheta_j} \sqrt{\frac{\mu_j}{\epsilon_j}} \tag{5.2}$$

when $\mathbf{E}$ is parallel to the interface, and

$$Z_j = \cos \vartheta_j \sqrt{\frac{\mu_j}{\epsilon_j}} \qquad (5.3)$$

when $\mathbf{H}$ is parallel to the interface.

Using impedance, the problem of wave reflection from a layer is solved in exactly the same way in both cases of polarization in electrodynamics, and in acoustics. We first consider the reflection of an electromagnetic wave when $\mathbf{E}$ is parallel to the interface, and consequently is perpendicular to the plane of incidence of the wave.

As a result of multiple reflections at the boundaries of the layer, two resulting waves with different directions of propagation, symmetric with respect to planes of $z = $ const (Fig. 18), will be developed. Therefore, the expression for the electric field in the layer is written in the form

$$E_2 = E_{2y} = [A \exp(-i\alpha_2 z) + B \exp(i\alpha_2 z)] \exp(i\sigma_2 x), \qquad (5.4)$$

where $\qquad \alpha_2 = k_{2z} = k_2 \cos \vartheta_2, \quad \sigma_2 = k_{2x} = k_2 \sin \vartheta_2,$

and A and B are, for the present, undetermined constants. As usual, the factor $e^{-i\omega t}$ has been suppressed. According to the definition of impedance, the tangential component of the magnetic field can be written

$$H_{2x} = \frac{1}{Z_2}[A \exp(-i\alpha_2 z) - B \exp(i\alpha_2 z)] \exp(i\sigma_2 x). \qquad (5.5)$$

We have taken account of the fact that the ratio E_{2y}/H_{2x} has different signs in the incident and reflected waves (see the remarks in §§ 2 and 3 in connection with the definition of impedance). At the boundary $z = 0$, the ratio E_{2y}/H_{2x} must be equal to the impedance of medium 1. Consequently we find

$$Z_2 \frac{A+B}{A-B} = Z_1, \quad \frac{B}{A} = \frac{Z_1 - Z_2}{Z_1 + Z_2}. \qquad (5.6)$$

Now let us determine the ratio E_{2y}/H_{2x} at the front side of the layer, i.e. at $z = d$. From Eqs. 5.4 and 5.5, we obtain

$$Z_{\text{in}} \equiv \left(\frac{E_{2y}}{H_{2x}}\right)_{z=d} \equiv \frac{A \exp(-i\alpha_2 d) + B \exp(i\alpha_2 d)}{A \exp(-i\alpha_2 d) - B \exp(i\alpha_2 d)} \cdot Z_2.$$

Substituting for B/A gives

$$Z_{\text{in}} = \frac{Z_1 - iZ_2 \tan \alpha_2 d}{Z_2 - iZ_1 \tan \alpha_2 d} Z_2. \qquad (5.7)$$

Since Z_{in} is equal to the ratio of E_{2y} to H_{2x} at the front side of the layer, it may properly be called the input impedance of the layer. Equation 5.7 is well known in transmission line theory. It is valid for any type of wave.

After the input impedance of the layer has been found, the problem of wave reflection from the layer is solved just as in the case of reflection from a single boundary, except that in place of the impedance of the reflecting medium, we must now substitute the input impedance of the layer.

The field of the incident and reflected waves in medium 3 will be

$$E_{3y} = \{C \exp\left[-i\alpha_3(z-d)\right] + D \exp\left[i\alpha_3(z-d)\right]\}.\exp\left(i\sigma_3 x\right),$$

$$H_{3x} = \frac{1}{Z_3} \{C \exp\left[-i\alpha_3(z-d)\right] - D \exp\left[i\alpha_3(z-d)\right]\} \exp\left(i\sigma_3 x\right). \tag{5.8}$$

At $z = d$, the ratio E_{3y}/H_{3x} must be equal to the input impedance of the layer, i.e.

$$\left(\frac{E_{3y}}{H_{3x}}\right)_{z=d} = Z_3 \frac{C+D}{C-D} Z_{\text{in}}; \ = Z_{\text{in}}$$

whence we obtain for the reflection coefficient

$$V = \frac{D}{C} = \frac{Z_{\text{in}} - Z_3}{Z_{\text{in}} + Z_3}, \tag{5.9}$$

or, if we substitute Eq. 5.7 for Z_{in},

$$V = \frac{(Z_1+Z_2)(Z_2-Z_3)\exp\left(-i\alpha_2 d\right) + (Z_1-Z_2)(Z_2+Z_3)\exp\left(i\alpha_2 d\right)}{(Z_1+Z_2)(Z_2+Z_3)\exp\left(-i\alpha_2 d\right) + (Z_1-Z_2)(Z_2-Z_3)\exp\left(i\alpha_2 d\right)}. \tag{5.10}$$

In the special case in which the media on both sides of the layer have the same physical properties, i.e. when $Z_3 = Z_1$, the expression for the reflection coefficient takes the form

$$V = \frac{Z_2^2 - Z_1^2}{Z_1^2 + Z_2^2 + 2iZ_1 Z_2 \cot \alpha_2 d}. \tag{5.11}$$

We shall now find the amplitude of the transmitted wave in medium 1. The electric field of this wave can be written in the form

$$E_{1y} = \mathscr{E} \exp\left(-i\alpha_1 z + i\sigma_1 x\right). \tag{5.12}$$

Using the last equation and Eq. 5.4, the condition that E_y be continuous at the boundary $z = 0$ gives

$$\mathscr{E} = A + B. \tag{5.13}$$

Similarly, using Eqs. 5.4 and 5.8, the condition that E_y be continuous at the boundary $z = d$ gives

$$C + D = A \exp\left(-i\alpha_2 d\right) + B \exp\left(i\alpha_2 d\right)$$

or

$$C(1 + V) = A \exp\left(-i\alpha_2 d\right) + B \exp\left(i\alpha_2 d\right). \tag{5.14}$$

Dividing Eq. 5.13 by Eq. 5.14, and using Eq. 5.10 for V and Eq. 5.6 for B/A, we obtain for the transmission coefficient

$$D = \frac{\mathcal{E}}{C} = \frac{4Z_1 Z_2}{(Z_1 - Z_2)(Z_2 - Z_3) \exp{(i\alpha_2 d)} + (Z_1 + Z_2)(Z_2 + Z_3) \exp{(-i\alpha_2 d)}}. \tag{5.15}$$

The expressions we have obtained for the reflection and transmission coefficients are valid both for electromagnetic waves, independently of their polarization, and for acoustic waves. The only difference in the treatment of these cases is the expression for the impedance. The derivation of Eqs. 5.10 and 5.15, given above, also remains valid. In particular, in the case of acoustics, it is only necessary to make the formal replacement of E_y by the acoustic pressure p, and H_x by the vertical component of the velocity v_z.

In the special case $d = 0$, Eqs. 5.10 and 5.15 become

$$V \to \frac{Z_1 - Z_3}{Z_1 + Z_3}, \tag{5.16}$$

$$D \to \frac{2Z_1}{Z_1 + Z_3}, \tag{5.17}$$

which are the expressions for the reflection and transmission coefficients of the boundary between media 1 and 3.

Let us also note that in the general case, Eqs. 5.10 and 5.15 for V and D can be written in the form

$$V = \frac{V_{23} + V_{12} \exp{(2i\alpha_2 d)}}{1 + V_{23} V_{12} \exp{(2i\alpha_2 d)}}, \tag{5.18}$$

$$D = \frac{4Z_1 Z_2}{(Z_1 + Z_2)(Z_2 + Z_3)} \cdot \frac{1}{\exp{(-i\alpha_2 d)} + V_{12} V_{23} \exp{(i\alpha_2 d)}}, \tag{5.19}$$

where

$$V_{12} = \frac{Z_1 - Z_2}{Z_1 + Z_2}, \quad V_{23} = \frac{Z_2 - Z_3}{Z_2 + Z_3} \tag{5.20}$$

are the reflection coefficients at the boundaries 2,1 and 3,2 respectively. It is easily proved by simple substitution that Eqs. 5.18 and 5.19 coincide with Eqs. 5.10 and 5.15, respectively.

2. Another derivation of the expression for the reflection coefficient

Expressions 5.18 and 5.19 for the reflection and transmission coefficients of a layer can be derived in another way by considering the multiple reflections at the boundaries of the layer. As an example of this approach, we shall derive the expression for the reflection coefficient, limiting ourselves to the case of normal incidence.

We consider an incident wave of unit amplitude. The resulting wave reflected from the layer may be regarded as a superposition of the following waves: 1. Waves reflected from the front surface of the layer (boundary between media 3 and 2); 2. Waves penetrating the front surface of the layer, passing through the layer, reflecting from the back surface of the layer, and finally passing through the layer again and leaving through its front surface; 3. Waves penetrating the layer, undergoing two reflections at the back surface and one at the front surface and then leaving the layer, etc.

The complex amplitudes (taking account of the phase factor) of these waves will be respectively,

$$V_{23}$$

$$D_{32} V_{12} D_{23} \exp(2ik_2 d)$$

$$D_{32} V_{12} V_{32} V_{12} D_{23} \exp(4ik_2 d) \tag{5.21}$$

where D_{32} denotes the transmission coefficient of the boundary between media 3 and 2 for a wave traveling from medium 3 into medium 2, and D_{23} represents the transmission coefficient for a wave traveling in the opposite direction.

The meaning of the individual factors in the last of expressions 5.21, for example, is as follows. The factor D_{32} characterizes the diminution of the wave as it passes from medium 3 to medium 2, V_{12} represents the diminution because of the subsequent reflection at the back surface of the layer, etc. Finally, the factor D_{23} takes account of the diminution as the wave leaves the layer in the opposite direction. The factor $\exp(4ik_2 d)$ takes account of the phase change of the wave (and the diminution which takes place if k_2 is complex) as it goes back and forth twice through the layer.

Summing up all the waves which form the reflected wave, we obtain

$$V_{23} + D_{32} V_{12} D_{23} \exp(2ik_2 d) + D_{32} V_{32} V_{12}^2 D_{23} \exp(4ik_2 d)$$
$$+ D_{32} V_{32}^2 V_{12}^3 D_{23} \exp(6ik_2 d) + \ldots$$

for the amplitude of the reflected wave.

Since the amplitude of the incident wave was taken as unity, the last expression is also V, the coefficient of reflection from the layer. Thus,

$$V = V_{23} + D_{32} D_{23} V_{12} \exp(2ik_2 d) \sum_{n=0}^{\infty} [V_{32} V_{12} \exp(2ik_2 d)]^n$$

$$= V_{23} + D_{32} D_{23} V_{12} \frac{\exp(2ik_2 d)}{1 - V_{32} V_{12} \exp(2ik_2 d)}.$$

Remembering that $V_{32} = -V_{23}$ (see Eq. 5.20), we have

$$D_{32} = \frac{2Z_3}{Z_3 + Z_2} = 1 + V_{23}, \quad D_{23} = 1 - V_{23}.$$

After some transformations, we obtain

$$V = \frac{V_{23} + V_{12} \exp(2ik_2 d)}{1 + V_{23} V_{12} \exp(2ik_2 d)}. \tag{5.22}$$

Exactly the same expression follows from Eq. 5.18 for this case since we have $\alpha_2 = k_2$ at normal incidence.

3. *Some special cases. Halfwave layer*

To begin with, we shall limit ourselves to normal incidence, and shall consider a layer with a thickness of an integral number of half-wavelengths:

$$d = m \frac{\lambda_2}{2}, \text{ i.e. } 2k_2 d = 2\pi m, \quad m = 1, 2, \ldots$$

According to Eq. 5.22, the coefficient of reflection from the layer will be

$$V = \frac{V_{23} + V_{12}}{1 + V_{23} V_{12}}. \tag{5.23}$$

Substituting for V_{23} and V_{12} from Eq. 5.20, we immediately obtain

$$V = \frac{Z_1 - Z_3}{Z_1 + Z_3} = V_{13}. \tag{5.24}$$

Thus, the halfwave layer has no effect on the incident wave (as if the layer were absent), and the reflection coefficient is equal to the usual coefficient of reflection from the boundary between media 3 and 1 just as if they were in direct contact with one another. In particular, if media 3 and 1 have the same properties ($Z_3 = Z_1$), the reflection coefficient is zero.

This result remains valid in the case of oblique incidence, but here the condition $2k_2 d = 2\pi m$ is replaced by the condition $2k_2 d \cos \vartheta_2 = 2\pi m$.

Quarterwave transmitting layer. Equation 5.22 for the reflection coefficient can be written in the form

$$V = \frac{(V_{23} + V_{12}) + i(V_{12} - V_{23}) \tan k_2 d}{(1 + V_{23} V_{12}) + i(V_{23} V_{12} - 1) \tan k_2 d}. \tag{5.25}$$

The square of the modulus of the reflection coefficient is then

$$\rho^2 = |V|^2 = \frac{(V_{23} + V_{12})^2 - 4V_{23} V_{12} \sin^2 k_2 d}{(1 + V_{23} V_{12})^2 - 4V_{23} V_{12} \sin^2 k_2 d}. \tag{5.26}$$

The function ρ^2 oscillates as d varies. By the usual methods, we find for the extrema

$$2k_2 d = \pi m, \tag{5.27}$$

where m is again an integer. For $V_{23} V_{12} > 0$, which is satisfied when the value of Z_2 lies between Z_3 and Z_1, we have maxima when m is even and minima when m is odd. But if Z_2 lies outside the interval Z_1, Z_3, we have minima when m is even, and maxima when m is odd.

Let us examine the first case ($V_{23} V_{12} > 0$). Here, the first minimum occurs at $m = 1$, i.e. when $2k_2 d = \pi$ or when $d = \lambda_2/4$. Thus, the least reflection will be observed when the thickness of the layer is equal to a quarter of a wavelength.

As follows from Eq. 5.26, the reflection coefficient will be

$$\rho_{\min} = \frac{V_{23} - V_{12}}{1 - V_{23} V_{12}}. \tag{5.28}$$

Reflection will be completely absent when $V_{23} = V_{12}$. Substitution of Eq. 5.20 gives

$$Z_2 = \sqrt{(Z_1 Z_3)}. \tag{5.29}$$

Thus, reflection at the boundary between any two media can be eliminated entirely by inserting a quarterwave layer between the two media, the layer having an impedance equal to the geometric mean of the impedances of the two media. A similar effect will be observed when the layer thickness is equal to $\frac{3}{4}\lambda$, $\frac{5}{4}\lambda$, etc.

This principle is applied in the improvement of the quality of optical systems ("reflection reduction"). Similarly, when two electrical transmission lines are joined, a coupling element is introduced with the appropriate impedance, creating an additional quarterwave phase change.

Behavior of the modulus and phase of the reflection coefficient as the thickness of the layer varies. The reflection coefficient (with respect to energy $R = |V|^2$) is shown in Fig. 19 as a function of the layer thickness for an electromagnetic wave reflecting from a layer situated on an air–water boundary ($\epsilon_3 = 1$, $\epsilon_1 = 81$). The three curves in the Figure correspond to different values of the dielectric constant of the layer. The Figure covers layer thicknesses from 0 to $\lambda_2/2$. As d is increased further, the curves repeat themselves. The phase change upon reflection (the phase of V) is found in the usual way. If the reflection coefficient is represented in the form $V = \rho e^{i\phi}$, where ρ is the modulus and ϕ is the phase, then from Eq. 5.25, we find

$$\tan \phi = \frac{(1 - V_{23}^2) \tan 2k_2 d}{1 + V_{23}^2 + \dfrac{V_{23}}{V_{12}} (1 + V_{12}^2) \dfrac{1}{\cos 2k_2 d}}. \tag{5.30}$$

The quantity ϕ is equal to the phase change upon reflection, and is shown in Fig. 20 as a function of the layer thickness, expressed in wavelengths, for the same three cases considered in Fig. 19.

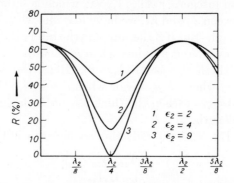

Fig. 19. The energy coefficient of reflection of an electromagnetic wave from a dielectric layer separating air and water, as a function of the layer thickness and for various values of its dielectric constant.

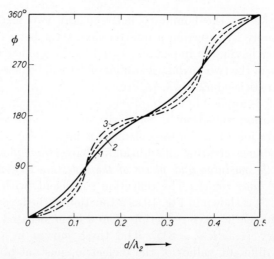

Fig. 20. The phase of the reflection coefficient under the same conditions as in Fig. 19. The curves are numbered to correspond with Fig. 19.

4. *The case of absorbing media*

When absorption is present, the reflection coefficient V and the transmission coefficient D will still be expressed by Eqs. 5.18 and 5.19, and the presence of absorption will be taken into account purely formally in that the wave numbers k_1, k_2 and k_3 in the various media will in general be complex.

In view of the relations for the angles of refraction

$$k_3 \sin \vartheta_3 = k_2 \sin \vartheta_2 = k_1 \sin \vartheta_1, \qquad (5.31)$$

the angles of refraction ϑ_2 and ϑ_1 will be complex, even if it is assumed that the angle of incidence ϑ_3 is real. The impedances Z_1, Z_2, Z_3 will be complex, and naturally the reflection coefficients V_{23} and V_{12} on both sides of the boundary will also be complex.

We use the notation

$$2\alpha_2 d = 2k_2 \cos \vartheta_2 d = \alpha + i\beta, \quad Z_j = \gamma_j + i\delta_j, \quad j = 1, 2, 3, \qquad (5.32)$$

$$V_{23} = \rho_{23}\, e^{i\phi_{23}}, \quad V_{12} = \rho_{12}\, e^{i\phi_{12}}. \qquad (5.33)$$

From Eq. 5.20 we obtain

$$\rho_{23}^2 = \frac{(\gamma_2 - \gamma_3)^2 + (\delta_2 - \delta_3)^2}{(\gamma_2 + \gamma_3)^2 + (\delta_2 + \delta_3)^2}, \quad \tan\phi_{23} = \frac{2(\delta_2 \gamma_3 - \delta_3 \gamma_2)}{\gamma_2^2 - \gamma_3^2 + \delta_2^2 - \delta_3^2} \qquad (5.34)$$

and similar expressions for ρ_{12}^2 and ϕ_{12} with appropriate subscripts.

Now, by means of elementary operations we obtain the modulus and phase of the reflection coefficient $V = \rho\, e^{i\phi}$ from Eq. 5.18:

$$\rho^2 = |V|^2 = \frac{\rho_{23}^2 + 2\rho_{12}\rho_{23}\, e^{-\beta} \cos(\phi_{12} - \phi_{23} + \alpha) + \rho_{12}^2\, e^{-2\beta}}{1 + \rho_{12}^2 \rho_{23}^2\, e^{-2\beta} + 2\rho_{12}\rho_{23}\, e^{-\beta} \cos(\phi_{12} + \phi_{23} + \alpha)}, \qquad (5.35)$$

$$\phi = A - B + \phi_{23},$$

$$A = \arctan \frac{\rho_{12}\, e^{-\beta} \sin(\phi_{12} - \phi_{23} + \alpha)}{\rho_{23} + \rho_{12}\, e^{-\beta} \cos(\phi_{12} - \phi_{23} + \alpha)}, \qquad (5.36)$$

$$B = \arctan \frac{\rho_{12}\rho_{23}\, e^{-\beta} \sin(\phi_{12} + \phi_{23} + \alpha)}{1 + \rho_{12}\rho_{23}\, e^{-\beta} \cos(\phi_{12} + \phi_{23} + \alpha)}.$$

In the same way we obtain the modulus and phase of the transmission coefficient $D = \eta\, e^{i\phi}$:

$$\eta^2 = \frac{16(\gamma_2^2 + \delta_2^2)(\gamma_1^2 + \delta_1^2)\, e^{-\beta/2}}{[(\gamma_1 + \gamma_2)^2 + (\delta_1 + \delta_2)^2]\,[(\gamma_2 + \gamma_3)^2 + (\delta_2 + \delta_3)^2]} \\ \times [1 + 2\rho_{12}\rho_{23}\, e^{-\beta} \cos(\phi_{12} + \phi_{23} + \alpha) + \rho_{12}^2 \rho_{23}^2\, e^{-2\beta}] \qquad (5.37)$$

$$\phi = \frac{\alpha}{2} + \arctan \frac{\gamma_1 \delta_2 + \delta_1 \gamma_2}{\gamma_1 \gamma_2 - \delta_1 \delta_2}$$

$$- \arctan \frac{(\gamma_2 + \gamma_3)(\delta_1 + \delta_2) + (\gamma_1 + \gamma_2)(\delta_2 + \delta_3)}{(\gamma_1 + \gamma_2)(\gamma_2 + \gamma_3) - (\delta_1 + \delta_2)(\delta_2 + \delta_3)} - B. \qquad (5.38)$$

The coefficient of reflection from an absorbing layer is again an oscillatory function of the layer thickness. However, as d increases, the amplitude of the oscillation decreases continuously, and, when d is sufficiently large, ρ becomes a constant, equal to the modulus of the reflection coefficient at the front surface of the layer.

In fact, when β is sufficiently great, all of the terms except for the first in the numerator and denominator of Eq. 5.35 can be neglected, and we obtain $\rho = \rho_{32}$. Physically, this means that when the layer is very thick, the waves are absorbed and do not reach the back surface of the layer.

Just as in the case of a nonabsorbing layer, the period of oscillation of ρ as d varies is approximately a half wavelength, provided that the absorption in the layer is not so great that it has a strong effect at distances of the order of λ_2.

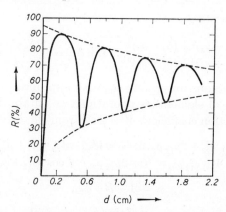

Fig. 21. The coefficient of reflection of an electromagnetic wave ($\lambda = 9.35$ cm) from a layer of water, as a function of the layer thickness, taking into account attenuation in the layer.

The experimental determination of the propagation velocity and of the absorption of sound in the medium of which the layer is composed can be based on an investigation of ρ as a function of d. The coefficient of reflection (with respect to energy) $R = \rho^2$ of an electromagnetic wave from a thin layer of water situated in air is shown in Fig. 21 as a function of the thickness of the layer. The wavelength in air is $\lambda_3 = 9.35$ cm, the wavelength in the layer is $\lambda_2 = 1.05$ cm, and $k_2 = (5.95 + 0.456i)$ cm^{-1}.

5. Wave "leakage" through a layer

We will consider a nonabsorbing layer in which the velocity of propagation is greater than in the medium through which the wave is incident. For an infinitely thick layer, total internal reflection would occur at sufficiently great angles of incidence. However, with a layer of finite thickness, complete reflection will not occur, and the wave will partially penetrate through the layer. This effect is completely analogous to the leakage of a particle through a potential barrier in quantum mechanics.

From Eq. 5.31 we have

$$\sin \vartheta_2 = \frac{k_3}{k_2} \sin \vartheta_3 = \frac{c_2}{c_3} \sin \vartheta_3.$$

When $\vartheta_3 > \arc \sin (c_3/c_2)$ (the angle of incidence is greater than the angle of total internal reflection) we obtain $\sin \vartheta_2 > 1$, whence ϑ_2 is a complex angle:

$$\cos \vartheta_2 = \pm i \sqrt{\{[(c_2/c_3) \sin \vartheta_3]^2 - 1\}}.$$

Thus, in both of Eqs. 5.10 and 5.18 for the reflection coefficient, the quantity $\alpha_2 = k_2 \cos \vartheta_2$ will be imaginary. The plus sign must be chosen in front of the radical in the expression for $\cos \vartheta_2$, if only to yield $V \to V_{23}$ in Eq. 5.18 as $d \to \infty$.

Substituting for $\cos \vartheta_2$ in Eq. 5.32, we obtain

$$2\alpha_2 d = i\beta = 2ik_2 d \sqrt{\{[(c_2/c_3) \sin \vartheta_3]^2 - 1\}}. \tag{5.39}$$

The impedance Z_2 will be purely imaginary, and according to Eqs. 5.1–5.3,

$$Z_2 = i\delta_2,$$

where $\quad \delta_2 = -\dfrac{\rho_2 c_2}{\sqrt{\{[(c_2/c_3) \sin \vartheta_3]^2 - 1\}}} \quad$ in acoustics;

$$\delta_2 = -\sqrt{(\mu_2/\epsilon_2)} \, \frac{1}{\sqrt{\{[(c_2/c_3) \sin \vartheta_3]^2 - 1\}}} \quad \text{in electrodynamics, when}$$

$\mathbf{E}$ is perpendicular to the plane of incidence;

$$\delta_2 = \sqrt{(\mu_2/\epsilon_2)} \sqrt{\{[(c_2/c_3) \sin \vartheta_3]^2 - 1\}} \quad \text{in electrodynamics, when}$$

$\mathbf{E}$ lies in the plane of incidence.

We shall consider in detail the case in which the media on both sides of the layer are identical. Then $Z_3 = Z_1$ are real, and Eq. 5.11 gives

$$V = \frac{Z_1^2 + \delta_2^2}{\delta_2^2 - Z_1^2 - 2iZ_1 \delta_2 \coth (\beta/2)}.$$

The modulus and phase of the reflection coefficient are then easily obtained:

$$\rho = \frac{Z_1^2 + \delta_2^2}{\sqrt{[(Z_1^2 - \delta_2^2)^2 + 4Z_1^2 \delta_2^2 \coth^2 (\beta/2)]}}, \tag{5.40}$$

$$\phi = \arc \tan \frac{2Z_1 \delta_2 \coth (\beta/2)}{\delta_2^2 - Z_1^2}. \tag{5.41}$$

We see from Eq. 5.40 that the modulus of the reflection coefficient is always less than unity, i.e. there is always leakage of the wave through the layer. As the thickness of the layer increases, the reflection coefficient increases. As $d \to \infty$ $(\beta \to \infty)$ it becomes equal to unity, i.e. the reflection becomes complete.

6. *The reflection and transmission coefficients for an arbitrary number of layers*

Let us suppose that between two semi-infinite media, which we denote by 1 and $n+1$, there are $n-1$ layers, denoted by $2, 3, ..., n$ (Fig. 22).

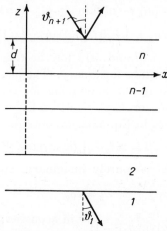

Fig. 22. Layer schematic used for calculation.

Let a plane wave be incident on the last layer at an angle of incidence ϑ_{n+1} and let the plane of incidence be the xz-plane. The z-axis will be normal to the boundaries of the layers. As a result of multiple reflections at the boundaries of the layers, two waves will exist, in each of the media, with the exception of medium 1. One of these waves propagates in the direction of positive z, and the other in the direction of negative z. In medium 1 there will be only the wave which has passed through the entire system of layers and propagates in the negative z direction. Our problem is to determine the amplitude of this transmitted wave and the amplitude of the reflected wave in medium $n+1$.

The method of calculation developed below will be valid for acoustics and for both types of polarization in electrodynamics. However, for definiteness, we shall start by considering the reflection of an electromagnetic wave in which the electric vector is perpendicular to the plane of incidence (along the y-axis).

We use the following notation: $\alpha_j = k_j \cos \vartheta_j$, $j = 1, 2, ..., n+1$, is the z-component of the wave vector in the jth layer; z_j is the coordinate of the boundary between the jth and the $(j+1)$st layers (see Fig. 22); $d_j = z_j - z_{j-1}$ is the thickness of the layer; $\phi_j = \alpha_j d_j$ is the phase change in the jth layer; $Z_j = \sqrt{(\mu_j / \epsilon_j)} \, (1/\cos \vartheta_j)$ is the normal impedance.

According to the definition of the impedance (see §5.1), the electric and magnetic fields in the jth medium are given by

$$E_{jy} = A_j \exp\left[-i\alpha_j(z-z_{j-1})\right] + B_j \exp\left[i\alpha_j(z-z_{j-1})\right],$$

$$H_{jx} = \frac{1}{Z_j}\{A_j \exp\left[-i\alpha_j(z-z_{j-1})\right] - B_j \exp\left[i\alpha_j(z-z_{j-1})\right]\}. \qquad (5.42)$$

$$j = 2, 3, \ldots, n+1.$$

In medium 1 we have

$$E_{1y} = A_1 \exp\left(-i\alpha_1 z\right),$$

$$H_{1x} = -\frac{A_1}{Z_1} \exp\left(-i\alpha_1 z\right).$$

The x and t dependence of all quantities is given by

$$\exp\left[i(k_{n+1} x \sin\vartheta_{n+1} - \omega t)\right]$$

which we omit for brevity. In the expressions written above, the coefficients A_j and B_j are the amplitudes of the direct and return waves in the jth medium. The amplitude A_{n+1} of the incident wave is assumed to be known. The remaining $2n$ amplitudes: $A_1, A_2, B_2, \ldots, A_n, B_n, B_{n+1}$ can be obtained from the $2n$ equations expressing the continuity of the tangential components of the field (E_y and H_x) across the boundaries.

In particular, on the boundary of the layers j and $j+1$ ($z = z_j$), these equations are written in the form:

continuity of E_y:

$$A_j \exp\left(-i\phi_j\right) + B_j \exp\left(i\phi_j\right) = A_{j+1} + B_{j+1}; \qquad (5.43)$$

continuity of H_x:

$$A_j \exp\left(-i\phi_j\right) - B_j \exp\left(i\phi_j\right) = \frac{Z_j}{Z_{j+1}}\left(A_{j+1} - B_{j+1}\right).$$

Letting the subscript j in the last equations vary from $j = 1$ to $j = n$ (with $B_1 = 0$), we obtain the required system of $2n$ equations for the determination of the amplitudes.

Methods for the investigation of this system of equations have been considered in a number of works (see, for example, Refs. 74, 102, 208). As a result, a general scheme has been obtained, in the form of which the solutions can be represented.

However, these solutions have an extremely unwieldy form and, if the number of layers exceeds 3, are almost useless for the analysis of the influence of one or another layer on the reflection coefficient.

We shall therefore follow a different path, namely, we shall develop a convenient method for the determination of the input impedance

$Z_{\text{in}}^{(n)}$ of a system of n layers, after which the reflection coefficient can be very simply determined.

The validity of the following discussion is independent of the form of the wave under consideration.

Let us suppose for the moment that we are considering the reflection of a wave from the boundary between media 1 and 2, and that medium 2 (see Fig. 22) stretches to infinity in the positive z direction. As we already know, the reflection coefficient can be found by using the formula

$$V = \frac{Z_{\text{in}} - Z_2}{Z_{\text{in}} + Z_2},\qquad(5.44)$$

where, in this case, Z_{in} is equal to Z_1—the impedance of the reflecting medium.

Now let us suppose that we have one layer (with subscript 2) and that medium 3 stretches to infinity. As has been shown above, the reflection from the layer is determined by the equation

$$V = \frac{Z_{\text{in}}^{(2)} - Z_3}{Z_{\text{in}}^{(2)} + Z_3},\qquad(5.45)$$

where

$$Z_{\text{in}}^{(2)} = \frac{Z_1 - iZ_2 \tan \phi_2}{Z_2 - iZ_1 \tan \phi_2} \cdot Z_2.\qquad(5.46)$$

The last equation is of great importance to us. It makes it possible to determine the impedance Z_{in} at the front (with respect to the incident wave) surface of the layer, if the impedance Z_1 at the back surface and the phase change ϕ_2 in the layer are known. Let us suppose further that there are now two layers with subscripts 2 and 3, and that the wave is incident from medium 4.

The impedance at the back surface of layer 3 will coincide with the input impedance $Z_{\text{in}}^{(2)}$ at the front surface of layer 2, because as the boundary of separation is crossed, the tangential components of the electric and magnetic fields (and in acoustics, the pressure and the normal component of the particle velocity), and hence, the impedances, remain unchanged. The input impedance at the front of layer 3 can now be calculated from Eq. 5.46 by replacing Z_2 by Z_3, Z_1 by Z_{in}, and ϕ_2 by ϕ_3. We then obtain the equation

$$Z_{\text{in}}^{(3)} = \frac{Z_{\text{in}}^{(2)} - iZ_3 \tan \phi_3}{Z_3 - iZ_{\text{in}}^{(2)} \tan \phi_3} \cdot Z_3.\qquad(5.47)$$

The input impedance of an arbitrary system of layers can be calculated by the successive application of Eq. 5.47. Thus, if we had found the input impedance $Z_{in}^{(n-1)}$ at the front surface of the $(n-1)$st layer, the addition of the nth layer results in the input impedance

$$Z_{in}^{(n)} = \frac{Z_{in}^{(n-1)} - iZ_n \tan\phi_n}{Z_n - iZ_{in}^{(n-1)} \tan\phi_n} \cdot Z_n \qquad (5.48)$$

for the entire system.

From the definition of $Z_{in}^{(n)}$—the input impedance at the front surface of the last layer—the reflection coefficient, equal to the amplitude ratio B_{n+1}/A_{n+1}, is given quite simply by the equation

$$V = \frac{Z_{in}^{(n)} - Z_{n+1}}{Z_{in}^{(n)} + Z_{n+1}}, \qquad (5.49)$$

which has the same form as the expression for the reflection coefficient from one boundary.

The calculation of the input impedance by means of the successive application of Eq. 5.48 is quite elementary when absorption is absent, i.e. when ϕ is real. When ϕ is complex, the calculation is more tedious. It is then convenient to use a special nomogram, [98] with the aid of which $Z_{in}^{(n)}$ can be determined from prescribed values of $Z_{in}^{(n-1)}$, Z_n and ϕ_n, in the most general case including the presence of absorption.

We shall now find the amplitude of the transmitted wave. We use boundary conditions (5.43) at the boundary between two arbitrary neighboring layers. Assuming that the input impedance $Z_{in}^{(j)}$ of the system of j layers is known, the field in the $(j+1)$st layer at the boundary with the jth layer will be given by

$$Z_{in}^{(j)} = \left(\frac{E_{y,j+1}}{H_{x,j+1}}\right)_{z_j} = \frac{A_{j+1} + B_{j+1}}{A_{j+1} - B_{j+1}} \cdot Z_{j+1}. \qquad (5.50)$$

Eliminating the ratios B_{j+1}/A_{j+1} and B_j/A_{j+1} from the three equations (5.43) and (5.50), we find

$$\frac{A_j}{A_{j+1}} = \frac{Z_j + Z_{in}^{(j)}}{Z_{j+1} + Z_{in}^{(j)}} \exp(i\phi_j).$$

Letting the subscript j take successively all values from $j = 1$ (here $d_1 = 0$) to $j = n$, and multiplying together all the equations thus obtained, we obtain an expression for the transmission coefficient

$$\frac{A_{n+1}}{A_1} = \prod_{j=1}^{j=n} \frac{Z_{j+1} + Z_{in}^{(j)}}{Z_j + Z_{in}^{(j)}} \exp(-i\phi_j). \qquad (5.51)$$

7. *The reflection coefficient for the special cases of two and three layers*

The method developed above for the determination of the reflection and transmission coefficients is convenient for numerical calculations.

However, it is difficult through the use of this method to estimate the influence on these coefficients of the parameters of the individual layers. Therefore, we shall find an explicit form for the expression for the reflection coefficient in the case of two and three layers. The single layer was considered above. It is not sensible to write out the expressions for a large number of layers because these expressions have an extremely unwieldy and unclear form.

Substituting Eq. 5.46 for $Z_{in}^{(2)}$ into Eq. 5.47, we find

$$Z_{in}^{(3)} = \frac{Z_1 Z_2 - Z_1 Z_3 \tan\phi_2 \tan\phi_3 - i(Z_2^2 \tan\phi_2 + Z_2 Z_3 \tan\phi_3)}{Z_2 Z_3 - Z_2^2 \tan\phi_2 \tan\phi_3 - i(Z_1 Z_3 \tan\phi_2 + Z_2 Z_1 \tan\phi_3)} \cdot Z_3 \qquad (5.52)$$

for the input impedance of two layers.

The reflection coefficient is found from the relation

$$V = \frac{Z_{in}^{(3)} - Z_4}{Z_{in}^{(3)} + Z_4}. \qquad (5.53)$$

We recall that, according to our notation, the subscripts 1 and 4 refer to the end media, and 2 and 3 to the layers. We now consider the case of three layers. Substituting Eq. 5.52 into Eq. 5.48, and, having set $n = 4$, we obtain for the input impedance of a system of three layers

$$Z_{in}^{(3)} = \frac{Z_1 Z_2 Z_3 - Z_1 Z_3^2 \delta_2 \delta_3 - Z_1 Z_3 Z_4 \delta_2 \delta_4 - Z_1 Z_2 Z_4 \delta_3 \delta_4}{Z_2 Z_3 Z_4 - Z_2^2 Z_4 \delta_2 \delta_3 - Z_2^2 Z_3 \delta_2 \delta_4 - Z_2 Z_3^2 \delta_3 \delta_4}$$

$$\frac{- i[Z_2^2 Z_3 \delta_2 + Z_2 Z_3^2 \delta_3 + Z_2 Z_3 Z_4 \delta_4 - Z_2^2 Z_4 \delta_2 \delta_3 \delta_4]}{- i[Z_1 Z_3 Z_4 \delta_2 + Z_1 Z_2 Z_4 \delta_3 + Z_1 Z_2 Z_3 \delta_4 - Z_1 Z_3^2 \delta_2 \delta_3 \delta_4]} \cdot Z_4, \qquad (5.54)$$

where we have used the notation

$$\delta_j = \tan\phi_j, \quad j = 2, 3, 4. \qquad (5.55)$$

The reflection coefficient is now written

$$V = \frac{Z_{in}^{(4)} - Z_5}{Z_{in}^{(4)} + Z_5}. \qquad (5.56)$$

In the special case in which the thicknesses of all the layers are zero or are an integral number of half wavelengths, i.e. $\phi_j = \pi N_j$ where N_j is an integer, we have $\delta_j = 0$ and $Z_{in}^{(4)} = Z_1$. As could be expected, the layers will have no effect on the reflection in this case, and the input impedance of the whole system will be equal to the impedance of medium 1, just as if the layers were not there.

Similarly, it can be verified that if one of the δ_j becomes zero, the corresponding impedance is eliminated from the formula, and the input impedance of the entire system will be the same as in the case of $(n-1)$ layers.

A number of numerical examples of reflection from two- and three-layered coverings will be considered in the next chapter in the sections concerned with the problem of the transmissivity of optical and acoustic systems.

§ 6. Elastic Waves in Solid Layered Media

1. *The reflection and transmission coefficients for an arbitrary number of layers*

The propagation of elastic waves in solid layers is of fundamental interest in seismology. and has been studied in Refs. 50, 53, 60, 61, 62, 188 and 193.

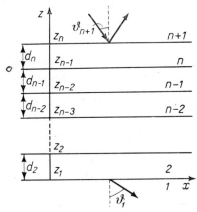

Fig. 23. Notation used in the calculation of the reflection and transmission coefficients for a system of solid layers.

We shall investigate the reflection of a plane wave from an arbitrary number of solid layers. The reflection and transmission coefficients will be determined by a method based on recurrence formulas connecting the wave amplitudes in neighboring layers.[198]

We shall use the notation previously employed in the investigation of reflection from a single solid–liquid interface.

Let us consider the nth layer, which may be located either within the system of layers or at the edge. We denote its thickness by d, and choose a coordinate system as illustrated in Fig. 23, with the origin at the boundary between the nth and the $(n-1)$st layer. As a result of reflections from the boundaries of the layer, a system of longitudinal and transverse waves will exist, propagating in the direction of positive and negative z. The potentials of the longitudinal and the transverse waves in the layer can then be written in the form

$$\phi = (\phi' \, e^{i\alpha z} + \phi'' \, e^{-i\alpha z}) \, e^{i(\sigma x - \omega t)},$$

$$\psi = (\psi' \, e^{i\beta z} + \psi'' \, e^{-i\beta z}) \, e^{i(\sigma x - \omega t)},$$

$$\alpha = \sqrt{(k_n^2 - \sigma^2)}, \quad \beta = \sqrt{(\kappa_n^2 - \sigma^2)} \qquad (6.1)$$

where $\sigma = k_n \sin \vartheta_n = k_{n+1} \sin \vartheta_{n+1}$ is the x component of the wave vector; σ determines the phase velocity along any boundary. As a result of the continuity conditions of the field at the boundaries, the phase velocity is the same for longitudinal and for transverse waves, and for all layers.

The components of the particle velocity v_x and v_z at any point in the layer are found from Eq. 4.2. The normal components of the stress tensor Z_z and the tangential components Z_x are obtained from Eq. 4.5, in which we must set $u_x = (i/\omega) \, v_x$, $u_z = (i/\omega) \, v_z$. We denote the values of v_x, v_z, Z_z, Z_x at the upper boundary of the nth layer (i.e. at $z = d$) by $v_x^{(n)}, v_z^{(n)}, Z_z^{(n)}, Z_x^{(n)}$. Straightforward calculations give (we use the notation $\alpha d = P, \beta \alpha = Q$): $\quad P = \alpha d \quad Q = \beta d$

$$v_x^{(n)} = [i\sigma \cos P(\phi' + \phi'') - \sigma \sin P(\phi' - \phi'') - i\beta \cos Q(\psi' - \psi'')$$
$$+ \beta \sin Q(\psi' + \psi'')] e^{i(\sigma x - \omega t)}. \qquad (6.2)$$

The formulas for $v_z^{(n)}, Z_z^{(n)}$ and $Z_x^{(n)}$ have a similar form. The only differences are the coefficients of $\phi' + \phi''$, $\phi' - \phi''$, $\psi' + \psi''$ and $\psi' - \psi''$.

All of these equations can be written in the form of the following table [the factor $\exp(\sigma x - \omega t)$ is omitted]:

$$
\begin{bmatrix}
v_x^{(n)} \\
v_z^{(n)} \\
Z_z^{(n)} \\
\dfrac{1}{2\mu} Z_x^{(n)}
\end{bmatrix}
$$

$$
=
\begin{bmatrix}
i\sigma \cos P & -\sigma \sin P & -i\beta \cos Q & \beta \sin Q \\
-\alpha \sin P & i\alpha \cos P & -\sigma \sin Q & i\sigma \cos Q \\
-\dfrac{i}{\omega}(\lambda k^2 + 2\mu\alpha^2) \cos P & \dfrac{1}{\omega}(\lambda k^2 + 2\mu\alpha^2) \sin P & -\dfrac{2i\mu}{\omega}\sigma\beta \cos Q & \dfrac{2\mu}{\omega}\sigma\beta \sin Q \\
\dfrac{\alpha\sigma}{\omega} \sin P & -\dfrac{i\alpha\sigma}{\omega} \cos P & \dfrac{1}{2\omega}(\sigma^2 - \beta^2) \sin Q & \dfrac{i}{2\omega}(\beta^2 - \sigma^2) \cos Q
\end{bmatrix}
$$

$$
\times
\begin{bmatrix}
\phi' + \phi'' \\
\phi' - \phi'' \\
\psi' - \psi'' \\
\psi' + \psi''
\end{bmatrix}. \qquad (6.3)
$$

To obtain the expression for the mth element of the column on the left hand side of the equation, we apply the rules of matrix algebra to the right hand side, i.e. the elements of the column on the right hand side are multiplied successively by the appropriate element of the mth row of the matrix, and the results are added. As an example, the reader should use Eq. 6.3 to obtain Eq. 6.2.

By setting $P = Q = 0$ in Eq. 6.3, we obtain a table which gives the analogous expressions at the other side of the layer (at $z = 0$). Since these components of the stress tensor and both components of the particle velocity remain unchanged as we cross the boundary, the same expression will hold for the quantities $v_x^{(n-1)}, \ldots, Z_x^{(n-1)}$ at the upper boundary of the $(n-1)$st layer (see Fig. 23). Thus, we will have

$$
\begin{bmatrix} v_x^{(n-1)} \\ v_z^{(n-1)} \\ Z_z^{(n-1)} \\ \dfrac{1}{2\mu} Z_x^{(n-1)} \end{bmatrix} = \begin{bmatrix} i\sigma & 0 & -i\beta & 0 \\ 0 & i\alpha & 0 & i\sigma \\ -\dfrac{i}{\omega}(\lambda k^2 + 2\mu\alpha^2) & 0 & -\dfrac{2i\mu}{\omega}\sigma\beta & 0 \\ 0 & -\dfrac{i\alpha\sigma}{\omega} & 0 & \dfrac{i}{2\omega}(\beta^2 - \sigma^2) \end{bmatrix} \begin{bmatrix} \phi' + \phi'' \\ \phi' - \phi'' \\ \psi' - \psi'' \\ \psi' + \psi'' \end{bmatrix}.
$$

$$(6.4)$$

We shall now find the recurrence formula connecting $v_x^{(n)}, \ldots, Z_x^{(n)}$ with $v_x^{(n-1)}, \ldots, Z_x^{(n-1)}$ in the neighboring layer. Solving the system of linear equations (6.4) for $\phi' + \phi'', \ldots, \psi' + \psi''$, we obtain

$$
\begin{bmatrix} \phi' + \phi'' \\ \phi' - \phi'' \\ \psi' - \psi'' \\ \psi' + \psi'' \end{bmatrix} = \begin{bmatrix} -\dfrac{2i\sigma}{\kappa^2} & 0 & \dfrac{i\omega}{\mu\kappa^2} & 0 \\ 0 & \dfrac{i(\sigma^2 + \beta^2)}{\alpha\kappa^2} & 0 & \dfrac{2i\omega\sigma}{\alpha\kappa^2} \\ \dfrac{i[k^2(\lambda/\mu) + 2\alpha^2]}{\beta\kappa^2} & 0 & \dfrac{i\sigma\omega}{\beta\mu\kappa^2} & 0 \\ 0 & -\dfrac{2i\sigma}{\kappa^2} & 0 & -\dfrac{2i\omega}{\kappa^2} \end{bmatrix} \begin{bmatrix} v_x^{(n-1)} \\ v_z^{(n-1)} \\ Z_z^{(n-1)} \\ \dfrac{1}{2\mu} Z_x^{(n-1)} \end{bmatrix}.
$$

$$(6.5)$$

Substituting Eqs. 6.5 for $\phi' + \phi'', \ldots, \psi' + \psi''$ into Eq. 6.3, we find, after some straightforward though tedious operations,

$$
\begin{bmatrix} v_x^{(n)} \\ v_z^{(n)} \\ Z_z^{(n)} \\ \dfrac{Z_x^{(n)}}{2\mu} \end{bmatrix} = \begin{bmatrix} a_{11} & a_{12} & a_{13} & a_{14} \\ a_{21} & a_{22} & a_{23} & a_{24} \\ a_{31} & a_{32} & a_{33} & a_{34} \\ a_{41} & a_{42} & a_{43} & a_{44} \end{bmatrix} \begin{bmatrix} v_x^{(n-1)} \\ v_z^{(n-1)} \\ Z_z^{(n-1)} \\ \dfrac{Z_x^{(n-1)}}{2\mu} \end{bmatrix}, \qquad (6.6)
$$

where $a_{11}, \ldots, a_{44}$ are given by the following expressions ($\sin \vartheta = \sigma/k$, $\sin \gamma = \sigma/\kappa$, c and b are the propagation velocities of the longitudinal and the transverse waves, respectively):

$$a_{11} = 2 \sin^2 \gamma \cos P + \cos 2\gamma \cos Q,$$

$$a_{12} = i(\tan \vartheta \cos 2\gamma \sin P - \sin 2\gamma \sin Q),$$

$$a_{13} = \frac{\sin \vartheta}{\rho c} (\cos Q - \cos P),$$

$$a_{14} = -2ib(\tan \vartheta \sin \gamma \sin P + \cos \gamma \sin Q),$$

$$a_{21} = i\left(\frac{b \cos \vartheta}{c \cos \gamma} \sin 2\gamma \sin P - \tan \gamma \cos 2\gamma \sin Q\right),$$

$$a_{22} = \cos 2\gamma \cos P + 2 \sin^2 \gamma \cos Q,$$

$$a_{23} = -\frac{i}{\rho c} (\cos \vartheta \sin P + \tan \gamma \sin \vartheta \sin Q),$$

$$a_{24} = 2b \sin \gamma (\cos Q - \cos P),$$

$$a_{31} = 2\rho b \sin \gamma \cos 2\gamma (\cos Q - \cos P),$$

$$a_{32} = -i\rho\left(\frac{c \cos^2 2\gamma}{\cos \vartheta} \sin P + 4b \cos \gamma \sin^2 \gamma . \sin Q\right),$$

$$a_{33} = \cos 2\gamma . \cos P + 2 \sin^2 \gamma . \cos Q,$$

$$a_{34} = 2i\rho b^2(\cos 2\gamma \tan \vartheta \sin P - \sin 2\gamma \sin Q),$$

$$a_{41} = -i\left(\frac{2}{c} \cos \vartheta \sin^2 \gamma \sin P + \frac{\cos^2 2\gamma}{2b \cos \gamma} \sin Q\right),$$

$$a_{42} = \frac{\sin \vartheta . \cos 2\gamma}{c} (\cos Q - \cos P),$$

$$a_{43} = \frac{i}{2\rho} \left(\frac{\sin 2\vartheta}{c^2} \sin P - \frac{\cos 2\gamma}{b^2} \tan \gamma . \sin Q\right),$$

$$a_{44} = 2 \sin^2 \gamma . \cos P + \cos 2\gamma . \cos Q.$$

The angles ϑ and γ can be expressed quite simply in terms of the angle of incidence ϑ_{n+1} by using Snell's law:

$$\frac{\sin \vartheta_{n+1}}{c_{n+1}} = \frac{\sin \vartheta}{c} = \frac{\sin \gamma}{b},$$

where ϑ, γ, c and b refer to an arbitrary layer. If $c/c_{n+1} > 1$, then we will have $\sin \vartheta > 1$ when $\sin \vartheta_{n+1} > c_{n+1}/c$, i.e. ϑ will be a complex angle. It will be convenient in calculations to set $\vartheta = (\pi/2) + i\xi$ (then $\sin \vartheta = \cosh \xi$, $\cos \vartheta = -i \sinh \xi$).

What has been said above also applies to the angle γ, if $b/c_{n+1} > 1$.

If v_x, v_z, Z_z, Z_x are given at the boundary of separation between media 1 and 2 [we denote them by $v_x^{(1)}, \ldots, Z_x^{(1)}$], then by successive application of Eqs. 6.6 we can find their values at the boundary between layers n and $n+1$, for arbitrary n.

The result of the successive application of Eq. 6.6 can be written symbolically in the form

$$
\begin{bmatrix} v_x^n \\ v_z^n \\ Z_z^{(n)} \\ \dfrac{1}{2\mu} Z_x^{(n)} \end{bmatrix}
= \begin{bmatrix} a_n \end{bmatrix} \begin{bmatrix} a_{n-1} \end{bmatrix} \cdots \begin{bmatrix} a_2 \end{bmatrix}
\begin{bmatrix} v_x^{(1)} \\ v_z^{(1)} \\ Z_z^{(1)} \\ \dfrac{1}{2\mu} Z_x^{(1)} \end{bmatrix}, \qquad (6.7)
$$

where $[a_k]$ denotes the matrix (of the type entering into Eq. 6.6) for the kth layer. Equation 6.7 can also be written in the form

$$
\begin{bmatrix} v_x^{(n)} \\ v_z^{(n)} \\ Z_z^{(n)} \\ \dfrac{1}{2\mu} Z_x^{(n)} \end{bmatrix}
= \begin{bmatrix} A_{11} & A_{12} & A_{13} & A_{14} \\ A_{21} & A_{22} & A_{23} & A_{24} \\ A_{31} & A_{32} & A_{33} & A_{34} \\ A_{41} & A_{42} & A_{43} & A_{44} \end{bmatrix}
\begin{bmatrix} v_x^{(1)} \\ v_z^{(1)} \\ Z_z^{(1)} \\ \dfrac{1}{2\mu} Z_x^{(1)} \end{bmatrix}, \qquad (6.8)
$$

where the resultant matrices $A_{11}, \ldots, A_{44}$ are the products of the matrices $[a_2], \ldots, [a_n]$.

If the bounding media 1 and $n+1$ are liquid, the tangential stresses $Z_x^{(1)}$ and $Z_x^{(n)}$ must vanish. But, according to Eq. 6.8, we have

$$
\frac{1}{2\mu} Z_x^{(n)} = A_{41} v_x^{(1)} + A_{42} v_z^{(1)} + A_{43} Z_z^{(1)} + A_{44} \frac{1}{2\mu} Z_x^{(1)}. \qquad (6.9)
$$

Therefore, using $Z_x^{(1)} = Z_x^{(n)} = 0$, we obtain the relation

$$
A_{41} v_x^{(1)} + A_{42} v_z^{(1)} + A_{43} Z_z^{(1)} = 0. \qquad (6.10)
$$

Using Eq. 6.8 to write the quantities $v_z^{(n)}$ and $Z_z^{(n)}$ in the form of linear combinations of the type (6.9), and expressing $v_x^{(1)}$ in terms of $v_z^{(1)}$ and $Z_z^{(1)}$ through Eq. 6.10, we obtain

$$
\begin{aligned}
v_z^{(n)} &= M_{22} v_z^{(1)} + M_{23} Z_z^{(1)}, \\
Z_z^{(n)} &= M_{32} v_z^{(1)} + M_{33} Z_z^{(1)},
\end{aligned} \qquad (6.11)
$$

where
$$
M_{22} = A_{22} - \frac{A_{21} A_{42}}{A_{41}}, \quad M_{23} = A_{23} - \frac{A_{21} A_{43}}{A_{41}},
$$

$$
M_{32} = A_{32} - \frac{A_{31} A_{42}}{A_{41}}, \quad M_{33} = A_{33} - \frac{A_{31} A_{43}}{A_{41}}. \qquad (6.12)
$$

3

We shall now find the reflection and transmission coefficients for the entire system of layers. We denote the total thickness by H, and place the origin of coordinates at the lower boundary of the first layer. We shall suppose that a sound wave is incident from the liquid with the subscript $n+1$ at an angle of incidence ϑ_{n+1}. The expression for the combined acoustic potential of the incident and reflected wave in this medium can be written in the form

$$\phi_{n+1} = \{\phi' \exp [i\alpha_{n+1}(z+H)] + \phi'' \exp [-i\alpha_{n+1}(z-H)]\} \exp [i(\sigma x - \omega t)],$$

$$\alpha_{n+1} = k_{n+1} \cos \vartheta_{n+1}, \tag{6.13}$$

whereas the potential of the transverse wave $\psi_{n+1} = 0$.

In the liquid on the other side of the system of solid layers, there will be only a transmitted sound wave

$$\phi_1 = \phi''' \exp (i\alpha_1 z) \exp [i(\sigma x - \omega t)], \quad \psi_1 = 0,$$

$$\alpha_1 = k_1 \cos \vartheta_1; \tag{6.14}$$

and the angles ϑ_{n+1} and ϑ_1 are connected by the relation

$$k_1 \sin \vartheta_1 = k_{n+1} \sin \vartheta_{n+1}.$$

Since the particle velocity is obtained from the displacement by multiplying by $-i\omega$, Eqs. 4.1 and 4.5 (for a liquid) give

$$v_z = \frac{\partial \phi}{\partial z},$$

$$Z_z = i\frac{\lambda}{\omega} \operatorname{div} \mathbf{v} = i\frac{\lambda}{\omega} \nabla^2 \phi = -i\frac{\lambda}{\omega} k^2 \phi = -i\rho\omega\phi. \tag{6.15}$$

In place of ϕ, we successively substitute ϕ_{n+1} and ϕ_1, setting $z = H$ and $z = 0$ as required. According to the continuity conditions, the values of the normal component of the particle velocity and the normal component of the stress tensor [$v_z^{(n)}$ and $Z_z^{(n)}$ respectively] at the upper boundary of layer n are equal to the same quantities in the liquid at $z = H$, and we obtain

$$v_z^{(n)} = i\alpha_{n+1}(\phi' - \phi''), \quad v_z^{(1)} = i\alpha_1 \phi''',$$

$$Z_z^{(n)} = -i\rho_{n+1}\omega(\phi' + \phi''), \quad Z_z^{(1)} = -i\rho_1 \omega\phi''', \tag{6.16}$$

where the common factor $e^{i(\sigma x - \omega t)}$ has been omitted for brevity.

Substituting Eq. 6.16 in Eq. 6.11, we obtain two equations for the determination of the reflection and the transmission coefficients. These equations yield

$$V = \frac{\phi''}{\phi'} = \frac{M_{32} - Z_1 M_{33} + (M_{22} - Z_1 M_{23}) \cdot Z_{n+1}}{M_{32} - Z_1 M_{33} - (M_{22} - Z_1 M_{23}) \cdot Z_{n+1}}, \tag{6.17}$$

where $Z_1 = \rho_1 \omega/\alpha$, $Z_{n+1} = \rho_{n+1}\omega/\alpha_{n+1}$ are the impedances of the liquid on one and the other side of the layer.

We define the transmission coefficient as the ratio of the acoustic pressures in the transmitted and the incident waves, i.e.

$$D = \rho_1 \phi''' / \rho_{n+1} \phi'.$$

The same equations then yield

$$D = \frac{2Z_1}{(M_{22} - Z_1 M_{23})Z_{n+1} - M_{32} + Z_1 M_{33}}. \tag{6.18}$$

Setting the shear wave velocity in the layers equal to zero, we obtain the previously studied case of reflection from liquid layers. We shall verify this for the simplest case of a single layer. Let a wave be incident from medium 3 onto layer 2, and let it penetrate partially into medium 1 (Fig. 24). Since there is only one layer, $A_{ik} = a_{ik}$. Since the layer is

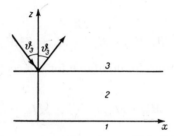

Fig. 24. Special case of a single layer.

assumed to be liquid, we must set $b = 0$ (and consequently $\kappa = \infty$, $\gamma = 0$) in the above formulas for a_{ik}. The table for the a_{ik} is then

$$
\begin{bmatrix}
\cos Q & i\tan\vartheta_2 \sin P & \dfrac{\sin\vartheta_2}{\rho_2 c_2}(\cos Q - \cos P) & 0 \\[2ex]
0 & \cos P & -\dfrac{i\cos\vartheta_2}{\rho_2 c_2}\sin P & 0 \\[2ex]
0 & -\dfrac{i\rho_2 c_2}{\cos\vartheta_2}\sin P & \cos P & 0 \\[2ex]
\infty & \dfrac{\sin\vartheta_2}{c_2}(\cos Q - \cos P) & \infty & \cos Q
\end{bmatrix}
$$

and, according to Eq. 6.12, we obtain

$$M_{22} = a_{22} = \cos P, \quad M_{23} = a_{23} = -\frac{i}{Z_2}\sin P,$$

$$M_{32} = a_{32} = -iZ_2 \sin P, \quad M_{33} = a_{33} = \cos P,$$

where $Z_2 = \rho_2 c_2/\cos\vartheta_2$.

Substituting these expressions into Eqs. 6.17 and 6.18, we obtain expressions coinciding with Eqs. 5.10 and 5.15, the only difference being the sign of α_2 as a result of the difference in the choice of the direction of the z-axis.

2. *Reflection from a solid plate*

We now consider the reflection from a solid plate 2 separating two liquid media 1 and 3. As previously, we will assume that the wave is incident on the plate from medium 3.

The reflection and transmission coefficients can be found from the general formulas (6.17) and (6.18), in which we must set $n = 2$. The quantities $M_{22}, \ldots$ in these formulas are found from relations (6.12), in which $A_{ik} = a_{ik}$.

We first consider the case in which the plate is bounded on both sides by the same liquid, i.e. $Z_3 = Z_1$. Then Eq. 6.18 for the transmission coefficient is written

$$D = \frac{2Z_1}{Z_1(M_{22} + M_{33}) - M_{32} - Z_1^2 M_{33}}. \qquad (6.19)$$

By the use of the values found above for the coefficients a_{ik}, Eq. 6.12 gives

$$M_{23} = -\frac{i}{Z_1 N}, \quad M_{22} + M_{33} = \frac{2M}{N}, \quad M_{32} = \frac{iZ_1}{N}(M^2 - N^2),$$

where

$$N = \frac{Z_2}{Z_1}\frac{\cos^2 2\gamma_2}{\sin P} + \frac{Z_{2t}}{Z_1}\frac{\sin^2 2\gamma_2}{\sin Q},$$

$$M = \frac{Z_2}{Z_1}\cos^2 2\gamma_2 \cot P + \frac{Z_{2t}}{Z_1}\sin^2 2\gamma_2 \cot Q,$$

$$Z_1 = \frac{\rho_1 c_1}{\cos \vartheta_1}, \quad Z_2 = \frac{\rho_2 c_2}{\cos \vartheta_2}, \quad Z_{2t} = \frac{\rho_2 b_2}{\cos \gamma_2}. \qquad (6.20)$$

As a result, the expression for the transmission coefficient is written

$$D = \frac{2N}{2M + i(N^2 - M^2 + 1)}. \qquad (6.21)$$

In exactly the same way, we obtain for the reflection coefficient,

$$V = \frac{i(N^2 - M^2 - 1)}{2M + i(N^2 - M^2 + 1)}. \qquad (6.22)$$

If the angle of incidence ϑ_3 does not exceed the angle of total internal reflection for longitudinal or for transverse waves at the liquid–solid interface, then the angles ϑ_2 and γ_2, and consequently M and N, will be real, and we obtain for the reflection and transmission coefficients (with respect to energy)

$$R = |V|^2 = \frac{(N^2 - M^2 - 1)^2}{4M^2 + (N^2 - M^2 + 1)^2},$$

$$T = |D|^2 = \frac{4N^2}{4M^2 + (N^2 - M^2 + 1)^2}. \tag{6.23}$$

Here, $R + T = 1$, which could also have been obtained from the law of conservation of energy.

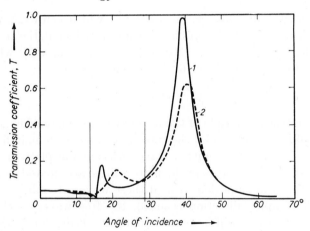

Fig. 25. Transmission coefficient of an aluminum plate, 2.5 mm thick, in water, at a frequency of 4.02×10^5 cps. 1. Theoretical curve. 2. Experimental curve.

When the properties of the liquids comprising media 3 and 1 are not identical, we obtain similarly

$$V = \frac{M(Z_1 - Z_3) + i[(N^2 - M^2)Z_3 - Z_1]}{M(Z_1 + Z_3) + i[(N^2 - M^2)Z_3 + Z_1]}, \tag{6.24}$$

$$D = \frac{2NZ_1}{M(Z_1 + Z_3) + i[(N^2 - M^2)Z_3 + Z_1]}, \tag{6.25}$$

$$Z_1 = \frac{\rho_1 c_1}{\cos \vartheta_1}, \quad Z_3 = \frac{\rho_3 c_3}{\cos \vartheta_3}. \tag{6.26}$$

Here, the transmission coefficient is defined as the ratio of the amplitude of the acoustic pressure in the transmitted wave to the analogous amplitude in the incident wave.

As an illustration, the theoretical and experimental curves of the (power) transmission coefficient T as a function of the angle of incidence are shown in Fig. 25 for an aluminum plate, 25 mm thick, in water.[190]

The acoustic frequency is 402,000 cps, which corresponds to an acoustic wavelength of $\lambda \approx 1.3$ cm in aluminum. The strong increase of the transmission in the region of $40°$ is explained by principles which will be considered in detail in the next Section. The lack of agreement in the positions of the peaks of the theoretical and experimental curves is due to the fact that attenuation was not taken into account in the theoretical calculations. The passage of sound through plates was also investigated in Ref. 111.

3. Free waves in a plate

By free waves in a plate we mean waves which can exist without external excitation, i.e. in the problem we have considered, the amplitude of the incident wave is zero. Clearly, this can occur only when the denominators in Eqs. 6.21 and 6.22 become zero, i.e.

$$2M + i(N^2 - M^2 + 1) = 0. \tag{6.27}$$

In this case, as the amplitude of the incident wave approaches zero, the amplitudes of the reflected and the transmitted waves remain finite. If the reflected and the transmitted waves leaving the plate are of the ordinary kind, the free waves in the plate will be attenuated as a result of the energy lost as radiation. But if these waves are inhomogeneous waves (with complex direction cosines), the free waves in the plate corresponding to them will be nonattenuating (absorption in the plate itself is neglected).

It is not difficult to see the analogy between the free waves in a plate and the surface waves on the interface between two media considered in §§ 3, 4. Also the occurrence of free waves in a plate is to some extent related to resonance effects.

Remembering the values of M and N (Eq. 6.20), Eq. 6.27 can be transformed into the form

$$(E + iZ_1)(F - iZ_1) = 0, \tag{6.28}$$

where
$$E = Z_2 \cos^2 2\gamma_2 \cot \frac{P}{2} + Z_{2t} \sin^2 2\gamma_2 \cot \frac{Q}{2},$$

$$F = Z_2 \cos^2 2\gamma_2 \tan \frac{P}{2} + Z_{2t} \sin^2 2\gamma_2 \tan \frac{Q}{2}. \tag{6.29}$$

The values of Z_1, Z_2 and Z_{2t} are given by Eq. 6.20. The quantities P and Q give the phase changes of longitudinal and transverse waves respectively, over the thickness of the plate. The subscript 2 refers to the plate, and the subscript 1 to the surrounding medium.

When the thickness and density of the plate, and the velocities of longitudinal and of transverse waves are given, only the angles ϑ_1, ϑ_2 and γ_2 remain undetermined in Eq. 6.28. However, since these angles are connected by Snell's law

$$\frac{\sin \vartheta_1}{c_1} = \frac{\sin \vartheta_2}{c_2} = \frac{\sin \gamma_2}{b_2}, \tag{6.30}$$

only one of these three angles is undetermined; the other two can be expressed in terms of it.

As the unknown angle to be determined by Eq. 6.28, we can choose, for example, ϑ_1. However, it will be more convenient in what follows to use Eq. 6.30 to define the quantity

$$v = \frac{c_1}{\sin \vartheta_1} = \frac{c_2}{\sin \vartheta_2} = \frac{b_2}{\sin \gamma_2}, \tag{6.31}$$

equal to the velocity with which each of the three waves moves along the plate.

Equation 6.28 separates into two equations

$$E + iZ_1 = 0,$$

$$F - iZ_1 = 0. \tag{6.32}$$

It can be shown[191] that the upper equation corresponds to the so-called symmetric waves, in which the mean plane separating the plate into two equal parts along its thickness remains at rest, and the remaining parts are displaced symmetrically with respect to this plane. The lower equation corresponds to "antisymmetric waves", a special case of which are flexural waves.

If the density of the surrounding medium is set equal to zero (whence $Z_1 = 0$) we obtain the following characteristic equation for free waves in the plate:

$$EF = 0. \tag{6.28a}$$

This equation has been investigated repeatedly.[126, 150, 191] Solving it, we find the velocity v as a function of ω

$$v = v(\omega),$$

i.e. the dispersion law. Since the solution of Eq. 6.32 is many-valued, the dispersion curve has many branches, each of which corresponds to a particular wave in the plate with a definite amplitude distribution across the thickness. The determination of the form of the dispersion curve is an extremely complicated problem, requiring lengthy calculations and involving the numerical solution of transcendental equations.

Curves calculated in this way[191] are shown in Fig. 26 for an aluminum plate. The ordinate represents the phase velocity along the plate (in m/sec), and the abscissa is the product of the frequency (in cps) and the plate thickness (in millimeters). The curves $a_0, a_1, a_2, \ldots$ correspond to antisymmetric waves, and the curves $s_0, s_1, s_2, \ldots$ to symmetric waves.

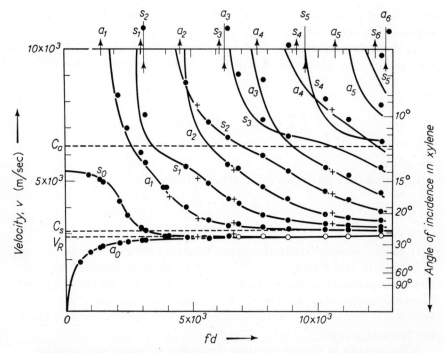

Fig. 26. Curves for the velocity of free waves in an aluminum plate in vacuum. The points were obtained from measurements of the transmission coefficient of the plate in xylene. The abscissa represents the product of the frequency in cps and the plate thickness in millimeters.

The curve a_0 refers to the usual flexural wave. For small values of the frequency–thickness product, this curves satisfies the $v \sim \sqrt{(\omega d)}$ dependence, i.e. the velocity is proportional to the square root of the frequency. This result is also obtained in the usual theory of flexural waves in thin plates.

As $\omega d \to \infty$, the curves a_0 and s_0 asymptotically approach straight lines corresponding to the velocity v_R of Rayleigh surface waves, while all the remaining curves approach straight lines corresponding to the velocity of transverse waves.

As the frequency is lowered, all the curves except a_0 and s_0 go toward infinity, asymptotically approaching vertical straight lines, corresponding to certain "limiting" frequencies. These frequencies are found from the relations:

$$\left.\begin{array}{l} \dfrac{\omega d}{2c_2} = \dfrac{\pi}{2}, \, 3\dfrac{\pi}{2}, \, \ldots \\[3mm] \dfrac{\omega d}{2b_2} = \pi, \, 2\pi, \, \ldots \end{array}\right\} \quad \text{for symmetric waves;}$$

$$\left.\begin{array}{l} \dfrac{\omega d}{2c_2} = \pi, \, 2\pi, \, \ldots \\[3mm] \dfrac{\omega d}{2b_2} = \dfrac{\pi}{2}, \, 3\dfrac{\pi}{2}, \, \ldots \end{array}\right\} \quad \text{for antisymmetric waves.}$$

Thus all the limiting frequencies correspond to plate thicknesses of an integral number of half wavelengths of longitudinal or transverse waves. The location of the limiting frequency for each curve is shown by an arrow at the upper edge of the Figure.

At lower frequencies and for a given form of oscillation of the plate, the characteristic velocity is imaginary. In this case there can be no wave propagating along the plate, but only a particle oscillation with the same phase along the plate and an exponentially decreasing amplitude. We shall encounter a similar phenomenon in § 25 in studying the propagation of waves in layers.

The fact that for waves in a plate in vacuum the velocity v must be purely real or purely imaginary also follows from energy considerations. In fact, since there is no energy radiated by the plate into the surrounding space, and absorption in the plate itself is neglected, the energy flux along the plate must be either constant (v real) or zero (v imaginary).

However, if the plate is surrounded by liquid, it is clear that waves propagating in the plate will lose energy in the form of radiation into the surrounding medium. Therefore, the waves will be attenuated, which corresponds to complex values of v. These complex values will be found from the complete equations (6.32).

In addition to these attenuating waves, there will be waves which excite inhomogeneous waves in the surrounding liquid, with amplitudes decreasing exponentially with distance from the plate. Since in this case the plate does not lose energy, the waves in the plate will be non-attenuating. A detailed investigation[191] shows that there will be two such waves. One of them is of the symmetric type; and the other is antisymmetric. The phase velocity of these waves is less than the velocity of sound in the liquid.

4. *The case of total transmission*

Using Eq. 6.23, we can easily obtain an expression for the reciprocal of the transmission coefficient T:

$$\frac{1}{T} = 1 + \frac{1}{4N^2}(1 + M^2 - N^2)^2; \tag{6.33}$$

using the notation (6.29), we can write (6.33) in the form

$$\frac{1}{T} = 1 + \left(\frac{Z_1^2 - EF}{\frac{1}{2}Z_1(E+F)}\right)^2. \tag{6.34}$$

The condition for total transmission ($T = 1$) is now written in the form

$$EF = Z_1^2. \tag{6.35}$$

This expression determines the angle of incidence ϑ_1 (or the phase velocity $v = c/\cos\vartheta$ along the plate) at which total transmission of the sound wave through the plate occurs.

Frequently, the ratio $\rho_1 c_1 / \rho_2 c_2$ of the wave impedances of the surrounding medium and the plate is very much less than unity. For example, for an aluminum plate in air it is approximately 2×10^{-5}, and for the same plate in water the ratio is 0.09. In these cases, Z^2 may be neglected as a first approximation, and Eq. 6.35 becomes

$$EF = 0. \tag{6.36}$$

This equation coincides with Eq. 6.28a for the phase velocity of free waves in a plate. Thus, we obtain the following important result: if the wave impedance of the surrounding medium is very much less than that of the plate, total transmission will occur at an angle of incidence such that the phase velocity of the incident wave along the plate coincides with the phase velocity of free waves in the plate. Cremer[101] has named this the "coincidence rule".

Using this rule and dispersion curves of the type shown in Fig. 26, we can find the velocity v at which total transmission occurs, at a given frequency and plate thickness. Then the corresponding values of the angle of incidence are determined by the relation $\cos\vartheta_1 = c_1/v$.† The positions of the experimentally observed transmission maxima for an aluminum plate in xylene are shown by the points in Fig. 26. We see

---

† It is useful to note that in the case of wave transmission through a liquid layer, the "coincidence rule" formulated above is not approximate, but is exact. In fact, according to Eq. 5.11, total transmission ($V = 0$) occurs when $\alpha_2 d = n\pi$, i.e. when $d \cos\vartheta_2 \approx n\lambda_2/2$, $n = 1, 2, \dots$. It can be shown that the same values for ϑ_2 can be obtained from Eq. 6.32 by setting $b_2 = 0$ therein, i.e. we pass from the case of a solid plate to the case of a liquid layer.

that these points are actually grouped around the various branches of the dispersion curve.

However, there are points and entire regions on the dispersion curves which do not refer to total transmission. Let us consider, for example, the case of normal incidence. Here, total transmission occurs at plate thicknesses satisfying the condition $\omega d/c_2 = \pi n$, where n is an integer, i.e. at the limiting frequencies of the dispersion curves which correspond to thickness resonances of the plate for longitudinal waves. Total transmission will not occur at the limiting frequencies satisfying the condition $\omega d/b_2 = \pi n$ and the corresponding resonances for shear waves. This is evident without any proof, since at normal incidence of the sound wave from the surrounding liquid shear waves will not be excited in the plate. For the same reason, free waves of this type in the plate will not radiate energy into the surrounding liquid.

Now, let a sound wave be incident on the plate at the angle of total internal reflection ($\sin \vartheta_1 = c_1/c_2$). Total transmission could be realized at points of intersection of the curves of family a, corresponding to anti-symmetric free waves, and the ordinate $v = c_2$ (see Fig. 26). The points of intersection of the curves of family s (symmetric waves) with this ordinate correspond to oscillations in which the points on the surface undergo only tangential displacements and, consequently, the waves in the plate are again decoupled from waves in the liquid. Under these conditions, the transmission through the plate is insignificant.

At angles of incidence lying between the angles of total internal reflection for longitudinal and transverse waves ($c_1/c_2 < \sin \vartheta_1 < c_1/b_2$), the longitudinal waves in the plate are inhomogeneous waves whose amplitudes decrease exponentially with distance from the boundary of the plate. If the plate is many wavelengths thick, energy transfer through the plate is accomplished principally through shear waves.

If the angle of incidence is greater than the angle of total internal reflection for shear waves ($c_1/c_2 < \sin \vartheta_1 < c_1/b_2$), total reflection will occur when the plate thickness is infinite.

With a plate of finite thickness we have two curves, a_0 and s_0, in this region of the dispersion curve diagram. A detailed investigation shows that in this case, if $\omega d/b_2$ is not too great, condition (6.35) for total transmission can be satisfied, and the intersection of the ordinate $v = c_1/\sin \vartheta_1$ with the curves a_0 and s_0 actually represents total transmission.[191] For large values of the product $\omega d/b_2$, condition (6.35) cannot be satisfied, and the transmission decreases rapidly with increasing d.

5. *Wave incidence at the angle of total internal reflection*

It is of particular interest to investigate the case in which the sound wave is incident on the plate at an angle ϑ_1, equal to the angle of total internal reflection ($\sin \vartheta_1 = c_1/c_2$). This case has a number of interesting features. In particular, it turns out that total reflection occurs at definite plate thicknesses. This effect arises as a result of a special interaction between the longitudinal and the transverse waves in the plate. If the plate is replaced by a liquid layer, in which transverse waves are absent, then, as we already know (see § 5), total reflection will not occur at finite plate thicknesses.

When $\sin \vartheta_1 = c_1/c_2$ we have: $\sin \vartheta_2 = (c_2/c_1) \sin \vartheta_1 = 1$, i.e. $\vartheta_2 = \pi/2$—a longitudinal wave is propagated parallel to the boundaries of the plate. Then in Eq. 6.29,

$$P = \frac{\omega d}{c_2} \cos \vartheta_2 \to 0, \quad Z_2 = \frac{\rho_2 c_2}{\cos \vartheta_2} \to \infty, \quad Z_2 \tan \frac{P}{2} \to \frac{\rho_2 \omega}{2} d.$$

As a result, we obtain

$$E \to \infty, \quad F \to \frac{\rho_2 \omega d}{2} \cos^2 2\gamma_2 + Z_{2t} \sin^2 2\gamma_2 \tan \frac{Q}{2}. \tag{6.37}$$

Equation 6.34 then takes the form

$$\frac{1}{T} = 1 + \frac{4}{Z_1^2} F^2 = 1 + \left(\frac{\rho_2 \omega d}{2} \cos^2 2\gamma_2 + Z_{2t} \sin^2 2\gamma_2 . \tan \frac{Q}{2} \right)^2 \frac{4}{Z_1^2}. \tag{6.38}$$

On the basis of this expression for the reciprocal of the transmissivity $1/T$, we consider two cases:

a. Total reflection. When

$$\frac{Q}{2} = (2n+1)\frac{\pi}{2}, \tag{6.39}$$

Eq. 6.38 gives $T = 0$, i.e. in this case the plate is completely opaque to sound. Remembering that $Q = (2\pi d/\lambda_t) \cos \gamma_2$, where λ_t is the wavelength of shear waves, condition 6.39 can be written

$$d = \tfrac{1}{2}(2n+1)\frac{\lambda_t}{\cos \gamma_2}. \tag{6.40}$$

Since $\lambda_t/\cos \gamma_2$ can be considered the effective wavelength of shear waves in the direction perpendicular to the boundaries of the plate, total reflection will occur when the plate thickness is an odd number of half wavelengths of the effective wavelength of shear waves.

These theoretical results are confirmed experimentally.[159] The abscissa in Fig. 27 represents the thickness of a glass plate (in mm), and the ordinate represents the angle of incidence of a sound wave at a frequency of 5×10^6 cps. The plate was immersed in transformer oil. The circles on the Figure indicate the thicknesses and angles of incidence at which maximum reflection occurred (minimum amplitude of the transmitted wave). The open circles correspond to weak maxima, the circles with crosses correspond to stronger maxima; finally, the dark circles correspond to total reflection.

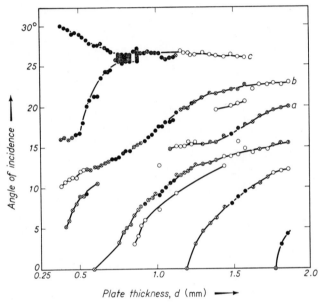

Fig. 27. The reflection maxima for a glass plate in oil, as a function of the angle of incidence and the plate thickness.

The angle of total internal reflection is approximately 13° for an oil–glass boundary. The thicknesses corresponding to total reflection are, from Eq. 6.40, $d = 0.42, 1.25, 2.09, \ldots$ mm.

We see from Fig. 27 that total reflection is actually observed at angles in the neighborhood of $\vartheta = 13°$ and in the region of thicknesses 0.42 and 1.25 mm. The value 2.09 mm lies at the limits of the thicknesses investigated. Thus, observing wave reflection from a plate at an angle equal to the angle of total internal reflection from an unbounded medium, and determining the plate thicknesses at which wave transmission through the plate does not occur, we can use Eq. 6.40 to determine the wavelength of the shear waves, and also their propagation velocity. We see from Fig. 27 that the region of total transmission is spread over a considerable interval around each of the values indicated above, so that the accuracy with which the shear wave velocity is determined is not too great.

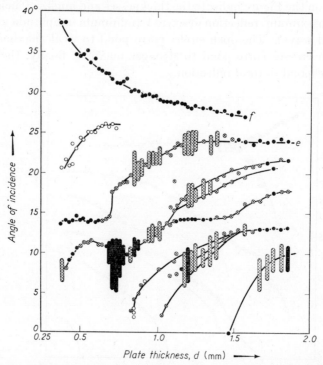

Fig. 28. The transmission maxima for a glass plate in oil, as a function of the angle of incidence and the plate thickness.

b. Total transmission. We obtain the condition for total transmission by setting $T = 1$ in Eq. 6.38. We then obtain a transcendental equation for the determination of the appropriate plate thicknesses

$$\omega d = \frac{2Z_{2t}}{\rho_2} \tan^2 2\gamma_2 \tan \frac{Q}{2}, \tag{6.41}$$

in which

$$Q = \frac{\omega d}{b_2} \cos \gamma_2.$$

The positions of the transmission maxima are shown in Fig. 28 for the same experimental conditions used in Fig. 27. The black circles correspond to very sharp transmission maxima (from 80 to 100 per cent transmission), the circles with crosses correspond to weaker maxima, and finally, the open circles correspond to very weak maxima.

The thicknesses corresponding to total transmission, given by Eq. 6.41 for the appropriate experimental conditions, are $d = 0.74$ and 1.49 mm. According to Fig. 28, total transmission at an angle of $13°$ will occur at thicknesses of $d = 0.79$ and 1.53 mm. The agreement with theory is quite satisfactory.

§ 7. WAVES IN FINELY LAYERED MEDIA

We consider the propagation of electromagnetic and elastic waves in a medium consisting of infinitely alternating layers of two different homogeneous and isotropic substances. If the thickness of the layers is sufficiently small with respect to the wavelength, this type of complex medium behaves on the whole as if it were homogeneous but anisotropic. Our problem is to define the electric and elastic parameters of this medium. This problem has been studied by various authors. We mention, for example, the investigations of Bruggeman.[96]

The elastic properties of solid finely layered media were studied by Tarkhov[76] and Riznichenko[65] in connection with problems of seismo-exploration.

M. L. Levin[56] considered in detail the propagation of an electromagnetic wave in a nonabsorbing, periodically layered medium, assuming propagation perpendicular to the layers. Quite recently, a very thorough and systematic exposition of the problem including electromagnetic as well as elastic waves was given by Rytov.[69, 70] He investigated general relations, valid for layers of any thickness. The results for a finely layered medium are obtained from these relations by a transition to a limit. Taking account of correction terms makes it possible to establish the conditions of applicability of the limiting results. The following exposition is based entirely on the work of Rytov.

The practical applications of the theory of wave propagation in finely layered media are many and varied. In seismo-exploration, one cannot avoid taking account of quasi-anisotropic layered media. Artificial anisotropic media for electromagnetic waves play a large role in engineering. Structures having the character of finely layered media are used in vibration isolation, and also in some areas of ultrasonics.

1. *The electromagnetic properties of finely layered media*

It is clear from the symmetry of the problem that, in order to determine the mean constants of the medium, it is sufficient to consider three cases of wave propagation: propagation parallel to the layers, with two polarizations (either the electric or the magnetic vector parallel to the layers), and propagation perpendicular to the layers. As previously, the z-axis will be directed perpendicular to the layers.

We consider *propagation along the x-axis with* **E** *directed along the y-axis.* Using Eqs. 2.2, we have the following equations in each of the layers for the nonzero components of the electromagnetic field $E_y = E$, H_x and H_z:

$$\frac{\partial E}{\partial z} = -ik\mu H_x, \quad \frac{\partial E}{\partial x} = ik\mu H_z, \quad \frac{\partial H_z}{\partial x} - \frac{\partial H_x}{\partial z} = ik\epsilon E, \qquad (7.1)$$

where $k = \omega/c$, and ϵ denotes the *complex* dielectric permeability of the medium (denoted in Eq. 2.2 by ϵ').† The permeabilities ϵ and μ vary periodically as a function of z, alternately taking the values ϵ_1, μ_1 in one layer and ϵ_2, μ_2 in the next.

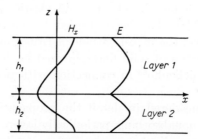

Fig. 29. The distribution of the electric and magnetic components of the field, over the width of the layers.

We seek a solution of Eq. 7.1 in the form

$$E = U(z)\,e^{iknx}, \quad H_x = V(z)\,e^{iknx}, \quad H_z = W(z)\,e^{iknx}, \qquad (7.2)$$

where U, V and W are for the time being unknown functions of z, and n is the index of refraction which we must determine, characterizing the propagation velocity of waves along the layers.

Substituting Eq. 7.2 in Eq. 7.1 gives

$$\frac{dU}{dz} = -ik\mu V, \quad nU = \mu W, \quad \frac{dV}{dz} - iknW = -ik\epsilon U. \qquad (7.3)$$

We select an arbitrary pair of neighboring layers and place the origin of coordinates on their interface. The layer with the constants ϵ_1 and μ_1 extends from $z = 0$ to $z = h_1$, and the layer with the constants ϵ_2 and μ_2 extends from $z = -h_2$ to $z = 0$ (Fig. 29).

---

† In comparing these equations (and all those following) with those of Rytov, we must remember that we have chosen the time factor in the form $e^{-i\omega t}$ whereas Rytov uses $e^{i\omega t}$. As a result, we always have the opposite sign in front of i.

The solutions of Eqs. 7.3 in each of the layers are

$$0 \leqslant z \leqslant h_1,$$

$$U_1 = A \cos \alpha_1 z + B \sin \alpha_1 z,$$

$$V_1 = \frac{\alpha}{ik\mu_1} (A \sin \alpha_1 z - B \cos \alpha_1 z),$$

$$W_1 = \frac{n}{\mu_1} (A \cos \alpha_1 z + B \sin \alpha_1 z),$$

$$\alpha_1 = k\sqrt{(n_1^2 - n^2)}, \quad n_1^2 = \epsilon_1 \mu_1; \tag{7.4}$$

$$-h_2 \leqslant z \leqslant 0,$$

$$U_2 = C \cos \alpha_2 z + D \sin \alpha_2 z,$$

$$V_2 = \frac{\alpha_2}{ik\mu_2} (C \sin \alpha_2 z - D \cos \alpha_2 z),$$

$$W_2 = \frac{n}{\mu_2} (C \cos \alpha_2 z + D \sin \alpha_2 z),$$

$$\alpha_2 = k\sqrt{(n_2^2 - n^2)}, \quad n_2^2 = \epsilon_2 \mu_2. \tag{7.5}$$

These solutions must be subjected to the four conditions of continuity and periodicity of E and H_x (i.e. U and V) with respect to z

$$U_1(0) = U_2(0), \qquad U_1(h_1) = U_2(-h_2),$$
$$V_1(0) = V_2(0), \qquad V_1(h_1) = V_2(-h_2). \tag{7.6}$$

With the help of Eqs. 7.4 and 7.5, Eq. 7.6 yields four homogeneous equations

$$C = A, \quad C \cos \alpha_2 h_2 - D \sin \alpha_2 h_2 = A \cos \alpha_1 h_1 + B \sin \alpha_1 h_1,$$

$$D = pB, \quad C \sin \alpha_2 h_2 + D \cos \alpha_2 h_2 = -p(A \sin \alpha_1 h_1 - B \cos \alpha_1 h_1), \tag{7.7}$$

where

$$p = \frac{\mu_2 \alpha_1}{\mu_1 \alpha_2}. \tag{7.8}$$

Setting the determinant of the system of equations (7.7) equal to zero, we obtain a dispersion equation, determining n as a function of k

$$(1 + p^2) \sin \alpha_1 h_1 \sin \alpha_2 h_2 + 2p(1 - \cos \alpha_1 h_1 \cos \alpha_2 h_2) = 0.$$

Solving this equation for p, we readily reduce it to the set of the following two equations

$$\frac{\tan(\tfrac{1}{2}\alpha_2 h_2)}{\tan(\tfrac{1}{2}\alpha_1 h_1)} = -p, \qquad \frac{\tan(\tfrac{1}{2}\alpha_2 h_2)}{\tan(\tfrac{1}{2}\alpha_1 h_1)} = -\frac{1}{p}. \tag{7.9}$$

Thus, waves of two types will propagate along the layers. Using Eqs. 7.7, we can show[69] that the first of Eqs. 7.9 corresponds to a wave in which E and H_z are distributed symmetrically with respect to the center in each layer, i.e. they are even functions of $z - h_1/2$ in layer 1 and $z + h_2/2$ in layer 2. However, the H_x component is odd. The components E_y and H_x are shown in Fig. 29 as a function of z. In what follows, we shall be interested in the electric and magnetic fields *averaged* over the period of the structure $d = h_1 + h_2$. Since the mean value of H_x is zero in the type of wave we are considering, the wave will have only the components E_y and H_z, i.e. it will be purely transverse.

In the same way it can be shown that the second of Eqs. 7.9 represents a wave of opposite character, in which H_x is even and E and H_z are odd with respect to the center of the layer. As a result, the average field will have only a longitudinal component H_x. However, it is not hard to see that this type of wave will be rapidly attenuated in the x-direction; consequently we shall not analyze it further. Actually, the space between the centers of the two layers, for example between $-h_2/2$ and $h_1/2$, can be considered as bounded by two perfectly conducting planes, since in the case we are considering, $E = H_z = 0$ on these planes. But it is well known that, in waveguides with perfectly conducting walls, waves with a tangential component of $\mathbf{E}$ (in our case the component $E_y = E$) are practically incapable of propagation, and are already strongly attenuated at a distance equal to the width of the waveguide (in our case $d/2$), if the wavelength is great in comparison with d. This result remains valid for any values of ϵ_1, μ_1 and ϵ_2, μ_2 for layers enclosed between perfectly conducting walls.

We shall therefore concentrate on the first of the two cases considered above, and denote the values of E_y and H_z averaged over the period d by $\bar{E}_y$ and $\bar{H}_z$. By definition, the relation

$$\frac{\bar{E}_y}{\bar{H}_z} = \sqrt{\frac{\mu^e}{\epsilon^e}}$$

must be observed, where μ^e and ϵ^e are "effective" values of μ and ϵ, and must be found. The mean values of the field can be found in an elementary way. Thus, for example,

$$\bar{E}_y = \frac{1}{h_1 + h_2} \left[\int_0^{h_1} U_1\, dz + \int_{-h_2}^0 U_2\, dz \right].$$

We substitute for U_1 and U_2 by using Eqs. 7.4 and 7.5, and after integration we use the relations (7.7) among the constants A, B, C and D. As a result we obtain

$$\frac{\bar{E}_y}{\bar{H}_z} = \sqrt{\frac{\mu^e}{\epsilon^e}} = \frac{\mu_1 \alpha_2^2 - \mu_2 \alpha_1^2}{n(\alpha_2^2 - \alpha_1^2)}.$$

Adding one more equation, $n = \sqrt{(\mu^e \epsilon^e)}$, we obtain the following expressions for ϵ^e and μ^e:

$$\epsilon^e = \frac{n^2(\alpha_2^2 - \alpha_1^2)}{\mu_1 \alpha_2^2 - \mu_2 \alpha_1^2}, \qquad \mu^e = \frac{\mu_1 \alpha_2^2 - \mu_2 \alpha_1^2}{(\alpha_2^2 - \alpha_1^2)}. \tag{7.10}$$

Here, n is a root of the first of Eqs. 7.9, which, incidentally, can also be written in the form

$$\frac{\alpha_2}{\mu_2} \tan \frac{\alpha_2 h_2}{2} = -\frac{\alpha_1}{\mu_1} \tan \frac{\alpha_1 h_1}{2}. \tag{7.11}$$

We now consider in more detail the relations we have obtained for the case in which the thicknesses h_1 and h_2 are small. Replacing the tangents in the last equation by their arguments, we obtain from Eq. 7.10

$$\epsilon^e = \bar{\epsilon}, \qquad \mu^e = \tilde{\mu}, \tag{7.12}$$

where

$$\left. \begin{array}{l} \bar{\epsilon} = \dfrac{h_1 \epsilon_1 + h_2 \epsilon_2}{h_1 + h_2} \quad \text{is the mean value of } \epsilon \text{ over a period,} \\[2mm] \dfrac{1}{\tilde{\mu}} = \overline{\left(\dfrac{1}{\mu}\right)} = \dfrac{h_1/\mu_1 + h_2/\mu_2}{h_1 + h_2} \text{ is the mean value of } 1/\mu \text{ over a period.} \end{array} \right\} \tag{7.13}$$

A more exact treatment, taking account of the cubic terms in the expansion of the tangents in Eq. 7.11, gives

$$\epsilon^e = \bar{\epsilon}\left[1 + \frac{k^2 h_1^2 h_2^2}{12(h_1 + h_2)^2} \frac{\bar{\mu}\tilde{\mu}}{\mu_1 \mu_2} (n_1^2 - n_2^2) \frac{\epsilon_1 - \epsilon_2}{\bar{\epsilon}}\right],$$

$$\mu^e = \tilde{\mu}\left[1 + \frac{k^2 h_1^2 h_2^2}{12(h_1 + h_2)^2} \frac{\bar{\mu}\tilde{\mu}^2}{\mu_1^2 \mu_2^2} (n_1^2 - n_2^2)(\mu_1 - \mu_2)\right], \tag{7.14}$$

where $n_1^2 \equiv \epsilon_1 \mu_1$, $n_2^2 \equiv \epsilon_2 \mu_2$.

The correction terms are of the order $k^2 d^2 = k^2(h_1 + h_2)^2$. Moreover, for given values of h_1 and h_2, the correction terms play a smaller role, the closer the parameters of one of the substances are to those of the other.

Propagation along the x-axis with **H** *directed along the y-axis.* The equations for the electromagnetic field, which in this case has the components E_x, E_z and $H_y = H$, are

$$\frac{\partial H}{\partial z} = ik\epsilon E_x, \qquad \frac{\partial H}{\partial x} = -ik\epsilon E_z, \qquad \frac{\partial E_x}{\partial z} - \frac{\partial E_z}{\partial x} = ik\mu H. \tag{7.15}$$

The components H and E_x are subject to the continuity and periodicity conditions at the boundaries between the layers. Clearly, all the formulas pertaining to the present case can be obtained from the formulas pertaining to the preceding case by replacing **H**, **E**, ϵ, μ by **E**, $-$**H**, μ, ϵ, respectively.

Averaging the expressions for the field components over the period of the structure, $d = h_1 + h_2$, yields a transverse wave, with components $\bar{H}_y$ and $\bar{E}_z$, satisfying Maxwell's equations

$$\frac{\partial \bar{H}_y}{\partial x} = -ik\epsilon^e \bar{E}_z, \quad \frac{\partial \bar{E}_z}{\partial x} = -ik\mu^e \bar{H}_y, \qquad (7.16)$$

where the effective values of ϵ and μ are given by the formulas

$$\epsilon^e = \frac{\epsilon_1 \alpha_2^2 - \epsilon_2 \alpha_1^2}{\alpha_2^2 - \alpha_1^2}, \quad \mu^e = n^2 \frac{\alpha_2^2 - \alpha_1^2}{\epsilon_1 \alpha_2^2 - \epsilon_2 \alpha_1^2}, \qquad (7.17)$$

obtained from Eqs. 7.10 by means of the substitutions indicated above. The equation for n is then

$$\frac{\alpha_2}{\epsilon_2} \tan \frac{\alpha_2 h_2}{2} = -\frac{\alpha_1}{\epsilon_1} \tan \frac{\alpha_1 h_1}{2}. \qquad (7.18)$$

When h_1 and h_2 are small, we obtain

$$\epsilon^e = \tilde{\epsilon}, \quad \mu^e = \bar{\mu}, \qquad (7.19)$$

where $$\frac{1}{\tilde{\epsilon}} = \left(\overline{\frac{1}{\epsilon}}\right) = \frac{h_1/\epsilon_1 + h_2/\epsilon_2}{h_1 + h_2}, \quad \bar{\mu} = \frac{h_1 \mu_1 + h_2 \mu_2}{h_1 + h_2}. \qquad (7.20)$$

Propagation along the z-axis. We choose $E_x = E$ and $H_y = H$ as the nonzero components of the field. These components satisfy the equations

$$\frac{dE}{dz} = ik\mu H, \quad \frac{dH}{dz} = ik\epsilon E. \qquad (7.21)$$

If $\epsilon(z)$ and $\mu(z)$ were continuous, we could immediately conclude [on the basis of Fatou's theorem[110]] that, as $d \to 0$, the solution of Eqs. 7.21 approaches the solution of the same equations, but with the averaged quantities $\tilde{\epsilon}$ and $\bar{\mu}$.†

A rigorous solution shows that this result remains valid even for the discontinuous ϵ and μ under consideration. Nevertheless, since we are interested in the magnitude of the correction to $\tilde{\epsilon}$ and $\bar{\mu}$ in the effective values ϵ^e and μ^e, we start with an exact calculation, following the method used by M. L. Levin.[56]

---

† Let us note (following S. M. Rytov[69]) the following important circumstance. Eliminating H (or E) from Eqs. 7.21, we obtain an equation of the second order for E (or H). The application of the Fatou theorem then gives $n^2 = \overline{\epsilon\mu}$ for the averaged index of refraction, whereas Eq. 7.21 gives $n^2 = \tilde{\epsilon}\bar{\mu}$. This difference stems from the fact that, to obtain the second order equation, one of Eqs. 7.21 must be differentiated, yet the derivative of the approximate solution is not equal to the derivative of the exact solution in the limit $d \to 0$. Therefore, in obtaining the approximate expressions for E and H, it is not valid to make the transition to an equation of the second order.

Keeping Floquet's theorem[114] in mind, we seek a solution of Eqs. 7.21 in the form

$$E = U(z)\exp(iknz), \quad H = V(z)\exp(iknz), \tag{7.22}$$

where U and V are periodic functions of z with the period d.

If we integrate Eqs. 7.21 separately for each layer and compare the results with Eqs. 7.22, we obtain

$$0 \leqslant z \leqslant h_1 \quad U_1 = \exp(-iknz)[A\exp(-i\alpha_1 z) + B\exp(i\alpha_1 z)],$$

$$V_1 = -\frac{\alpha_1}{k\mu_1}\exp(-iknz)[A\exp(-i\alpha_1 z) - B\exp(i\alpha_1 z)],$$

$$\alpha_1 = kn_1 = k\sqrt{(\epsilon_1\mu_1)},$$

$$-h_2 \leqslant z \leqslant 0 \quad U_2 = \exp(-iknz)[C\exp(-i\alpha_2 z) + D\exp(i\alpha_2 z)],$$

$$V_2 = -\frac{\alpha_2}{k\mu_2}\exp(-iknz)[C\exp(-i\alpha_2 z) - D\exp(i\alpha_2 z)],$$

$$\alpha_2 = kn_2 = k\sqrt{(\epsilon_2\mu_2)}. \tag{7.23}$$

Imposing the conditions of continuity and periodicity on U and V, which are still written in the form (7.6), we obtain four homogeneous equations for the constants of integration:

$$C + D = A + B,$$

$$\exp(iknh_2)[C\exp(i\alpha_2 h_2) + D\exp(-i\alpha_2 h_2)]$$
$$= \exp(-iknh_1)[A\exp(-i\alpha_1 h_1) + B\exp(i\alpha_1 h_1)],$$

$$C - D = p(A - B),$$

$$\exp(iknh_2)[C\exp(i\alpha_2 h_2) - D\exp(-i\alpha_2 h_2)]$$
$$= p\exp(-iknh_1)[A\exp(-i\alpha_1 h_1) - B\exp(i\alpha_1 h_1)],$$

$$p = \frac{\alpha_1\mu_2}{\alpha_2\mu_1} = \sqrt{\frac{\epsilon_1\mu_2}{\epsilon_2\mu_1}}. \tag{7.24}$$

Equating the determinant of system (7.24) to zero yields the dispersion equation, determining n

$$\cos knd = \cos\alpha_1 h_1 \cos\alpha_2 h_2 - \frac{1+p^2}{2p}\sin\alpha_1 h_1 \sin\alpha_2 h_2. \tag{7.25}$$

We again determine the mean values of the fields $\bar{E}_x = \bar{E}$ and $\bar{H}_y = \bar{H}$ over a period of the structure. Requiring that the mean fields satisfy Maxwell's equations

$$\frac{\partial\bar{E}_x}{\partial z} = ik\mu^e\bar{H}_y, \quad \frac{\partial\bar{H}_y}{\partial z} = ik\epsilon^e\bar{E}_x,$$

we must determine ϵ^e and μ^e from the conditions

$$n = \sqrt{(\epsilon^e \mu^e)}, \quad \frac{\bar{E}_x}{\bar{H}_y} = \frac{\bar{U}}{\bar{V}} = \sqrt{\frac{\mu^e}{\epsilon^e}},$$

whence

$$\epsilon^e = n \frac{\bar{H}_y}{\bar{E}_x}, \quad \mu^e = n \frac{\bar{E}_x}{\bar{H}_y}. \tag{7.26}$$

After some manipulation, we obtain

$$\epsilon^e = \bar{\epsilon} \left[1 + \frac{ikh_1 h_2}{4d} \frac{\mu_1 \epsilon_2 - \mu_2 \epsilon_1}{\sqrt{(\bar{\epsilon}\bar{\mu})}} \right]$$

$$\mu^e = \bar{\mu} \left[1 - \frac{ikh_1 h_2}{4d} \frac{\mu_1 \epsilon_2 - \mu_2 \epsilon_1}{\sqrt{(\bar{\epsilon}\bar{\mu})}} \right]. \tag{7.27}$$

We thus see that the correction to the approximate quantity $n^2 = \bar{\epsilon}\bar{\mu}$ is quadratic.

Let the thicknesses of the layers h_1 and h_2 be sufficiently small so that the correction terms in Eq. 7.27 (and in the analogous expressions obtained above) can be neglected. Combining all the cases considered above, we obtain the following system of Maxwell's equations for the averaged fields in a "quasi-anisotropic" medium:

$$\frac{\partial \bar{E}_y}{\partial x} = ik\tilde{\mu}\bar{H}_z, \quad \frac{\partial \bar{H}_z}{\partial x} = ik\bar{\epsilon}\bar{E}_y,$$

$$\frac{\partial \bar{H}_y}{\partial x} = -ik\bar{\epsilon}\bar{E}_z, \quad \frac{\partial \bar{E}_z}{\partial x} = -ik\bar{\mu}\bar{H}_y,$$

$$\frac{\partial \bar{E}_x}{\partial z} = ik\bar{\mu}\bar{H}_y, \quad \frac{\partial \bar{H}_y}{\partial z} = ik\bar{\epsilon}\bar{E}_x.$$

This is a unique system of Maxwell's equations for a medium in which the tensors for the dielectric constant ϵ^e and the magnetic permeability μ^e are simply degenerate, having the same principal axes and the following principal values:

$$\epsilon_1^e = \epsilon_2^e = \bar{\epsilon} = \frac{h_1 \epsilon_1 + h_2 \epsilon_2}{h_1 + h_2}, \quad \epsilon_3^e = \tilde{\epsilon} = \frac{\epsilon_1 \epsilon_2 (h_1 + h_2)}{h_1 \epsilon_2 + h_2 \epsilon_1},$$

$$\mu_1^e = \mu_2^e = \bar{\mu} = \frac{h_1 \mu_1 + h_2 \mu_2}{h_1 + h_2}, \quad \mu_3^e = \tilde{\mu} = \frac{\mu_1 \mu_2 (h_1 + h_2)}{h_1 \mu_2 + h_2 \mu_1}. \tag{7.28}$$

Thus, the complex medium under consideration has the property of a monoaxial crystal, with the optic axis perpendicular to the layers.

In conclusion, we recall that in the expressions obtained above $\epsilon_1, \mu_1, \epsilon_2, \mu_2$ will be complex if the layers are absorbing.

2. *Elastic properties of finely layered media*

It will be shown below that, as regards its elastic properties, a finely layered medium is analogous to a crystal with hexagonal symmetry, i.e. it is characterized by five elastic constants. Four of these constants can be found by investigating the propagation of longitudinal and transverse waves along and perpendicular to the layers. The fifth constant can be found only by solving the problem of wave propagation at an angle of incidence (with respect to the z-axis) other than zero or $\pi/2$.

We decompose the total particle displacement $\mathbf{U}$ in any layer into an irrotational and a solenoidal part

$$\mathbf{U} = \mathbf{u} + \mathbf{v} \qquad (7.29)$$

where, in accordance with Eq. 4.1, $\mathbf{u} = \operatorname{grad}\phi$, $\mathbf{v} = \operatorname{curl}\psi$, ϕ and ψ are the scalar and vector potentials; $\mathbf{u}$ and $\mathbf{v}$ will satisfy the equations

$$\ddot{\mathbf{u}} - c^2 \operatorname{grad}\operatorname{div}\mathbf{u} = 0, \quad \operatorname{curl}\mathbf{u} = 0,$$

$$\ddot{\mathbf{v}} - b^2 \operatorname{curl}\operatorname{curl}\mathbf{v} = 0, \quad \operatorname{div}\mathbf{v} = 0, \qquad (7.30)$$

where the compressional and the shear wave velocities are determined by Eqs. 4.4.

Let us consider *propagation parallel to the layers*, for definiteness, along the x-axis. Clearly, a shear wave with displacements U_y parallel to the layers is autonomous, i.e. it is not coupled to a compressional wave. On the other hand, a transverse displacement of the layer (U_z) is generally coupled with a longitudinal displacement U_x. We first consider the latter case. The displacement vector $\mathbf{U}$ is given by the components $[U_x(x,z),\ 0,\ U_z(x,z)]$. For harmonic waves of frequency ω, Eqs. 7.30 take the form

$$\frac{\partial^2 \mathbf{u}}{\partial x^2} + \frac{\partial^2 \mathbf{u}}{\partial z^2} + k^2 \mathbf{u} = 0, \quad \frac{\partial u_z}{\partial x} = \frac{\partial u_x}{\partial z},$$

$$\frac{\partial^2 \mathbf{v}}{\partial x^2} + \frac{\partial^2 \mathbf{v}}{\partial z^2} + \kappa^2 \mathbf{v} = 0, \quad \frac{\partial v_x}{\partial x} = -\frac{\partial v_z}{\partial z},$$

$$k = \frac{\omega}{c}, \quad \kappa = \frac{\omega}{b}. \qquad (7.31)$$

The solutions we are seeking of these equations will represent traveling waves in the x-direction and standing waves in the z-direction. The wave velocity along the x-axis will be denoted by c_{xz} and the wave number by $a = \omega/c_{xz}$. Now the solution for u_x in layer 1 can be written in the form

$$u_x = P(z)\,e^{iax}, \quad P(z) = A\cos\alpha\left(z - \frac{h_1}{2}\right) + B\sin\alpha\left(z - \frac{h_1}{2}\right), \qquad (7.32)$$

$$\alpha^2 = k^2 - a^2,$$

where A and B are for the present undetermined constants. In the expression for $P(z)$, we separated the symmetric and the antisymmetric parts with respect to the center of the layer $z = h_1/2$.

From the upper right hand equation in (7.31), we obtain

$$u_z = \frac{P'(z)}{ia}\,e^{iax}. \qquad (7.33)$$

In the same way we obtain for v_x and v_z

$$v_z = Q(z)\,e^{iax}, \quad v_x = -\frac{Q'(z)}{ia}\,e^{iax}, \qquad (7.34)$$

where $\quad Q(z) = C\cos\beta\left(z - \frac{h_1}{2}\right) + D\sin\beta\left(z - \frac{h_1}{2}\right), \quad \beta^2 = \kappa^2 - a^2.$ $\quad (7.35)$

To simplify the notation, we use a bar to denote quantities referring to the second layer. Thus, in the second layer we will have

$$\bar{u}_x = \bar{P}(z)\,e^{iax}, \quad \bar{u}_z = \frac{1}{ia}\,\bar{P}'(z)\,e^{iax},$$

$$\bar{v}_x = -\frac{1}{ia}\,\bar{Q}'(z)\,e^{iax}, \quad \bar{v}_z = \bar{Q}(z)\,e^{iax}, \qquad (7.36)$$

where $\quad \bar{P}(z) = \bar{A}\cos\bar{\alpha}\left(z + \frac{h_2}{2}\right) + \bar{B}\sin\bar{\alpha}\left(z + \frac{h_2}{2}\right), \quad \bar{\alpha}^2 = \bar{k}^2 - a^2,$

$$\bar{Q}(z) = \bar{C}\cos\bar{\beta}\left(z + \frac{h_2}{2}\right) + \bar{D}\sin\bar{\beta}\left(z + \frac{h_2}{2}\right), \quad \bar{\beta}^2 = \bar{\kappa}^2 - a^2. \qquad (7.37)$$

The components of the stress tensor Z_z and Z_x are expressed in terms of the displacement components through Eqs. 4.5. It must be remembered that the displacement components were denoted by the small letters u_x and u_z in Eqs. 4.5 but are now denoted by U_x and U_z. Using relation (7.29), and also expressions (7.32)–(7.34) for u_x, u_z, v_x and v_z, we find

$$Z_x = 2\mu\left(P' - \frac{a^2 - \beta^2}{2ia}\,Q\right)e^{iax}, \quad Z_z = \left(\frac{\lambda k^2 + 2\mu a^2}{-ia}\,P + 2\mu Q'\right)e^{iax}. \qquad (7.38)$$

The components $\bar{Z}_x$ and $\bar{Z}_z$ in layer 2 can be expressed similarly. The deformation components $U_x = u_x + v_x$, $U_z = u_z + v_z$ and the components of the stress tensor must be continuous at the boundary between the layers, $z = 0$, i.e.

$$U_x(0) = \bar{U}_x(0), \quad Z_x(0) = \bar{Z}_x(0),$$

$$U_z(0) = \bar{U}_z(0), \quad Z_z(0) = \bar{Z}_z(0). \tag{7.39}$$

The equality of these same quantities at the boundaries $z = h_1$ and $z = -h_2$ follow from the periodicity conditions, i.e.

$$U_x(h_1) = \bar{U}_x(-h_2), \quad Z_x(h_1) = \bar{Z}_x(-h_2),$$

$$U_z(h_1) = \bar{U}_z(-h_2), \quad Z_z(h_1) = \bar{Z}_z(-h_2). \tag{7.40}$$

Substitution of the appropriate expressions into Eqs. 7.39 and 7.40 yields eight equations for the constants $A, B, \ldots, \bar{D}$, which divide into two independent groups, associated with two types of waves:

a. *Waves, longitudinal in the mean* ($B = C = \bar{B} = \bar{C} = 0$). It is clear from Eqs. 7.36 and 7.37 that in this case, the longitudinal displacements u_x and v_x are even (symmetrical) with respect to the centers of the layers, whereas the transverse displacements u_z and v_z are odd (antisymmetrical). Therefore, if we are interested in displacements averaged over a period of the structure, waves of this type will be longitudinal, accompanied only by compression, and not shear. Equations 7.39 and 7.40 yield the following equations for the constants A, D and $\bar{A}$, $\bar{D}$:

$$A i a \cos\frac{\alpha h_1}{2} - D\beta \cos\frac{\beta h_1}{2} = \bar{A} i a \cos\frac{\bar{\alpha} h_2}{2} - \bar{D}\beta \cos\frac{\beta h_2}{2},$$

$$A(\lambda k^2 + 2\mu\alpha^2)\cos\frac{\alpha h_1}{2} - 2 D i a \mu\beta \cos\frac{\beta h_1}{2}$$
$$= \bar{A}(\bar{\lambda}\bar{k}^2 + 2\bar{\mu}\bar{\alpha}^2)\cos\frac{\bar{\alpha} h_2}{2} - 2\bar{D} i a \bar{\mu}\beta \cos\frac{\beta h_2}{2},$$

$$A\alpha \sin\frac{\alpha h_1}{2} - D i a \sin\frac{\beta h_1}{2} = -\bar{A}\bar{\alpha} \sin\frac{\bar{\alpha} h_2}{2} + \bar{D} i a \sin\frac{\beta h_2}{2},$$

$$A 2 i a \mu\alpha \sin\frac{\alpha h_1}{2} + \mu(a^2 - \beta^2) D \sin\frac{\beta h_1}{2}$$
$$= -\bar{A} 2 i a \bar{\mu}\bar{\alpha} \sin\frac{\bar{\alpha} h_2}{2} - \bar{\mu}(a^2 - \beta^2) \bar{D} \sin\frac{\beta h_1}{2}.$$

Setting the determinant of this system equal to zero, we obtain a dispersion equation, determining the wave number a, and consequently the propagation velocity $c_{xx} = \omega/a$:

$$4(\mu - \bar{\mu})^2 X\bar{X} + \omega^2\rho\left[\frac{\omega^2\rho}{a^2} - 4(\mu - \bar{\mu})\right]\bar{X}\tan\frac{\beta h_1}{2}$$

$$+ \omega^2\bar{\rho}\left[\frac{\omega^2\bar{\rho}}{a^2} + 4(\mu - \bar{\mu})\right]X\tan\frac{\bar{\beta}h_2}{2} - \frac{\omega^4\rho\bar{\rho}}{a^2}\left(Y\tan\frac{\bar{\beta}h_2}{2} + \bar{Y}\tan\frac{\beta h_1}{2}\right) = 0,$$

$$(7.41)$$

where for brevity we have introduced the notation

$$X = a^2\tan\frac{\beta h_1}{2} + \alpha\beta\tan\frac{\alpha h_1}{2}, \quad Y = a^2\tan\frac{\beta h_1}{2} - \bar{\alpha}\beta\tan\frac{\alpha h_2}{2},$$

$$\bar{X} = a^2\tan\frac{\bar{\beta}h_2}{2} + \bar{\alpha}\bar{\beta}\tan\frac{\bar{\alpha}h_2}{2}, \quad \bar{Y} = a^2\tan\frac{\bar{\beta}h_2}{2} - \alpha\bar{\beta}\tan\frac{\alpha h_1}{2}.$$

Equation 7.41 determines a for any values of the parameters of the layers in the periodic structure. The transition to a finely layered medium is accomplished by replacing all tangents by their arguments. This simplifies the dispersion equation considerably and yields the following expression for the square of the compressional wave velocity c_{xx}:

$$c_{xx}^2 = \frac{1 + 4s\bar{s}\dfrac{(\mu - \bar{\mu})[\lambda + \mu - (\bar{\lambda} + \bar{\mu})]}{(\lambda + 2\mu)(\bar{\lambda} + 2\bar{\mu})}}{\bar{\rho}\left(\dfrac{s}{\lambda + 2\mu} + \dfrac{\bar{s}}{\bar{\lambda} + 2\bar{\mu}}\right)}, \quad (7.42)$$

where s and $\bar{s}$ denote the relative thicknesses of the layers, and $\bar{\rho}$ denotes the mean density of the medium:

$$s = \frac{h_1}{h_1 + h_2}, \quad \bar{s} = \frac{h_2}{h_1 + h_2}, \quad \bar{\rho} = s\rho + \bar{s}\bar{\rho}. \quad (7.43)$$

The conditions which must be imposed on the layer thicknesses in order that Eq. 7.42 be valid can be obtained by keeping the cubic terms in the expansion of the tangents (see Eq. 7.1). However, we shall not consider this question further.

b. *Waves, transverse in the mean* $(A = D = \bar{A} = \bar{D} = 0)$. According to Eqs. 7.36 and 7.37, the transverse displacements u_z and v_z are even with respect to the centers of the layers, and the longitudinal displacements u_x and v_x are odd. Therefore, the deformations averaged over a period of the structure will be purely shear. Rather than write out the equations for B, C and $\bar{B}$, $\bar{C}$, and the resulting dispersion equation for arbitrary layer thicknesses, we shall simply give the limiting value of the wave velocity for the case of thin layers:

$$c_{xz}^2 = \frac{1}{\tilde{\rho}(s/\mu + \bar{s}/\bar{\mu})}. \tag{7.44}$$

We now consider propagation along the x-axis of a purely shear wave for which only one displacement component $U_y = v_y$ is nonzero. In layer 1 we have

$$v_y = Q(z)\,e^{iax},$$

where $\quad Q(z) = C\cos\beta\left(z - \frac{h_1}{2}\right) + D\sin\beta\left(z - \frac{h_1}{2}\right), \quad \beta^2 = \kappa^2 - a^2,$

and in layer 2 we have

$$\bar{v}_y = \bar{Q}(z)\,e^{iax}, \quad \bar{Q}(z) = \bar{C}\cos\bar{\beta}\left(z + \frac{h_2}{2}\right) + \bar{D}\sin\bar{\beta}\left(z + \frac{h_2}{2}\right), \quad \bar{\beta}^2 = \bar{\kappa}^2 - a^2.$$

The only nonzero component of the stress tensor will be Y_z. The continuity and periodicity conditions on v_y and Y_z yield four equations for C, D and $\bar{C}$, $\bar{D}$. Setting the determinant of this system equal to zero, we obtain the dispersion equation

$$p\left(\tan^2\frac{\beta h_1}{2} + \tan^2\frac{\bar{\beta} h_2}{2}\right) + (1 + p^2)\tan\frac{\beta h_1}{2}\tan\frac{\bar{\beta} h_2}{2} = 0,$$

where

$$p = \frac{\bar{\mu}\bar{\beta}}{\mu\beta}.$$

Hence, for thin layers, i.e. when the tangents are replaced by their arguments, the square of the velocity $c_{xy} = \omega/a$ is

$$c_{xy}^2 = \frac{s\mu + \bar{s}\bar{\mu}}{\tilde{\rho}}. \tag{7.45}$$

Propagation perpendicular to the layers. When a longitudinal wave with a nonzero displacement component $U_z = u_z = P(z)\,e^{iaz}$, and a transverse wave with a displacement parallel to the layers (for example, $U_x = v_x = Q(z)\,e^{iaz}$) propagate along the z-axis, they can be investigated independently. According to Floquet's theorem, $P(z)$ and $Q(z)$ must be periodic functions with the period of the structure $d = h_1 + h_2$. The conditions of periodicity (at $z = h_1$ and $z = -h_2$) and of continuity (at $z = 0$) must be satisfied by the amplitudes U_z and Z_z of the compressional wave, and the amplitudes U_x and Z_x of the shear wave. In the first case the dispersion equation has the form

$$\cos a(h_1 + h_2) = \cos kh_1\cos\bar{k}h_2 - \frac{1 + p^2}{2p}\sin kh_1\sin\bar{k}h_2,$$

where

$$p = \frac{(\bar{\lambda} + 2\bar{\mu})\,\bar{k}}{(\lambda + 2\mu)\,k}.$$

In the second case the dispersion equation has the same form, but with k and $\bar{k}$ in the arguments of the cosines and sines replaced by κ and $\bar{\kappa}$, respectively, and with

$$p = \frac{\bar{\mu}\bar{\kappa}}{\mu\kappa}.$$

For thin layers, replacing the sines by their arguments and limiting the cosines to quadratic terms, we obtain for compressional waves

$$c_{zz}^2 = \frac{1}{\bar{\rho}[s/(\lambda + 2\mu) + \bar{s}/(\bar{\lambda} + 2\bar{\mu})]}. \tag{7.46}$$

As could be expected, the shear wave velocity c_{zx} turns out to be equal to c_{xz} (see Eq. 7.44).

3. *The finely layered medium as a crystal with hexagonal symmetry*

It was indicated above that, as regards its elastic properties, a finely layered medium is similar to a crystal with hexagonal symmetry. The free energy per unit volume of this type of crystal is written in the form†

$$F = \frac{1}{2}\lambda_1 l_{zz}^2 + \frac{\lambda_2}{2}(l_{xx} + l_{yy})^2 + \lambda_3 l_{zz}(l_{xx} + l_{yy})$$
$$+ 2\lambda_4(l_{xy}^2 - l_{xx}l_{yy}) + 2\lambda_5(l_{xz}^2 + l_{yz}^2), \tag{7.47}$$

where l_{ik} are the components of the deformation tensor

$$l_{ik} = \frac{1}{2}\left(\frac{\partial U_i}{\partial x_k} + \frac{\partial U_k}{\partial x_i}\right).$$

The medium under consideration is isotropic in the xy-plane. Complete isotropy occurs when

$$\lambda_1 = \lambda_2 = \lambda + 2\mu, \quad \lambda_3 = \lambda, \quad \lambda_4 = \lambda_5 = \mu,$$

where λ and μ are the Lamé coefficients. The constants λ_1, λ_2, λ_4, and λ_5 in Eq. 7.47 can be expressed in terms of the propagation velocities found above for the various types of waves. Let us take, for example, a longitudinal wave propagating in the x-direction, with velocity previously denoted by c_{xx}. The only nonzero component of the deformation tensor will be l_{xx}, and the free energy is written

$$F = \frac{\lambda_2}{2}l_{xx}^2, \quad l_{xx} = \frac{\partial U_x}{\partial x}.$$

---

† See, for example, Ref. 10. The system of notation for the elastic constants is somewhat different here.

We shall have exactly the same expressions in the isotropic case, where $\lambda + 2\mu$ enters in place of λ_2. In the same way as for an isotropic medium (using the expression for F) we obtain $c_{xx}^2 = \lambda_2/\tilde{\rho}$ for the wave velocity, where $\tilde{\rho}$ is the density of the medium. The same approach can be used for the velocities of the other waves. As a result, we obtain

$$c_{xx}^2 = \frac{\lambda_2}{\tilde{\rho}}, \quad c_{xz}^2 = c_{zx}^2 = \frac{\lambda_5}{\tilde{\rho}}, \quad c_{xy}^2 = \frac{\lambda_4}{\tilde{\rho}}, \quad c_{zz}^2 = \frac{\lambda_1}{\tilde{\rho}}, \tag{7.48}$$

where the first subscript indicates the propagation direction, and the second indicates the direction of the displacement vector $\mathbf{U}$. Using Eqs. 7.42, 7.44, 7.45 and 7.46 for c_{xx}, we find the connection between crystal constants $\lambda_1, \lambda_2, \lambda_4, \lambda_5$, and the characteristics of the finely layered medium:

$$\frac{1}{\lambda_1} = \frac{s}{\lambda + 2\mu} + \frac{\bar{s}}{\bar{\lambda} + 2\bar{\mu}}, \quad \lambda_2 = \lambda_1 \left[1 + s\bar{s} \frac{(\mu - \bar{\mu})[\lambda + \mu - (\bar{\lambda} + \bar{\mu})]}{(\lambda + 2\mu)(\bar{\lambda} + 2\bar{\mu})} \right],$$

$$\lambda_4 = s\mu + \bar{s}\bar{\mu}, \quad \frac{1}{\lambda_5} = \frac{s}{\mu} + \frac{\bar{s}}{\bar{\mu}}. \tag{7.49}$$

The constant λ_3 is still undetermined. In order to determine this constant, we must study the propagation of elastic waves at an arbitrary angle of incidence with respect to the z-axis. As shown in Ref. 70 we obtain

$$\lambda_3 = -\lambda_5 + \sqrt{\left\{ (\lambda_1 - \lambda_5)^2 + 4s\bar{s}\lambda_1^2 \lambda_5 \left(\frac{1}{\bar{\mu}} - \frac{1}{\mu} \right) \right.}$$

$$\left. \times \left(\frac{\lambda + \mu}{\lambda + 2\mu} - \frac{\bar{\lambda} + \bar{\mu}}{\bar{\lambda} + 2\bar{\mu}} \right) \left[\frac{s(\lambda + \mu)}{\lambda + 2\mu} + \frac{\bar{s}(\bar{\lambda} + \bar{\mu})}{\bar{\lambda} + 2\bar{\mu}} \right] \right\}. \tag{7.50}$$

Knowing all five effective elastic constants, we have the characteristics of the elastic properties of the crystal completely at our disposal, and can calculate the propagation velocities of waves with arbitrary propagation directions and polarizations.

4. Wave absorption in finely layered media

In a number of applications, such as the design of vibration isolation materials and artificial anisotropic media in acoustics, it is important to know not only the propagation velocities of waves of various types, but also, their absorption coefficients. This problem is solved quite simply if the condition for thin layers is also satisfied when attenuation is taken into account. Equations 7.49 and 7.50 remain valid, except that the real coefficients λ and μ must be replaced by complex coefficients because of the introduction of the bulk and shear viscosities ζ and η:

$$\lambda + \tfrac{2}{3}\mu \to \lambda + \tfrac{2}{3}\mu - i\omega\zeta, \quad \mu \to \mu - i\omega\eta,$$

so that $\lambda \to \lambda - i\omega(\zeta - \tfrac{2}{3}\eta), \quad \lambda + 2\mu \to \lambda + 2\mu - i\omega(\gamma + \tfrac{4}{3}\eta).$

According to Eq. 7.48, the complex wave numbers for the various types of waves $(a_{ik} = \omega/c_{ik})$ are written

$$a_{xx}^2 = \frac{\omega^2 \tilde{\rho}}{\lambda_2}, \quad a_{xz}^2 = a_{zx}^2 = \frac{\omega^2 \tilde{\rho}}{\lambda_5}, \quad a_{xy}^2 = \frac{\omega^2 \tilde{\rho}}{\lambda_4}, \quad a_{zz}^2 = \frac{\omega^2 \tilde{\rho}}{\lambda_1}. \quad (7.51)$$

Substituting for $\lambda_1, \lambda_2, \lambda_4, \lambda_5$, from Eqs. 7.49, where λ, μ and $\bar{\lambda}, \bar{\mu}$ are the complex Lamé coefficients in each of the layers, and setting $a_{ik} = a_{ik}^0 + i\gamma_{ik}$, we obtain the absorption coefficients γ_{ik}. When the viscosity is sufficiently small so that we need consider only terms of the first order in η and ζ, the expressions for the absorption coefficients have the following form:

$$\gamma_{xx} = \frac{\omega^2 \lambda_1}{2} \sqrt{\left(\frac{\tilde{\rho}}{\lambda_2}\right)} \left\{ \frac{s(\zeta + \tfrac{4}{3}\eta)}{(\lambda + 2\mu)^2} + \frac{\bar{s}(\bar{\zeta} + \tfrac{4}{3}\bar{\eta})}{(\bar{\lambda} + 2\bar{\mu})^2} \right.$$

$$\left. + \left(\frac{1}{\lambda_1} - \frac{1}{\lambda_2}\right) \left[\frac{\eta - \bar{\eta}}{\mu - \bar{\mu}} + \frac{\zeta - \bar{\zeta} + \tfrac{1}{2}(\eta - \bar{\eta})}{\lambda + \mu - \bar{\lambda} - \bar{\mu}} - \frac{\zeta + \tfrac{4}{3}\eta}{\lambda + 2\mu} - \frac{\bar{\zeta} + \tfrac{4}{3}\bar{\eta}}{\bar{\lambda} + 2\bar{\mu}} \right] \right\},$$

$$\gamma_{xz} = \gamma_{zx} = \frac{\omega^2}{2} \sqrt{(\tilde{\rho}\lambda_5)} \left(\frac{s\eta}{\mu^2} + \frac{\bar{s}\bar{\eta}}{\bar{\mu}^2}\right), \quad \gamma_{xy} = \frac{\omega^2}{2\lambda_4} \sqrt{\left(\frac{\tilde{\rho}}{\lambda_4}\right)} (s\eta + \bar{s}\bar{\eta}),$$

$$\gamma_{zz} = \frac{\omega^2}{2} \sqrt{(\tilde{\rho}\lambda_1)} \left[\frac{s(\zeta + \tfrac{4}{3}\eta)}{(\lambda + 2\mu)^2} + \frac{\bar{s}(\bar{\zeta} + \tfrac{4}{3}\bar{\eta})}{(\bar{\lambda} + 2\bar{\mu})^2} \right], \quad (7.52)$$

where $\lambda_1, \lambda_2, \lambda_4, \lambda_5$ are real constants, determined by Eqs. 7.49; absorption is neglected.

The case in which layers of a nonlossy material (for example, a metal) alternate with layers of a viscous liquid deserves attention. Considering only the shear viscosity in the liquid (layer 2), we have

$$\bar{\lambda} = \bar{\rho}\bar{c}^2 + \frac{2i\omega\bar{\eta}}{3}, \quad \bar{\mu} = -i\omega\bar{\eta},$$

where $\bar{c}$ is the velocity of sound in the liquid. For brevity, we use the notation

$$\bar{\rho}\bar{c}^2 = \bar{K}, \quad \omega\bar{\eta} = \bar{\eta}_1.$$

Then, according to Eq. 7.49,

$$\frac{1}{\lambda_1} = \frac{s}{\lambda + 2\mu} + \frac{\bar{s}}{\bar{K} - \frac{4}{3}i\bar{\eta}_1}; \quad \lambda_2 = \lambda_1\left\{1 + 4s\bar{s}\frac{(\mu + i\bar{\eta}_1)(\lambda + \mu - \bar{K} + \frac{1}{3}i\bar{\eta}_1)}{(\lambda + 2\mu)(\bar{K} - \frac{4}{3}i\bar{\eta}_1)}\right\},$$

$$\lambda_4 = s\mu - i\bar{s}\bar{\eta}_1, \quad \frac{1}{\lambda_5} = \frac{s}{\mu} + \frac{i\bar{s}}{\bar{\eta}_1}. \tag{7.53}$$

Now, we can use Eqs. 7.51 to obtain the propagation velocities and the absorption coefficients of the various waves. Thus, for shear waves we obtain

$$c_{xz}^2 = \frac{2\mu}{s\bar{\rho}\{\sqrt{[1 + (\bar{s}\mu/s\bar{\eta}_1)^2]} + 1\}}, \quad \gamma_{xz}^2 = \frac{s\omega^2\,\bar{\rho}}{2\mu}\{\sqrt{[1 + (\bar{s}\mu/s\bar{\eta}_1)^2]} - 1\},$$

$$c_{xy}^2 = \frac{2(s^2\,\mu^2 + \bar{s}^2\,\bar{\eta}_1^2)}{s\bar{\rho}\mu\{\sqrt{[1 + (\bar{s}\bar{\eta}_1/s\mu)^2]} + 1\}}, \quad \gamma_{xy}^2 = \frac{s\omega^2\,\bar{\rho}\mu}{2(s^2\,\mu^2 + \bar{s}^2\,\bar{\eta}_1^2)}\{\sqrt{[1 + (\bar{s}\bar{\eta}_1/s\mu)^2]} - 1\}. \tag{7.54}$$

If $\bar{\eta}_1 = \omega\bar{\eta}$ is sufficiently small with respect to μ, these equations can be simplified:

$$c_{xz}^2 = \frac{2\omega\bar{\eta}}{\bar{s}\bar{\rho}}, \quad c_{xy}^2 = \frac{s\mu}{\bar{\rho}}, \quad \gamma_{xz}^2 = \frac{\omega\bar{\rho}\bar{s}}{2\bar{\eta}}, \quad \gamma_{xy}^2 = \frac{\omega^4\,\bar{\rho}\bar{s}^2\,\bar{\eta}^2}{2\bar{s}^3\,\mu^3}. \tag{7.55}$$

It is essential to note that as $\bar{\eta}$ decreases, the velocity c_{xz} decreases, and the absorption γ_{xz} increases (however, we must remember that the thickness of the liquid layers must be small in comparison with the wavelength of viscosity waves). On the other hand, when $\bar{\eta}$ is small, the velocity c_{xy} is independent of $\bar{\eta}$, and the absorption coefficient γ_{xy} decreases with decreasing $\bar{\eta}$.

5. *Alternation of solid and liquid layers*

In considering the alternation of solid and liquid layers in the preceding paragraph, we assumed that the condition that the layers be thin in comparison with the wavelengths of compressional and of shear waves was satisfied by both the solid and the liquid layers. However, if the viscosity of the liquid is small, it is also of interest to examine the case in which the liquid layer thickness is small compared to the wavelength of compressional waves, but is considerably greater than the wavelength of shear (viscous) waves. Under these conditions we need consider, as a first approximation, only longitudinal waves in the liquid, and require that the tangential stresses be zero at the boundary between the liquid and the solid. The wave attenuation due to the viscosity of the liquid can then be found from the discontinuity in tangential velocity. A layered medium of this type will have some interesting characteristics. In particular, shear waves will not be able to propagate perpendicular to the layers.

In the liquid we will have only longitudinal waves with the displacements $\mathbf{U}[\bar{U}_x, \bar{U}_y, \bar{U}_z]$. The pressure p is then given by the relation $p = -\bar{K}\,\mathrm{div}\,\mathbf{U}$, where $\bar{K} = \bar{\rho}\bar{c}^2$. The remaining notation is the same as in the preceding paragraph.

At the boundary $z = 0$ we have the following conditions: the normal components of the velocity and the stress tensor are continuous, and the tangential components of the stress tensor are zero:

$$U_z(0) = \bar{U}_z(0), \quad Z_z(0) = -p(0), \quad Z_x(0) = Z_y(0) = 0. \qquad (7.56)$$

The values of the corresponding quantities at the boundaries $z = h_1$ and $z = -h_2$ are related by the periodicity conditions

$$U_z(h_1) = \bar{U}_z(-h_2), \quad Z_z(h_1) = -p(-h_2), \quad Z_x(h_1) = Z_y(h_1) = 0. \qquad (7.57)$$

It is not hard to see that if we consider longitudinal waves propagating perpendicular to the layers (along the z-axis), we shall find nothing particularly different from the case of alternating solid layers, examined above. For c_{zz}, we obtain an expression similar to Eq. 7.46:

$$c_{zz}^2 = \frac{1}{\tilde{\rho}[s/(\lambda + 2\mu) + \bar{s}/\bar{K}]}, \qquad (7.58)$$

i.e. $\lambda + 2\mu$ is simply replaced by $\bar{K}$, the compressibility of the liquid.

In the remaining cases, the results will naturally be quite different from those obtained earlier.

We again begin by examining wave propagation along the layers (along the x-axis) and consider first waves in which the displacement lies in the xz-plane. The displacements and stresses in the solid layers are given as previously by Eqs. 7.32–7.35 and 7.38; in the liquid layers these are

$$\bar{U}_x = \bar{P}(z)\,e^{iax}, \quad \bar{U}_z = \frac{\bar{P}'(z)}{ia}\,e^{iax}, \quad p = \frac{\bar{\rho}\omega^2}{ia}\,\bar{P}(z)\,e^{iax},$$

where $\quad \bar{P}(z) = A\cos\bar{\alpha}\left(z + \frac{h_2}{2}\right) + B\sin\bar{\alpha}\left(z + \frac{h_2}{2}\right), \quad \bar{\alpha}^2 = \bar{k}^2 - a^2, \quad \bar{k} = \frac{\omega}{\bar{c}}.$

Equations 7.56 and 7.57 yield six equations for the constants A, B, C, D, $\bar{A}$, $\bar{B}$, which can again be divided into two groups. In accordance with this, we will again consider two cases:

a. *Waves, longitudinal in the mean* $(B = C = \bar{B} = 0)$. The boundary conditions give

$$A\alpha \sin\frac{\alpha h_1}{2} - Dia \sin\frac{\beta h_1}{2} + \bar{A}\bar{\alpha}\sin\frac{\bar{\alpha}h_2}{2} = 0,$$

$$-A2ia\alpha \sin\frac{\alpha h_1}{2} + D\gamma \sin\frac{\beta h_1}{2} = 0,$$

$$A\mu\gamma \cos\frac{\alpha h_1}{2} - D2ia\mu\beta \cos\frac{\beta h_1}{2} - \bar{A}\bar{\rho}\omega^2 \cos\frac{\bar{\alpha}h_2}{2} = 0,$$

$$\gamma = \kappa^2 - 2a^2.$$

The dispersion equation is then

$$\mu\bar{\alpha}\tan\frac{\bar{\alpha}h_2}{2}\left(4a^2\alpha\beta\tan\frac{\alpha h_1}{2} + \gamma^2\tan\frac{\beta h_1}{2}\right)$$

$$+ \bar{\rho}\omega^2\alpha(\gamma + 2a^2)\tan\frac{\alpha h_1}{2}\tan\frac{\beta h_1}{2} = 0. \quad (7.59)$$

For thin layers, replacing the tangents by their arguments, and making the substitutions $\mu = \omega^2\rho/\kappa^2$, $\alpha^2 = k^2 - a^2$, $\bar{\alpha}^2 = \bar{k}^2 - a^2$, we obtain

$$\bar{s}\rho(\bar{k}^2 - a^2)[4a^2(k^2 - \kappa^2) + \kappa^4] + s\bar{\rho}\kappa^4(k^2 - a^2) = 0.$$

If we introduce the notation

$$x = \left(\frac{a}{\kappa}\right)^2, \quad \xi = \frac{s\tilde{\rho}}{\bar{s}\rho}, \quad \eta = \left(\frac{k}{\kappa}\right)^2, \quad \zeta = \left(\frac{\bar{k}}{\kappa}\right)^2, \quad (7.60)$$

the previous equation takes the form

$$4(1-\eta)x^2 - [4\zeta(1-\eta) + \xi + 1]x + \xi\eta + \zeta = 0,$$

whence

$$x_{1,2} = \frac{4\zeta(1-\eta) + \xi + 1 \pm \sqrt{\{[4\zeta(1-\eta) + \xi - 1]^2 + 4\xi[1 - 4\eta(1-\eta)]\}}}{8(1-\eta)}. \quad (7.61)$$

It is not hard to see that when $\eta < 1$ (which is the only case of real interest), both roots x_1 and x_2 are real and positive, i.e. there are two compressional wave velocities c_{xx}.

Let us consider some special cases. When $\eta = \frac{1}{2}$ $[c = \sqrt{(2)}b]$ we obtain

$$x_1 = \zeta + \frac{\xi}{2}, \quad c_{xx}^{(1)} = \frac{\bar{c}c}{\sqrt{(\bar{c}^2 + c^2)}}, \quad A:D:\bar{A} = 1:\sqrt{\left(\frac{\xi + 2\zeta}{\xi + 2\zeta - 2}\right)}:\frac{\rho}{\bar{\rho}},$$

$$x_2 = \frac{1}{2}, \quad c_{xx}^{(2)} = c, \quad A:D:\bar{A} = 1:0:0.$$

4

Thus, the second root corresponds to a purely compressional wave in the solid layers, unaccompanied by shear because of the choice of a special relation between the velocities c and b. Under these conditions, the liquid layers remain at rest.

In the next special case we will set $\xi \to 0$, which corresponds to making the solid layers thinner, or to increasing their density. We then have

$$x_1 = \zeta, \quad c_{xx}^{(1)} = \bar{c}, \quad A : D : \bar{A} = 0 : 0 : 1,$$

$$x_2 = \frac{1}{4(1-\eta)}, \quad c_{xx}^{(2)} = \frac{2b\sqrt{(c^2 - b^2)}}{c},$$

$$A : D : \bar{A} = \frac{\eta}{1-2\eta} : \frac{-1}{\sqrt{(4\eta - 3)}} : \frac{2h_1(1-\eta)}{h_2[4\zeta(1-\eta) - 1]},$$

where x_1 corresponds to a purely compressional wave in the liquid with fixed solid layers.

Finally, as ξ increases, which corresponds to increasing the density of the liquid, or to making its layers thinner (however, their thickness must always be large in comparison with the wavelength of viscous waves), we obtain

$$x_1 = \frac{\xi}{4(1-\eta)}, \quad c_{xx}^{(1)} = \frac{2b}{c}\sqrt{\left(\frac{c^2 - b^2}{\xi}\right)}, \quad A : D : \bar{A} = 1 : 1 : \frac{2(1-\eta)\rho}{\bar{\rho}},$$

$$x_2 = \eta, \quad c_{xx}^{(2)} = c, \quad A : D : \bar{A} = 1 : 0 : 0.$$

Thus, as ξ increases, the velocity $c_{xx}^{(1)}$ decreases. The root x_2 yields compressional waves in the solid layers.

b. *Waves, transverse in the mean* $(A = D = \bar{A})$. In this case, the transverse displacement in the solid layer is even with respect to the center of the layer, and the longitudinal displacement is odd, i.e. bending of the solid layers must occur. The dispersion equation has the form

$$\mu\bar{\alpha}\left(4a^2\,\alpha\beta \tan\frac{\beta h_1}{2} + \gamma^2 \tan\frac{\alpha h_1}{2}\right) + \bar{\rho}\omega^2\,\alpha(\gamma + 2a^2)\tan\frac{\bar{\alpha}h_2}{2} = 0.$$

If, in the transition to thin layers, we retain only terms of the first order in h_1 and h_2, the condition $\tilde{\rho} = s\rho + \bar{s}\bar{\rho} = 0$ is not satisfied. Thus, this expansion is not sufficiently accurate. Taking account of the cubic terms in the expansion of the tangents, the equation is written

$$\kappa^2\{h_1^2\,[x(3-4x) + \eta(2x-1)^2] + h_2^2\,\xi_1(\zeta - x)\} + 12(1+\xi_1) = 0,$$

where $\xi_1 = \bar{s}\rho/s\rho$; in addition, the notation of (7.60) has been used.

When κh_1 and κh_2 are small, one of the roots of the last equation is positive and very large. In particular, when $h_1 = h_2$, the root is equal to

$$x \approx \frac{1}{\kappa h_1} \sqrt{\left(\frac{3(1+\xi_1)}{1-\eta}\right)}. \tag{7.62}$$

It can be verified that terms of the fifth and higher orders in the expansion of the tangents can be neglected, so that the last solution is completely correct. In the limit of very thin layers, we obtain

$$c_{xz} = 0. \tag{7.63}$$

Let us now examine the propagation along the x-axis of a purely shear wave with displacements in the y-direction. In this case the solid layers do not interact with the liquid layers or with one another, but slip along the fixed liquid as the wave propagates. The displacements in the solid are

$$U_y = v_y = \left[C \cos \beta\left(z - \frac{h_1}{2}\right) + D \sin \beta\left(z - \frac{h_1}{2}\right)\right] e^{iax},$$

whereas in the liquid $\bar{U}_y = 0$. The boundary conditions reduce to the vanishing of the tangential stress Z_y on both sides of the solid layer, whence we obtain two equations for the constants C and D

$$C \sin \frac{\beta h_1}{2} + D \cos \frac{\beta h_1}{2} = 0,$$

$$- C \sin \frac{\beta h_1}{2} + D \cos \frac{\beta h_1}{2} = 0.$$

The condition that these equations have a solution is $\sin \beta h_1 = 0$, i.e. $\beta h_1 = n\pi$, and consequently

$$a = \sqrt{\left[\kappa^2 - \left(\frac{n\pi}{h_1}\right)^2\right]}.$$

As a result we have obtained the wave number for wave propagation in a layer with perfectly reflecting boundaries (see § 26). In the limiting case of thin layers ($\kappa h \ll 1$), only the wave $n = 0$ will propagate along the layers, with velocity

$$c_{xy} = b, \tag{7.64}$$

i.e. the velocity of shear waves. In this wave the displacement is constant over the thickness of the layer $U_y = C^{\kappa x}$. Calculating its attenuation (with respect to the losses introduced by the viscosity of the liquid η as the solid layers slide along the liquid layers), we find

$$\gamma_{xy} = \frac{1}{2h_1} \sqrt{\frac{\bar{\rho}\omega\eta}{\rho\mu}}$$

for the amplitude absorption coefficient.

Finally, in the propagation of shear waves along the z-axis, interaction between the solid and the liquid layers is again absent, and so we must set $c_{zx} = 0$.

§ 8. THE REFLECTION OF BOUNDED BEAMS

So far, we have considered the reflection of plane waves from layered media, or their penetration through layered media. However, under practical and experimental conditions, we are always concerned with

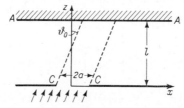

Fig. 30. Beam formed by a plane wave incident on an aperture in a screen.

more or less bounded beams rather than with infinitely extended plane waves. It is therefore necessary to examine the peculiarities that arise in the reflection of bounded beams. Our method of investigation will be based on the expansion of the bounded beam into an infinite set of plane waves, since we have already studied the reflection of plane waves in detail.[44]

1. *The representation of a bounded beam as a superposition of plane waves*

Let us suppose that a bounded beam is created as a result of the passage of a plane wave through a slit CC in a screen, as shown in Fig. 30. We will further suppose that the beam is incident on the boundary AA of a layered-inhomogeneous medium, and we shall investigate the reflection from this boundary. The geometrical boundaries of the beam are shown in Fig. 30 by the dashed lines. Actually, the beam will become somewhat diffused as a result of diffraction.

The angle of incidence of the beam on the boundary will be denoted by ϑ_0. For simplicity, we consider the two dimensional problem, i.e. the slit in the screen extends to infinity, and the plane of incidence is perpendicular to the axis of the slit. Under these conditions, and with a coordinate system as in Fig. 30, the y-coordinate need not be considered.

The z-axis is perpendicular to the plane of the screen, the x-axis is perpendicular to the axis of the slit, and the origin of coordinates is located on the axis of the slit. The width of the slit is $2a$, and its distance from the reflecting plane is l.

Between the screen and the reflecting plane, the field of the beam must satisfy the wave equation. In the plane of the screen, the field must satisfy boundary conditions which, assuming the slit width to be large compared to the wavelength,† we shall give in an approximate form following Kirchhoff, namely, we suppose that:

(1) the field is zero on the back side of the screen;

(2) in the plane of the slit the field is the same as if the screen were not there.

The field of the incident beam is determined completely by the wave equation and the above boundary conditions.

In order to represent the beam as a superposition of plane waves, it is expedient to write the field in the plane $z = 0$ (a function of x) as a Fourier integral.

The field of the incident plane wave in front of the screen is assumed to be of the form

$$\psi = \exp\left[i(\alpha x + \gamma z) - i\omega t\right], \tag{8.1}$$

where

$$\alpha = k \sin \vartheta_0, \quad \gamma = k \cos \vartheta_0. \tag{8.2}$$

After passage through the slit, we have at $z = 0$ (omitting the factor $e^{-i\omega t}$ for brevity)

$$\psi(x) = e^{i\alpha x} \text{ for } -a \lesssim x \lesssim a, \quad \psi(x) = 0 \text{ for } |x| > a. \tag{8.3}$$

We generalize the problem somewhat by allowing an arbitrary amplitude distribution over the cross section of the beam in the plane $z = 0$. Therefore, in what follows, we suppose that the function $\psi(x)$ has the form

$$\psi(x) = F(x)\,e^{i\alpha x}, \quad -\infty < x < \infty. \tag{8.4}$$

The function $F(x)$ in the situation described by Fig. 30 corresponds to a variable transparency of the screen as a function of x. In order for the concept of a beam to have meaning, the beam width must be large compared to the wavelength. This reduces to the requirement, essential in what follows, that $F(x)$ be a slowly varying function (changes only slightly over a distance of the order of a wavelength).

---

† In the exact theory of diffraction at a slit it is shown that Kirchhoff's assumptions are valid only when the angle of incidence is not too great. However, for our present purposes this is not too important.

4*

We represent the field in the plane $z = 0$ as a Fourier integral

$$\psi(x) = \int_{-\infty}^{+\infty} \Phi(p)\,e^{ipx}\,dp. \tag{8.5}$$

The function $\Phi(p)$ is determined by the well known formula

$$\Phi(p) = \frac{1}{2\pi}\int_{-\infty}^{+\infty}\psi(x)\,e^{-ipx}\,dx = \frac{1}{2\pi}\int_{-\infty}^{+\infty}F(x)\,e^{i(\alpha-p)x}\,dx. \tag{8.6}$$

Thus, for example, in the case of a beam formed by a wave passing through a slit, we have

$$F(x) = 1, \quad -a < x < a,$$

$$F(x) = 0, \quad |x| > a, \tag{8.7}$$

which gives $\qquad \Phi(p) = \dfrac{1}{2\pi}\displaystyle\int_{-a}^{+a}e^{i(\alpha-p)x}\,dx = \dfrac{\sin(\alpha-p)a}{\pi(\alpha-p)}. \tag{8.8}$

It is useful to note that the function $\Phi(p)$ is of unit order of magnitude only for small differences $(\alpha-p)$, satisfying the condition $a(\alpha-p)\lesssim 1$. When $a(\alpha-p)\gg 1$, the function $\Phi(p)$ will be very small.

Using the function $\psi(x)$, given in the form (8.4) and (8.5) and characterizing the field in the plane of the screen, we construct the following function of the two variables x and z:

$$\psi(x, z) = \int_{-\infty}^{+\infty}\Phi(p)\,e^{i(px+\mu z)}\,dp, \tag{8.9}$$

where $\mu = \sqrt{(k^2 - p^2)}$.

This function will also describe the field of the incident beam between the plane of the screen and the reflecting plane because:

(1) it satisfies the wave equation

$$\frac{\partial^2\psi}{\partial x^2} + \frac{\partial^2\psi}{\partial z^2} + k^2\psi = 0, \tag{8.10}$$

since the integrand satisfies this equation;

(2) at $z = 0$ it is transformed into the function $\psi(x)$ given by Eq. 8.5, i.e. it satisfies the boundary condition.

Since the solution of the wave equation subject to these boundary conditions is unique, there can be no other expression describing the beam.

The exponent in the integrand in Eq. 8.9 for any given p represents a plane wave propagating at an angle

$$\vartheta = \arcsin\frac{p}{k} \tag{8.11}$$

with respect to the z-axis. Thus, each component of the expansion of the field in the plane $z = 0$ in the Fourier integral extends through space as a separate plane wave. Equation 8.9 will also be the expansion (that we are seeking) of the incident beam into plane waves.

When $p > k$, the angle ϑ will be complex according to Eq. 8.11, i.e. the expansion will also contain inhomogeneous plane waves (see § 1). Their amplitude will decrease exponentially with distance from the plane of the screen. This is evident from Eq. 8.9 since, when $p > k$, the exponent in the integrand takes the form $\exp[ipx - \sqrt{(p^2 - k^2)}z]$.

However, in Eq. 8.9, the most important rôle will be played by those plane waves traveling in approximately the same direction as the plane wave incident on the slit. In fact, as we have already noted, the function $\Phi(p)$ will have a significant value only when

$$(\alpha - p)a \lesssim 1.$$

Since $\alpha = k \sin \vartheta_0$ and $p = k \sin \vartheta$, the last condition can be written in the form $ak(\sin \vartheta_0 - \sin \vartheta) \lesssim 1$, or, in view of the closeness of ϑ_0 and ϑ,

$$\vartheta_0 - \vartheta \lesssim \frac{1}{ak \cos \vartheta_0}. \tag{8.12}$$

The quantity $a \cos \vartheta_0$ is equal to the width of the normal cross section of the beam (whereas a is the width of the cross section in the plane $z = 0$). Since this width must be much greater than a wavelength (otherwise the concept of a beam loses meaning), we have, consequently, $ak \cos \vartheta \gg 1$ and

$$\vartheta_0 - \vartheta \ll 1. \tag{8.13}$$

2. *The field of the reflected beam. The displacement of a beam upon reflection*

We will let $V(\vartheta)$ denote the reflection coefficient of a plane wave incident on a reflecting boundary at the angle ϑ. This quantity will be complex; the square of its absolute value gives the reflection coefficient with respect to energy, and its argument gives the phase change undergone by the wave upon reflection. We write $V(\vartheta)$ in the form

$$V(\vartheta) = \rho(p)\, e^{i\phi(p)}, \tag{8.14}$$

where $\rho(p)$ and $\phi(p)$ are functions of p, and consequently are functions of the angle of incidence ϑ.

By setting $z = l$ in Eq. 8.9, we obtain the field created on the boundary by the incident beam. As a result, at the boundary we have

$$\psi_{\text{inc}}(x) = \int_{-\infty}^{+\infty} \Phi(p)\, e^{i(px + \mu l)}\, dp. \tag{8.15}$$

If we multiply the integrand in Eq. 8.15, which represents a plane wave with a variable angle of incidence, by the reflection coefficient $V(\vartheta)$, we obtain the field of the reflected beam. Hence,

$$\psi_{\text{refl}}(x) = \int_{-\infty}^{+\infty} \Phi(p)\,\rho(p) \exp\left[i\phi(p) + i(px + \mu l)\right] dp. \qquad (8.16)$$

Substituting the value of $\Phi(p)$ from Eq. 8.6 into Eq. 8.16, the field of the incident and reflected beams can be written in another form:

$$\psi_{\text{inc}}(x) = \frac{1}{2\pi} \iint_{-\infty}^{+\infty} F(\xi) \exp\left[i(\alpha - p)\,\xi + i(px + \mu l)\right] dp\,d\xi, \qquad (8.17)$$

$$\psi_{\text{refl}}(x) = \frac{1}{2\pi} \iint_{-\infty}^{+\infty} F(\xi)\,\rho(p) \exp\left[i\phi(p)\right]$$
$$\times \exp\left[i(\alpha - p)\,\xi + i(px + \mu l)\right] dp\,d\xi. \qquad (8.18)$$

In what follows, we shall consider the case in which the absolute value of the reflection coefficient does not vary very much within the limits of the important angles of incidence. Thus, one of the most significant special cases is that of total internal reflection from an interface where the absolute value of the reflection coefficient is constant and equal to unity for all angles greater than the angle of total internal reflection.

Under these conditions, the quantity $\rho(p)$ can be taken outside the integral sign at the value $p = \alpha$. Furthermore, we introduce a new quantity

$$\Omega = p - \alpha \qquad (8.19)$$

in Eqs. 8.17 and 8.18. Only values of the functions corresponding to small Ω will play an essential rôle in the integrals. We therefore expand the function $\phi(p)$ in a power series in Ω, i.e. we write

$$\phi(p) = \phi(\alpha) + \phi'(\alpha)\,\Omega + \tfrac{1}{2}\phi''(\alpha)\,\Omega^2 + \ldots, \qquad (8.20)$$

where

$$\phi'(\alpha) = \left(\frac{\partial \phi}{\partial p}\right)_{p=\alpha}, \qquad (8.21)$$

and limit ourselves (for the time being) to terms of first degree in Ω. Then, taking account of the value of μ, Eqs. 8.17 and 8.18 can be written in the form

$$\psi_{\text{inc}}(x) = \exp(i\alpha x) f_{\text{inc}}(x),$$

$$f_{\text{inc}}(x) = \frac{1}{2} \iint_{-\infty}^{+\infty} F(\xi) \exp\left\{i\Omega(x - \xi) + i\sqrt{[k^2 - (\Omega + \alpha)^2]}\,l\right\} d\xi\,d\Omega, \qquad (8.22)$$

$$\psi_{\text{refl}}(x) = \exp(i\alpha x) f_{\text{refl}}(x),$$

$$f_{\text{refl}}(x) = \frac{V(\vartheta_0)}{2\pi} \iint_{-\infty}^{+\infty} F(\xi) \exp\left\{i\Omega[x + \phi'(\alpha) - \xi]\right.$$
$$\left. + i\sqrt{[k^2 - (\Omega + \alpha)^2]}\,l\right\} d\xi\,d\Omega. \qquad (8.23)$$

where $f_{\mathrm{inc}}(x)$ and $f_{\mathrm{refl}}(x)$ are functions describing the form of the beam.

Comparing Eqs. 8.22 and 8.23, we find

$$f_{\mathrm{refl}}(x) = f_{\mathrm{inc}}[x + \phi'(\alpha)] \cdot V(\vartheta_0). \qquad (8.24)$$

Hence, upon reflection, the beam is displaced along the boundary by an amount

$$\Delta = -\phi'(\alpha) = -\left(\frac{\partial \phi}{\partial p}\right)_{p=\alpha}. \qquad (8.25)$$

The physical meaning of Δ will become clear after we examine the following simple phenomenon.

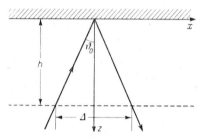

Fig. 31. The case of a perfectly reflecting boundary.

Let a plane wave be incident on a boundary $z = 0$ (Fig. 31) whose reflection coefficient is equal to unity (the boundary of a perfect reflector). Identifying the xz-plane with the plane of incidence, the incident wave can be written as previously in the form $\exp[i(\alpha x - \gamma z) - i\omega t]$ and the reflected wave in the form $\exp[i(\alpha x + \gamma z) - i\omega t]$, where $\alpha = k \sin \vartheta_0$ and $\gamma = k \cos \vartheta_0$.

At $z = 0$, both waves have the same amplitude and phase. In some plane $z = h$, the incident and the reflected waves will have the expressions $\exp[i(\alpha x - \gamma h) - i\omega t]$ and $\exp[i(\alpha x + \gamma h) - i\omega t]$. The ratio of the reflected and the incident waves

$$V = e^{2i\gamma h} \qquad (8.26)$$

can be considered as the reflection coefficient in the plane $z = h$. The modulus of this coefficient is unity, and the phase $\phi = 2\gamma h = 2\sqrt{(k^2 - \alpha^2)}\,h$ corresponds to the phase change as the wave travels from the plane $z = h$ to the plane $z = 0$ and back.

In the general case, with an arbitrary angle of incidence, the phase ϕ will be

$$\phi = 2h\sqrt{(k^2 - p^2)}, \qquad (8.27)$$

where p, as previously, is the projection of the wave vector on the x-axis ($p = k \sin \vartheta$).

In this case, the displacement Δ as given by Eq. 8.25 is

$$\Delta = -\left(\frac{\partial\phi}{\partial p}\right)_{p\,=\,\alpha} = 2h\frac{\alpha}{\gamma} = 2h\tan\vartheta_0. \tag{8.28}$$

The quantity Δ is shown in Fig. 31. It is equal to the horizontal displacement of the ray as it travels from the plane $z = h$ to the plane $z = 0$ and back.

Thus, in this case, the ray displacement calculated by Eq. 8.25 has a very simple and natural significance. Of course, the displacement could have been calculated by using ray theory, without resorting to Eq. 8.25. However, the theory developed above is of fundamental value under conditions in which the ray concept is inapplicable. We shall examine cases of this type below.

We should also note that the theory of beam displacement upon reflection is completely analogous to the theory of the propagation of a quasi-monochromatic pulse in a dispersive medium (see, for example, Ref. 4).

3. Total internal reflection of a beam

According to Eq. 8.25, we expect that the beam displacement upon reflection is greatest when the phase of the reflection coefficient changes most rapidly with the angle. One such case is the incidence of a beam on the boundary between two homogeneous media at an angle somewhat greater than the limiting angle of total internal reflection. As is seen in Fig. 8 (the limiting angle here is $56°$), the phase of the reflection coefficient changes very rapidly under these conditions.

We shall therefore assume that the propagation velocity in the reflecting medium is greater than in the medium through which the wave is incident, and that the angle of incidence is greater than the limiting angle of total internal reflection. Then total reflection will occur, and the reflection coefficient can be written in the form $V = e^{i\phi}$.

As an example, we examine the reflection of an electromagnetic wave whose electric vector is perpendicular to the plane of incidence. With $\mu_1 = \mu$ and $\sin\vartheta > n$, Eq. 2.16 gives

$$V_\perp = \frac{\cos\vartheta - i\sqrt{(\sin^2\vartheta - n^2)}}{\cos\vartheta + i\sqrt{(\sin^2\vartheta - n^2)}} = e^{i\phi}, \quad \phi = -2\arctan\sqrt{\left(\frac{p^2 - k^2 n^2}{k^2 - p^2}\right)}, \tag{8.29}$$

where $p = k\sin\vartheta$ and ϑ is the angle of incidence. Now, using Eq. 8.25 for the ray displacement, we obtain the expression

$$\Delta_\perp = \frac{\lambda}{\pi}\frac{\tan\vartheta_0}{\sqrt{(\sin^2\vartheta_0 - n^2)}}, \tag{8.30}$$

where, as previously, ϑ_0 is the angle between the axis of the beam and the normal to the boundary.

Similarly, if the electric vector lies in the plane of incidence, Eq. 2.18 gives

$$\Delta_\| = \frac{\lambda}{\pi n^2} \frac{(1-n^2) \tan \vartheta_0}{\sqrt{(\sin^2 \vartheta_0 - n^2) \left[\cos^2 \vartheta_0 + (1/n^4)(\sin^2 \vartheta_0 - n^2)\right]}}.$$

If we take into account that a significant effect occurs only when $\sin \vartheta_0$ is close to n, the previous equation can be written in a simpler form:

$$\Delta_\| = \frac{\lambda}{\pi n^2} \frac{\tan \vartheta_0}{\sqrt{(\sin^2 \vartheta_0 - n^2)}}. \tag{8.31}$$

In these formulas, λ is the wavelength in the medium through which the incident wave travels, and n is the index of refraction at the boundary.

We see that the displacement of the beam is greater, the closer its angle of incidence is to the limiting angle of total internal reflection. As $\sin \vartheta_0 \to n$, Eqs. 8.30 and 8.31 give $\Delta_\perp \to \infty$ and $\Delta_\| \to \infty$.

An experimental determination of the displacement of a light beam upon total internal reflection at a glass–air boundary was performed by German investigators.[124, 125] Although the displacement at each reflection may be many wavelengths, multiple reflections were used in the experiments in order to make the displacements readily observable (the displacement increases proportionately to the number of reflections).

The theoretical calculations were confirmed completely by the experiments.

In the case of the reflection of acoustic waves from the interface between two liquid media, the reflection coefficient will be given by Eq. 3.14. Under conditions of total internal reflection ($\sin \vartheta_0 > n$), the reflection coefficient can be written in the form $V = e^{i\phi}$, where the phase ϕ is given correctly by Eq. 3.20. When $\sin \vartheta_0$ is close to n, we again use Eq. 8.25, and obtain for the displacement

$$\Delta = \frac{\lambda}{\pi m} \frac{\tan \vartheta_0}{\sqrt{(\sin^2 \vartheta_0 - n^2)}}. \tag{8.31a}$$

4. The reflection of acoustic beams from solids and from plates

When an acoustic wave is incident from a liquid on a solid, total internal reflection will occur when $\sin \vartheta > c/b_1$, where c is the acoustic velocity in the liquid, and b_1 is the shear wave velocity in the solid. Near the limiting angle of total reflection, just as in optics, considerable displacement of the beam will occur. This displacement can be calculated from Eq. 8.25, using Eq. 4.32 for the reflection coefficient and writing it in the form $V = e^{i\phi}$.

However, in addition to the above case, considerable displacement also occurs when the angle of incidence is such that the phase velocity of the wave along the boundary coincides with the velocity of Rayleigh waves on the boundary of the solid. This angle will be somewhat greater than the angle of total internal reflection. In the neighborhood of this angle, which is determined by Eq. 4.48 and the relation $\sin \vartheta = (c/b_1) \sin \gamma_1$, the reflection coefficient will be a rapidly varying function, as is clear from Eq. 4.25. In particular, it can be shown that in the neighborhood of this angle, the phase of the wave will change by approximately 2π. In this case,[192] the displacement is

$$\Delta = \frac{2\lambda}{\pi} \frac{\rho_1}{\rho} \sqrt{\left[\frac{r(r-s)}{s(s-1)}\right]} \frac{1 + 6s^2(1-q) - 2s(3-2q)}{s-q}, \qquad (8.32)$$

where λ is the wavelength in the liquid, $s = (b_1/v_R)^2$, and v_R is the velocity of Rayleigh waves. The remaining notation is the same as in § 4 (see Eq. 4.50).

The quantity s is a function of Poisson's ratio, and varies between 1.1 and 1.3 for various solids. For aluminum, for example, $s = 1.15$.

Substitution of numerical values into Eq. 8.32 gives:

(a) $\Delta = 24.4\lambda$ for a water–aluminum boundary;

(b) $\Delta = 33.4\lambda$ for a xylene–aluminum boundary.

The angles of incidence at which these displacements occur are determined from the relation $\sin \vartheta_R = c/v_R$, and are $\vartheta_R = 31°$ and $\vartheta_R = 27°$, respectively.

The displacement Δ is sufficiently great to be easily measured. Shadow photographs of the total internal reflection of an ultrasonic beam (1.6×10^7 cps) from a xylene–aluminum boundary are shown in Fig. 32 for three angles of incidence, increasing from left to right.[192] The angle of incidence is ϑ_R in the center picture; the displacement of the reflected beam with respect to the incident beam is quite evident. This displacement is not present in the end pictures, where the angles of incidence differ somewhat from ϑ_R. The beam width is 12 mm, the wavelength in xylene at this frequency is 0.08 mm, and the displacement, in agreement with the calculated value, is 2.7 mm.

The photographs shown in Fig. 33 refer to the same conditions as in Fig. 32, but at a lower frequency, 5.5×10^6 cps ($\lambda = 0.24$ mm). The displacement is greater here since according to the theory it is proportional to the wavelength. The beam is displaced by almost its entire width. However, because of the increase of the wavelength, the boundaries of the beam are more diffuse.

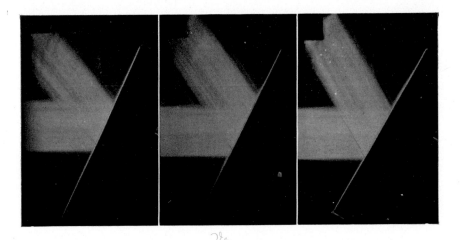

Fig. 32. The displacement of an ultrasonic beam of frequency 1.6×10^7 cps upon reflection at a xylene–aluminum interface.

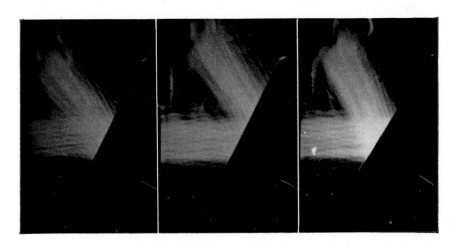

Fig. 33. The same as in Fig. 32, but at the frequency 5.5×10^6 cps.

4**

The presence of displacement is a sufficiently good indication of the angle ϑ_R to be used for an accurate determination of the Rayleigh wave velocity and, consequently, for the determination of the elastic parameters of the medium.

A very large displacement of an acoustic beam can arise upon reflection from a plate because in this case the phase of the reflection coefficient varies particularly rapidly with angle (see, for example, Fig. 34, showing the phase of the reflection coefficient as a function of the angle of incidence for the reflection of an acoustic wave from an aluminum plate in air). Here, the transmitted beam is also displaced with respect to the incident beam.

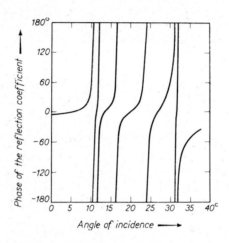

Fig. 34. The phase of the coefficient of reflection of a sound wave from an aluminum plate in air (calculated). The plate thickness is 3.1 mm, and the frequency is 3.35×10^6 cps.

Two cases of reflection of an ultrasonic beam of frequency 16.3×10^6 cps from an aluminum plate 0.43 mm thick in xylene are shown in Fig. 35.[191] The angles of incidence are somewhat different in the two cases. In one case we have almost total reflection, and in the other case we have almost total penetration of the plate.

The reflected and transmitted beams are displaced by several plate thicknesses. It is also quite evident that the cross section of the reflected beam is not uniform. We will consider this question further in the next section.

5. *The energy distribution in the cross section of the reflected beam*

In order to investigate the "internal structure" of the reflected beam, i.e. the energy distribution over its cross section, terms of the second

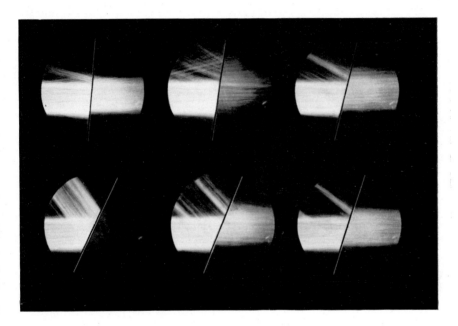

Fig. 35. The reflection of an ultrasonic beam of frequency 1.63×10^7 cps from an aluminum plate 0.43 mm thick, in xylene.

order must be taken into account in Eq. 8.20 for the phase of the reflection coefficient.† Then, in place of Eq. 8.23, we obtain the following expression for the field of the reflected beam

$$\psi_{\text{refl}}(x) = \frac{1}{2\pi} V(\vartheta_0) e^{i\alpha x} \int\!\!\int_{-\infty}^{+\infty} F(\xi) \exp\{i\Omega[x + \phi'(\alpha) - \xi]$$
$$+ i\sqrt{[k^2 - (\Omega + \alpha)^2]}\, l + \tfrac{1}{2} i\phi''(\alpha)\Omega^2\}\, d\xi\, d\Omega. \quad (8.33)$$

We treat the factor $\exp\{i\sqrt{[k^2 - (\Omega + \alpha)^2]}\,l\}$ as a slowly varying function of Ω and take it outside the integral sign at the value corresponding to $\Omega = 0$. For this to be valid, l must not be too large. Leaving this factor under the integral for large l would make it possible to take account of the spreading of the beam because of diffraction as it propagates from the slit to the reflecting surface. This spreading has been quite thoroughly investigated in diffraction theory, and we shall not take it into account here.

We introduce a new variable η through the relation

$$\phi''\left(\Omega + \frac{x + \phi' - \xi}{\phi''}\right)^2 = \pi\eta^2. \quad (8.34)$$

Carrying out the integration in Eq. 8.33 with respect to η, and using

$$\int_{-\infty}^{+\infty} \exp\left(\pm i\frac{\pi}{2}\eta^2\right) d\eta = 1 \pm i, \quad (8.35)$$

we obtain

$$\psi_{\text{refl}}(x) = \frac{1+i}{2} V(\vartheta_0) \frac{1}{\sqrt{(\pi\phi'')}} \exp\{i[\alpha x + \sqrt{(k^2 - \alpha^2)}\, l]\}$$
$$\times \int_{-\infty}^{+\infty} \exp\left[-i\frac{(x + \phi' - \xi)}{2\phi''}\right] F(\xi)\, d\xi. \quad (8.36)$$

Introducing the further notation

$$\frac{x + \phi' - \xi}{\sqrt{(\pi\phi'')}} = u, \quad (8.37)$$

we obtain

$$\psi_{\text{refl}}(x) = \frac{1+i}{2} V(\vartheta_0) \exp\{i[\alpha x + \sqrt{(k^2 - \alpha^2)}\, l]\}$$
$$\times \int_{-\infty}^{+\infty} \exp\left(-i\frac{\pi}{2}u^2\right) F[x + \phi' + \sqrt{(\pi\phi'')}u]\, du. \quad (8.38)$$

---

† The problem we are now considering is formally analogous to the problem of the change of form of a wave packet during propagation in a dispersive medium (see Ref. 4), on which we will base our approach.

When ϕ'' is sufficiently small (and if we take account of Eq. 8.35), we again obtain Eq. 8.24, giving displacement of the beam without changing its form.

Further analysis of Eq. 8.38 without neglecting terms containing $\sqrt{(\pi\phi'')}$ is possible only when a definite amplitude distribution over the cross section of the beam is given, i.e. when the form of the function F is given. We shall assume here that the function F is given by relations (8.7), i.e. that the width of the incident beam is $2a$ and the intensity is constant over the cross section. Then the integral in Eq. 8.38 can be taken between the limits (u_1, u_2), determined from the relations

$$x + \phi' + \sqrt{(\pi\phi'')} \cdot u_{1,\,2} = \mp a \qquad (8.39)$$

or $\qquad u_1 = -\dfrac{1}{\sqrt{(\pi\phi'')}}\,(a+x+\phi'), \quad u_2 = \dfrac{1}{\sqrt{(\pi\phi'')}}\,(a-x-\phi'), \qquad (8.40)$

since the function F is zero outside these limits.

Since $F = 1$ within the limits of integration, Eq. 8.38 gives

$$\psi_{\text{refl}}(x) = \frac{1+i}{2}\,V(\vartheta_0)\exp\left\{i[\alpha x + \sqrt{(k^2-\alpha^2)}\,l]\right\}\int_{u_1}^{u_2}\exp\left(-i\frac{\pi}{2}n^2\right)du, \quad (8.41)$$

which can be written in the form

$$\psi_{\text{refl}}(x) = \frac{1+i}{2}\,V(\vartheta_0)\exp\left\{i[\alpha x + \sqrt{(k^2-\alpha^2)}\,l]\right\}[f(u_2)-f(u_1)], \qquad (8.42)$$

where $\qquad f(u_1) = \displaystyle\int_0^{u_1}\exp\left(-i\frac{\pi}{2}u^2\right)du = C(u_1) - iS(u_1) \qquad (8.43)$

and similarly for $f(u_2)$. The quantities C and S are Fresnel integrals.[16] Since the functions C and S are oscillating functions of their arguments, the intensity of the reflected beam will not vary monotonically over the cross section. This behavior is confirmed experimentally (see, for example, Fig. 35).

6. The energy flux in total internal reflection

Beam displacement upon reflection can also be obtained by considering the energy flux in the incident and reflected beams.[119, 190] The lines of energy flux (Umov-Poynting vector) in the total internal reflection of an unbounded plane wave were first studied by Eikhenval'd.[80] He showed that these lines enter the reflecting medium and then leave it at a distance $\lambda/2$ along the boundary. However, it is difficult to compare these results with a definitive experiment since the measurement of the energy flux density at every point of space requires the separation of a beam bounded by a diaphragm.

We assume from the very beginning that a bounded beam is incident on the interface. For definiteness we consider the acoustic case, although the electromagnetic case is no different in principle.

The energy flux density in any direction is given by the Umov vector

$$\mathbf{I} = p\mathbf{v}, \tag{8.44}$$

where p is the acoustic pressure, and $\mathbf{v}$ is the particle velocity in the acoustic wave. If p and $\mathbf{v}$ are expressed in terms of the acoustic potential ϕ, we obtain

$$\mathbf{I} = -\rho \frac{\partial \phi}{\partial t} \operatorname{grad} \phi. \tag{8.45}$$

We write ϕ in the form

$$\phi = \psi(x, y, z) \, \mathrm{e}^{-i\omega t}. \tag{8.46}$$

Since we are using complex notation, we must remember, as has been mentioned previously, that ϕ in Eq. 8.45 refers to the real part of the potential.

We find the time average of the energy flux, which reduces to the time average of the quantity $(\partial/\partial t)\,(\operatorname{Re}\phi)\operatorname{grad}(\operatorname{Re}\phi)$ (the symbol Re denotes the real part). We then have

$$\operatorname{Re}\phi = \tfrac{1}{2}(\phi + \phi^*) = \tfrac{1}{2}(\psi\,\mathrm{e}^{-i\omega t} + \psi^*\,\mathrm{e}^{i\omega t}),$$

$$\operatorname{grad}(\operatorname{Re}\phi) = \tfrac{1}{2}(\operatorname{grad}\psi\,\mathrm{e}^{-i\omega t} + \operatorname{grad}\psi^*\,\mathrm{e}^{i\omega t}), \tag{8.47}$$

$$\frac{\partial}{\partial t}(\operatorname{Re}\phi) = \frac{i\omega}{2}(\psi^*\,\mathrm{e}^{i\omega t} - \psi\,\mathrm{e}^{-i\omega t}). \tag{8.48}$$

Multiplying Eqs. 8.47 and 8.48 and taking the time average we obtain

$$\bar{\mathbf{I}} = \frac{i\omega\rho}{4}(\psi\operatorname{grad}\psi^* - \psi^*\operatorname{grad}\psi). \tag{8.49}$$

If we apply the last formula to the calculation of the energy flux in a plane wave, and remember that

$$\psi(x, y, z) = \psi_0\,\mathrm{e}^{i\mathbf{k}\cdot\mathbf{r}},$$

we obtain $\qquad \operatorname{grad}\psi = i\mathbf{k}\psi, \quad \operatorname{grad}\psi^* = -i\mathbf{k}\psi^*,$

and Eq. 8.49 gives $\qquad \bar{\mathbf{I}} = \frac{\omega\rho}{2}\psi_0^2\,\mathbf{k}. \tag{8.50}$

Hence, using $\omega\rho\psi_0 = |p|, |\mathbf{k}| = \omega/c$, we obtain Eq. 3.28 for the absolute value of the average flux.

We examine the projection of the average energy flux onto the normal to the interface in the case of a bounded beam.

From Eq. 8.49, we have

$$\bar{I}_l = \frac{i\omega\rho}{2}\left(\psi\frac{\partial\psi^*}{\partial z} - \psi^*\frac{\partial\psi}{\partial z}\right). \qquad (8.51)$$

According to Eqs. 8.14–8.16, the total field of the incident and reflected beams at the interface can be written in the form

$$\psi(x) = \psi_{\text{inc}}(x) + \psi_{\text{refl}}(x) = \int_{-\infty}^{+\infty} A(p)\left[1 + V(p)\right]e^{ipx}\,dp, \qquad (8.52)$$

where
$$A(p) = \Phi(p)\exp\left[i\sqrt{(k^2 - p^2)}\,l\right], \qquad (8.53)$$

and $V(p)$ is the reflection coefficient of a plane wave. The parameter p is connected with the angle of incidence at the boundary by the relation $p = k\sin\vartheta$. The quantity $\partial\psi/\partial z$ is obtained by multiplying the integrand in the expression for the incident beam (see Eq. 8.9) by $i\mu$, and the integrand for the reflected beam by $-i\mu$. As a result, similar to Eq. 8.52, we obtain

$$\frac{\partial\psi}{\partial z} = \int_{-\infty}^{+\infty} iA(p)\left[1 - V(p)\right]e^{ipx}\,\mu\,dp. \qquad (8.54)$$

Now we substitute Eqs. 8.52 and 8.54 into Eq. 8.51, and obtain

$$\bar{I}_z = \frac{\omega\rho}{4}\left\{\iint_{-\infty}^{+\infty}\left[A(p')A^*(p)e^{ix(p'-p)} + A^*(p')A(p)e^{-ix(p'-p)}\right]\mu\,dp'\,dp\right.$$

$$-\iint_{-\infty}^{+\infty}\mu[A(p')A^*(p)V(p')V^*(p)e^{ix(p'-p)}$$

$$+A^*(p')A(p)V^*(p')V(p)e^{-ix(p'-p)}]\,dp'\,dp$$

$$+\iint_{-\infty}^{+\infty}\mu[A(p')A^*(p)(V(p') - V^*(p))e^{ix(p'-p)}$$

$$\left. +A^*(p')A(p)(V^*(p') - V(p))e^{-ix(p'-p)}]\,dp'\,dp\right\}. \qquad (8.55)$$

Thus, the expression for $\bar{I}_z$ is split into three terms: $I_{\text{I}} + I_{\text{II}} + I_{\text{III}}$. The first of these, I_{I} (the first integral in Eq. 8.55), contains quantities referring only to the incident beam, and thus corresponds to the energy flux in the incident beam. Similarly, I_{II} (the second integral) gives the energy flux in the reflected beam and therefore has the opposite sign of I_{I}. Finally, in I_{III} (the third integral), quantities referring to the incident and reflected waves are mixed. This last term characterizes the non-additivity of the energy fluxes in the incident and reflected waves.

In the case of total reflection, which we are currently considering, $\rho(p)$ is equal to unity in Eq. 8.14 for the reflection coefficient. We again expand $\phi(p)$ in powers of Ω (Eq. 8.20), taking account of terms up to the first degree in Ω. We recall that the validity of this expansion depends on the smallness of the angular spread of the beam. As a result, after some straightforward transformations, we obtain

$$I_{\text{I}}(x) = \frac{\omega\rho}{2} k \cos\vartheta_0 \iint_{-\infty}^{+\infty} A(p') A^*(p) e^{ix(p'-p)} \, dp' \, dp,$$

$$I_{\text{II}}(x) = -\frac{\omega\rho}{2} k \cos\vartheta_0 \iint_{-\infty}^{+\infty} A(p') A^*(p) e^{i(x-\Delta)(p'-p)} \, dp' \, dp = -I_{\text{I}}(x-\Delta),$$

$$I_{\text{III}}(x) = \frac{i\omega\rho}{2} \sin\phi_0 \iint_{-\infty}^{+\infty} (\mu'-\mu) A(p') A^*(p) e^{i(x-\frac{1}{2}\Delta)(p-p)} \, dp' \, dp,$$

where ϕ_0 denotes the value of the phase of the reflection coefficient $\phi(p)$ at $p = \alpha = k \sin\vartheta_0$, that is, at an inclination corresponding to the axis of the beam; Δ is still given by Eq. 8.25.

In the transformations of the expressions for I_{I} and I_{II}, the function μ was treated as a slowly varying function and was taken outside the integral with the value $\mu = k \cos\vartheta_0$. However, this could not be done in the expression for I_{III} since then, I_{III} would become identically zero. Therefore, we make a series expansion of the difference

$$\mu' - \mu = \sqrt{(k^2 - p'^2)} - \sqrt{(k^2 - p^2)}$$

in the neighborhood of the point $p = \alpha$. We then have

$$\mu' - \mu = \left[\frac{\partial \sqrt{(k^2 - p'^2)}}{\partial p'}\right]_{p'=\alpha} (p' - \alpha) - \left[\frac{\partial \sqrt{(k^2 - p^2)}}{\partial p}\right]_{p=\alpha} (p - \alpha)$$

$$= -\tan\vartheta_0 (p' - p).$$

We finally obtain $\quad I_{\text{III}} = -\dfrac{\sin\phi_0 \tan\vartheta_0}{k \cos\vartheta_0} \dfrac{\partial}{\partial x} I_{\text{I}}\left(x - \dfrac{\Delta}{2}\right).$ (8.56)

For aid in visualization, the component parts of the energy flux, I_{I}, I_{II}, I_{III}, are shown schematically in Fig. 36 as functions of the coordinate x.

We see that I_{II} is the same energy flux as I_{I} in the incident beam, except that it has the opposite sign and is displaced along the boundary by the distance Δ, which signifies the displacement of the reflected beam with respect to the incident beam. Inside the cross section of the beam, the resultant energy flux $\bar{I}_z = I_{\text{I}} + I_{\text{II}} + I_{\text{III}}$ is zero, because the fluxes of the incident and reflected beams cancel one another. However, at the edges of the beam, energy flows into the reflecting medium on one side and returns on the other (Fig. 37).

The total quantity of energy taking part in this "exchange", i.e. entering the reflecting medium per unit time, is obtained by integrating $\bar{I}_z$ from $x = -\infty$ to $x = x_m$, where x_m is the center of the beam (the beam width is considered large with respect to Δ):

$$\int_{-\infty}^{x_m} \bar{I}_z \, dx = \int_{-\infty}^{x_m} (I_{\mathrm{I}} + I_{\mathrm{II}} + I_{\mathrm{III}}) \, dx = I_{\mathrm{I}}(x_m) \left(\Delta - \frac{\sin \phi_0 \tan \vartheta_0}{k \cos \vartheta_0} \right). \quad (8.57)$$

Since the beam width is assumed to be sufficiently large compared to the wavelength, we can approximate $I_{\mathrm{I}}(x_m)$ by the normal component of the energy flux density in a plane wave with the same amplitude as that in the center of the beam.

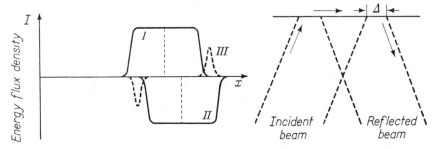

Fig. 36. The distribution of the component parts of the energy flux normal to the interface.

Fig. 37. Illustration of the energy flux upon total internal reflection of a beam.

It is interesting to note that as the angle of incidence approaches the limiting angle of total internal reflection ($\sin \vartheta_0 \to n$), the value of the total flux increases rapidly since, according to Eqs. 8.30 and 8.31, the beam displacement increases. It can also be shown that under these conditions the wave penetrates more deeply into the reflecting medium.

7. *Reflection from an inhomogeneous medium*

The study of the reflection of a bounded beam from a medium with continuously varying properties is of great interest. Here, the application of Eq. 8.25 for the displacement leads to interesting results, supplementing the usual ray representations of wave refraction.[39]

We shall consider a concrete example in which the index of refraction is the following function of the z-coordinate:

$$n = 1 \text{ for } z \leqslant 0,$$
$$n = \sqrt{(1 - az)} \text{ for } z > 0, a > 0. \quad (8.58)$$

It will be seen below that the results obtained here are of more general significance.

We suppose that a bounded beam is incident on the boundary $z = 0$ from a homogeneous medium ($z < 0$). In order to find the form and location of the reflected beam, we must, as we have seen, know the reflection coefficient $V(\vartheta)$ of a plane wave from this interface as a function of the angle of incidence. This function is found in § 15 and is given by Eq. 15.15. To continue the analysis, it is expedient to express the Bessel functions in this formula in terms of Airy functions in accordance with the relations[23]

$$v(-t) = \tfrac{1}{3}\sqrt{(\pi t)}\,[J_{-\frac{1}{3}}(w_0) + J_{\frac{1}{3}}(w_0)],$$

$$v'(-t) = -\tfrac{1}{3}\sqrt{(\pi t)}\,[J_{-\frac{2}{3}}(w_0) - J_{\frac{2}{3}}(w_0)],$$

$$t = \left(\frac{3}{2}w_0\right)^{\frac{2}{3}} = \left(\frac{k_0}{a}\right)^{\frac{2}{3}}\cos^2\vartheta. \tag{8.59}$$

As a result, if we write

$$V(\vartheta) = \exp[i\phi(\vartheta)], \tag{8.60}$$

we obtain
$$\phi(\vartheta) = -2\arctan\left[\sqrt{t}\,\frac{v(-t)}{v'(-t)}\right] - \pi. \tag{8.61}$$

Before proceeding to the analysis of this expression in the general case, it is convenient to examine the case $t \gg 1$. As can be seen from Eq. 8.59, this condition will be satisfied if:

(a) $k_0/a \gg 1$, i.e. the index of refraction does not change much over a single wavelength;

(b) $\cos\vartheta$ is not too small (the angle of incidence is not too close to grazing).

When $t \gg 1$, we have the following asymptotic expressions:[23]

$$v(-t) = t^{-\frac{1}{4}}\sin\left(w_0 + \frac{\pi}{4}\right), \quad v'(-t) = -t^{\frac{1}{4}}\cos\left(w_0 + \frac{\pi}{4}\right). \tag{8.62}$$

Substituting into Eq. 8.61, we obtain

$$\phi = \frac{4k_0}{3a}\cos^3\vartheta - \frac{\pi}{2}. \tag{8.63}$$

It is interesting to note that this expression for the phase, except for the term $\pi/2$ (which, with the assumptions we have made, is small compared with the first term), can be obtained on the basis of geometrical optics. According to these representations, if a ray is incident on the interface at an angle ϑ, its direction at an arbitrary point of the inhomogeneous medium is determined by the angle $\vartheta(z)$, which is found from the relation

$$n(z)\sin\vartheta(z) = \sin\vartheta. \tag{8.64}$$

But at the point $z = z_m$, determined by the equation

$$n(z_m) = \sin \vartheta, \tag{8.65}$$

the ray is horizontal, and then turns in the opposite direction (Fig. 38).

The phase change as the wave travels from the boundary $z = 0$ to the plane $z = z_m$ and back is evidently

$$\phi = 2 \int_0^{z_m} k_z \, dz = 2k_0 \int_0^{z_m} n(z) \cos \vartheta(z) \, dz.$$

This is also the phase of the reflection coefficient.

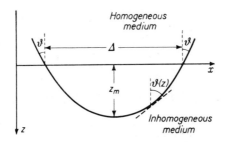

Fig. 38. The ray picture of total reflection from an inhomogeneous medium.

Taking relation (8.64) into account, we obtain

$$\phi = 2k_0 \int_0^{z_m} \sqrt{[n^2(z) - \sin^2 \vartheta]} \, dz. \tag{8.66}$$

In the case under consideration, in which the function $n(z)$ is given for $z > 0$ by the second of Eqs. 8.58, we find

$$\phi = 2k_0 \int_0^{z_m} \sqrt{(\cos^2 \vartheta - az)} \, dz, \tag{8.67}$$

or, after the replacement $\zeta = \cos^2 \vartheta - az$,

$$\phi = \frac{2k_0}{a} \int_0^{\cos^2 \vartheta} \zeta^{\frac{1}{2}} \, d\zeta = \frac{4k_0}{3a} \cos^3 \vartheta, \tag{8.68}$$

which agrees with Eq. 8.63 to within the term $\pi/2$.

It is also of interest to calculate the beam displacement upon reflection in the geometric optics approximation. Expressing ϑ in terms of p by the relation

$$p = k_0 \sin \vartheta, \tag{8.69}$$

we obtain

$$\phi = \frac{4k_0}{3a} \left(1 - \frac{p^2}{k_0^2}\right)^{\frac{3}{2}}. \tag{8.70}$$

Then, remembering that $\alpha = k_0 \sin \vartheta_0$, where ϑ_0 is the angle of incidence of the beam, Eq. 8.25 gives

$$\Delta = \frac{2 \sin 2\vartheta_0}{a}. \tag{8.71}$$

The same expression for the displacement can be obtained by calculating the path of the ray in Fig. 38, and taking account of Eqs. 8.64 and 8.58. We shall let the reader verify this for himself.

We now proceed to the consideration of the general case. According to Eqs. 8.59 and 8.69, we have

$$t = \left(\frac{k_0}{a}\right)^{\frac{2}{3}} \left(1 - \frac{p^2}{k_0^2}\right). \tag{8.72}$$

Moreover, remembering that $\alpha = k_0 \sin \vartheta_0$, we obtain

$$\left(\frac{d\phi}{dp}\right)_{p=\alpha} = -\frac{2 \sin 2\vartheta_0}{a} - \frac{2 \tan \vartheta_0}{a^{\frac{1}{3}} k_0^{\frac{2}{3}}} \cdot \frac{vv'}{v'^2 + tv^2}, \tag{8.73}$$

where
$$v \equiv v(-t), \quad v' \equiv \frac{\partial}{\partial t} v(-t).$$

In the differentiation, we also take account of the differential equation for the Airy function

$$v''(t) = tv(t). \tag{8.74}$$

The beam displacement upon reflection, Δ, is now determined by Eq. 8.25. The expression thus obtained for the displacement is of interest only when the angle of incidence ϑ_0 is close to $\pi/2$, since otherwise, it is convenient to use the simpler formula (8.71), obtained in the geometric optics approximation. In place of the angle of incidence ϑ_0 we introduce the angle $\alpha_0 = \pi/2 - \vartheta_0$, and assuming that $\alpha_0 \ll 1$, we obtain

$$\left(\frac{a^2}{4\lambda_0}\right)^{\frac{1}{3}} \Delta = 2^{\frac{1}{3}} \sqrt{(-t)} \left[1 + \frac{1}{2t} \frac{v(t)\, v'(t)}{v'^2(t) - tv^2(t)}\right], \tag{8.75}$$

where now the sign for t is changed

$$t \approx \frac{-\alpha_0^2}{(a\lambda_0)^{\frac{2}{3}}}, \quad \lambda_0 \equiv \frac{1}{k_0}. \tag{8.76}$$

The quantity $(a^2/4\lambda_0)^{\frac{1}{3}}\Delta$ is plotted as a function of t in Fig. 39 (solid line). This curve can be used for practical calculations.

It is interesting to compare the value of Δ we have just obtained with the approximate value (8.71), obtained by the use of ray representations.

We write Eq. 8.71 in the form

$$\left(\frac{a^2}{4\lambda_0}\right)^{\frac{1}{3}}\Delta \approx 2^{\frac{2}{3}}(-t)^{\frac{1}{2}}. \qquad (8.77)$$

This function is shown in Fig. 39 by the dotted line.

Comparing the solid and dotted curves, we see that the ray theory gives an approximately correct result when

$$|t| \geqslant 1, \quad \text{or} \quad \alpha_0 \geqslant (a\lambda_0)^{\frac{1}{3}}. \qquad (8.78)$$

When this condition is not satisfied, the predictions of the exact and the ray theories are quite different.

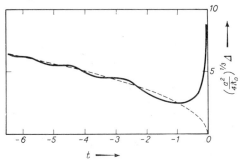

Fig. 39. Beam displacement at reflection, as a function of the parameter t.

It is important to note that in the important region for $|t| < 1$, Eq. 8.75 for the displacement is valid not only when the index of refraction $n(z)$ is given by the second of Eqs. 8.58, but also whenever the expansion of the function $n(z)$ in powers of z has a linear term, i.e. when (for small az) we can set

$$n(z) = 1 - \frac{az}{2}, \qquad (8.79)$$

where a is some parameter.

This broad applicability of the results we have obtained is due to the following: whatever the form of the function $n(z)$, at small grazing angles the reflection process takes place in a layer for which $az \ll 1$, and consequently, expansion (8.79) is valid.

We shall now show, without attempting to be rigorous, that at large grazing angles, at which the wave during reflection reaches deeper layers of the inhomogeneous medium [where Eq. 8.79 cannot be satisfied], we can use the ray representation. Then Δ can be calculated by elementary methods for any form of $n(z)$.

At the smallest grazing angles for which the ray theory is still applicable, we can, in accord with Eq. 8.78, take

$$\alpha_{\text{lim}} = (a\lambda_0)^{\frac{1}{3}} \ll 1. \qquad (8.80)$$

A ray, incident on the interface between a homogeneous and an inhomogeneous medium at this grazing angle, will penetrate into the inhomogeneous medium to a depth z_m, determined by the relation (see Eq. 8.65)

$$n(z_m) = \cos\alpha_{\lim} \approx 1 - \tfrac{1}{2}\alpha_{\lim}^2.$$

Using expansion (8.79) for $n(z)$, the last equation gives

$$\tfrac{1}{2}az_m = \tfrac{1}{2}\alpha_{\lim}^2 \ll 1.$$

At smaller grazing angles, where the ray theory is already inapplicable, the penetration depth into the inhomogeneous medium will in any case be no greater,† so that the condition $az \ll 1$ is satisfied, which completes the proof.

The problem of the total reflection from an interface between a homogeneous and an inhomogeneous medium was also solved by Gans[121] (see also Ref. 4, § 19). He calculated the path of the ray in the neighborhood of the turning point in the inhomogeneous medium, and found that at the turning point itself the ray is broken. It seems to us that his result has no physical meaning since he traced the path of the ray into regions where the ray theory is inapplicable. We shall discuss this erroneous result of Gans in more detail in § 17.

§ 9. THE REFLECTION OF PULSES

1. *General relations. The law of conservation of the integrated pulse*

We shall examine the reflection of a pulse with a plane front from a plane boundary. We expand the pulse into harmonic waves, which impinge on the boundary at the same angle of incidence as the pulse.

Again, for definiteness, we shall assume that the plane of incidence of the pulse is the xz-plane (Fig. 40). Then, in the general expression for a pulse (1.1), we will have $n_x = \sin\vartheta$, $n_y = 0$, $n_z = \cos\vartheta$, where ϑ is the angle of incidence. Expanding the incident pulse into harmonic waves, according to Eq. 1.3a, we will have the form

$$F(\xi) = \frac{1}{2}\int_0^\infty \Phi(\omega)\,e^{i\omega\xi}\,d\omega + \text{c.c.} \tag{9.1}$$

$$\xi = \frac{x\sin\vartheta + z\cos\vartheta}{c} - t, \tag{9.2}$$

where the symbol c.c. is used for brevity to denote the complex conjugate of the preceding term.

---

† Generally, this requires a proof based on the wave theory. The reader may carry out the proof for $n(z)$ given by Eq. 8.58 by using the expressions given in § 15 for the field in the inhomogeneous medium.

In the acoustic case, F will represent the acoustic pressure. In the electromagnetic case, $F(\xi)$ will represent the electric field E_y when $\mathbf{E}$ is perpendicular to the plane of incidence, and will represent the magnetic field H_y when $\mathbf{H}$ is perpendicular to the plane of incidence.

Again denoting the reflection coefficient of a plane harmonic wave by V, we write the reflected pulse in the form

$$F_{\text{refl}}(\xi^-) = \frac{1}{2} \int_0^\infty V\,\Phi(\omega)\,e^{i\omega\xi^-}\,d\omega + \text{c.c.}, \quad \xi^- = \frac{x\sin\vartheta - z\cos\vartheta}{c} - t. \quad (9.3)$$

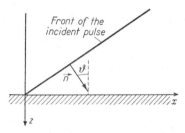

Fig. 40. A pulse with a plane front incident on a plane interface.

In general, V will be complex, and will be a function of the frequency ω (for example, in reflection from a layer). Therefore, the form of the reflected pulse may be completely different from that of the incident pulse.

The form of the pulse remains unchanged only when V is real and is independent of ω. In fact, in this case $V^* = V$, where V can be taken outside the integral sign as a constant, and from Eq. 9.3 we obtain Eq. 9.1, multiplied by the reflection coefficient, the only difference being that ξ enters in one case, and ξ^- in the other.

However, if we have total internal reflection from the interface between two homogeneous media, where V, although independent of frequency, is complex, the function $F_{\text{refl}}(\xi^-)$ will be essentially different from $F_{\text{refl}}(\xi)$, i.e. the form of the pulse will change. We shall examine this case in detail below.

The transmitted pulse is written by analogy with Eq. 9.3 as

$$F_{\text{tr}}(\xi_1) = \frac{1}{2} \int_0^\infty D\Phi(\omega)\,e^{i\omega\xi_1}\,d\omega + \text{c.c.}, \quad \xi_1 = \frac{x\sin\vartheta_1 + z\cos\vartheta_1}{c_1} - t, \quad (9.4)$$

where c_1 is the propagation velocity in the medium into which the wave penetrates, ϑ_1 is the angle of refraction, connected with the angle of incidence by the relation $\sin\vartheta_1 = (c_1/c)\sin\vartheta$, and D is the transmission coefficient.

We shall now prove an extremely interesting theorem which can be formulated as follows: *the total (integrated) pulse at any point in the upper medium is equal to the total pulse at any point in the lower medium.*

This theorem may be called the law of conservation of the total pulse. It is expressed mathematically by the identity

$$\int_{-\infty}^{+\infty} (F + F_{\text{refl}})\, dt = \int_{-\infty}^{+\infty} F_{\text{tr}}\, dt, \tag{9.5}$$

the validity of which is independent of the form of the incident pulse and the points of space at which the integrals on the right and left hand sides of the equation are evaluated.†

Thus, for example, in the total internal reflection of a pulse, the maximum value of F_{tr} in the pulse will (as we shall see below) decrease with depth as it penetrates into the reflecting medium. However, the pulse will stretch out in time in such a way that the area given by the integral on the right hand side of Eq. 9.5 will remain constant at any distance from the boundary.

In order to convince ourselves of the validity of Eq. 9.5, we begin by examining the expression for the integrated value of the incident pulse.

Remembering that according to Eq. 1.4 we have

$$\Phi^*(\omega) = \Phi(-\omega), \tag{9.6}$$

we can write Eq. 9.1 for the incident pulse in the form

$$F(\xi) = \frac{1}{2} \int_0^\infty \Phi(\omega)\, e^{i\omega\xi}\, d\omega + \frac{1}{2} \int_0^\infty \Phi(-\omega)\, e^{-i\omega\xi}\, d\omega,$$

or, replacing $-\omega$ by ω in the second integral and combining the two integrals, we obtain

$$F(\xi) = \frac{1}{2} \int_{-\infty}^{+\infty} \Phi(\omega)\, e^{i\omega\xi}\, d\omega. \tag{9.7}$$

The time integral between infinite limits is equivalent to the integral with respect to ξ. Therefore,

$$\int_{-\infty}^{+\infty} F(\xi)\, dt = \int_{-\infty}^{+\infty} F(\xi)\, d\xi = \frac{1}{2} \iint_{-\infty}^{+\infty} \Phi(\omega)\, e^{i\omega\xi}\, d\xi\, d\omega. \tag{9.8}$$

But, as is well known,[26]

$$\int_{-\infty}^{+\infty} e^{i\omega\xi}\, d\xi = 2\pi\delta(\omega), \tag{9.9}$$

---

† See Ref. 44. This law was also formulated in Ref. 113 for some restricted cases.

where $\delta(\omega)$ is the Dirac function, equal to zero everywhere except at the point $\omega = 0$ at which it becomes infinite. The Dirac δ-function also has the following properties:

$$\int_{-\infty}^{+\infty} \delta(\omega)\,d\omega = 1, \quad \int_{-\infty}^{+\infty} f(\omega)\,\delta(\omega)\,d\omega = f(0), \tag{9.10}$$

where $f(\omega)$ is an arbitrary continuous function. Then Eq. 9.8 can be written

$$\int_{-\infty}^{+\infty} F(\xi)\,d\xi = \pi \int_{-\infty}^{+\infty} \Phi(\omega)\,\delta(\omega)\,d\omega = \pi\Phi(0). \tag{9.11}$$

This result could have been foreseen, since it is known that the area of a curve is given by the constant in the Fourier series or integral expansion of the curve, i.e. the component corresponding to the frequency $\omega = 0$.

Each of the incident plane waves will give, according to Eq. 9.3, the reflected wave

$$\tfrac{1}{2} V(\omega)\,\Phi(\omega)\,e^{-i\omega\xi^-} + \text{c.c.},$$

which, if we use the notation

$$V(\omega)\,\Phi(\omega) = A(\omega) + iB(\omega), \tag{9.12}$$

can be written as

$$\tfrac{1}{2} A(\omega)\,(e^{-i\omega\xi^-} + e^{-i\omega\xi^-}) - B(\omega)\sin\omega\xi^-.$$

In the integration with respect to ξ^-, the last term yields zero since it is an odd function of ξ^-, and the first term, just as for the incident wave, yields

$$\pi A(0), \quad \text{i.e.} \quad \pi \,\mathrm{Re}\,[V(0)\,\Phi(0)].$$

A similar expression, with V replaced by D, is obtained for the integral of the refracted pulse at any point of space.

Thus, for the proof of Eq. 9.5, it is sufficient to convince ourselves of the validity of the equation

$$\Phi(0) + \mathrm{Re}\,[V(0)\,\Phi(0)] = \mathrm{Re}\,[D(0)\,\Phi(0)].$$

However, according to Eq. 9.6, we have $\Phi^*(0) = \Phi(0)$, i.e. $\Phi(0)$ is real. Therefore, the last equation reduces to

$$1 + \mathrm{Re}\,V(0) = \mathrm{Re}\,D(0). \tag{9.13}$$

It is not difficult to show that the more general relation

$$1 + V(0) = D(0) \tag{9.14}$$

holds. Taking the real part of Eq. 9.14, we obtain Eq. 9.13.

In the case of reflection from the interface between two homogeneous media, the coefficients V and D are independent of frequency, and the relation $1 + V = D$ follows directly from the continuity conditions

5

satisfied by the acoustic pressure or the appropriate component of the electromagnetic field (see §§ 2 and 3). Under more complicated conditions, when the reflection occurs at a layer or a set of layers, as $\omega \to 0$ (i.e. for infinite wavelength), the entire set of layers will act like a concentrated system having no effect on the reflection process, and reflection will take place just as if the media separated by the set of layers were in direct contact with one another. For the case of one layer, this can be seen from Eqs. 5.10 and 5.15, which show that $\omega \to 0$ ($k_2 \to \infty$) is equivalent to the case $d \to 0$, i.e. to the elimination of the influence of the layer.

Another proof of the law (9.5) is also of interest. We will give this proof for the case of an acoustic pulse.†

First of all, it is clear without proof that at the two points A and A' (Fig. 41) lying in a plane parallel to the reflecting boundary (in this case,

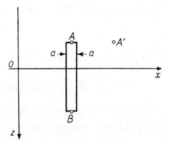

Fig. 41. Concerning the proof of the law of conservation of the total pulse.

in a plane $z = $ const), the integrated pulse will be the same since both points are located similarly with respect to the reflecting plane. Therefore, it is sensible to compare the integrated pulses only for points A and B lying on the same normal to the boundary.

Let us consider an elementary cylinder with infinitesimal ends, at which the points A and B are located. This cylinder was at rest before the arrival of the wave, and remains at rest after the wave passes. This means that the total impulse it receives is zero, i.e. the integral of the pressure acting on its surface, taken over the entire surface and from $t = -\infty$ to $t = +\infty$, is equal to zero. The forces acting on the lateral surface (the arrows in Fig. 41) mutually cancel in view of the symmetry of the points lying on the lateral surface. Therefore, the integrated impulses acting on the ends of the cylinder, i.e. the expressions

$$dS \int_{-\infty}^{+\infty} p_A(t)\, dt \quad \text{and} \quad dS \int_{-\infty}^{+\infty} p_B(t)\, dt,$$

† The possibility of this proof was indicated to the author by M. A. Isakovich.

where dS is the area of the end of the cylinder, and $p_A(t)$ and $p_B(t)$ are the acoustic pressures at the points A and B as functions of time, must be equal. Then

$$\int_{-\infty}^{+\infty} p_A(t)\,dt = \int_{-\infty}^{+\infty} p_B(t)\,dt,$$

which was to be proved.

Among the general laws satisfied during the reflection of a pulse of arbitrary form is, of course, the law of conservation of energy, which is written in the form

$$S_z + S_z^{\text{refl}} = S_z^{\text{tr}}, \tag{9.15}$$

where S_z, S_z^{refl} and S_z^{tr} denote the integrated components of the energy flux along the z-axis for the incident, reflected and refracted waves, respectively. S_z and S_z^{refl} naturally have different signs.

In the case of an acoustic pulse, we have

$$S_z = \int_{-\infty}^{+\infty} p v_z\,dt, \tag{9.16}$$

where p and v_z are the acoustic pressure and the z-component of the particle velocity in the incident wave. S_z^{refl} and S_z^{tr} are written similarly.

It is not difficult to see that Eq. 9.15 is valid for the pulse since it is satisfied for each of the harmonic waves into which the pulse may be expanded. At total internal reflection, $S_z^{\text{tr}} = 0$, since the refracted pulse propagates along the interface. For the proof of Eq. 9.15 in this case, see Ref. 13.

2. *The change of pulse form upon total internal reflection from a boundary between two homogeneous media*

We shall examine this problem for some special pulse forms. We begin with a pulse whose form is given by the function

$$F(\xi) = \frac{\epsilon}{\epsilon^2 + \xi^2}, \tag{9.17}$$

where ϵ is a parameter characterizing the width of the pulse. The form of this pulse is shown in Fig. 42. It is easy to verify that its expansion in a Fourier integral has the form

$$F(\xi) = \frac{1}{2}\int_0^\infty \exp\left(-\epsilon\omega + i\omega\xi\right)d\omega + \text{c.c.} \tag{9.18}$$

or

$$F(\xi) = \frac{1}{2}\int_{-\infty}^{+\infty} \exp\left(-\epsilon|\omega| + i\omega\xi\right)d\omega. \tag{9.19}$$

Thus, if we represent Eq. 9.19 in the general form (9.7), we must set

$$\Phi(\omega) = \exp(-\epsilon|\omega|). \tag{9.20}$$

We note that, as $\epsilon \to 0$, the pulse under consideration becomes a pulse characterized by the δ-function, as can be seen immediately by comparing Eqs. 9.9 and 9.19.

Remembering the definition of ξ, the incident pulse, we can write Eq. 9.17 in the form

$$F(x, z, t) = \frac{\epsilon}{\epsilon^2 + \left(\dfrac{x \sin \vartheta + z \cos \vartheta}{c} - t\right)^2}. \tag{9.21}$$

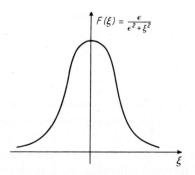

Fig. 42. The pulse form used in the calculations.

If the usual kind of reflection, and not total internal reflection, occurs, then, as we have seen, the form of the reflected and the refracted pulses will be the same as the form of the incident pulse, i.e.

$$F_{\text{refl}}(x, z, t) = V \cdot \frac{\epsilon}{\epsilon^2 + \left(\dfrac{x \sin \vartheta + z \cos \vartheta}{c} - t\right)^2}, \tag{9.22}$$

$$F_{\text{tr}}(x, z, t) = D \cdot \frac{\epsilon}{\epsilon^2 + \left(\dfrac{x \sin \vartheta_1 + z \cos \vartheta_1}{c_1} - t\right)^2}, \tag{9.23}$$

where V and D are the reflection and transmission coefficients.

In the case of total internal reflection, V and D are complex. For definiteness, we shall consider the acoustic case. Then (see Eq. 3.19)

$$V = \frac{m \cos \vartheta - is}{m \cos \vartheta + is}, \quad D = 1 + V, \quad s = \sqrt{(\sin^2 \vartheta - n^2)}. \tag{9.24}$$

In electrodynamics, the reflection coefficient is expressed by analogous formulas (see Eqs. 2.16 and 2.18). We set $V = A + iB$, where

$$A = \frac{m^2 \cos^2 \vartheta - s^2}{m^2 \cos^2 \vartheta + s^2}, \quad B = -\frac{2sm \cos \vartheta}{m^2 \cos^2 \vartheta + s^2}. \tag{9.25}$$

Then, taking account of Eqs. 9.3 and 9.20, we can write the reflected pulse in the form

$$F_{\text{refl}}(\xi^-) = \frac{A}{2} \int_{-\infty}^{+\infty} \exp\left(-\epsilon |\omega| + i\omega\xi^-\right) d\omega$$

$$- B \int_0^{\infty} \exp\left(-\epsilon\omega\right) \sin \omega\xi^- d\omega. \tag{9.26}$$

Both integrals in the last expression can be evaluated without difficulty; as a result we obtain (remembering the definition of ξ^-)

$$F_{\text{refl}} = A \frac{\epsilon}{\epsilon^2 + [(x \sin \vartheta - z \cos \vartheta)/c - t]^2}$$

$$- B \frac{(x \sin \vartheta - z \cos \vartheta)/c - t}{\epsilon^2 + [(x \sin \vartheta - z \cos \vartheta)/c - t]^2}, \tag{9.27}$$

or, as $\epsilon \to 0$,

$$F_{\text{refl}} = \pi A \delta\left(\frac{x \sin \vartheta - z \cos \vartheta}{c} - t\right) - B \frac{1}{(x \sin \vartheta - z \cos \vartheta)/c - t}. \tag{9.28}$$

Thus, the reflected pulse consists of two parts, one of which corresponds to a pulse of the same form as the incident pulse, and the other to a pulse with a modified form.

Let us investigate the pulse penetrating into the lower medium. In Eq. 9.4 we have for this pulse

$$\xi_1 = \frac{x \sin \vartheta_1}{c_1} - t + \frac{is}{c} z. \tag{9.29}$$

As a result, using Eq. 9.20, we obtain

$$F_{\text{tr}} = \frac{D}{2} \int_0^{\infty} \exp\left(-\epsilon\omega - \frac{\omega s}{c} z\right) \exp\left[i\omega\left(\frac{x \sin \vartheta_1}{c_1} - t\right)\right] d\omega + \text{c.c.},$$

or, for brevity, using the notation

$$\frac{x \sin \vartheta_1}{c_1} - t = g, \quad \epsilon + \frac{s}{c} z = h, \tag{9.30}$$

and substituting the value of the integral

$$\int_0^\infty \exp\left(-h\omega - ig\omega\right) d\omega = \frac{1}{h+ig}, \tag{9.31}$$

we have

$$F_{\text{tr}} = \frac{1}{2}\left(\frac{D}{h-ig} + \frac{D^*}{h+ig}\right). \tag{9.32}$$

Using $D = 1 + V$ and Eq. 9.25, the last expression can be transformed into the form

$$F_{\text{tr}} = \frac{(1+A)h - Bg}{h^2 + g^2}, \tag{9.33}$$

or, expressing h and g in terms of the coordinates and time:

$$F_{\text{tr}} = \frac{(1+A)\left(\epsilon + \frac{s}{c}z\right) - B\left(\frac{x\sin\vartheta_1}{c_1} - t\right)}{\left(\epsilon + \frac{s}{c}z\right)^2 + \left(\frac{x\sin\vartheta_1}{c_1} - t\right)^2}. \tag{9.34}$$

In the case $\epsilon = 0$, where the incident pulse is described by a δ-function, Eq. 9.34 gives

$$F_{\text{tr}} = \frac{-B\left(\frac{x\sin\vartheta_1}{c_1} - t\right) + \frac{s}{c}z(1+A)}{\left(\frac{s}{c}z\right)^2 + \left(\frac{x\sin\vartheta_1}{c_1} - t\right)^2}. \tag{9.35}$$

We see that, as regards its form, the refracted pulse has essentially nothing in common with the incident pulse. Furthermore, in Eqs. 9.34 and 9.35, the time t enters only in the combination $(x\sin\vartheta_1/c_1) - t$. Hence, in the lower medium, the pulse propagates along the interface with the velocity $c_1/\sin\vartheta_1$, which according to the law of refraction is equal to $c/\sin\vartheta$, the propagation velocity of the pulse in the upper medium along the interface.

It is interesting to note that on the straight line

$$B\left(\frac{x\sin\vartheta_1}{c_1} - t\right) + A\frac{s}{c}z = 0 \tag{9.36}$$

the field of the refracted pulse is zero, and has different signs on the opposite sides of this line. The incident, reflected and refracted pulses at the instant $t = 0$ are shown schematically in Fig. 43, under the assumption that the incident pulse is given by a δ-function. In Fig. 43, AA is the boundary between the media, OB is the front of the incident pulse, given by the equation

$$\frac{x\sin\vartheta + z\cos\vartheta}{c} - t = 0,$$

and OD is the front of the part of the reflected pulse which corresponds to the first term in Eq. 9.28. The second term in Eq. 9.28 is shown schematically by the solid line shading on one side of OD and the dotted line shading on the other side of OD. The dotted line shading corresponds to the negative field, and the solid line shading to the positive field. The field decreases with distance from the line OD, corresponding to the decreasing density of the shading.

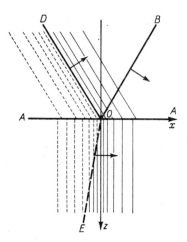

Fig. 43. Schematic illustration of the reflection and refraction of a δ-pulse.

The field of the refracted pulse has a finite value everywhere except at the origin O. It is zero on the line OE, and has different signs on the opposite sides of the line. The propagation directions of the incident, reflected and refracted pulses are shown by arrows.

Let us note that, according to Eq. 9.35, the z-dependence of the field of the refracted (transmitted) pulse at $x = c_1 t/\sin \vartheta_1$ will be

$$F_{\text{tr}} = \frac{(1+A)c}{sz}.$$

Thus, the field decreases with distance from the boundary, not exponentially as in the case of a harmonic plane wave, but inversely proportionately to the distance from the boundary.

Cross sections of the field that is shown schematically in Fig. 43 are pictured in Figs. 44 and 45[113] across planes of $x = \text{const}$, for a δ-function pulse incident on a boundary characterized by the parameters $n = c/c_1 = 0.26$, $m = 0.07$, which corresponds in acoustics, for example, to a boundary of air ($c = 343$ m/sec, $\rho = 1.2 \times 10^{-3}$ gm/cm³) and hydrogen ($c_1 = 1305$ m/sec, $\rho_1 = 8.4 \times 10^{-5}$ gm/cm³), at a temperature of 20° and at normal pressure. The angle of total internal reflection for this case is 15.25°. In the case shown in Figs. 44 and 45, the angle of incidence is $\vartheta = 60°$.

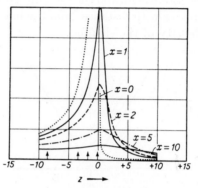

Fig. 44. Cross sections of the field shown schematically in Fig. 43, over planes of $x = $ const (const > 0).

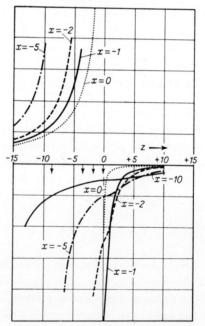

Fig. 45. Cross sections of the field shown schematically in Fig. 43, over planes of $x = $ const (const < 0).

The ordinates represent either the acoustic pressure or the electric field intensity in relative units. The functions F_{refl} and F_{tr}, describing the reflected and the refracted (transmitted) pulses, are homogeneous functions of x and z at $t = 0$, according to Eqs. 9.28 and 9.35, i.e. changing the scale of x and z changes nothing but the scale of the ordinates. Therefore, we can also express x and z in relative units, as was done in Figs. 44 and 45. The incident pulse and the part of the field of the reflected pulse which is described by a δ-function are not shown in the Figures. The values of the z-coordinate at which the front of these pulses intersect the corresponding planes of $x = $ const are shown by arrows on the abscissas.

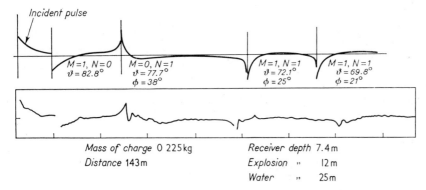

Fig. 46. The theoretical (upper) and experimental (lower) curves of the change of form of an exponential pulse as it undergoes successive reflections at the water surface and at the bottom.

Studies of the change of the pulse form upon total internal reflection for other forms of the incident pulse are available in the literature. Thus, Friedlander[120] investigated the reflection and refraction of a "plateau" pulse, i.e. the field in the pulse is constant over the time interval (t_1, t_2) and is zero outside this interval.

Arons and Yennie[13] studied the reflection of an exponential pulse, given by the equation

$$F(\xi) = \begin{cases} 0 & \text{for} \quad \xi < 0, \\ F_0 e^{-\lambda \xi} & \text{for} \quad \xi \geqslant 0. \end{cases} \qquad (9.37)$$

This function gives a fair description of the form of the pulse in a shock wave generated by an underwater explosion. The authors compared their theoretical conclusions with experimental results obtained by recording an explosion pulse in a layer of water bounded above by its free surface and below by the bottom (packed sea sand $\rho_1 \approx 2.7$, $c_1 \approx 1800$ m/sec, and an angle of total internal reflection at the boundary with water of $58°$).

The theoretically calculated form of the time dependence of the acoustic pressure is shown at the top in Fig. 46, and the experimental result is shown in the bottom part of the Figure. Here M is the number of reflections from the surface of the water, N is the number of reflections from the bottom, ϑ is the angle of incidence of the pulse at the bottom, and ϕ is the phase change of a plane harmonic wave upon reflection.

We see that following the direct pulse is a pulse reflected from the surface of the water ($M = 1, N = 0$), which has the same form as the incident pulse, but the opposite sign. The next pulse to arrive at the receiver is the one reflected from the bottom ($M = 0, N = 1$), the form of which is completely different from that of the incident pulse. Then come pulses which were reflected once from the surface of the water and once from the bottom (with different orders of occurrence of the reflections). We see that the experimental results are generally in fair agreement with the theoretical curve. Detailed agreement could not be expected in view of the idealized form of the incident pulse.

Chapter II

SOME APPLICATIONS OF THE THEORY OF PLANE WAVE PROPAGATION IN LAYERED MEDIA

The theory developed in Chapter I has numerous applications. We shall consider only a few of these applications, without going into great detail, since the specific literature regarding each of the problems we shall examine is voluminous (particularly as regards the reflection reduction of optical systems). We shall try to present the fundamental theoretical treatment from a unified point of view, and shall discuss experimental verifications as far as possible.

We shall also undertake, in this chapter, to use some of the methods and representations which are common to various branches of physics. An example of such a common representation is the concept of impedance which is now becoming quite useful in optics.

§ 10. The Reflection Reduction of Optical Systems

By the reflection reduction of optical systems we mean, as is well known, the lowering of the reflection coefficient at the air–glass boundaries by depositing thin layers of various materials on the glass. The greater the number of layers that could be deposited on the glass, the greater the spectral range and the range of angles of incidence over which increased transmissivity of the boundary could be attained. However, so far, no more than three layers have been tried, either in practice or experimentally. Below, we shall investigate deposits of one, two and three layers, emphasizing only the fundamental theoretical results and their experimental verification, without going into the numerous details which the reader may find in a monograph devoted to this problem[5] and also in a review article.[67]

We shall also find it useful to refer to Refs. 60, 149, 187 and 194, which are concerned with these problems, and use the apparatus of transmission line and four-terminal network theory, developed originally in radio engineering.

Recently, the problem of increasing the transparency of interfaces to ultrasonic waves has become of interest. This is due to the growing application of liquid and solid lenses used for focusing sound in connection with ultrasonic flaw detection and the problem of "seeing" into optically opaque media[145] with ultrasonic waves. In their general form, the results obtained below will also apply to the "reflection reduction" of acoustical systems.

Also of great interest is the problem of increasing the transparency of radomes to centimeter waves and the transparency of sonar housings to ultrasound. The theoretical aspects of this problem can be solved by the methods presented below.

1. Single-layer coating

The theory of wave reflection from a plane-parallel layer separating two media was considered in § 5. By substituting

$$\exp\left(\pm i\alpha_2 d\right) = \cos\alpha_2 d \pm i\sin\alpha_2 d$$

in Eq. 5.10, and carrying out some straightforward transformations, the reflection coefficient can be written in the form

$$V = \frac{Z_2(Z_1 - Z_3)\cos\alpha_2 d - i(Z_2^2 - Z_1 Z_3)\sin\alpha_2 d}{Z_2(Z_1 + Z_3)\cos\alpha_2 d - i(Z_2^2 + Z_1 Z_3)\sin\alpha_2 d}, \tag{10.1}$$

where d is the layer thickness, $\alpha_2 = k_2\cos\vartheta_2$ is the projection of the wave vector in the layer on the normal to the boundary, $\alpha_2 d$ is the phase change of the wave over the thickness of the layer, Z_j $(j = 1, 2, 3)$ is the impedance defined by Eqs. 5.1–5.3, and finally, ϑ_j is the angle between the normal to the wave front and the normal to the boundary in each of the media.

The subscripts 3, 2, 1 refer respectively to the medium through which the incident wave travels, the layer, and the medium into which the wave is transmitted.

In the case of normal incidence of an electromagnetic wave $(\vartheta_j = 0)$, where it is assumed that the magnetic permeability $\mu_j = 1$, we have $Z_j = 1/n_j$ for both cases of polarization, where n_j is the index of refraction of the jth medium. Then Eq. 10.1 can be written in the form

$$V = \frac{n_2(n_3 - n_1)\cos\left(2\pi d/\lambda_2\right) - i(n_1 n_3 - n_2^2)\sin\left(2\pi d/\lambda_2\right)}{n_2(n_3 + n_1)\cos\left(2\pi d/\lambda_2\right) - i(n_1 n_3 + n_2^2)\sin\left(2\pi d/\lambda_2\right)}, \tag{10.2}$$

where λ_2 is the wavelength in the layer.

The reflection coefficient V is zero when the following two conditions are satisfied:

$$n_2 = \sqrt{(n_1 n_3)}, \quad d = \tfrac{1}{4}\lambda_2(2k+1), \quad k = 0, 1, 2, \ldots, \tag{10.3}$$

i.e. when the layer thickness is an odd number of quarter wavelengths, and the index of refraction of the layer is the geometric mean of the indices of refraction of the media separated by the layer.

In the case of oblique incidence, conditions (10.3) must be replaced by the more general conditions

$$Z_2 = \sqrt{(Z_1 Z_3)}, \quad d \cos \vartheta_2 = \tfrac{1}{4}\lambda_2(2k+1). \tag{10.4}$$

Thus, at a given frequency and angle of incidence, it is always possible in principle to select a layer that would eliminate reflection at the boundary.

The quality of a covering is usually measured by the magnitude of the reflection coefficient with respect to energy, $R = |V|^2$. From Eq. 10.1 we obtain

$$R = \frac{Z_2^2(Z_1-Z_3)^2 \cos^2\alpha_2\, d + (Z_2^2-Z_1 Z_3)^2 \sin^2\alpha_2\, d}{Z_2^2(Z_1+Z_3)^2 \cos^2\alpha_2\, d + (Z_2^2+Z_1 Z_3)^2 \sin^2\alpha_2\, d}. \tag{10.5}$$

At oblique incidence, the reflection coefficient will depend on the polarization of the wave. If we let $R_\perp$ and $R_\parallel$ denote the reflection coefficients for light polarized perpendicular and parallel, respectively, to the plane of incidence, then, as is well known, the reflection coefficient for unpolarized light is

$$R = \tfrac{1}{2}(R_\perp + R_\parallel). \tag{10.6}$$

The reflection coefficient R is shown as a function of the angle of incidence in Fig. 47,[5] for the case in which the index of refraction of the glass is $n = 1.52$, and the index of refraction and the thickness of the reflection reduction layer are chosen according to conditions (10.3), i.e. $n_2 = \sqrt{n_1}$ ($n_3 = 1$ for air) and $d = \tfrac{1}{4}\lambda_2$.

The single-layer covering has the following disadvantages:

1. When the index of refraction of the optical components is low, the index of refraction of the layer must not be too far from unity. Thus, to obtain reflection reduction for mirror glass ($n = 1.52$), the index of refraction of the layer must be $n_2 = \sqrt{n_1} = 1.23$. However, it is difficult to find a solid transparent substance which has this index of refraction and is also practical technically.

The substances used in practice for reflection-reducing coverings have large indices of refraction, and the quality of transmission augmentation is thereby reduced.

The theoretically calculated reflection coefficient R is shown in Fig. 48 as a function of the angle of incidence for the case of glass with an index of refraction of $n_1 = 1.52$ covered by a layer of titanium dioxide (TiO_2) with an index of refraction of $n = 2.2$ and a thickness of $d/\lambda_2 = 0.034$.[5]

2. Using a single-layer covering, it is very difficult to obtain good quality reflection reduction over a broad spectral region, since the reflection coefficient increases comparatively rapidly for wavelengths differing from the optimum wavelength for which the layer was designed.

The reflection coefficient as a function of the wavelength is shown in Fig. 49 for the case $n_1 = 1.52$, $n_2 = \sqrt{n_1} = 1.233$ for various layer thicknesses, corresponding to total transmission at wavelengths of $\lambda = 0.50\mu$,

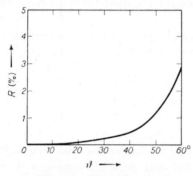

Fig. 47. The reflection coefficient as a function of the angle of incidence for glass covered by a single reflection-reducing layer.

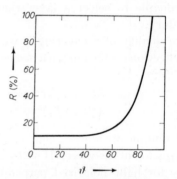

Fig. 48. The reflection coefficient as a function of the angle of incidence for glass covered by a reflection-reducing layer of titanium dioxide ($n_2 = 2.2$) of thickness 0.034λ.

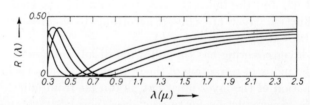

Fig. 49. The reflection coefficient as a function of the wavelength for reflection reduction by layers of various thicknesses.

0.58μ, 0.70μ and 0.80μ.[5] The variation with wavelength of the indices of refraction of the glass and the layer were not taken into account in the calculations since this effect does not change the results significantly.

2. *Double-layer coating*

The coefficient of reflection from two layers, separating two different media, is given by Eqs. 5.52 and 5.53. We will limit ourselves to the case of normal incidence. Then $Z_j = 1/n_j$, $j = 1, 2, 3$; $Z_4 = 1/n_4 = 1$

(n_4 is the index of refraction of air). The reflection coefficient with respect to energy $R = |V|^2$ is obtained from Eqs. 5.52 and 5.53:

$$R = \left| \frac{[n_2 n_3(1-n_1) - (n_2^2 - n_1 n_3^2) \tan \phi_2 \tan \phi_3]}{[n_2 n_3(1+n_1) - (n_2^2 + n_1 n_3^2) \tan \phi_2 \tan \phi_3]} \right.$$

$$\left. \frac{-i[n_3(n_1 - n_2^2) \tan \phi_2 + n_2(n_1 - n_3^2) \tan \phi_3]}{-i[n_3(n_1 + n_2^2) \tan \phi_2 + n_2(n_1 + n_3^2) \tan \phi_3]} \right|^2 , \qquad (10.7)$$

where n_1, n_2, n_3 are indices of refraction of the components whose transmissivities are to be increased, the lower and the upper layer, respectively; $\phi_2 = (2\pi/\lambda_2) d_2$, $\phi_3 = (2\pi/\lambda_3) d_3$ are the phase changes in the lower and upper layer.

When both layers are a quarter wavelength in thickness

$$(\tan \phi_2 \to \tan \phi_3 \to \infty),$$

Eq. 10.7 gives
$$R = \left(\frac{n_2^2 - n_1 n_3^2}{n_2^2 + n_1 n_3^2} \right)^2 . \qquad (10.8)$$

Thus, in this case the reflection vanishes when

$$n_2^2 = n_1 n_3^2. \qquad (10.9)$$

This shows that the selection of materials for a double-layer coating may be made over a comparatively broad range, and is much more easily accomplished than in the case of a single layer.

However, a vanishing reflection coefficient can be attained not only with quarter-wave layers. Equating the real and imaginary parts of Eq. 10.7 to zero, we find that reflection will be absent when the following conditions are satisfied:

$$\tan \phi_3 = -\frac{n_3(n_1 - n_2^2)}{n_2(n_1 - n_3^2)} \tan \phi_2, \qquad (10.10)$$

$$\tan \phi_2 = \left[\frac{n_2^2(n_1 - 1)(n_1 - n_3^2)}{(n_2^2 - n_1 n_3^2)(n_1 - n_2^2)} \right]^{\frac{1}{2}} . \qquad (10.11)$$

If the expression in square brackets in the last equation is positive, then $\tan \phi_2$ will be real, and it will be possible to choose layer thicknesses, different from a quarter wavelength, which will yield total transmission.

The indices of refraction of all substances which can be deposited as layers for reflection reduction of glass are considerably greater than $\sqrt{1.5}$ or $\sqrt{1.6}$. We shall therefore assume that $n_2^2 > n_1, n_3^2 > n_1$ in Eq. 10.11. Then the expression in the square brackets will be positive when $n_2^2 > n_1 n_3^2$.

If the index of refraction of the glass is given, there are many possible combinations of n_2 and n_3 which lead to the absence of reflection. For example, in a practical case, we may have $n_1 = 1.75$ and $n_2 = 2.0$. Then n_3 must be greater than 1.51, which is not difficult to accomplish. When $n_2^2 = n_1 n_3^2$, we return to the case of quarter-wave layers.

We shall now present some experimental data pertaining to double-layer coatings, and compare[154] them with theoretical results.

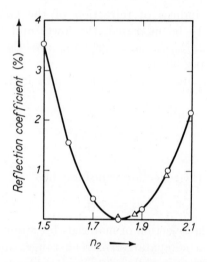

Fig. 50. The minimum value of the reflection coefficient as a function of the index of refraction of the internal layer in double-layer reflection reduction of glass (using quarter-wave layers). The index of refraction of the upper layer is 1.47 and that of the glass is 1.525. ○. Theoretical values. △. Experimental values.

Let the glass whose reflection is to be reduced have an index of refraction of $n_1 = 1.525$. The smallest index of refraction of the layer, which can be obtained without difficulty, remembering that the layer must be stable and must not scatter light, is approximately 1.47. Setting $n_1 = 1.525$ and $n_3 = 1.47$, Eq. 10.10 gives $n_2 = 1.81$ for the index of refraction of the interior layer.

It is of interest to see how the reflection coefficient R varies as n_2 departs from the above value. This is shown in Fig. 50, in which the abscissa represents n_2 and the ordinate represents R. Both layers are a quarter wavelength, so that, as n_2 varies, the layer thickness d_2 must also vary. The solid curve represents the theoretical results, and the triangles indicate the experimental points. There is evidently excellent agreement between theory and experiment.

The advantage of a double layer over a single layer is quite evident in Fig. 51, in which the index of refraction of the glass is represented by the abscissa, and the reflection coefficient is represented by the ordinate. The dotted curve shows the reflection reduction of a single layer with an index of refraction $n_2 = 1.38$ (MgF$_2$). The solid curve shows the double-layer transmission increase with quarter-wave layers, the indices of refraction being 1.80 and 1.47 for the interior and the

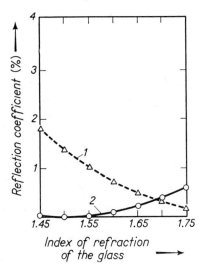

Fig. 51. Theoretical values of the minimum reflection coefficient obtained in the reflection reduction of glass by a single layer of MgF$_2$ ($n = 1.38$) and by two quarter-wave layers ($n = 1.80$ and $n_3 = 1.47$). 1. Single-layer reflection reduction. 2. Two-layer reflection reduction ($n_2 = 1.38$).

exterior layers, respectively. It is clear from Fig. 51 that by using quarter-wave layers of substances with these indices of refraction, transmission increases can be accomplished for all sorts of glass from $n_1 = 1.50$ to $n_1 = 1.70$, with the reflection coefficient remaining below 0.5%.

An important characteristic of a coating is the extent of the spectral range over which transmission increase is realized. Theoretical and experimental curves of the reflection coefficient as a function of the wavelength (of light) are shown in Fig. 52 for the case of double-layer reflection reduction with quarter-wavelength layers. The agreement between theory and experiment is satisfactory. The lack of complete agreement may be due to inaccuracy in the determination of the layer thicknesses, and also to not taking into account the variation of the index of refraction with wavelength.

Figure 53 is similar to Fig. 52, except that the thickness of the upper layer is not one-quarter, but rather three-quarters of a wavelength. The agreement between theory and experiment is again good. We should note that in this case the spectral band over which the transmission increase is realized is narrower than in the preceding case.

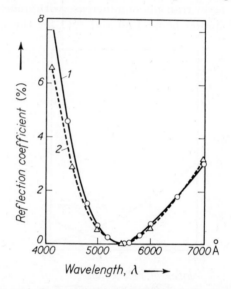

Fig. 52. Spectral characteristics of reflection reduction by two quarter-wave layers. The indices of refraction of the glass, the inner and the outer layer are 1.525, 1.80 and 1.47, respectively. 1. Theoretical curve. 2. Experimental curve.

As was indicated above, a transmission increase can be accomplished not only with quarter-wave layers. In Fig. 54, we show the spectral characteristics of a covering with the following properties: interior layer, $d_2 = 0.385\lambda_2, n_2 = 2.0$; exterior layer, $d_3 = 0.190\lambda_3, n_3 = 1.47$; the index of refraction of the glass is $n_1 = 1.53$.

It is completely unnecessary to require that the reflection coefficient be zero at some wavelength. In Fig. 55, we show the spectral characteristics of a coating with the following properties: interior layer, $d_2 = 0.20\lambda_2$, $n_2 = 1.47$; exterior layer, $d_3 = 0.350\lambda_3, n_3 = 1.87$; the index of refraction of the glass is $n_1 = 1.53$. It is clear from the Figure that the reflection coefficient is very small over the entire visible spectrum. The reflection coefficient, measured with a photoelement having a sensitivity curve approximately the same as that of the eye, is equal to 0.53%. Note that the reflection coefficient does not become zero at any wavelength, although its lowest value is only 0.04%.

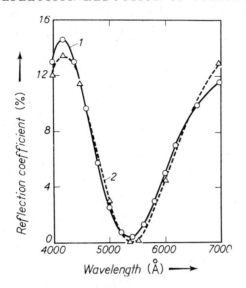

Fig. 53. The spectral characteristics of double-layer reflection reduction when the thickness of the upper layer is three-quarters of a wavelength, and that of the lower layer is one-quarter of a wavelength. The indices of refraction of the glass, the inner and the outer layer are 1.53, 1.87 and 1.47, respectively. 1. Theoretical curve. 2. Experimental curve.

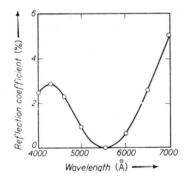

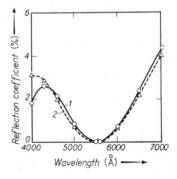

Fig. 54. The theoretical spectral characteristics of a double-layer coating with the parameters: internal layer, $n_2 = 2.0$, $d_2 = 0.385\lambda_2$; external layer, $n_3 = 1.47$, $d_3 = 0.190\lambda_3$; glass, $n_1 = 1.53$.

Fig. 55. The spectral characteristics of a double-layer coating for which the reflection coefficient is comparatively small over the entire visible spectrum but is not zero at any wavelength. Parameters of the coating: outer layer, $d_3 = 0.350\lambda_3$, $n_3 = 1.87$; inner layer, $d_2 = 0.20\lambda_2$, $n_2 = 1.47$; glass, $n_1 = 1.53$. 1. Theoretical curve. 2. Experimental curve.

3. Triple-layer coating

The use of triple-layer coatings permits a further broadening of the spectral region over which the reflection is sufficiently small to approach the realization of "invisible glass".

The reflection coefficient from a system of three layers is given by Eqs. 5.56 and 5.54. We shall again limit ourselves to the case of normal incidence, where $Z_j = 1/n_j$, $j = 1, 2, 3, 4$. Furthermore, $Z_5 = 1/n_5 = 1$, since the medium through which the incident wave travels is air. If, moreover, we use the notation

$$\delta_j = \tan \phi_j,$$

where $\phi_j = 2\pi d_j/\lambda_j$ is the phase change in the jth layer ($\lambda_j = \lambda/n_j$ is the wavelength in this layer), the expression for the reflection coefficient with respect to energy $R = |V|^2$ can be written in the form

$$R = \left| \frac{X-1}{X+1} \right|^2, \tag{10.12}$$

where

$$X = \frac{[n_2 n_3 n_4 - n_2^2 n_4 \delta_2 \delta_3 - n_2^2 n_3 \delta_2 \delta_4 - n_2 n_3^2 \delta_3 \delta_4]}{[n_1 n_2 n_3 n_4 - n_1 n_3^2 n_4 \delta_2 \delta_3 - n_1 n_3 n_4^2 \delta_2 \delta_4 - n_1 n_2 n_4^2 \delta_3 \delta_4]}$$
$$\frac{- i[n_1 n_3 n_4 \delta_2 + n_1 n_2 n_4 \delta_3 + n_1 n_2 n_3 \delta_4 - n_1 n_3^2 \delta_2 \delta_3 \delta_4]}{- i[n_2^2 n_3 n_4 \delta_2 + n_2 n_3^2 n_4 \delta_3 + n_2 n_3 n_4^2 \delta_4 - n_2^2 n_4^2 \delta_2 \delta_3 \delta_4]}. \tag{10.13}$$

The interior layer in contact with the glass is denoted by the subscript 2, the middle layer by the subscript 3, and the external layer by the subscript 4.

Of all the possible combinations of layer thicknesses and indices of refraction leading to vanishing reflection for a given wavelength, we shall examine only two:

1. Each layer has an optical thickness equal to a quarter wavelength ($n_j d_j = \lambda_j/4$). Then $\delta_2 = \delta_3 = \delta_4 = \infty$; dividing Eq. 10.13 by the product $\delta_2 \delta_3 \delta_4$, we obtain

$$X = \frac{n_1 n_3^2}{n_2^2 n_4^2}.$$

The coefficient of reflection is then

$$R = \left(\frac{n_1 n_3^2 - n_2^2 n_4^2}{n_1 n_3^2 + n_2^2 n_4^2} \right)^2. \tag{10.14}$$

Thus, the reflection coefficient is zero when

$$\sqrt{(n_1)}\, n_3 = n_2 n_4. \tag{10.15}$$

2. The end layers are quarter-wave, and the middle layer is half-wave. In this case, $\delta_2 = \delta_4 = \infty$, $\delta_3 = 0$. Dividing the numerator and denominator in Eq. 10.13 by $\delta_2 \delta_4$, and then setting $\delta_3 = 0$, we obtain $X = n_2^2/n_1 n_4^2$, and consequently,

$$R = \left(\frac{n_2^2 - n_1 n_4^2}{n_2^2 + n_1 n_4^2}\right)^2. \tag{10.16}$$

Thus, in this case, $R = 0$ when

$$n_2^2 = n_1 n_4^2. \tag{10.17}$$

The index of refraction of the middle layer does not enter into the condition for the vanishing of reflection. This could have been expected beforehand, since it was shown in the preceding chapter that a half-wave layer is equivalent to a layer of zero thickness.

The spectral characteristics, calculated from Eqs. 10.19 and 10.13, for two concrete cases of a triple-layer coating[154] are shown in Figs. 56 and 57. In the first case, all three layers are quarter-wave, and in

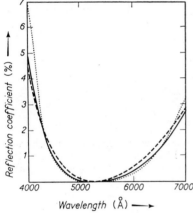

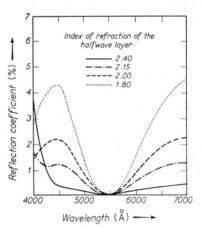

Fig. 56. The theoretical spectral characteristics of triple-layer reflection reduction for various indices of refraction of the layers: $- - -$ $n_1 = 1.53$, $n_2 = 1.70$, $n_3 = 2.02$, $n_4 = 1.47$; ——— $n_1 = 1.53$, $n_2 = 1.80$, $n_3 = 2.14$, $n_4 = 1.47$; .... $n_1 = 1.53$, $n_2 = 2.00$, $n_3 = 2.38$, $n_4 = 1.47$. All layers are quarter-wave.

Fig. 57. The theoretical spectral characteristics of a triple-layered coating (the end layers are quarter-wave, and the middle layer is half-wave). The indices of refraction are: glass, 1.53; lower layer, 1.80; upper layer, 1.47; middle layer, various values.

the second case, the middle layer is half-wave. The various curves in the Figures refer to different values of the indices of refraction. In the second case, the index of refraction of the middle layer plays an important rôle at all wavelengths, except for the wavelength at which it is a half-wave layer.

We see that a triple-layer coating in which all the layers are quarter-wave yields a broader minimum for the reflection coefficient than the double-layer coating considered previously. However, the broadest minimum is obtained when the end layers are quarter-wave, and the middle layer is half-wave (Fig. 57). Moreover, the broadest minimum is obtained when the index of refraction of the half-wave layer is of the order 2.3–2.4. It is difficult to find a suitable substance with such a high index of refraction since we must at the same time require that absorption in the layer be negligible.

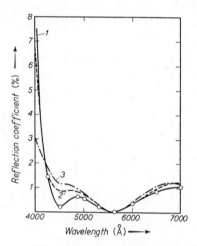

Fig. 58. Comparison of the theoretical and experimental spectral characteristics of a triple-layer coating (the end layers are quarter-wave, and the middle layer is half-wave). 1. Experimental curve. 2. Theoretical curve taking dispersion into account. 3. Theoretical curve without taking account of dispersion.

The theoretical and experimental results for the coefficient of reflection from glass covered by a triple-layer coating in which the middle layer is half-wave and the end layers are quarter-wave are shown in Fig. 58. The dependence of the index of refraction of the glass and of each of the layers on the wavelength is shown in Fig. 59. Two theoretical curves are given in Fig. 58. One of these was obtained without taking into account the variation of the indices of refraction of the glass and the layers with wavelength; the other curve takes this effect into account. The two theoretical curves do not differ significantly from one another, and follow the experimental curve satisfactorily.

For comparison, the experimental characteristics of typical single-, double- and triple-layer coatings are shown in Fig. 60 for glass with an index of refraction $n_1 = 1.53$. The curve for the single-layer coating is based on a layer of MgF_2. The advantage of a triple-layer coating is clear from the Figure.

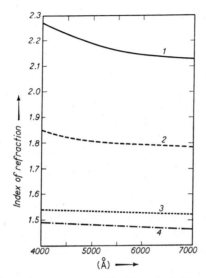

Fig. 59. The dispersion curves used in the calculations of Fig. 58. 1. Middle layer. 2. Lower layer. 3. Glass. 4. Upper layer.

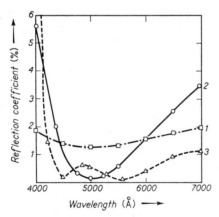

Fig. 60. Spectral characteristics of typical single-layer (curve *1*), double-layer (curve *2*) and triple-layer (curve *3*) coatings.

4. *Reflection reduction of acoustical systems*

Formulas for the calculation of the coefficient of reflection of acoustic waves from systems of layers, deposited on solid surfaces to increase their acoustic transparency, are easily obtained from the general formulas of § 5, Chap. I, for the case of normal incidence. A detailed theoretical investigation of the double-layer and triple-layer acoustic coating has not as yet been carried out. Some results concerning this problem have been published by B. D. Tartakovskii.[75]

At oblique incidence, the picture becomes much more complicated since shear waves come into play.

§ 11. INTERFERENCE LIGHTFILTERS

1. *General considerations*

Recently, layered, or so-called interference, lightfilters have found widespread application in optics. The behavior of these filters is a result of the interference of the waves reflected from the layers of which the filter is constructed. Similar principles are used in electrical filters[29] and in mechanical filters in acoustics. This type of filter was also used during the last war for ultrashort radiowaves.

In comparison with the usual absorption lightfilters, the interference lightfilters have the advantage that the loss of light in the pass band is considerably less. The position of the pass band in interference lightfilters may easily be varied.

On the other hand, in comparison with monochromators of various kinds, the interference lightfilters have the advantage that the cross section of the beam being filtered may be quite large.

The fundamental physical problems arising in the analysis of the behavior of lightfilters are examined in the special literature.[129, 179] These problems are discussed particularly thoroughly in the review article of G. V. Rozenberg.[67]

Here, we shall dwell only on the fundamental theoretical results concerning lightfilters, working both in reflected and transmitted light, and shall compare these results with experiment.

The simplest form of an interference filter working in reflected light is shown in Fig. 61. In this type of filter, a good reflecting metallic mirror *aa* is coated with a layer of dielectric *baab* which in turn is covered by a thin semitransparent metallic layer *bb*.

When a wave is normally incident on a filter of this type, total reflection occurs in the spectral regions for which the distance between the mirror and the metallic layer is an integral number of half wavelengths

$$\Delta = \frac{\lambda}{2}, 2\frac{\lambda}{2}, 3\frac{\lambda}{2}, \dots \tag{11.1}$$

where λ is the wavelength in the dielectric. On the other hand, reflection will be a minimum (practically zero) if the condition

$$\Delta = \frac{\lambda}{4}, 3\frac{\lambda}{4}, 5\frac{\lambda}{4}, \ldots \qquad (11.1a)$$

is fulfilled.

In fact, if we assume that the mirror is a perfect conductor, the intensity of the electric field will be zero at the boundary aa and at distances from it which are multiples of $\lambda/2$, as a result of the interference

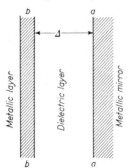

Fig. 61. Simple form of a reflection interference lightfilter.

of the direct and the reflected waves. If the metallic layer is located at one of these positions, it will absorb practically no energy from the electromagnetic wave, and the wave will be completely reflected. This condition is realized exactly when condition (11.1) is satisfied. On the other hand, if condition (11.1a) is satisfied, the metallic layer will be at an antinode of the electric field, and the energy absorption from the wave will be maximum. Waves of this type will be almost completely absorbed by the filter.

The simplest form of filter for transmitted light is shown in Fig. 62. It consists of two thin metallic layers aa and bb with a layer of dielectric between them. This system yields maximum transmission for waves for which the thickness of the dielectric layer is an integral number of half wavelengths. However, under these conditions, the transmissivity is not complete, but reaches 30–40% in practical configurations. That total transmission cannot be attained will be clear from the following con-siderations. Wave absorption in the lightfilter is due to the ohmic losses of currents induced in the metallic layers. This loss will be absent only if the metallic layers are in locations at which the electric field is always zero. Clearly, for this to occur, a standing wave must be produced in the filter, with nodes of the electric field at the positions of metallic layers. However, such a standing wave can be produced only when the amplitude of the reflected wave is equal to that of the incident wave. Under these conditions, we would obtain total reflection instead of total transmission.

The width of the pass band of the filter may be of the order of 100 Å.

The filter shown in Fig. 62 has minimum transmissivity (in practice, less than 1%) for waves satisfying the condition $\Delta = \lambda/4, 3\lambda/4, \ldots$ Under these conditions, if one of the metallic layers is at a node of the electric field, the other must be at an antinode, and as a result the absorption will be a maximum.

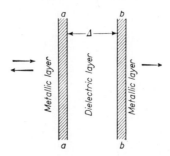

Fig. 62. Schematic illustration of an interference lightfilter for transmitted light.

The reflection coefficient of a light wave incident on the filter can be found by a complete solution of the problem of the electromagnetic field in each of the layers of the filter, taking account of the appropriate conditions at the boundaries. However, we shall follow a different and, in the present case, a simpler path.

2. Theory of the simplest reflection filter

The wave reflected from the filter can be represented as a superposition of plane waves $E_1 + E_2 + E_3 + \ldots$ (Fig. 63), where:

$E_1 = V_{31} E_0$ results from the reflection of the incident wave from the metallic layer (V_{31} is the corresponding reflection coefficient);

$E_2 = D_{13} e^{2i\phi} V_{43} D_{31} E_0$ results when the wave penetrates through the metallic layer (transmission coefficient D_{13}), traverses the dielectric layer of thickness Δ (the phase change is $\phi = kn_3 \Delta \cos \vartheta_3$, where k is the wave number in vacuum, n_3 is the index of refraction of the dielectric, ϑ_3 is the angle of refraction of the wave in the dielectric), is reflected from the metallic mirror (reflection coefficient V_{43}), traverses the dielectric layer again and, finally, penetrates through the metallic layer from the dielectric side (transmission coefficient D_{31});

$E_3 = D_{13} e^{4i\phi} V_{43}^2 V_{13} D_{31} E_0$ results from penetration through the metallic layer, quadruple traversal of the dielectric layer, double reflection from the metallic mirror, single reflection from the metallic layer from the dielectric side (reflection coefficient V_{13}), and return through the metallic layer, etc.

Reflection from a metallic layer and transmission through the layer can also be considered as consisting of multiple reflections (see § 5). However, we will characterize these processes by the resultant coefficients V_{31}, D_{31}, V_{13} and D_{13}.

Summing over all the waves, we obtain for the reflection coefficient of the filter:

$$\frac{E}{E_0} = V_{31} + D_{13}\,e^{2i\phi}\,V_{43}\,D_{31} + D_{13}\,e^{4i\phi}\,V_{43}^2\,V_{13}\,D_{31} + D_{13}\,e^{6i\phi}\,V_{43}^3\,V_{13}^2\,D_{31} + \ldots$$

$$= V_{31} + D_{13}\,D_{31}\,V_{43}\,e^{2i\phi}\sum_{s=0}^{\infty}e^{2is\phi}\,V_{43}^s\,V_{13}^s.$$

Hence, using the formula for a geometric progression, we obtain

$$\frac{E}{E_0} = \frac{V_{31} + V_{43}\,e^{2i\phi}\,(D_{13}\,D_{31} - V_{13}\,V_{31})}{1 - V_{43}\,V_{13}\,e^{2i\phi}}. \tag{11.2}$$

The reflection and transmission coefficients of the metallic layer, V_{31} and D_{13}, are equal, respectively, to V and D given by Eqs. 5.35–5.38. The coefficients V_{13} and D_{31} corresponding to a wave traveling in the reverse direction are obtained from these formulas by interchanging the subscripts 1 and 3.

Equation 11.2 can be simplified by limiting the generality of the investigation.

We shall assume that complete reflection occurs at the metallic mirror, i.e. $V_{43} = -1$. The minus sign results from the boundary condition that the tangential component of the electric field be zero at the surface of the mirror.

An important area of application of reflecting interference filters is the infrared region of the spectrum, where the selection of usual absorption filters is very difficult or even impossible. In this region of the spectrum, the thickness of the metallic layer may be considered small with respect to the wavelength. The index of refraction of the metal will be $n^2 \approx 2i\sigma/\nu$, where σ is the conductivity of the metal in Gaussian units, and ν is the frequency. Under these conditions, $|n|^2 \gg 1$.

Thus, if we limit ourselves for the time being to the case of normal incidence, the reflection coefficient of the filter will be given approximately by[†]

$$R = \left|\frac{E}{E_0}\right|^2 = \frac{(1-f)^2 + n_3^2\cot^2(2\pi n_3\,\Delta/\lambda)}{(1+f)^2 + n_3^2\cot^2(2\pi n_3\,\Delta/\lambda)}, \tag{11.3}$$

[†] This formula can be obtained from Eqs. 11.2, 5.10 and 5.15 with the conditions $\phi_2 = \alpha_2 l \approx n_2(2\pi/\lambda)\,l \ll 1$. Compare also Ref. 129 where the expression for f is written somewhat differently because of the particular system of units employed.

where $f = (4\pi\sigma/c)\,l$, c and λ are the velocity and wavelength of light in vacuum, and the meanings of l and n_3 are clear from Fig. 63.

As is seen in the last formula, the reflection coefficient becomes zero when the following two conditions are satisfied:

$$f = 1, \quad \Delta = (2m-1)\frac{\lambda}{4n_3}, \quad m = 1, 2, \ldots \tag{11.4}$$

On the other hand, when the condition

$$\Delta = \frac{m\lambda}{2n_3} \tag{11.5}$$

is satisfied, the reflection coefficient is unity.

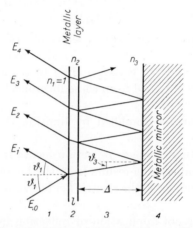

Fig. 63. For the calculation of the reflection coefficient of a filter.

The reflection coefficient of the filter as a function of the frequency is shown in Fig. 64 for $f = 0.5, 1, 2$, assuming that $n_3 = 1.3$. The symbol ν_0 denotes the frequency for which $\Delta = \lambda_0/4n_3$.

Analysis of Eq. 11.3 shows that the form of the curves of R as a function of frequency depends on n_3. As n_3 increases, the reflection maxima become broader, and the minima more narrow.

When light strikes the filter at oblique incidence, we have formulas similar to Eq. 11.3 for the various polarizations

$$R_\perp = \frac{[1-(f/p)]^2 + (q/p)^2 \cot^2[2\pi q(\Delta/\lambda)]}{[1+(f/p)]^2 + (q/p)^2 \cot^2[2\pi q(\Delta/\lambda)]}, \tag{11.6}$$

$$R_\parallel = \frac{(1-fp)^2 + (n_3\,p/q)^2 \cot^2[2\pi q(\Delta/\lambda)]}{(1+fp)^2 + (n_3\,p/q)^2 \cot^2[2\pi q(\Delta/\lambda)]}, \tag{11.7}$$

where the subscripts $\perp$ and $\parallel$ denote respectively $\mathbf{E}$ perpendicular and parallel to the plane of incidence. Furthermore,

$$p = \cos \vartheta_1,$$

$$q = n_3 \cos \vartheta_3 = \sqrt{(n_3^2 - \sin^2 \vartheta_1)}.$$

In a number of cases, the reflection coefficients $R_\perp$ and $R_\parallel$ can differ considerably. In particular, if the filter is designed so that $R_\perp = 0$ at an angle of incidence of $\vartheta_1 = 60°$ (for which we must have $f = \frac{1}{2}$), we will have $R_\parallel = [(1 - fp)/(1 + fp)]^2 = 0.36$, i.e. $R_\parallel$ will be quite noticeable. A filter of this type can be used to obtain polarized light.

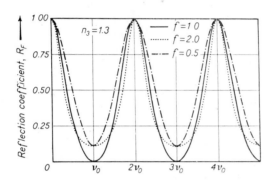

Fig. 64. The reflection coefficient of a filter as a function of the frequency.

3. *Theory of the simplest transmission filter*

Just as in the preceding paragraph, we sum the successive waves traversing the dielectric layer various numbers of times and obtain an expression for the amplitude E_{tr} of the wave transmitted through the system consisting of two thin metallic layers and a dielectric layer (Fig. 62):

$$E_{tr} = \frac{D_{13} D_{31} e^{i\phi}}{1 - R_{13}^2 e^{2i\phi}} E_0, \tag{11.8}$$

where E_0 is the amplitude of the incident wave, and D_{31} is the transmission coefficient of the metallic layer for a wave traveling from air through the layer and into the dielectric. In the general case of a wave of arbitrary polarization, this coefficient is given by Eqs. 5.15, 5.37 and 5.38; D_{13} is the analogous coefficient for a wave traveling in the reverse direction, and is obtained from the expression for D_{31} by interchanging the subscripts 1 and 3.

The reflection coefficient R_{13} at the metallic layer on the dielectric side is given by Eqs. 5.10, 5.35 and 5.36. The quantity ϕ is the phase change across the dielectric layer.

We introduce the notation

$$D_{31} = \eta_{31}\,e^{i\Phi_{31}}, \quad D_{13} = \eta_{13}\,e^{i\Phi_{13}}, \quad R_{13} = \rho_{13}\,e^{i\phi_{13}}. \tag{11.9}$$

Then Eq. 11.8 yields the following expression for the transmission coefficient of the filter:

$$T = \left|\frac{E_{\mathrm{tr}}}{E_0}\right|^2 = \frac{\eta_{31}^2\,\eta_{13}^2}{(1-\rho_{13}^2)^2 + 4\rho_{13}^2 \sin^2(\phi+\phi_{13})}. \tag{11.10}$$

Maximum transmission will occur for frequencies satisfying the condition

$$\phi + \phi_{13} = m\pi, \quad m = 1, 2, \ldots$$

At these frequencies

$$T = T_{\max} = \frac{\eta_{31}^2\,\eta_{13}^2}{(1-\rho_{13}^2)^2}. \tag{11.11}$$

On the other hand, minimum transmission occurs when

$$\phi + \phi_{13} = (2m-1)\,\pi/2,$$

and the corresponding value is

$$T_{\min} = \frac{\eta_{31}^2\,\eta_{13}^2}{(1+\rho_{13}^2)^2}. \tag{11.12}$$

We define the pass band width W as twice the wavelength interval between the transmission maximum and the point at which the transmission is one half of maximum.

To calculate this width, we remember that in Eq. 11.10

$$\phi = \frac{2\pi}{\lambda}\,n_3\,\Delta\cos\vartheta_3,$$

where n_3 is the index of refraction of the dielectric layer, Δ is its thickness, and ϑ_3 is the angle of refraction in this layer. Then Eq. 11.10 yields the following expression for the width of the pass band:

$$W = \frac{2\lambda_{\max}}{m\pi - \phi_{13} + \arcsin^2[(1-\rho_{13}^2)/2\rho_{13}]}\,\arcsin\frac{1-\rho_{13}^2}{2\rho_{13}}, \tag{11.13}$$

where $\lambda_{\max}$ is the wavelength corresponding to maximum transmission. In the derivation of Eq. 11.13, it was assumed that within the limits of the pass band, the quantities ρ_{13}, ϕ_{13}, η_{13} and η_{31} are constant.

It is clear from Eq. 11.13 that the greater the coefficient of reflection from the metallic layer, the smaller the width of the pass band. However, as the reflection coefficient is increased, the transmissivity of the filter decreases. Therefore, in practice we must find an optimum value of the reflection coefficient which provides the required pass band width, and at the same time allows satisfactory transmission.

It is usually best to choose a transmission maximum between 10 and 40%. The pass band width is then between 50 and 300 Å, depending on the order of interference m and the region of the spectrum. Under these conditions, the transmission minimum between neighboring pass bands can be less than 0.25%.

The experimental transmission curve of a filter of the type we are considering[127] is shown in Fig. 65. The values of the transmission minima between the pass bands $m = 2$ and $m = 3$, and between $m = 3$ and $m = 4$, are 0.03 and 0.15%, respectively.

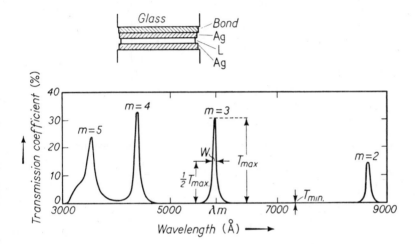

Fig. 65. The transmission characteristics of the simple interference filter in transmitted light.

4. More complicated filters

By increasing the number of layers in the filter, we widen the possibilities for selecting a filter with the necessary characteristics. In particular, we will be able to suppress undesirable bands of transmission and reflection.

The characteristics of a simple reflection filter consisting of a layer of MgF_2, fixed on an aluminum mirror and covered with a thin layer of the same metal,[67] are shown in Fig. 66. Curves indicating the reflecting power of the aluminum mirror and the thin aluminum layer used in the filter are also shown in Fig. 66.

The characteristics of a reflecting filter consisting of two layers of dielectric and two metallic layers are shown in Fig. 67. We see that the second reflection maximum in the long wavelength region, which is quite strong in the case of the simple filter (Fig. 66), is now cut down considerably.

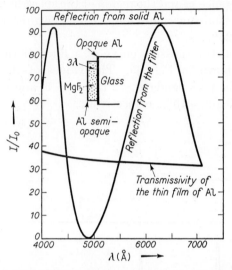

Fig. 66. A typical spectral characteristic of a simple reflecting filter.

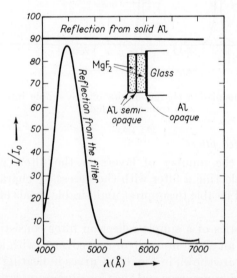

Fig. 67. The spectral characteristic of a double reflecting filter.

Another type of filter which has been successfully developed is one consisting of a number of layers of dielectric with layers of low and high indices of refraction alternating with one another. Metallic layers are absent. The advantage of this type of filter is the absence of energy loss through absorption. The best calculational approach to such multi-layered systems is the impedance method developed in § 5.

The filter characteristics shown in Fig. 68 are those of a multilayered filter without metallic layers[127]. The distinctive feature of this filter is that it transmits the entire visible region of the spectrum, and at the same time reflects the spectral region containing the thermal radiation maximum of the majority of powerful sources of light. This type of filter is of great practical value.

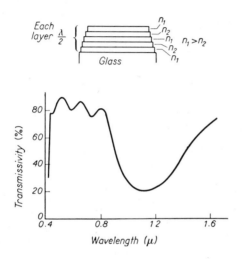

Fig. 68. Transmission filter reflecting the infrared rays.

§ 12. LAYERED SOUND INSULATORS

1. *Theory of the layered sound insulator*

Light and effective sound insulating partitions can be made by using layered construction. This type of partition consists of alternating layers of an impermeable (to steady air flow) material and a porous material, the layers also being separated in some cases by air spaces. Structures of this kind are many and varied.

In this paragraph, we shall investigate the theory of the sound insulation of a structure (Fig. 69) consisting of:

(1) a layer of impermeable material of mass σ_1 cm^{-2},

(2) an air space of thickness l,

(3) a layer of porous material of thickness d_1,

(4) an impermeable layer of mass σ_2 cm^{-2},

(5) a layer of porous material of thickness d_2.

6

Structures of this kind have been studied in detail[86] in connection with the sound insulation of aircraft cabins. In this case, the impermeable layer σ_1 is the Dural cabin wall.

In this Section, we shall limit ourselves to the case of normal wave incidence. We let p_0 denote the acoustic pressure in front of the layered structure (the sum of the pressures in the incident and the reflected waves), and let p_5 denote the acoustic pressure in the wave transmitted through the structure (see Fig. 69). The sound insulation of the structure will be defined as the ratio p_0/p_5, and our problem is to find this ratio.

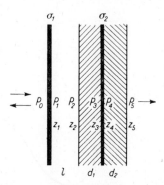

Fig. 69. Schematic illustration of a sound insulating structure of a general form.

First of all, we shall find the relations between the successive acoustic pressures $p_0, p_1, ..., p_5$. Thus, for example, the relation between p_0 and p_1 is determined from the equation of motion of layer σ_1, considered as an inert mass (the thickness of this layer is assumed to be small compared to the wavelength), i.e.

$$p_0 - p_1 = \sigma_1 \frac{dv_1}{dt},$$

where v_1 is the velocity of the layer.

Remembering that $(1/p_1)(dv_1/dt) = -i\omega\, v_1/p_1$, and that by definition the impedance is $Z_1 = p_1/v_1$, this equation can be written in the form

$$\frac{p_0}{p_1} = 1 - i\omega \frac{\sigma_1}{Z_1}. \tag{12.1}$$

The acoustic pressures p_3 and p_4 are similarly related.

We now find the relation between the pressures p_1 and p_2 at points separated by the air layer. The pressure in the layer (dropping the factor $e^{-i\omega t}$ as usual) can be written in the form

$$p = a_1 e^{bz} + a_2 e^{-bz}, \tag{12.2}$$

where z is a coordinate directed perpendicular to the plane layers, and b is the propagation constant, equal to $i\omega/c$ in nonabsorbing media, where c is the acoustic velocity.

Equation 12.2 can also be written in the form

$$p = 2A \cosh(bz + \epsilon), \tag{12.3}$$

where A and ϵ are constants, connected with a_1 and a_2 by the relations

$$a_1 = A e^{\epsilon}, \quad a_2 = A e^{-\epsilon}.$$

We locate the coordinate origin in Eq. 12.3 at a point where the impedance is assumed to be known, and denote the value of the impedance, corresponding to the point $z = 0$, by Z_T. It is not difficult to express the constant ϵ in terms of Z_T. In fact, remembering that in a plane monochromatic wave the particle velocity is expressed in terms of the pressure by the formula (see § 3)

$$\mathbf{v} = \frac{1}{i\omega\rho} \operatorname{grad} p, \tag{12.4}$$

Eq. 12.3 yields

$$Z_T = \left(\frac{p}{v}\right)_{z=0} = i\omega\rho \left(\frac{p}{dp/dz}\right)_{z=0} = \frac{i\omega\rho}{b \tanh \epsilon}.$$

But $i\omega\rho/b \equiv Z_0$ is the characteristic impedance of the medium. Therefore

$$\epsilon = \operatorname{arc} \tanh \frac{Z_0}{Z_T}. \tag{12.5}$$

Applying Eq. 12.3 to the air layer, and assuming that the point $z = 0$ corresponds to its boundary with the sound insulating layer d_1, we find

$$\left. \begin{array}{l} p_2 = 2A \cosh \epsilon_1, \quad p_1 = 2A \cosh\left(i\frac{\omega}{c}l + \epsilon_1\right), \\[2mm] \quad \epsilon_1 = \operatorname{arc} \tanh \frac{\rho c}{Z_2}. \end{array} \right\} \tag{12.6}$$

where

In the derivation, we took into account that for the air layer

$$b = i\frac{\omega}{c}, \quad Z_0 = \rho c.$$

From Eq. 12.6, we obtain

$$\frac{p_1}{p_2} = \frac{\cosh\left[(i\omega/c)l + \epsilon_1\right]}{\cosh \epsilon_1}. \tag{12.7}$$

By analogy with Eqs. 12.7 and 12.1, we have further

$$\frac{p_2}{p_3} = \frac{\cosh (bd_1 + \epsilon_2)}{\cosh \epsilon_2}, \quad \epsilon_2 = \text{arc tanh} \frac{Z_0}{Z_3}, \tag{12.8}$$

$$\frac{p_3}{p_4} = 1 - \frac{i\omega\sigma_2}{Z_4}, \tag{12.9}$$

$$\frac{p_4}{p_5} = \frac{\cosh (bd_2 + \epsilon_4)}{\cosh \epsilon_4}, \quad \epsilon_4 = \text{arc tanh} \frac{Z_0}{Z_5}, \tag{12.10}$$

where $Z_1, Z_2, ..., Z_5$ are the impedances in the planes indicated in Fig. 69, and, in particular, $Z_5 = \rho c$ is the wave impedance of air; Z_0 and b are respectively the characteristic impedance and the propagation constant of the porous material. Here, Z_0 and b are generally complicated functions of the parameters characterizing the porous material: the specific resistance R_1, the density ρ_m and to a lesser extent the structure factor $k,$[27] the porosity Y, and the bulk moduli K and Q of the air in the material and the material itself, respectively. We limit ourselves to a soft porous material (the ratio K/Q is greater than, say, 20), in which a wave traveling along the skeleton will be very rapidly attenuated and can be neglected. Only under these conditions can the material be characterized by a single propagation constant b.

The relations between the impedances Z_1 and Z_2, Z_2 and Z_3, etc., at points separated by layers, were considered in § 5 (see, for example, Eq. 5.48). However, it will be convenient for us here to present these relations in a somewhat different form.

As we have seen above, the acoustic pressure in the layer, on one side of which (at $z = 0$) the impedance is Z_T, is given by Eqs. 12.3 and 12.5. For the impedance on the other side of the layer (at $z = l$), we obtain

$$Z = i\omega\rho \left(\frac{p}{dp/dz}\right)_{z=l} = \frac{i\omega\rho}{b} \coth (bl + \epsilon) = Z_0 \coth (bl + \epsilon). \tag{12.11}$$

In the case of an air layer

$$Z_0 = \rho c, \quad b = i\frac{\omega}{c}.$$

We are now able to write down the entire system of impedances required for calculations. First of all, we have

$$\left. \begin{array}{l} Z_5 = \rho c, \\ Z_4 = Z_0 \coth (bd_2 + \epsilon_4). \end{array} \right\} \tag{12.12}$$

Furthermore, since the particle velocities are the same on both sides of the impermeable layer σ_2, the ratio of the impedances on the two sides of this layer is equal to the ratio of the acoustic pressures, i.e. according to Eq. 12.9,

$$Z_3 = \frac{p_3}{p_4} Z_4 = Z_4 - i\omega\sigma_2. \tag{12.13}$$

Finally, by analogy with Eq. 12.12,

$$Z_2 = Z_0 \coth(bd_1 + \epsilon_2), \quad Z_1 = \rho c \coth[i(\omega/c)l + \epsilon_1]. \tag{12.14}$$

Multiplying Eqs. 12.1 and 12.7–12.10, we obtain the ratio p_1/p_5. Taking Eqs. 12.12–12.14 into account, this ratio can be written in the following form, which is most convenient for calculations

$$\frac{p_0}{p_5} = \cosh bd_2(X_1 \cosh bd_1 + X_2 \sinh bd_1)$$
$$+ \sinh bd_2(X_3 \cosh bd_1 + X_4 \sinh bd_1), \tag{12.15}$$

where

$$X_1 = A - i\left(B + A\frac{\omega\sigma_2}{\rho c}\right), \qquad X_2 = \frac{AZ_0}{\rho c} - \frac{B\omega\sigma_2}{Z_0} - i\frac{B\rho c}{Z_0},$$

$$X_3 = \frac{AZ_0}{\rho c} - i\left(\frac{B\rho c}{Z_0} + \frac{A\omega\sigma_2}{Z_0}\right), \quad X_4 = A - \frac{\omega\sigma_2 \rho c}{Z_0^2}B - iB \tag{12.16}$$

and finally

$$A = \cos\frac{\omega l}{c} + \frac{\omega\sigma_1}{\rho c}\sin\frac{\omega l}{c}, \quad B = -\sin\frac{\omega l}{c} + \frac{\omega\sigma_1}{\rho c}\cos\frac{\omega l}{c}. \tag{12.17}$$

It is of interest to examine a number of sound insulating structures, which are obtained from the structure we are now considering by choosing special values for the parameters.

We will investigate five structures, denoting them by the numbers 0 to IV, where the higher numbers correspond to more complicated structures. The structure we considered above will be denoted by the number V as the most complicated; the remaining structures are then obtained by assigning the following values to the parameters (Fig. 70):

structure IV, $d_2 = 0$, $d_1 = d$,

structure III, $d_1 = 0$, $d_2 = d$,

structure II $\sigma_2 = 0$, $d_2 = 0$, $d_1 = d$,

structure I $l = 0$, $\sigma_2 = 0$, $d_2 = 0$, $d_1 = d$,

structure 0 $d_1 = 0$, $d_2 = 0$.

Thus, structure 0 consists of two impermeable walls, separated by an air space, structure I is a wall covered by a layer of porous material, etc.

Formulas for the sound insulation of all the above structures are obtained from Eqs. 12.15–12.17.

Structure IV: $\quad \dfrac{p_0}{p_5} = X_1 \cosh bd + X_2 \sinh bd.$ $\hfill$ (12.18)

Structure III: $\quad \dfrac{p_0}{p_5} = X_1 \cosh bd + X_3 \sinh bd.$ $\hfill$ (12.19)

Structure II: $\quad \dfrac{p_0}{p_5} = (A - iB)\cosh bd + \left(\dfrac{AZ_0}{\rho c} - iB\dfrac{\rho c}{Z_0}\right)\sinh bd.$ $\hfill$ (12.20)

Structure I: $\quad \dfrac{p_0}{p_5} = \left(1 - \dfrac{i\omega\sigma_1}{\rho c}\right)\cosh bd + \left(\dfrac{Z_0}{\rho c} - \dfrac{i\omega\sigma_1}{Z_0}\right)\sinh bd.$ $\hfill$ (12.21)

Structure 0: $\quad \dfrac{p_0}{p_5} = X_1 = A - i\left(B + \dfrac{A\omega\sigma_2}{\rho c}\right).$ $\hfill$ (12.22)

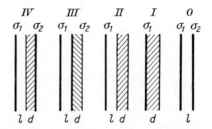

Fig. 70. Sound insulating structures of various special forms.

It is also of interest to consider a structure consisting of two identical impermeable layers separated by a sound-absorbing material. Then $d_2 = 0$, $l = 0$, $\sigma_1 = \sigma_2 = \sigma$, and

$$\frac{p_0}{p_5} = \left(1 - 2i\frac{\omega\sigma}{\rho c}\right)\cosh bd + \left(\frac{Z_0}{\rho c} - \frac{\omega^2\sigma^2}{Z_0\,\rho c} - \frac{i\omega\sigma}{Z_0}\right)\sinh bd. \qquad (12.23)$$

Curves of sound insulation vs frequency, calculated with the formulas obtained above, are shown in Figs. 71 and 72 for structure 0 (double wall) and structure V.

We see that at certain frequencies the addition of the second wall actually reduces the sound insulation (see Fig. 71). It is not hard to show that these frequencies correspond to the resonant frequencies of the system consisting of the two layers with an air space between them, and loaded on both sides by air.

The sound insulation curves for the double wall also show quite a number of strong peaks corresponding to local failures of the sound insulation. These peaks are observed at the frequencies for which the air space between the walls is an integral number of half wavelengths. It can be shown that at these frequencies, the sound insulation is equal to

$$10 \log \left| \frac{p_0}{p_5} \right|^2 = 10 \log \left[1 + \frac{\omega^2 (\sigma_1 + \sigma_2)^2}{\rho^2 c^2} \right], \qquad (12.24)$$

i.e. in this case the two walls act as a single wall with the combined mass. This could have been expected from the general theory developed in § 5, where it was shown that a layer whose thickness is equal to an integral number of half wavelengths has no influence on the transmission and reflection coefficients of the system, and is equivalent in this regard to a layer of zero thickness†.

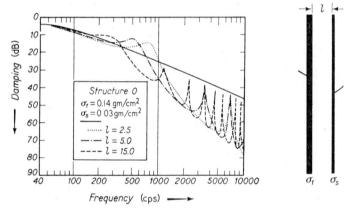

Fig. 71. The sound insulation of a double wall (in decibels). The solid line shows the sound insulation of a single wall with a surface density of σ_1.

In the intervals between the peaks, the sound insulation is equal to

$$10 \log \left| \frac{p_0}{p_5} \right|^2 = 10 \log \left(1 + \frac{\omega^2 \sigma_1^2}{\rho^2 c^2} \right) + 10 \log \left(1 + \frac{\omega^2 \sigma_2^2}{\rho^2 c^2} \right), \qquad (12.25)$$

i.e. in this case the effect of both walls is added independently.

---

† In the case of a single wall, we express the sound insulation by the formula $10 \log [1 + (\omega^2 \sigma^2 / \rho^2 c^2)]$, where σ is the wall density per square cm. This differs from the formula which is usually employed, $10 \log [1 + (\omega^2 \sigma^2 / 4 \rho^2 c^2)]$. The difference is due to the fact that we define the sound insulation of the wall as the ratio (expressed in decibels) of the acoustic pressures in front of the wall and behind it, whereas the sound insulation is usually defined as the ratio (expressed in decibels) of the acoustic pressures at a single point behind the wall when the wall is present and when it is absent. In particular, when $\omega \sigma / \rho c \gg 1$, almost total reflection occurs at the wall, the pressure in front of the wall is doubled, and the two definitions of sound insulation differ by 6 db. Our definition seems the more natural.

In Fig. 72, corresponding to a comparatively complex sound insulating structure containing layers of porous materials, the peaks are not as strong because of damping in the layers.

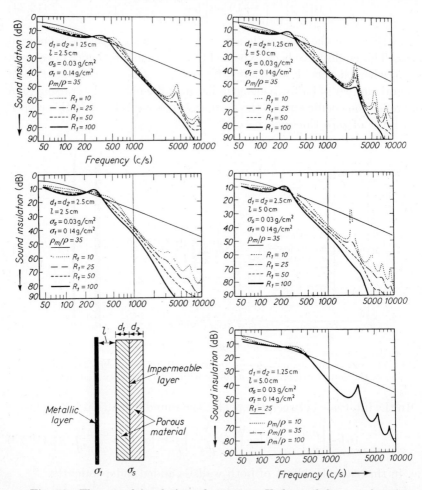

Fig. 72. The sound insulation of structure V (lower left) as a function of frequency for various values of the structure parameters. On all of the Figures, the solid line with the greatest slope represents the sound insulation of a single wall with a density of σ_1 cm^{-2}.

In these figures, ρ_m denotes the density of the porous material, ρ is the density of air, and R_1 is the specific (dyne-sec/cm^3) resistance of the material and at all frequencies is close to the resistance in steady flow. The results of Ref. 85 were used to determine the propagation constant b. It is also shown in Ref. 85 that, for soft porous materials with the porosity Y greater than 95%, the characteristic impedance and the constant b are related by[†]

$$Z_0 = -\frac{iKb}{\omega Y}.$$

2. Comparison with experiment. Analysis of other cases

In the experimental verification of the theory of layered sound insulation, it is necessary to take a number of precautions. In particular it is very important to eliminate the possibility of sound transmission through a layer resulting from the excitation of flexural waves in the layer which will then radiate sound on the other side of the layer.

This elimination was accomplished by building a layer out of a set of thinner metallic layers which had sheets of mica (of density 0.04 gm/cm²) bonded to them. Flexural waves were thereby highly attenuated. In this way, layers were made with total densities from 0.16 to 1.6 gm/cm². Experiment showed that only in this way could a single wall be obtained which would yield the sound insulation corresponding to its mass.

Moreover, it was also necessary to eliminate waves propagating along the air gap between layers σ_1 and d_1. This was accomplished by placing sound absorbing material around the edges of the air gap. The reader will find a detailed description of the experimental setup by referring to the original works.[85, 86]

The theory developed above is well confirmed by experiment. The experimental and theoretical sound insulation curves are shown in Figs. 73 and 74 for a number of structures. The abscissas represent the frequency in cps, and the ordinates represent (in decibels) the excess of the sound insulation over that of a single wall with a surface density σ_1. Four different structures are represented in Fig. 73, while Fig. 74 is concerned with structures III and V. The notation for the parameters of the structures is the same as above.

The sound absorbing material used in the structures was made of a type of rock wool.

It is clear from the graphs that the general trends of the theoretical and experimental curves are in good agreement.

A detailed study of the frequency characteristics of the sound insulation of the various structures showed that:

(a) to provide good sound insulation at the lower frequencies, the total thickness of the structure had to be a maximum;

[†] We use the time factor $\exp(-i\omega t)$, whereas in Refs. 85 and 86, the factor $\exp(i\omega t)$ is used. As a result, we always have the opposite sign in front of i.

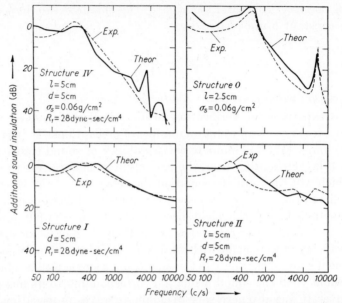

Fig. 73. Comparison of the experimental and theoretical results for the sound insulation of various structures as a function of frequency. The ordinate represents the excess sound insulation afforded by the structure in comparison with the sound insulation of a single wall with a density of σ_1/cm^2.

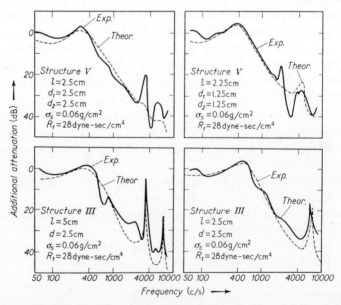

Fig. 74. The same as Fig. 73, but for other parameters of the structures.

(b) at frequencies of 1000–5000 cps, which are the most important in the problem of the sound insulation of aircraft cabins, structures III, IV and V gave approximately the same results.

In a number of cases, it is essential to determine the mean sound insulation for various angles of incidence of the sound wave on the wall. In its theoretical aspects, this problem involves no fundamental difficulties. The formulas obtained above are generalized to apply to an arbitrary angle of incidence (which is not particularly difficult if the results of § 5 are taken into account) and the results are then averaged over the angles.

In practice, this averaging is reduced to a numerical integration. This problem has not yet been investigated sufficiently thoroughly, although a partial investigation has been performed by London.[155, 156]

We should also mention some older work on layered sound insulation. The sound insulation of two identical partitions separated by an air space was considered in Ref. 99. The sound insulation of a system of partitions, each of which has a natural frequency, was examined in Ref. 140. Ref. 100 is also of interest.

CHAPTER III

PLANE WAVES IN LAYERED-INHOMOGENEOUS MEDIA

CHAPTER I was concerned with the theory of plane waves in homogeneous media separated from one another by plane boundaries. The properties of the media changed discontinuously across these boundaries. In the present chapter, we shall study a medium whose properties vary continuously along one axis of a rectangular coordinate system (the z-axis, for example) but do not change in planes perpendicular to this axis. Media of this type will be called layered-inhomogeneous. The earth's atmosphere, the sea, and the core of the earth are approximately layered-inhomogeneous media since their properties (velocities of elastic and of electromagnetic waves, density, etc.) vary considerably along the vertical, whereas the variations along the horizontal are much less noticeable.

Our problem is the study of acoustic and electromagnetic fields in layered-inhomogeneous media. Of particular importance is the determination of the reflection coefficient from the various kinds of inhomogeneities in the atmosphere, which have a layered character. The scattering of waves on irregular inhomogeneities (for example, on clouds in the atmosphere, etc.) will not be considered.

The results obtained below can be used directly to determine the reflection of ultrashort radiowaves and microwaves from layered inhomogeneities in the troposphere, and are also applicable to the reflection of radiowaves from ionospheric layers. However, certain effects peculiar to the latter case (influence of the magnetic field, nonlinear effects, absorption, etc.) will not be considered. The reader will find discussions of these effects in the literature on the propagation of radiowaves in the ionosphere.

All the results obtained below are directly applicable to sound propagation in the atmosphere, and also in the sea.

§ 13. EQUATIONS OF ELECTROMAGNETIC AND ACOUSTIC FIELDS IN AN INHOMOGENEOUS MEDIUM

1. *The equations of the electromagnetic field*

Equations 2.2 remain valid for an inhomogeneous medium; however, the complex dielectric constant ϵ' and magnetic permeability μ will be

168

functions of position. Also, the two equations in (2.1) which contain div **E** and div **H** are replaced by

$$\operatorname{div}(\epsilon' \mathbf{E}) = 0, \quad \operatorname{div}(\mu \mathbf{H}) = 0. \tag{13.1}$$

It is sufficient for our purposes to consider $\mu = 1$ throughout space.

Then Eqs. 2.2 yield
$$\nabla \times \nabla \times \mathbf{E} = \frac{\omega^2}{c^2} \epsilon' \mathbf{E}, \tag{13.2}$$

or
$$\nabla^2 \mathbf{E} + \frac{\omega^2}{c^2} \epsilon' \mathbf{E} - \operatorname{grad} \operatorname{div} \mathbf{E} = 0. \tag{13.3}$$

This equation can be represented in a somewhat different form. In fact, the first of Eqs. 13.1 gives

$$\epsilon' \operatorname{div} \mathbf{E} + \mathbf{E} \operatorname{grad} \epsilon' = 0.$$

Using the last equation to eliminate div **E** from Eq. 13.2, we obtain

$$\nabla^2 \mathbf{E} + \frac{\omega^2}{c^2} \epsilon' \mathbf{E} + \operatorname{grad}\left(\mathbf{E}\, \frac{1}{\epsilon'} \operatorname{grad} \epsilon'\right) = 0. \tag{13.4}$$

Thus, when ϵ' varies sufficiently slowly in space, the last term can be neglected, and the equation for the electric field reduces to the usual wave equation with a variable wave number k

$$\nabla^2 \mathbf{E} + k^2(x, y, z)\, \mathbf{E} = 0,$$

$$k^2(x, y, z) \equiv \frac{\omega^2}{c^2} \epsilon'(x, y, z). \tag{13.5}$$

However, the careless suppression of the small term in the differential equation (13.4) may sometimes lead to a large error in the final result. It is therefore of interest to consider methods by which we can reduce the exact equation (13.4) to a wave equation. This is accomplished by using the scalar and vector potentials ϕ and **A**.

  As is well known, the electric and magnetic fields are expressed in terms of the potentials ϕ and **A** by the equations (for $\mu = 1$)†

$$\mathbf{E} = -\operatorname{grad} \phi - \frac{1}{c}\frac{\partial \mathbf{A}}{\partial t} = -\operatorname{grad} \phi + \frac{i\omega}{c} \mathbf{A},$$

$$\mathbf{H} = \nabla \times \mathbf{A}. \tag{13.6}$$

Equations 2.2 and 13.1 will be satisfied if ϕ and **A** are subjected to the condition

$$\phi = -\frac{ic}{\omega\epsilon'} \operatorname{div} \mathbf{A}, \tag{13.7}$$

---

† In a homogeneous medium, the vector potential **A** coincides, to within the constant factor $-i\omega\epsilon'/c$, with the Hertz vector **Π** used in the following chapters.

and if **A** satisfies the equation

$$\nabla^2 \mathbf{A} + k^2 \mathbf{A} - \frac{2}{k} \operatorname{div} \mathbf{A} \operatorname{grad} k = 0. \tag{13.8}$$

In what follows, we shall consider a layered medium in which the wave number k is a function only of the z-coordinate. Furthermore, since we are considering a plane problem, the y-axis may be directed in such a way that **A** will be independent of y. We shall examine two cases in detail:

a. *The vector* **A** *is directed along the z-axis* $(A_x = A_y = 0)$. Equations 13.6 and 13.7 give

$$E_x = \frac{i\omega}{ck^2} \frac{\partial^2 A_z}{\partial x \, \partial z}, \quad E_y = 0, \quad E_z = \frac{i\omega}{c}\left[A_z + \frac{\partial}{\partial z}\left(\frac{1}{k^2}\frac{\partial A_z}{\partial z}\right)\right],$$

$$H_x = 0, \quad H_y = -\frac{\partial A_z}{\partial x}, \quad H_z = 0. \tag{13.9}$$

Thus, we have the case of waves polarized in the plane of incidence. From Eq. 13.8, the equation for A_z is

$$\nabla^2 A_z + k^2 A_z - \frac{1}{k^2}\frac{\partial A_z}{\partial z}\frac{\partial k^2}{\partial z} = 0. \tag{13.10}$$

In place of A_z, we introduce the new function

$$\psi = \frac{A_z}{k}. \tag{13.11}$$

Using Eq. 13.10, the equation for the new function is

$$\nabla^2 \psi + \left(k^2 + \frac{k''}{k} - \frac{2k'^2}{k^2}\right)\psi = 0, \tag{13.12}$$

where the primes on k denote derivatives with respect to z.

We have thus obtained a wave equation with an "effective" wave number squared equal to

$$K^2 = k^2 + \frac{k''}{k} - \frac{2k'^2}{k^2}. \tag{13.13}$$

b. *The vector* **A** *is directed along the y-axis* $(A_x = A_z = 0)$. A wave equation for A_y is obtained directly from Eq. 13.8:

$$\nabla^2 A_y + k^2 A_y = 0. \tag{13.14}$$

The components of the electromagnetic field are obtained from Eqs. 13.6 and 13.7 (note that here, $\phi = 0$):

$$E_x = 0, \quad E_y = \frac{i\omega}{c} A_y, \quad E_z = 0,$$

$$H_x = -\frac{\partial A_y}{\partial z}, \quad H_y = 0, \quad H_z = \frac{\partial A_y}{\partial x}. \tag{13.15}$$

This is the case of waves polarized perpendicular to the plane of incidence.

2. *The equations of the acoustic field in an inhomogeneous medium*

The fundamental equations of the acoustic field in an inhomogeneous medium have the form[9]

$$\frac{\partial p}{\partial t} + \rho c^2 \operatorname{div} \mathbf{v} = 0,$$

$$\frac{\partial \mathbf{v}}{\partial t} + \frac{1}{\rho} \operatorname{grad} p = 0, \tag{13.16}$$

where p is the acoustic pressure, $\mathbf{v}$ is the particle velocity in the wave, ρ is the density, and c is the velocity of sound. In the general case, ρ and c are functions of position. The first of Eqs. 13.6 is the equation of continuity, the second is the Euler equation.

Remembering that $\partial/\partial t = -i\omega$, and eliminating $\mathbf{v}$ from Eq. 13.16 we obtain

$$\rho \operatorname{div} \left(\frac{1}{\rho} \operatorname{grad} p \right) + k^2 p = 0, \tag{13.17}$$

or

$$\nabla^2 p + k^2 p - \frac{1}{\rho} \operatorname{grad} \rho \operatorname{grad} p = 0. \tag{13.18}$$

Just as in the electromagnetic case, this equation can be reduced to a wave equation. For this purpose, in place of p, we introduce the new function ψ, defined by[87]

$$\psi = \frac{p}{\sqrt{\rho}}. \tag{13.19}$$

After some straightforward transformations, we obtain the wave equation for ψ

$$\nabla^2 \psi + K^2(x, y, z) \psi = 0, \tag{13.20}$$

where

$$K^2(x, y, z) = k^2 + \frac{1}{2\rho} \nabla^2 \rho - \frac{3}{4} \left(\frac{1}{\rho} \operatorname{grad} \rho \right)^2.$$

§ 14. WAVE REFLECTION FROM AN INHOMOGENEOUS LAYER OF THE SIMPLEST FORM

1. *Statement of the problem*

We consider wave reflection from an inhomogeneous layer. To obtain a rigorous solution of this problem, we must solve the equations obtained in the previous section, subject to the appropriate boundary conditions. Such rigorous solutions in closed form are known only for a few forms of the function $k(z)$, characterizing the z-dependence of the propagation velocity. We shall examine one such case in order to exhibit the fundamental features of reflection from inhomogeneous layers. The type of reflecting layer under examination was first studied by Epstein.[109] However, we shall generalize his calculations somewhat, and consider an arbitrary angle of incidence rather than normal incidence only.

The mathematical problem reduces to the solution of the wave equation

$$\nabla^2 \psi + k^2(z)\psi = 0. \tag{14.1}$$

As previously, we take the xz-plane as the plane of incidence. Then ψ will be a function only of x and z. The meaning of this function will be different in the various cases. We again consider three cases:

(a) The vector $\mathbf{E}$ is perpendicular to the plane of incidence ($E_x = E_z = 0$). Comparing Eqs. 13.14 and 13.15, we see that E_y will satisfy an equation of the form (14.1). Therefore, in this case, $\psi \equiv E_y$.

(b) The vector $\mathbf{E}$ is parallel to the plane of incidence. Comparing Eqs. 13.11 and 13.12, and assuming that the function varies sufficiently slowly so that the terms involving k' and k'' in Eq. 13.12 can be neglected, we obtain an equation of the form (14.1), in which $\psi = A_z/k(z)$. Furthermore, according to Eq. 13.9, we have $H_y = -(\partial A_z/\partial x)$. As will be shown below, differentiation with respect to x corresponds to multiplication by a constant. Therefore, in this case we can take $\psi \equiv H_y/k(z)$.

(c) The acoustic case. It is clear from Eq. 13.19 that in this case we must set $\psi \equiv p/\sqrt{\rho}$, where p is the acoustic pressure. We again assume that the parameters vary sufficiently slowly so that we can neglect terms in Eq. 13.20 which depend on spatial derivatives of the density ρ.

Thus, in the three cases considered above, we have, respectively,

$$\text{(a)} \quad \psi \equiv E_y, \quad \text{(b)} \quad \psi \equiv \frac{H_y}{k(z)}, \quad \text{(c)} \quad \psi \equiv \frac{p}{\sqrt{\rho}}. \tag{14.2}$$

By expanding $\nabla^2 \psi$, Eq. 14.1 is written

$$\frac{\partial^2 \psi}{\partial x^2} + \frac{\partial^2 \psi}{\partial z^2} + k^2(z)\psi = 0. \tag{14.3}$$

We seek a solution of this equation in the form

$$\psi(x, z) = X(x) . Z(z). \qquad (14.3a)$$

Then Eq. 14.3 becomes

$$-\frac{1}{X} . \frac{\partial^2 X}{\partial x^2} = \frac{1}{Z} \frac{\partial^2 Z}{\partial z^2} + k^2.$$

Since the left hand side of the above equation cannot depend on z, and the right hand side cannot depend on x, the equality can hold only if both sides are constant. We denote this constant by a^2. Then we obtain the following equations for $Z(z)$ and $X(x)$:

$$\frac{\partial^2 Z}{\partial z^2} + [k^2(z) - a^2]Z = 0, \qquad (14.4)$$

$$\frac{\partial^2 X}{\partial x^2} + a^2 X = 0. \qquad (14.5)$$

Equation 14.5 is satisfied by

$$X = A\,e^{iax} + B\,e^{-iax}. \qquad (14.6)$$

The constant a has a simple physical meaning. Far from the layer, where the medium is homogeneous, the incident plane wave can be written in the form

$$\psi(x, z) = C \exp\left[ik_0(x \sin \vartheta_0 + z \sin \vartheta_0)\right], \qquad (14.7)$$

where k_0 is the wave number in the homogeneous medium through which the incident wave travels, and ϑ_0 is the angle of incidence. The reflected wave will be

$$CV \exp\left[ik_0(x \sin \vartheta_0 - z \cos \vartheta_0)\right], \qquad (14.8)$$

where V is the reflection coefficient. Thus, the x-dependence of the field outside the layer will be described by the function $\exp(ik_0 \sin \vartheta_0 z)$. We could have chosen a wave propagating in the opposite direction with respect to x, in which case we would have the exponent $\exp(-ik_0 \sin \vartheta_0 z)$. However, this will have no influence on the results.

Comparing Eqs. 14.7 and 14.8 with 14.6, we are led to the following conclusions:

(1) we must set $B = 0$;

(2) remembering that a is a constant, and taking account of its value outside the layer, we must set

$$a = k_0 \sin \vartheta_0. \qquad (14.9)$$

Substituting the solution (14.6) into Eq. 14.3a, and letting the constant A be part of the function $Z(z)$, which for the present is still unknown, we obtain

$$\psi(x, z) = Z(z)\, e^{iax}, \tag{14.10}$$

where $Z(z)$ is found from Eq. 14.4, and a is given by Eq. 14.9.

Below, we shall examine the case in which the form of the function $k(z)$ is such that Eq. 14.4 can be transformed into a hypergeometric equation.

2. *The hypergeometric equation*

The hypergeometric equation for the function $F(\xi)$ is written in the form

$$\frac{d^2 F}{d\xi^2} - \frac{(\alpha + \beta + 1)\,\xi - \gamma}{\xi(1 - \xi)} \cdot \frac{dF}{d\xi} - \frac{\alpha\beta}{\xi(1 - \xi)}\, F = 0, \tag{14.11}$$

where α, β and γ are parameters.

By simple, although somewhat tedious, manipulations, it can be shown that, by means of the replacements

$$F = r(z)\, Z, \quad \xi = P(z), \tag{14.12}$$

where $P(z)$ is an arbitrary function, and

$$r(z) = r_0\, \xi^{-\gamma/2}(1 - \xi)^{\frac{\gamma - \alpha - \beta - 1}{2}} \left(\frac{d\xi}{dz}\right)^{\frac{1}{2}}, \tag{14.13}$$

Eq. 14.11 is reduced to the wave equation

$$\frac{d^2 Z}{dz^2} + g(z)\, Z = 0. \tag{14.14}$$

In Eq. 14.13, r_0 is a constant.

The function $g(z)$ is given by the expression

$$g(z) = \left\{\frac{1}{2}\frac{d^2}{dz^2}\left(\ln \frac{dP}{dz}\right) - \frac{1}{4}\left[\frac{d}{dz}\left(\ln \frac{dP}{dz}\right)\right]^2\right\}$$
$$- \left(\frac{d}{dz}\ln P\right)^2 \left\{K_1 + K_2 \frac{P}{1 - P} + K_3 \frac{P}{(1 - P)^2}\right\}, \tag{14.15}$$

where

$$4K_1 = \gamma(\gamma - 2),$$

$$4K_2 = 1 - (\alpha - \beta)^2 + \gamma(\gamma - 2),$$

$$4K_3 = (\alpha + \beta - \gamma)^2 - 1. \tag{14.16}$$

By choosing various forms of the function $P(z)$, we obtain various forms of $g(z)$. Comparing Eqs. 14.14 and 14.4, we see that, having chosen a definite form for the function $g(z)$, we can find the solution of our problem for a layer characterized by the function

$$k^2(z) = g(z) + a^2. \tag{14.17}$$

In particular, at normal incidence $(a = 0)$, the square of the wave number $k^2(z)$ will coincide with the function $g(z)$.

To continue the analysis further, we shall prescribe the most simple form for the function $P(z)$, namely, we set

$$P(z) = -e^{mz}. \tag{14.18}$$

Then, using Eqs. 14.15 and 14.17, the square of the wave number in the layer becomes

$$k^2(z) = a^2 - m^2\left(K_1 + \frac{1}{4}\right) + m^2 K_2 \frac{e^{mz}}{1+e^{mz}} + m^2 K_3 \frac{e^{mz}}{(1+e^{mz})^2}.$$

It is convenient to express the constants K_1, K_2 and K_3 in terms of k_0 and two new constants M and N as follows:

$$K_1 = (a^2 - k_0^2)\frac{1}{m^2} - \frac{1}{4}, \quad K_2 = -\frac{k_0^2}{m^2}N, \quad K_3 = -\frac{4k_0^2}{m^2}M. \tag{14.19}$$

The physical meaning of the constants M and N will be explained below. Now we have

$$\frac{k^2(z)}{k_0^2} = 1 - N\frac{e^{mz}}{1+e^{mz}} - 4M\frac{e^{mz}}{(1+e^{mz})^2}. \tag{14.20}$$

The quantity $n^2 = k^2(z)/k_0^2$, i.e. the square of the index of refraction of the medium as a function of z, is shown graphically in Fig. 75 for two cases: $M = 0$, $N \neq 0$ and $M \neq 0$, $N = 0$. We see that in the first case, Eq. 14.20 describes a "transition" layer in which the index of refraction varies smoothly from the value $n^2 = 1$ at large negative values of z to the value $n^2 = 1 - N$ at large positive values of z. In the second case, we have a "symmetrical" layer in which the index of refraction is an even function of z. It is equal to unity at large distances from the mean plane $z = 0$, and differs most greatly from unity at $z = 0$, where it is $n^2 = 1 - M$. Rather than plot n^2 along the abscissa in Fig. 75, it is convenient to use $(1 - n^2)/N$ for the transitional layer, and $(1 - n^2)/M$ for the symmetrical layer.

The ordinate represents the combination $z/S\lambda_0 = mz/4\pi$, where

$$S \equiv \frac{2k_0}{m} \tag{14.21}$$

is usually called the relative thickness of the layer. Simple calculations show that in the case of the symmetrical layer, the thickness of the layer, defined as the distance along the z-axis between the points on both sides of the middle of the layer at which $[1 - n^2(z)]/M$ is equal to one half of its maximum value (at $z = 0$), is

$$l = 0.28\lambda_0 S. \tag{14.22}$$

It now remains for us to connect the parameters S, M and N which characterize the layer, with the parameters α, β and γ of the hypergeometric equation.

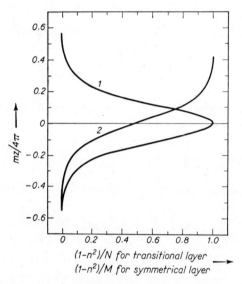

Fig. 75. The forms of the symmetrical and the transitional layers. 1. Curve for the symmetrical layer, $N = 0$. 2. Curve for the transitional layer, $M = 0$.

Solving the system of equations (14.16) for α, β and γ, we obtain

$$\alpha = \tfrac{1}{2}\{1 + \sqrt{(1 + 4K_1)} + \sqrt{(1 + 4K_3)} - \sqrt{[1 + 4(K_1 - K_2)]}\},$$
$$\beta = \tfrac{1}{2}\{1 + \sqrt{(1 + 4K_1)} + \sqrt{(1 + 4K_3)} + \sqrt{[1 + 4(K_1 - K_2)]}\},$$
$$\gamma = 1 + \sqrt{(1 + 4K_1)}. \tag{14.23}$$

On the other hand, taking account of Eq. 14.19, the first of Eqs. 14.23 gives

$$\sqrt{(1 + 4K_1)} = \frac{2}{m}\sqrt{(a^2 - k_0^2)} = iS \cos\vartheta_0. \tag{14.24}$$

Similarly, from the other two of Eqs. 14.23, we obtain

$$\sqrt{[1 + 4(K_1 - K_2)]} = iS\sqrt{(\cos^2\vartheta_0 - N)},$$
$$\sqrt{(1 + 4K_3)} = \sqrt{(1 - 4S^2 M)} \equiv 2(d_2 + id_1), \tag{14.25}$$

where d_2 and d_1 denote the real and imaginary parts, respectively, of $\frac{1}{2}\sqrt{(1-4S^2\,M)}$.

We now obtain

$$\alpha = \tfrac{1}{2}+d_2+i\,\frac{S}{2}\,[\cos\vartheta_0-\sqrt{(\cos^2\vartheta_0-N)}]+id_1,$$

$$\beta = \tfrac{1}{2}+d_2+i\,\frac{S}{2}\,[\cos\vartheta_0+\sqrt{(\cos^2\vartheta_0-N)}]+id_1,$$

$$\gamma = 1+iS\cos\vartheta_0. \tag{14.26}$$

We now go on to the analysis of the solutions of the hypergeometric equation (14.11), and to the determination of the reflection and transmission coefficients of the layer.

3. The reflection and transmission coefficients of the layer

It is shown in courses of mathematical analysis (see, for example, Ref. 22) that Eq. 14.11 can be satisfied by the following "hypergeometric series":

$$F_1 = F(\alpha,\beta,\gamma,\xi) = 1+\frac{\alpha\beta}{\gamma}\,\xi+\frac{\alpha(\alpha+1)\,\beta(\beta+1)}{1\cdot 2\gamma(\gamma+1)}\,\xi^2$$

$$+\frac{\alpha(\alpha+1)\,(\alpha+2)\,\beta(\beta+1)\,(\beta+2)}{1\cdot 2\cdot 3\gamma(\gamma+1)\,(\gamma+2)}\,\xi^3+\dots \tag{14.27}$$

This series converges for $|\xi|<1$. It can also be shown that a second, linearly independent, solution of the equation, convergent for $|\xi|<1$, is the series

$$F_2 = \xi^{1-\gamma}\,F(\alpha-\gamma+1,\beta-\gamma+1,2-\gamma,\xi). \tag{14.28}$$

Equation 14.11 has three singular points: $\xi = 0,\ 1,\ \infty$. There are accordingly three pairs of linearly independent solutions, and each pair of solutions converges in the neighborhood of "its own" singular point. Thus, the solutions which are convergent around the point $\xi = 0$ are the series

$$F_1 = F(\alpha,\beta,\gamma,\xi),\quad F_2 = \xi^{1-\gamma}\,F(\alpha-\gamma+1,\beta-\gamma+1,2-\gamma,\xi), \tag{14.29}$$

as already indicated above.

Around the point $\xi = 1$,

$$F_3 = F(\alpha,\beta,\alpha+\beta-\gamma+1,1-\xi),$$

$$F_4 = (1-\xi)^{\gamma-\alpha-\beta}\,F(\gamma-\alpha,\gamma-\beta,\gamma-\alpha-\beta+1,1-\xi). \tag{14.30}$$

Around the point $\xi = \infty$,

$$F_5 = \xi^{-\alpha} F\left(\alpha, \alpha-\gamma+1, \alpha-\beta+1, \frac{1}{\xi}\right),$$

$$F_6 = \xi^{-\beta} F\left(\beta, \beta-\gamma+1, \beta-\alpha+1, \frac{1}{\xi}\right). \tag{14.31}$$

Each of these expressions is an analytic function, being a solution of the hypergeometric equation over the entire region in which the series converge.

By the method of analytic continuation, each of these solutions can be continued beyond the region of convergence of the series corresponding to it. Then, in the new region, we shall have three solutions, one continued from another region, and two given by one of Eqs. 14.29–14.31. Since there can be only two linearly independent solutions in each region, there must be a linear relation with constant coefficients among the three solutions.

In what follows, we shall use only one relation of this kind. Thus, it turns out[22] that if the solution

$$F_5 = \xi^{-\alpha} F\left(\alpha, \alpha-\gamma+1, \alpha-\beta+1, \frac{1}{\xi}\right), \tag{14.32}$$

valid in the region $|\xi| > 1$, is analytically continued into the region $|\xi| < 1$, then in the latter region we obtain a new solution which can be expressed in terms of F_1 and F_2 through the linear combination

$$(-1)^{-\alpha} \frac{\Gamma(\alpha-\beta+1)\,\Gamma(1-\gamma)}{\Gamma(1-\beta)\,\Gamma(1+\alpha-\gamma)} F_1 + (-1)^{\gamma-1-\alpha} \frac{\Gamma(\alpha-\beta+1)\,\Gamma(\gamma-1)}{\Gamma(\gamma-\beta)\,\Gamma(\alpha)} F_2, \tag{14.33}$$

where Γ is the gamma-function.

This result has a simple physical meaning. In order to reveal this meaning, we use Eqs. 14.13 and 14.14 to go from the solution of the hypergeometric equation F to the solution of the wave equation Z. Since, in the present case,

$$\xi = -e^{mz}, \quad \frac{d\xi}{dz} = m\xi, \tag{14.34}$$

we obtain

$$Z = r_0^{-1} m^{-\frac{1}{2}} \xi^{\frac{1}{2}(\gamma-1)} (1-\xi)^{\frac{1}{2}(1+\alpha+\beta-\gamma)} F. \tag{14.35}$$

We shall now consider asymptotic expressions for the solutions of the wave equation Z, corresponding to F_5, F_1 and F_2, for the cases:

(a) $z \to \infty$, which corresponds to $\xi \to -\infty$,

(b) $z \to -\infty$, which corresponds to $\xi \to 0$.

In case (a), we can use expression (14.31) for F_5. Moreover, when $|\xi| \to \infty$, only the first term, equal to unity, remains in the expression for the hypergeometric series F. As a result, taking account of the relation (14.34) between ξ and z we obtain

$$z \to \infty, \quad Z \to r_0^{-1} m^{-\frac{1}{2}} (-1)^{\frac{1}{2}(\gamma-1)-\alpha} \exp[\tfrac{1}{2}m(\beta-\alpha)z]. \qquad (14.36)$$

Similarly, in case (b), we can obtain asymptotic expressions for the solutions of the wave equation, corresponding to the solutions of the hypergeometric equation F_1 and F_2.

The asymptotic expression for the linear combination of F_1 and F_2, entering into Eq. 14.33, can be obtained immediately. Again, using Eqs. 14.35 and 14.34, and in place of F, substituting successively F_1 and F_2 which according to Eq. 14.29 approach 1 and $\xi^{1-\gamma}$ as $\xi \to 0$, we obtain

$$r_0^{-1} m^{-\frac{1}{2}} (-1)^{\frac{1}{2}(\gamma-1)-\alpha} \left\{ \frac{\Gamma(\alpha-\beta+1)\,\Gamma(1-\gamma)}{\Gamma(1-\beta)\,\Gamma(1+\alpha-\gamma)} \exp[\tfrac{1}{2}m(\gamma-1)z] \right.$$

$$\left. + \frac{\Gamma(\alpha-\beta+1)\,\Gamma(\gamma-1)}{\Gamma(\gamma-\beta)\,\Gamma(\alpha)} \exp[-\tfrac{1}{2}m(\gamma-1)z] \right\}. \qquad (14.37)$$

for the combination (14.33) as $z \to -\infty$. This expression, valid for $z \to -\infty$, is the analytic continuation of Eq. 14.36, valid for $z \to \infty$.

Let the parameters of the hypergeometric equation α, β and γ be chosen such that $(\beta-\alpha)/i$ and $(\gamma-1)/i$ are real and positive. Then Eq. 14.36 represents a plane nonattenuating wave, propagating in the direction $z \to \infty$. But Eq. 14.37 will represent a combination of two plane waves, one of which propagates in the direction $z = -\infty$ (the second term in the brackets), while the other propagates in the opposite direction.

Taking into account the connection indicated above between the solutions of the wave and hypergeometric equations, we conclude that the solutions of the hypergeometric equation F_1, F_2 and F_5 found above enable us to find the solution of the wave equation over the entire range of z from $-\infty$ to $+\infty$. Moreover, as $z \to \infty$, this solution degenerates into a plane wave (Eq. 14.36), and as $z \to -\infty$, the solution degenerates into a combination of two plane waves, propagating in opposite directions (Eq. 14.37). It follows from this that, sufficiently far from the layer, where the medium can be considered homogeneous, Eq. 14.37 can be treated as the superposition of the incident and reflected waves, and Eq. 14.36 can be treated as the transmitted wave. Comparing the coefficients of the exponential factors, we obtain

$$V = \frac{\Gamma(\gamma-1)\,\Gamma(1-\beta)\,\Gamma(1+\alpha-\gamma)}{\Gamma(1-\gamma)\,\Gamma(\gamma-\beta)\,\Gamma(\alpha)}, \qquad (14.38)$$

for the reflection coefficient, and

$$D = \frac{\Gamma(1-\beta)\,\Gamma(1+\alpha-\gamma)}{\Gamma(1-\gamma)\,\Gamma(\alpha-\beta+1)}, \tag{14.39}$$

for the transmission coefficient of the layer. Thus, we had to use only the limiting values of the hypergeometric series. By using these series with an arbitrary value of the variable, we could have calculated the field at any point in the layer. However, in what follows, we shall have no need for such calculations.

4. The reflection coefficient at a transitional layer

We shall now analyze the expression for the reflection coefficient (Eq. 14.38) in the case of a transitional layer, i.e. in the case $M = 0$, $N \neq 0$. With $M = 0$, Eqs. 14.25 and 14.26 give

$$d_1 = 0, \quad d_2 = \tfrac{1}{2}, \quad \alpha = 1 + (i/2)\,S[\cos\vartheta_0 - \sqrt{(\cos^2\vartheta_0 - N)}],$$

$$\beta = 1 + (i/2)\,S[\cos\vartheta_0 + \sqrt{(\cos^2\vartheta_0 - N)}], \quad \gamma = 1 + iS\cos\vartheta_0. \tag{14.40}$$

In this case, Eq. 14.38 takes the form

$$V = \frac{\Gamma(iS\cos\vartheta_0)\,\Gamma\{-i(S/2)\,[\cos\vartheta_0 + \sqrt{(\cos^2\vartheta_0 - N)}]\}}{\Gamma(-iS\cos\vartheta_0)\,\Gamma\{i(S/2)\,[\cos\vartheta_0 - \sqrt{(\cos^2\vartheta_0 - N)}]\}}$$

$$\times \frac{\Gamma\{1 - i(S/2)\,[\cos\vartheta_0 + \sqrt{(\cos^2\vartheta_0 - N)}]\}}{\Gamma\{1 + i(S/2)\,[\cos\vartheta_0 - \sqrt{(\cos^2\vartheta_0 - N)}]\}}. \tag{14.41}$$

A simpler expression is obtained for the modulus of the reflection coefficient $\rho \equiv |V| \equiv \sqrt{(VV^*)}$. We use the well known relation

$$\Gamma(z)\,\Gamma(1-z) = \frac{\pi}{\sin\pi z}. \tag{14.42}$$

It is necessary to distinguish the two cases

$$\cos^2\vartheta_0 < N \quad \text{and} \quad \cos^2\vartheta_0 > N.$$

In the first case we have $\sqrt{(\cos^2\vartheta_0 - N)} = i\sqrt{(N - \cos^2\vartheta_0)}$. Multiplying Eq. 14.41 by its complex conjugate, we obtain $\rho = 1$, i.e. we obtain total reflection. Proceeding in the same way, the modulus of the reflection coefficient in the second case is

$$\rho = \frac{\sinh\{(\pi S/2)\,[\cos\vartheta_0 - \sqrt{(\cos^2\vartheta_0 - N)}]\}}{\sinh\{(\pi S/2)\,[\cos\vartheta_0 + \sqrt{(\cos^2\vartheta_0 - N)}]\}}. \tag{14.43}$$

It was shown above (see Sec. 2) that $1 - N = n_\infty^2$, where n_∞ is the index of refraction of the medium far from the layer on the side opposite the incident wave (at $z = \infty$). Therefore, the condition $\cos^2\vartheta_0 < N$ can be written in the form

$$\sin\vartheta_0 > n_\infty. \tag{14.44}$$

When the gradient of the index of refraction in the layer is very small, the ray representations are applicable, and inequality (14.44) can be treated as a condition under which the incident ray turns around at some height in the layer and returns into the medium from which it came (see § 8).

If we let the thickness S of the transitional layer approach zero, Eq. 14.43 yields the Fresnel formula for the reflection coefficient of an electromagnetic wave polarized perpendicular to the plane of incidence (see § 2):

$$V = \frac{\cos\vartheta_0 - \sqrt{(n^2 - \sin^2\vartheta_0)}}{\cos\vartheta_0 + \sqrt{(n^2 - \sin^2\vartheta_0)}}, \qquad (14.45)$$

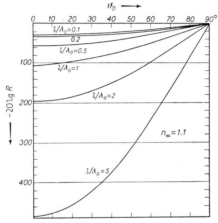

Fig. 76. The reflection coefficient at a transitional layer as a function of the angle of incidence, and for various thicknesses of the layer $n_\infty = 1.1$.

where the subscript ∞ on n is no longer needed. As $S \to 0$, we do not obtain the correct value for the reflection coefficient of an electromagnetic wave polarized in the plane of incidence, or for the reflection coefficient of a sound wave. The reason for this is as follows: in this case the gradient of the index of refraction approaches infinity, and in the reduction of the field equations to the wave equation, the "effective" wave number given by Eqs. 13.13 and 13.20 also approaches infinity, as a result of which the method of solution we have chosen becomes inapplicable. The case of small layer thickness will be examined in the next paragraph for a layer of more general form than that considered here; we shall therefore dwell no longer on this question.

The coefficient of reflection ρ from a transitional layer is shown in Figs. 76 and 77 as a function of the angle of incidence ϑ_0 for $n_\infty = 0.9$ and $n_\infty = 1.1$, and for various values of l/λ_0, where l is the effective thickness of the layer as given by Eq. 14.22. We recall that the wave is assumed to be incident from $z = -\infty$, where $n = 1$ and the wavelength is λ_0. The abscissas represent the angle of incidence, and the ordinates represent the reflection coefficient in decibels. We see that as the thickness of the layer is increased, the reflection coefficient falls sharply. Thus, at normal incidence, and with $n_\infty = 1.1$, as the thickness of the layer l is increased from $0.1\lambda_0$ to $5\lambda_0$, the amplitude of the reflected wave decreases from -27 db ($\rho = 4 \times 10^{-2}$) to -486 db ($\rho = 5 \times 10^{-25}$). The greater the thickness of the layer, the more sharply does the reflection

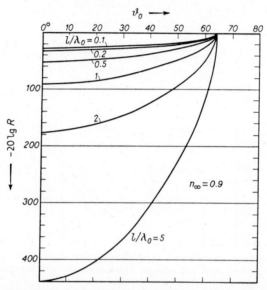

Fig. 77. The reflection coefficient at a transitional layer as a function of the angle of incidence, for various thicknesses of the layer $n_\infty = 0.9$.

coefficient depend on the angle of incidence. At $\vartheta_0 \to 90°$, we have $\rho \to 1$ ($\log \rho \to 0$). When $n_\infty = 0.9$, total internal reflection occurs for $\vartheta_0 > 64° \, 10'$.

In many practical cases, the results obtained above lead to a correct qualitative estimate of the dependence of the reflection coefficient on the layer thickness, the angle of incidence, the index of refraction on one or the other side of the layer, etc.

However, it must be kept in mind that for other forms of the function $n(z)$ in the layer, the reflection coefficient may have some irregularities. It is therefore of interest to compare the reflection coefficient ρ for two forms of the transitional layer, at normal incidence.

Using the relation $N = 1 - n_\infty^2$, Eq. 14.43 will have the form (at normal incidence, $\vartheta_0 = 0$)

$$\rho = \frac{\sinh\left[(\pi S/2)(1 - n_\infty)\right]}{\sinh\left[(\pi S/2)(1 + n_\infty)\right]}. \tag{14.46}$$

We shall investigate the reflection from a layer of thickness L, in which the index of refraction, for $-L/2 \leqslant z \leqslant L/2$, varies according to the law

$$n = \frac{A}{A + (z/L + 1/2)}, \tag{14.47}$$

where

$$A = \frac{n_1}{1 - n_1},$$

and n_1 is the index of refraction for $z \geqslant L/2$. The index of refraction on the incident side of the layer ($z \leqslant 0$) will, as previously, be taken equal to unity.

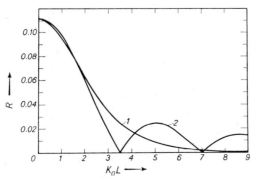

Fig. 78. Comparison of the reflection coefficients as functions of the layer thickness for two forms of transitional layers. The indices of refraction $n(z)$ for the two layers are shown in Fig. 79.

The coefficient of reflection at a layer of this type can be calculated comparatively easily for the case of normal incidence. The result is (Refs. 152 and 15, Vol. I, § 148b)

for $k_0 LA < 1/2$:

$$\rho = \frac{\sinh(\mu \ln n_1)}{\sqrt{[\sinh^2(\mu \ln n_1) + 4\mu^2]}}, \quad \mu = \sqrt{[\tfrac{1}{4} - (k_0 LA)^2]}, \tag{14.48}$$

for $k_0 LA > 1/2$

$$\rho = \frac{\sin(m \ln n_1)}{\sqrt{[\sin^2(m \ln n_1) + 4m^2]}}, \quad m = \sqrt{[(k_0 LA)^2 - \tfrac{1}{4}]};$$

where $k_0 = 2\pi/\lambda_0$, and λ_0 is the wavelength in the medium through which the incident wave travels.

The reflection coefficients given by Eq. 14.46 (curve 1) and Eqs. 14.48 (curve 2) are shown in Fig. 78. In both cases, the index of refraction of the medium on the opposite side of the layer with respect to the incident wave was taken as 0.8 (n_1 in Eq. 14.48 and n_∞ in Eq. 14.46). The relation between the thicknesses of the layers was chosen such that $S = 2.09L/\lambda_0$, where S characterizes the thickness of the transitional layer we previously examined. The form of the function $n(z)$ is shown for both cases in Fig. 79.

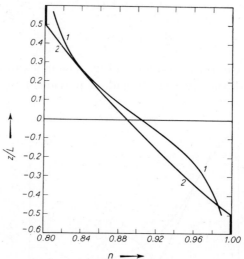

Fig. 79. Comparison of the functions $n(z)$ for two forms of layers.

We see that in the case of the smooth transitional layer (curve 1 in Fig. 79), the reflection coefficient decreases monotonically as the thickness of the layer increases. But in the case of a transitional layer that is bounded in thickness (curve 2 in Fig. 79), the reflection coefficient becomes zero at points satisfying the equation $m \ln n_1 = \nu\pi$, where ν is an integer (see Eq. 14.48), and goes through a maximum approximately half-way between such points.

At very great layer thicknesses, Eqs. 14.46 and 14.48 give

$$\frac{\pi S}{2} \gg 1, \quad \rho \approx \exp\left(-\pi S n_\infty\right), \tag{14.46a}$$

and

$$k_0 LA \gg 1, \quad \rho \approx \frac{1}{2m}\sin\left(m \ln n_1\right), \tag{14.48a}$$

respectively. Thus, in one case, the reflection coefficient decreases exponentially with increasing layer thickness, and in the other case, it is inversely proportional to the thickness of the layer.

Finally, we note that we have said nothing about the transmission coefficient of the layer D. However, there is no need for such a discussion since, in the majority of cases, only $|D|^2$ is of interest, and can be obtained from the law of conservation of energy after ρ has been determined.

5. The reflection coefficient at a symmetrical layer

We now analyze Eq. 14.38 for the reflection coefficient in the case of a symmetrical layer, given by Eq. 14.20 with $N = 0$, $M \neq 0$. The functional form of the index of refraction $n(z)$ is shown in Fig. 75.

Setting $N = 0$ in Eqs. 14.26, substituting the resulting values of α, β and γ into Eq. 14.38, and using Eq. 14.42, we obtain the following expression for the reflection coefficient:

$$V = \frac{\Gamma(iS \cos \vartheta_0)}{\Gamma(-iS \cos \vartheta_0)} \, \Gamma(\tfrac{1}{2} - d_2 - id_1 - iS \cos \vartheta_0)$$

$$\times \Gamma(\tfrac{1}{2} + d_2 + id_1 - iS \cos \vartheta_0) \frac{1}{\pi} \cos \pi (d_2 + id_1). \qquad (14.49)$$

We consider two cases:

(a) $\qquad\qquad\qquad\qquad 4S^2 M > 1.$

In this case, according to Eq. 14.25, we have $d_2 = 0$, $d_1 = \tfrac{1}{2}\sqrt{(4S^2 M - 1)}$. Using Eq. 14.42 once again, we obtain

$$\rho^2 = VV^* = \frac{\cosh^2 \pi d_1}{\cosh \pi (d_1 + S \cos \vartheta_0) \cosh \pi (d_1 - S \cos \vartheta_0)} \qquad (14.50)$$

for the square of the modulus of the reflection coefficient.

(b) $\qquad\qquad\qquad\qquad 4S^2 M < 1.$

In this case, $d_1 = 0, d_2 = \tfrac{1}{2}\sqrt{(1 - 4S^2 M)}$, and proceeding as previously, Eq. 14.49 gives

$$\rho^2 = VV^* = \frac{\cos^2 \pi d_2}{\cos \pi (d_2 + iS \cos \vartheta_0) \cos \pi (d_2 - iS \cos \vartheta_0)}, \qquad (14.51)$$

or $\qquad \rho^2 = \dfrac{\cos^2 \pi d_2}{\cos^2 \pi d_2 \cosh^2 (\pi S \cos \vartheta_0) + \sin^2 \pi d_2 \sinh^2 (\pi S \cos \vartheta_0)}.$

In these formulas, M is expressed in terms of the index of refraction n_1 at the center of the layer, through the formula

$$M = 1 - n_1^2.$$

The index of refraction is assumed to be unity sufficiently far from the layer, on both sides, and is most different from unity at the center of the layer.

As $M \to 0$, the layer vanishes. Under these conditions, we have $d_2 \to \frac{1}{2}$, and according to Eq. 14.51, $\rho \to 0$ as could be expected.

Let us note that if the propagation velocity in the layer is less than in the media on both sides of it, i.e. $n_1 > 1$, then M will be negative, and the reflection coefficient will be given by Eq. 14.51 for any layer thickness S. It is not hard to see that as the thickness of the layer approaches ∞, the formula gives $\rho \to 0$, i.e. reflection vanishes. This is a perfectly natural result, since the gradient of the index of refraction approaches zero as the layer becomes infinitely extended. Under these conditions, the geometrical optics approximation becomes applicable (see the following Paragraph).

Matters become somewhat more complicated in the limit $S \to 0$ when $n_1 < 1$, i.e. when the velocity in the layer is greater than in the adjoining homogeneous media. In this case, $M > 0$, and when the layer is sufficiently thick, we must use Eq. 14.50. We distinguish two cases:

$$(1) \ d_1 > S \cos \vartheta_0 \quad \text{and} \quad (2) \ d_1 < S \cos \vartheta_0. \tag{14.52}$$

In the first case, as $S \to \infty$, we have

$$\cosh \pi d_1 \to \tfrac{1}{2} \exp (\pi d_1),$$
$$\cosh \pi(d_1 + S \cos \vartheta_0) \to \tfrac{1}{2} \exp [\pi(d_1 + S \cos \vartheta_0)],$$
$$\cosh \pi(d_1 - S \cos \vartheta_0) \to \tfrac{1}{2} \exp [\pi(d_1 - S \cos \vartheta_0)]. \tag{14.53}$$

As a result, Eq. 14.50 gives

$$\rho \to 1,$$

i.e. we have total reflection.

In the second case, the first two of Eqs. 14.53 remain unchanged, while the third is written

$$\cosh \pi(d_1 - S \cos \vartheta_0) \to \tfrac{1}{2} \exp [\pi(S \cos \vartheta_0 - d_1)].$$

As a result, as $S \to 0$, Eq. 14.50 gives

$$\rho \to 0.$$

Remembering that

$$d_1 = \tfrac{1}{2} \sqrt{(4S^2 M - 1)} \to S M^{\frac{1}{2}} = S \sqrt{(1 - n_1^2)},$$

Eq. 14.52 is written

$$(1) \ \sin \vartheta_0 > n_1, \quad (2) \ \sin \vartheta_0 < n_1.$$

The first of these conditions indicates that, as a result of refraction, the ray corresponding to the incident plane wave is reversed in the layer, and returns into the medium from which it came. The second condition corresponds to the case in which the ray, although deflected in the layer, still leaves the layer at the other side.

In the first case, wave transmission through the layer occurs, similar to the tunnel effect in quantum mechanics in which particles pass through

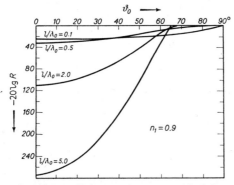

Fig. 80. The reflection coefficient at a symmetrical layer as a function of the angle of incidence, for various thicknesses of the layer $n_1 = 0.9$.

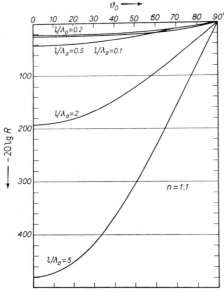

Fig. 81. The reflection coefficient at a symmetrical layer as a function of the angle of incidence, for various thicknesses of the layer $n_1 = 1.1$.

potential barriers. As the thickness of the layer is increased, this kind of transmission occurs less readily which explains the result

$$\rho \to 1 \quad \text{as} \quad S \to \infty.$$

This concludes our general analysis of Eqs. 14.50 and 14.51. These expressions are sufficiently simple to allow the reader to obtain the numerical value of the reflection coefficient rapidly under any conditions.

In Figs. 80 and 81, R is shown as a function of the angle of incidence ϑ_0, for various layer thicknesses l, and for $n_1 = 0.9$ and $n_1 = 1.1$†.

Equations 14.50 and 14.51 are of great value in the study of the reflection of radiowaves from ionospheric layers. Although the true form of the dielectric constant $\epsilon = k^2(z)/k_0^2$ in the ionosphere may be essentially different from the form of Eq. 14.20, the results we obtain give a qualitative picture of the phenomenon which is useful in understanding the rôle of the ionosphere.

When an electromagnetic wave is incident normally on an ionospheric layer, the square of the index of refraction at the center of the layer n_1^2 is[1]

$$n_1^2 = 1 - \frac{f_k^2}{f^2}, \tag{14.54}$$

where f is the frequency, f_k is the so-called critical frequency of the layer, given in terms of the maximum concentration of electrons in the layer N_m by the formula

$$f_k^2 = \frac{N_m e^2}{\pi m}, \tag{14.55}$$

and m and e are the mass and charge of the electron. We are neglecting absorption in the ionosphere.

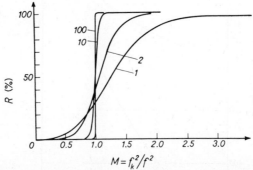

Fig. 82. The energy reflection coefficient $R = \rho^2$ at a simple ionospheric layer as a function of frequency, for various thicknesses of the layer.

We thus have

$$M = 1 - n_1^2 = \frac{f_k^2}{f^2}. \tag{14.56}$$

The reflection coefficient at normal incidence as a function of f_k^2/f^2 is shown in Fig. 82 for various values of S.

We see that when $S \geqslant 100$, which, according to Eq. 14.22 corresponds to layer thicknesses $l \geqslant 28\lambda_0$, the wave is almost totally reflected from

---

† For given values of n_1 and ϑ, there is a value of the ratio l/λ at which the reflection coefficient is a maximum. This can be shown by using Eq. 14.51.

the layer if its frequency is less than the critical frequency, and is almost completely transmitted through the layer if its frequency is greater than critical. As the layer thickness is decreased, the width of the frequency range corresponding to the transition zone between total reflection and total transmission increases.

§ 15. Wave Reflection from an Inhomogeneous Halfspace

1. *General relations*

We consider the case in which a homogeneous halfspace $z < 0$ is in contact with an inhomogeneous halfspace $z > 0$. The index of refraction n in the homogeneous halfspace will be taken as unity. In the inhomogeneous halfspace, the index of refraction will be assumed to vary in a definite way with z. In particular, we assume that the square of the index of refraction (and consequently, the dielectric constant in the electromagnetic case) is a linear function of z.

Thus,

$$z \leqslant 0, \quad n = 1,$$

$$z \geqslant 0, \quad n^2 = 1 \pm az. \tag{15.1}$$

We will assume that a is always positive. The plus and minus signs in front of az correspond, respectively, to an index of refraction which increases or decreases with distance from the boundary.

Let a plane wave be incident on the boundary $z = 0$ from the homogeneous medium $z < 0$. The total field in this halfspace is the sum of the incident and reflected waves, and is written (see Eqs. 14.7 and 14.8)

$$\psi(x, z) = Z(z) \exp (ik_0 \sin \vartheta_0 x), \tag{15.2}$$

where

$$Z(z) = [\exp (ik_0 z \cos \vartheta_0) + V(\vartheta_0) \exp (-ik_0 z \cos \vartheta_0)], \tag{15.3}$$

with ϑ_0 the angle of incidence, and $V(\vartheta_0)$ the reflection coefficient. Our problem is to find $V(\vartheta_0)$. The amplitude of the incident wave is assumed to be unity.

The problem we are now considering is of great importance in the study of radiowave propagation in the ionosphere and troposphere, and is also important in the theory of sound wave propagation in the atmosphere and in the sea.

We seek an expression for the field in the inhomogeneous medium $z > 0$ of the form (15.2), where the function $Z(z)$ will satisfy the equation (see § 14)

$$\frac{d^2 Z}{dz^2} + k_0^2 [n^2(z) - \sin^2 \vartheta_0] Z = 0. \tag{15.4}$$

7

Taking Eq. 15.1 into account, and in place of z introducing the new variable

$$\zeta = \cos^2 \vartheta_0 \pm az,\qquad\qquad (15.5)$$

the last equation can be written in the form

$$\frac{d^2 Z}{d\zeta^2} + \left(\frac{k_0}{a}\right)^2 \zeta Z = 0.\qquad\qquad (15.6)$$

The solutions of this equation are cylindrical functions of the 1/3 order (see Ref. 3, p. 109). It will be convenient to treat the different signs in Eq. 15.5 separately.

2. The case of a decreasing index of refraction

Here we have $\zeta = \cos^2 \vartheta_0 - az$, so that, at $z = z_m = \cos^2 \vartheta_0 / a$, the variable ζ becomes zero. Under these conditions, the solution of Eq. 15.6 can be written in the general form

$$z < z_m, \quad Z(z) = w^{\frac{1}{3}}[AH^{(1)}_{\frac{1}{3}}(w) + BH^{(2)}_{\frac{1}{3}}(w)], \qquad w \equiv \tfrac{2}{3}(k_0/a)\,\zeta^{\frac{3}{2}},$$

$$z > z_m, \quad Z(z) = w_1^{\frac{1}{3}}[CH^{(1)}_{\frac{1}{3}}(iw_1) + DH^{(2)}_{\frac{1}{3}}(iw_1)], \quad w_1 \equiv \tfrac{2}{3}\cdot(k_0/a)\,(-\zeta)^{\frac{3}{2}}.$$

$$(15.7)$$

We note that as $z \to \infty$,

$$w_1 \to \infty, \quad |H^{(2)}_{\frac{1}{3}}(iw_1)| \sim \sqrt{\left(\frac{2}{\pi w_1}\right)}\, e^{w_1} \to \infty.$$

From physical considerations we must require that the field remain finite everywhere, and in particular, as $z \to \infty$. Therefore, the constant D must be set equal to zero.

Furthermore, the function $Z(z)$ and its derivative dZ/dz must be continuous at $z = 0$ as well as at $z = z_m$. This gives four equations for the determination of the three remaining constants A, B and C, and the reflection coefficient V.

When z is close to z_m, the arguments of the Hankel functions in Eq. 15.7 are small, and we can expand the cylindrical functions in powers of their arguments. These calculations are performed in the next Paragraph. In particular, using Eqs. 16.53 and 16.54, it can be shown that the continuity conditions on Z and dZ/dz at $z = z_m$ lead to the two relations

$$A = B e^{i(\pi/3)}, \quad C = i\, e^{i(\pi/3)}B.\qquad\qquad (15.8)$$

We now return to the continuity conditions at $z = 0$. Using Eq. 15.3, the first of Eqs. 15.7, and Eq. 15.8, these conditions can be written:

1. Continuity of $Z(z)$:

$$1 + V = B\{w^{\frac{1}{3}}[e^{i(\pi/3)}H^{(1)}_{\frac{1}{3}}(w) + H^{(2)}_{\frac{1}{3}}(w)]\}_{z=0}.\qquad\qquad (15.9)$$

2. Continuity of dZ/dz:

$$ik_0 \cos \vartheta_0 (1 - V) = B \left\{ \frac{d}{dz} [w^{\frac{1}{2}}(e^{i(\pi/3)} H_{\frac{1}{3}}^{(1)}(w) + H_{\frac{1}{3}}^{(2)}(w))] \right\}_{z=0}.$$

In differentiating with respect to z, we must remember that, according to the definition of w, we obtain

$$\left(\frac{dw}{dz}\right)_{z=0} = \frac{k_0}{a}\left(\zeta^{\frac{1}{2}}\frac{d\zeta}{dz}\right)_{z=0} = -k_0 \cos \vartheta_0. \tag{15.10}$$

Moreover (see Ref. 16, p. 359),

$$\left\{\frac{d}{dw}[w^{\frac{1}{2}}H_{\frac{1}{3}}^{(1)}(w)]\right\}_{z=0} = [w^{\frac{1}{2}}H_{-\frac{2}{3}}^{(1)}(w)]_{z=0} = w_0^{\frac{1}{2}}H_{-\frac{2}{3}}^{(1)}(w_0), \tag{15.11}$$

where w_0 is the value of w at $z = 0$:

$$w_0 = \tfrac{2}{3}.(k_0/a) \cos^3 \vartheta_0. \tag{15.12}$$

Equation 15.11 is also valid for Hankel functions of the second kind. Dividing the second of Eqs. 15.9 by the first, we obtain

$$\frac{V-1}{V+1} = i\frac{e^{i(\pi/3)}H_{-\frac{2}{3}}^{(1)}(w_0) + H_{-\frac{2}{3}}^{(2)}(w_0)}{e^{i(\pi/3)}H_{\frac{1}{3}}^{(1)}(w_0) + H_{\frac{1}{3}}^{(2)}(w_0)}. \tag{15.13}$$

We shall find it convenient to express the Hankel functions in terms of Bessel functions using the following relations (see Ref. 3, § 3.61)

$$H_p^{(1)}(x) = \frac{i}{\sin \pi p}[e^{-p\pi i}J_p(x) - J_{-p}(x)],$$

$$H_p^{(2)}(x) = \frac{-i}{\sin \pi p}[e^{p\pi i}J_p(x) - J_{-p}(x)]. \tag{15.14}$$

Successively setting $p = \frac{1}{3}$ and $p = -\frac{2}{3}$, and using Eqs. 15.13 and 15.14, we obtain

$$V = \frac{J_{-\frac{2}{3}} - J_{\frac{2}{3}} + i(J_{\frac{1}{3}} + J_{-\frac{1}{3}})}{J_{-\frac{2}{3}} - J_{\frac{2}{3}} - i(J_{\frac{1}{3}} + J_{-\frac{1}{3}})}. \tag{15.15}$$

The argument in all of the Bessel functions appearing here is w_0, so that, for example, $J_{\frac{1}{3}}$ denotes $J_{\frac{1}{3}}(w_0)$, etc.

It is immediately evident from Eq. 15.15 that the modulus of the reflection coefficient is equal to unity, i.e. we have total reflection.

A more complete investigation of the reflection coefficient V was given in § 8, Section 7. It was shown there that when the condition $(k_0/a)^{\frac{1}{3}} \cos \vartheta_0 \gg 1$ is satisfied, the expression for V can be obtained more easily by the use of geometrical optics.

3. *The case of an increasing index of refraction*

We now consider the case in which the plus sign is chosen in Eqs. 15.1 and 15.5. In this case, the variable ζ does not vanish for any positive value of z, and the field in the halfspace $z > 0$ can be written in the form

$$z > 0, \quad Z(z) = A w^{\frac{1}{3}} H_{\frac{1}{3}}^{(1)}(w), \quad w = \frac{2k_0}{3a} \zeta^{\frac{3}{2}}. \tag{15.16}$$

We might expect another term, containing $H_{\frac{1}{3}}^{(2)}$, to enter. However, this term must be rejected in order that Z remain finite as $z \to \infty$.

As previously, the field in the homogeneous medium ($z < 0$) is the sum of the incident and reflected waves, and is given by Eq. 15.3.

The continuity conditions on Z and dZ/dz at the boundary $z = 0$ give

$$1 + V = A w_0^{\frac{1}{3}} H_{\frac{1}{3}}^{(1)}(w_0),$$

$$ik_0 \cos \vartheta_0 (1 - V) = A \left\{ \frac{\partial}{\partial w} [w^{\frac{1}{3}} H_{\frac{1}{3}}^{(1)}(w)] \frac{\partial w}{\partial z} \right\}_{z=0}, \tag{15.17}$$

where w_0 is given by Eq. 15.12.

Using Eqs. 15.10 and 15.11, Eq. 15.17 gives

$$\frac{1 - V}{1 + V} = -i \frac{H_{-\frac{2}{3}}^{(1)}(w_0)}{H_{\frac{1}{3}}^{(1)}(w_0)}, \tag{15.18}$$

whence
$$V = \frac{i H_{\frac{1}{3}}^{(1)}(w_0) - H_{-\frac{2}{3}}^{(1)}(w_0)}{i H_{\frac{1}{3}}^{(1)}(w_0) + H_{-\frac{2}{3}}^{(1)}(w_0)}. \tag{15.19}$$

It is convenient to consider two extreme cases:

(a) $w_0 \equiv \frac{2}{3}(k_0/a) \cos^3 \vartheta_0 \gg 1$, which corresponds to the case of a slowly varying index of refraction (a/k_0 is small), and angles ϑ_0 not too close to $\pi/2$.

(b) $w_0 \ll 1$, which is realized at angles ϑ_0 close to $\pi/2$ (grazing incidence).

In the first case it is convenient to use asymptotic representations of the Hankel functions for large values of the arguments

$$H_{\frac{1}{3}}^{(1)}(w_0) \approx \sqrt{\left(\frac{2}{\pi w_0}\right)} \exp\left[i(w_0 - \tfrac{5}{12}\pi)\right]\left(1 + \frac{5}{72 i w_0} + \ldots\right),$$

$$H_{-\frac{2}{3}}^{(1)}(w_0) \approx \sqrt{\left(\frac{2}{\pi w_0}\right)} \exp\left[i(w_0 + \tfrac{1}{12}\pi)\right]\left(1 - \frac{7}{72 i w_0} + \ldots\right). \tag{15.20}$$

Substituting these expressions into Eq. 15.18, and neglecting quantities of the second degree and higher in $1/w_0$, we obtain

$$\frac{1 - V}{1 + V} = 1 + \frac{i}{6 w_0},$$

whence, to within the same approximation,

$$V = -\frac{i}{12w_0} = -\frac{ia}{8k_0 \cos^3 \vartheta_0}. \tag{15.21}$$

If we substitute $k_0 = 2\pi/\lambda_0$, $a = 2(dn/dz)_{z=0}$, and compute the modulus of the reflection coefficient, we obtain

$$|V| = \frac{\lambda_0}{8\pi \cos^3 \vartheta_0} \left(\frac{dn}{dz}\right)_{z=0}. \tag{15.22}$$

Thus, the reflection coefficient increases with increasing wavelength and angle of incidence.

In case (b) ($w_0 \ll 1$), it is convenient to express the Hankel functions in terms of Bessel functions according to Eq. 15.14, and then to use the expansion of the Bessel function in powers of its argument

$$J_p(w) = \sum_{m=0}^{\infty} \frac{(-1)^m (w/2)^{p+2m}}{m! \, \Gamma(p+m+1)}. \tag{15.23}$$

As a result of some straightforward calculations, which we do not reproduce here, we obtain

$$V = -\left\{1 - 2\gamma\left(\frac{k_0}{3a}\right)^{\frac{1}{3}} \cos \vartheta_0 [\sqrt(3) - i]\right\}, \quad \text{where} \quad \gamma \equiv \frac{\Gamma(\frac{1}{3})}{\Gamma(\frac{2}{3})} = 1.978; \tag{15.24}$$

and as $\vartheta_0 \to \pi/2$, we have $V \to -1$.

§ 16. WAVES IN AN ARBITRARY LAYERED MEDIUM

1. *The field in a layered-inhomogeneous medium in the geometrical optics approximation*

As previously, we let the properties of the medium be functions of z only. We also assume for the time being that the properties of the medium change only slightly over a distance of the order of a wavelength. In this case, wave propagation will be described by Eq. 14.1, where, in general, the wave number $k(z)$ will be a complex function (its imaginary part will correspond to absorption in the medium). The value of the function $\psi(x, z)$ in Eq. 14.1 is given by Eqs. 14.2.

As was shown above, the solution of Eq. 14.1 can be represented in the form (14.10), where the function $Z(z)$ satisfies Eq. 14.4, and a is a constant whose physical meaning was discussed.

For simplicity, we restrict ourselves for the present to normal incidence, in which case we have the equation

$$\frac{d^2\psi}{dz^2} + k^2(z)\,\psi = 0. \tag{16.1}$$

We recall that to make the transition to oblique incidence, we must replace $k^2(z)$ by $k^2(z) - a^2$, as is evident from a comparison of Eqs. 16.1 and 14.4. Furthermore, the solution must contain an additional x-dependent factor e^{iax} or e^{-iax} depending on the propagation direction.

Were k constant, the solutions of Eq. 16.1 would be plane waves e^{ikz} and e^{-ikz} (the factor $e^{-i\omega t}$ is omitted everywhere), propagating in the positive and negative z-directions, respectively. When $k(z)$ is variable, but changes sufficiently slowly, Eq. 16.1 can be satisfied by the solution

$$\psi(z) = \Psi(z)\, e^{i\phi(z)}. \qquad (16.2)$$

In the case of a homogeneous medium, Ψ and $d\phi/dz$ are constants (ϕ is proportional to z). In the present case of slowly varying $k(z)$, these functions will also be slowly varying.

Substituting Eq. 16.2 into Eq. 16.1, we obtain

$$\Psi'' + 2i\phi'\,\Psi' + i\phi''\,\Psi + [k^2(z) - (\phi')^2]\,\Psi = 0, \qquad (16.3)$$

where the prime indicates differentiation with respect to z.

We will let L denote the distance over which the functions Ψ and ϕ' vary significantly. The assumption that the variation is sufficiently slow corresponds to the condition $L \gg \lambda$. For an estimate of the orders of magnitude of the derivatives, we have the relations $\Psi' \sim \Psi/L$, $\Psi'' \sim \Psi/L^2$, $\phi'' \sim \phi'/L$. Using these relations, and also $k = 2\pi/\lambda$, we obtain the following estimates for the various terms in Eq. 16.3:

$$\Psi'' \sim \frac{\Psi}{L^2}, \qquad (\alpha)$$

$$2i\phi'\,\Psi' + i\phi''\,\Psi \sim \frac{\phi'\,\Psi}{L}, \qquad (\beta)$$

$$[k^2(z) - (\phi')^2]\,\Psi \sim \frac{\Psi}{\lambda^2}. \qquad (\gamma)$$

Since term (α) is small in comparison with the remaining terms, it can be neglected. On the other hand, since terms (β) and (γ) have different orders of smallness, they must be equal to zero separately. As a result, we obtain

$$(\phi')^2 = k^2(z),$$

$$2\phi'\Psi' + \phi''\Psi = 0. \qquad (16.4)$$

Hence, we find, first of all,

$$\phi' = \pm\, k(z), \quad \phi = \pm \int_{z_0}^{z} k(z)\, dz, \qquad (16.5)$$

where z_0 is an arbitrary constant.

Substituting for ϕ', the second of Eqs. 16.4 becomes

$$\frac{\psi'}{\psi} = -\frac{k'}{2k}.$$

The solution of this equation is

$$\psi = \frac{C}{\sqrt{[k(z)]}},$$

where C is an arbitrary constant. As a result, taking account of both signs in Eq. 16.5, the complete solution of Eq. 16.2 is

$$\psi(z) = \frac{1}{\sqrt{[k(z)]}} \left\{ C_1 \exp\left[i \int_{z_0}^{z} k(z)\,dz \right] + C_2 \exp\left[-i \int_{z_0}^{z} k(z)\,dz \right] \right\}. \quad (16.6)$$

The first and second terms in the brackets correspond to waves propagating in the positive and negative z-directions, respectively. These waves propagate through the medium independently of one another, and to within the present approximation, there is no reflection.†

Equation 16.6 is called the geometrical optics approximation, and is also sometimes called the approximate solution of Wentzel-Kramers-Brillouin (abbreviated as WKB).

The meaning of solution (16.6) can be elucidated by simple physical considerations. Let us consider the first term in Eq. 16.6. The integral

$$\int_{z_0}^{z} k(z)\,dz$$

in the exponent is the phase change as the wave travels from an arbitrary point z_0 to the point of observation z. If instead of z_0, we choose any other point z_0', nothing changes except for an additional constant factor

$$\exp\left[i \int_{z_0'}^{z_0} k(z)\,dz \right],$$

in the solution which can be included in the arbitrary constant C_1.

The z-dependence of the wave amplitude in Eq. 16.6 is given by the factor $1/\sqrt{[k(z)]}$. This dependence can be explained on the basis of the law of conservation of energy. In fact, since there is no reflected wave, the magnitude of the time average of the energy flux (Umov-Poynting vector) must be the same for all z. For electromagnetic waves, the

---

† The above derivation of Eq. 16.6 cannot be considered completely rigorous since the neglect of even small terms in the differential equation can lead to significant differences in the solution. It can be shown, however (see below, § 16, Section 3), that a rigorous derivation yields the same result.

Umov-Poynting vector, as is well known, is given by the expression $(c/4\pi)\,\mathbf{E}\times\mathbf{H}$, and for acoustic waves the Umov vector is $p\mathbf{v}$, where p is the acoustic pressure and $\mathbf{v}$ is the particle velocity.

In accordance with Eq. 14.2, we will set

$$\psi = E_y \text{ in the electromagnetic case}$$

$$\psi = p/\sqrt{\rho} \text{ in acoustics.}$$

Then for waves propagating along the z-axis, remembering that

$$H_x \sim \frac{\partial E_y}{\partial z} \sim \frac{\partial \psi}{\partial z} \quad \text{and} \quad v_z \sim \frac{1}{\rho}\frac{\partial p}{\partial z} \sim \frac{1}{\sqrt{\rho}}\frac{\partial \psi}{\partial z},$$

the average energy flux in both the electromagnetic case and in acoustics is, to within a constant factor, equal to

$$\overline{\psi \frac{\partial \psi}{\partial z}}^t,$$

where the bar denotes the time average. As is well known, when ψ is written in complex form, the last expression can be written in the form

$$\psi^* \frac{d\psi}{dz}.$$

We shall now show that, in the geometrical optics approximation, this quantity is independent of z. Again limiting ourselves to the first term in Eq. 16.6 and differentiating with respect to z, we obtain

$$\frac{d\psi}{dz} = \frac{C_1}{\sqrt{k}}\left(ik - \frac{1}{2k}\frac{dk}{dz}\right)\exp\left(i\int_{z_0}^z k\,dz\right).$$

In the derivation of Eq. 16.6, we assumed that the properties of the medium changed only slightly over the distance of a single wavelength. Therefore, the second term is negligible compared with the first, and can be neglected. As a result, we obtain

$$\psi^* \frac{d\psi}{dz} = |C_1|^2,$$

i.e. the energy flux is in fact constant.

Generalizing the solution (16.6) to an arbitrary angle of incidence by the method indicated at the beginning of this Paragraph, we obtain

$$\psi(x,z) = \frac{1}{\sqrt{[k(z)\cos\vartheta(z)]}}\left\{C_1\exp\left[i\int_{z_0}^z k(z)\cos\vartheta(z)\,dz\right]\right.$$

$$\left. + C_2\exp\left[-i\int_{z_0}^z k(z)\cos\vartheta(z)\,dz\right]\right\}\exp(iax), \quad (16.7)$$

where
$$k(z)\cos\vartheta(z) = \sqrt{[k^2(z) - a^2]}. \qquad (16.8)$$

The quantity $a \equiv k_0 \sin\vartheta_0$ is the projection of the wave vector on the x-axis, and is independent of z.

Now, remembering the meaning of the function ψ, given by Eqs. 14.2, we obtain:

(a) when $\mathbf{E}$ is perpendicular to the plane of incidence:

$$E_y = \sqrt{\left(\frac{1}{k\cos\vartheta}\right)}\left[C_1\exp\left(i\int_{z_0}^z k\cos\vartheta\,dz\right)\right.$$
$$\left. + C_2\exp\left(-i\int_{z_0}^z k\cos\vartheta\,dz\right)\right]\exp(iax); \qquad (16.9)$$

(b) when $\mathbf{E}$ is parallel to the plane of incidence:

$$H_y = \sqrt{\left(\frac{k}{\cos\vartheta}\right)}\left[C_3\exp\left(i\int_{z_0}^z k\cos\vartheta\,dz\right)\right.$$
$$\left. + C_4\exp\left(-i\int_{z_0}^z k\cos\vartheta\,dz\right)\right]\exp(iax). \qquad (16.10)$$

In the latter case, the components of the electric field can be found from the equation curl $\mathbf{H} = -i(\omega/c)\epsilon'\,\mathbf{E}$, which, using the notation

$$k_0 = \frac{\omega}{c}, \quad \epsilon' = n^2, \quad k = k_0 n$$

gives
$$E_x = -\frac{ik_0}{k^2}\cdot\frac{\partial H_y}{\partial z}, \quad E_y = 0, \quad E_z = \frac{ik_0}{k^2}\cdot\frac{\partial H_y}{\partial x}.$$

In differentiating Eq. 16.10 with respect to z, the slowly varying factors $k(z)$ and $\cos\vartheta(z)$ can be considered constant. Then only the exponential will be subject to differentiation, and we obtain

$$E_x = k_0\sqrt{\left(\frac{\cos\vartheta}{k}\right)}\exp(iax)\left[C_3\exp\left(i\int_{z_0}^z k\cos\vartheta\,dz\right)\right.$$
$$\left. - C_4\exp\left(-i\int_{z_0}^z k\cos\vartheta\,dz\right)\right], \qquad (16.11)$$

$$E_z = -\frac{k_0\sin\vartheta}{\sqrt{(k\cos\vartheta)}}\exp(iax)\left[C_3\exp\left(i\int_{z_0}^z k\cos\vartheta\,dz\right)\right.$$
$$\left. + C_4\exp\left(-i\int_{z_0}^z k\cos\vartheta\,dz\right)\right], \qquad (16.11a)$$

where, in the expression for E_z, we have used the relation

$$k_0\sin\vartheta_0 = k(z)\sin\vartheta(z),$$

which is easily obtained from Eq. 16.8.

(c) in acoustics:

$$p = \sqrt{\left(\frac{\rho c}{\cos \vartheta}\right)} \left[C_5 \exp \left(i \int_{z_0}^{z} k \cos \vartheta \, dz \right) \right.$$
$$\left. + C_6 \exp \left(-i \int_{z_0}^{z} k \cos \vartheta \, dz \right) \right], \quad (16.12)$$

where $\rho = \rho(z)$ is the density, and $c = \omega/k(z)$ is the sound velocity.

In view of the great importance of solution (16.7), we shall demonstrate one more simple derivation of this equation. This derivation will apply directly to oblique incidence. The advantage of this derivation is that it shows clearly how the ray representations arise from the results we obtained above.

2. Another derivation of the geometrical optics approximation

We imagine a medium consisting of a large number of thin, plane-parallel, homogeneous layers in contact with one another, situated as shown in Fig. 83. In passing from layer to layer, the properties of the

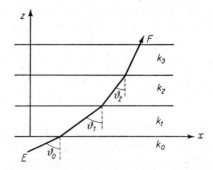

Fig. 83. Regarding the derivation of the geometrical optics approximation.

medium change discontinuously. However, letting the thickness of the layers approach zero while their number approaches infinity, we obtain a layered-inhomogeneous medium with continuously varying parameters.

Now, we shall suppose that there is a plane wave in layer 0, immediately adjacent to the plane $z = 0$, and that the normal to the wave front makes an angle ϑ_0 with the z-axis. We denote the amplitude of this wave by A_0. Upon striking the boundary with layer 1, the wave is refracted and partially reflected. As shown in §§ 2 and 3, the reflection coefficient is given by

$$V = \frac{Z_1 - Z_0}{Z_1 + Z_0},$$

where Z_j is the normal impedance in the jth layer (in this case, $j = 0, 1$), defined for the various types of waves by Eqs. 5.1–5.3.

In this Section, the function ψ describing the wave, will be assumed to represent the acoustic pressure in acoustics, and the tangential component of the electric field in electrodynamics, i.e. it will represent the quantity which must vary continuously across the boundaries between the layers. In view of this requirement, the field of the direct-plus-the-reflected waves on one side of the boundary must be equal to the field of the transmitted wave on the other side, or

$$1 + V = W,$$

where W is the transmission coefficient, equal numerically to the amplitude of the transmitted wave if the amplitude of the incident wave is taken as unity. Using the expression for V, we obtain

$$W = \frac{2Z_1}{Z_0 + Z_1}. \tag{16.13}$$

In view of the assumption that Z_0 and Z_1 differ only slightly from one another, we can set $V \approx 0$, i.e. we can neglect the reflection,† and write the expression for W in the form

$$\frac{A_1}{A_0} \approx 1 + \frac{\Delta Z_0}{2Z_0}, \quad \text{where} \quad \Delta Z_0 \equiv Z_1 - Z_0. \tag{16.14}$$

Passing through layer 1, the wave is refracted at the boundary between layers 1 and 2, and penetrates into layer 2. The amplitude A_2 in this layer is connected with A_1 by the relation

$$\frac{A_2}{A_1} = \frac{2Z_2}{Z_1 + Z_2} \approx 1 + \frac{\Delta Z_1}{2Z_1}, \tag{16.15}$$

etc.

The wave striking layer n will have the amplitude A_n, the value of which is obtained by multiplying equations 16.14, 16.15, and those with higher values of the index j. As a result

$$A_n = A_0\left(1 + \frac{\Delta Z_0}{2Z_0}\right)\left(1 + \frac{\Delta Z_1}{2Z_1}\right)\cdots\left(1 + \frac{\Delta Z_{n-1}}{2Z_{n-1}}\right). \tag{16.16}$$

This expression can be written in the form

$$A_n = A_0 \exp\left[\sum_{s=0}^{n-1} \ln\left(1 + \frac{\Delta Z_s}{2Z_s}\right)\right],$$

---

† In the next paragraph we will show that single, double, etc. reflections at the boundaries of the layers correspond to the second, third, etc. approximations in the solution of the problem.

or, using the relation

$$\ln\left(1+\frac{\Delta Z_s}{2Z_s}\right) \approx \frac{\Delta Z_s}{2Z_s},$$

and replacing the sum by an integral, we can obtain

$$A_n = A_0 \exp\left(\int_{s=0}^{s=n-1} \frac{dZ_s}{2Z_s}\right) = A_0 \exp\left[\frac{1}{2}\ln\frac{Z_{n-1}}{Z_0}\right] = A_0\sqrt{\frac{Z_{n-1}}{Z_0}}. \quad (16.17)$$

Thus, the wave amplitude at an arbitrary point is

$$A_z = A_0\sqrt{\frac{Z(z)}{Z_0}}, \quad (16.18)$$

where $Z(z)$ is the normal impedance at the point z, and A_0 and Z_0 are the wave amplitude and impedance, respectively, at $z = 0$.

We will now consider the phase of the wave. We will suppose that in the initial layer, the plane wave had the phase factor

$$\exp\left[ik_0(x\sin\vartheta_0 + z\cos\vartheta_0)\right].$$

Upon refraction at the interfaces, the form of the x-dependence of the phase does not change, and the term $ik_0 x\sin\vartheta_0$ in the exponent remains unchanged (see §§ 2, 3 and 5). This is explained physically by the fact that the phase velocity of the wave along the interface does not change as the boundary is crossed. Otherwise the continuity conditions on the field could not be satisfied.

On the other hand, the phase velocity of a wave along the z-axis will change from layer to layer. In particular, the phase change in the jth layer will be $k_{zj}\Delta z_j$, where Δz_j is the thickness of the jth layer, and k_{zj} is the z-component of the wave vector in the jth layer. Clearly, in the limit, as the layer thickness approaches zero and the number of layers approaches infinity, the total phase change as the wave propagates from the point $x = z = 0$ to the point x, z will be

$$\phi(x, z) = \int_0^z k_z(z)\,dz + k_0\sin\vartheta_0 . x, \quad (16.19)$$

where $$k_z(z) = k(z)\cos\vartheta(z). \quad (16.20)$$

We have considered only waves propagating in the positive z-direction.

With a set of a large number of thin homogeneous layers, the angle ϑ_j in each of the layers ($j = 0, 1, 2, \ldots$) is determined by the relation

$$k_0\sin\vartheta_0 = k_1\sin\vartheta_1 = \ldots = k_n\sin\vartheta_n.$$

In the limit, in the case of continuously varying parameters of the medium, this relation is written in the form

$$k(z)\sin\vartheta(z) = k_0 \sin\vartheta_0 = \text{const.} \qquad (16.21)$$

The complete expression for the field of the wave at the point x, z, taking account of both amplitude and phase, is

$$\psi(z) = A(z)\,e^{i\phi(z)} = A_0 \sqrt{\left(\frac{Z(z)}{Z_0}\right)} \exp\left[i\int_{z_0}^z k_z(z)\,dz + ik_0\sin\vartheta_0 . x\right]; \qquad (16.22)$$

where $\psi(z)$ is either the tangential component of the electric field or the acoustic pressure. Using the value indicated above for the impedance, and taking into account that

$$\sqrt{[\epsilon(z)]} = \frac{k(z)}{k_0},$$

we obtain (combining all the constants into one constant C):

(a) when the vector $\mathbf{E}$ is perpendicular to the plane of incidence ($E_x = E_z = 0$),

$$E_y(z) = \frac{C}{\sqrt{(k\cos\vartheta)}}\exp\left(\int_0^z k\cos\vartheta dz + ik_0\sin\vartheta_0 . x\right); \qquad (16.23)$$

(b) when the vector $\mathbf{E}$ lies in the plane of incidence ($E_y = 0$);

$$E_x(z) = C\sqrt{\left(\frac{\cos\vartheta}{k}\right)}\exp\left(i\int_0^z k\cos\vartheta dz + ik_0\sin\vartheta_0 . x\right); \qquad (16.24)$$

(c) in acoustics,

$$p(z) = C\left(\frac{\rho c}{\cos\vartheta}\right)^{\frac{1}{2}}\exp\left(i\int_0^z k\cos\vartheta dz + ik_0\sin\vartheta_0 . x\right). \qquad (16.25)$$

Equations 16.23–16.25 coincide to within a constant factor with the first terms of Eqs. 16.9, 16.11 and 16.12, if we set $z_0 = 0$ in the latter three equations. Thus, the present derivation yields the same results as the previous one. The second term is absent in these expressions because we considered only one wave, propagating in the positive z-direction.

It is useful to note that everything that has been presented above can also be considered from the point of view of ray representations. In the limit, the broken line EF in Fig. 83, which in each layer is normal to the wave front, becomes a continuous curve, which is called the ray.†

---

† Usually, the curve whose tangent at every point is in the direction of the energy flux is called the ray. However, we are not considering anisotropic media for the present, so that the directions of the energy flux and the normal to the wave front coincide.

The direction of the ray at any point is determined by Eq. 16.21. As is well known, the wave amplitude at any point can also be determined by using the ray picture. To do this, a so-called ray tube must be constructed, and its transverse cross section must be evaluated at each point. Then the energy of the wave, which to within a constant factor is equal to the square of the amplitude, will be inversely proportional to the cross sectional area of the ray tube at any point. We already used this method for determining the wave amplitude in § 3, Section 3 in the investigation of the refraction of a sound wave. We shall again have recourse to this method in § 39.

3. *Geometrical optics as the first approximation to the exact solution*[91, 93]

It can be shown that geometrical optics is only the first approximation to the exact solution of the problem. Since the successive approximations have a clear physical meaning, we shall go into this question in some detail. For definiteness, we consider the propagation of an electromagnetic wave in which the vector **E** is parallel to the plane of incidence $(E_y = H_x = H_z = 0)$.

If we set $\mu = 1$ and use the notation $\omega/c = k_0$, $\epsilon' = n^2$, Maxwell's equations (2.1) and (2.2) for this case can be written as four scalar equations:

$$\text{(a)} \quad \frac{\partial H_y}{\partial z} = ik_0 n^2 E_x, \qquad \text{(c)} \quad -\left(\frac{\partial E_z}{\partial x} - \frac{\partial E_x}{\partial z}\right) = ik_0 H_y, \qquad (16.26)$$

$$\text{(b)} \quad \frac{\partial H_y}{\partial x} = -ik_0 n^2 E_z, \qquad \text{(d)} \quad \frac{\partial E_x}{\partial x} + \frac{\partial E_z}{\partial z} = 0. \qquad (16.27)$$

We then seek a solution of these equations in the form

$$E_x = \sqrt{\left(\frac{\cos \vartheta}{n}\right)} \exp\left(ik_0 \sin \vartheta_0 . x\right) \left[\chi_1 \exp\left(ik_0 \int_0^z n \cos \vartheta \, dz\right)\right.$$
$$\left. - \chi_2 \exp\left(-ik_0 \int_0^z n \cos \vartheta \, dz\right)\right],$$

$$E_z = -\frac{\sin \vartheta}{\sqrt{(n \cos \vartheta)}} \exp\left(ik_0 \sin \vartheta_0 . x\right) \left[\chi_1 \exp\left(ik_0 \int_0^z n \cos \vartheta \, dz\right)\right.$$
$$\left. + \chi_2 \exp\left(-ik_0 \int_0^z n \cos \vartheta \, dz\right)\right],$$

$$H_y = \sqrt{\left(\frac{n}{\cos \vartheta}\right)} \exp\left(ik_0 \sin \vartheta_0 . x\right) \left[\chi_1 \exp\left(ik_0 \int_0^z n \cos \vartheta \, dz\right)\right.$$
$$\left. + \chi_2 \exp\left(-ik_0 \int_0^z n \cos \vartheta \, dz\right)\right],$$

$$E_y = H_x = H_z = 0, \qquad (16.28)$$

where $\chi_1 = \chi_1(z)$, $\chi_2 = \chi_2(z)$ are two, as yet unknown, functions of z. The angle $\vartheta = \vartheta(z)$ in Eq. 16.28 is determined from Eq. 16.21. If we let $n = n(z)$ be the index of refraction with respect to the point $z = 0$, Eq. 16.21 can be written $n \sin \vartheta = \sin \vartheta_0$, whence

$$\cos \vartheta = \frac{1}{n} \sqrt{(n^2 - \sin^2 \vartheta_0)}. \tag{16.29}$$

Substituting Eq. 16.28 into Eqs. 16.26 and 16.27, it can be shown[91] that these equations will be satisfied if the functions χ_1 and χ_2 satisfy the following system of two equations:

$$\frac{d\chi_1}{dz} = \frac{1}{2}\left[\frac{d}{dz}\ln\left(\frac{\cos\vartheta}{n}\right)\right]\exp\left(-2ik_0\int_0^z n\cos\vartheta dz\right)\cdot\chi_2,$$

$$\frac{d\chi_2}{dz} = \frac{1}{2}\left[\frac{d}{dz}\ln\left(\frac{\cos\vartheta}{n}\right)\right]\exp\left(2ik_0\int_0^z n\cos\vartheta dz\right)\cdot\chi_1. \tag{16.30}$$

Since it is assumed that the parameters of the medium vary slowly, the right hand sides of these equations will be small because the quantity in the square brackets containing a derivative with respect to z will be small. It is therefore convenient to solve the system of equations by the method of successive approximations. In the zeroth approximation, we set the right hand sides equal to zero, and obtain

$$\chi_1^{(0)} = C_1, \quad \chi_2^{(0)} = C_2. \tag{16.31}$$

Substituting these values of χ_1 and χ_2 into Eq. 16.28, we obtain a solution in the geometrical optics approximation. In this solution, the term containing $\chi_1^{(0)}$ describes a wave propagating in the positive z-direction ("direct" wave), and the term containing $\chi_2^{(0)}$ describes a wave propagating in the negative z-direction ("return" wave). Under these conditions, Eqs. 16.28 for E_x and E_z coincide exactly with Eqs. 16.11 and 16.11a obtained earlier.

For convenience in future calculations, we shall write Eqs. 16.30 in the form

$$\frac{d\chi_1}{dz} = \epsilon\lambda_1(z)\cdot\chi_2(z),$$

$$\frac{d\chi_2}{dz} = \epsilon\lambda_2(z)\cdot\chi_1(z), \tag{16.32}$$

where ϵ is a small parameter. It is now not difficult to obtain a solution of Eqs. 16.32 in the form of convergent series of the following kind:

$$\chi_1^m = \chi_1^{(0)} + \epsilon\chi_1^{(1)} + \epsilon^2\chi_1^{(2)} + \cdots,$$

$$\chi_2^m = \chi_2^{(0)} + \epsilon\chi_2^{(1)} + \epsilon^2\chi_2^{(2)} + \cdots \tag{16.33}$$

Substituting these expressions into Eqs. 16.32 and equating the coefficients of like powers of ϵ, we find that the successive approximations are related among themselves by the equations

$$\frac{d\chi_1^m}{dz} = \lambda_1(z) \cdot \chi_2^{(m-1)}(z),$$

$$\frac{d\chi_2^m}{dz} = \lambda_2(z) \cdot \chi_1^{(m-1)}(z). \tag{16.34}$$

When $m = 1$, this system of equations signifies the following: at every point of space, a return wave of the zeroth approximation is the origin of a direct wave of the first order $\chi_1^{(1)}$, and vice versa. In other words, as a result of the inhomogeneity of the medium, a wave of the zeroth approximation undergoes reflection at every point of space and gives rise to a wave of the first approximation propagating in the opposite direction.

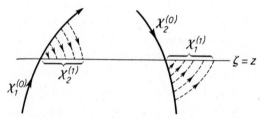

Fig. 84. The generation of secondary waves by primary waves.

We will denote the variable of integration corresponding to the z-coordinate by ζ. It is clear that at a given z, a secondary return wave $\chi_2^{(1)}$ is obtained as a result of the reflection of a primary direct wave $\chi_1^{(0)}$ at all points of the inhomogeneous medium for which ζ varies from z to ∞. On the other hand, a direct secondary wave $\chi_1^{(1)}$ is obtained as the result of the reflection of a wave $\chi_2^{(0)}$ at all levels for which $-\infty < \zeta < z$. This is shown schematically in Fig. 84. These considerations refer to the generation of waves of any order m.

In complete accordance with these graphic considerations, the solutions of Eqs. 16.34 can be written in the form

$$\chi_1^{(m)}(z) = \int_{-\infty}^{z} \lambda_1(\zeta)\,\chi_2^{(m-1)}(\zeta)\,d\zeta,$$

$$\chi_2^{(m)}(z) = \int_{\infty}^{z} \lambda_2(\zeta)\,\chi_1^{(m-1)}(\zeta)\,d\zeta. \tag{16.35}$$

Setting $m = 1, 2, \ldots$, in these equations, using Eq. 16.31, and substituting the resulting expressions for $\chi_1^{(1)}, \chi_2^{(1)}, \chi_1^{(2)}, \chi_2^{(2)}$, etc. into

Eq. 16.33, we obtain the solutions of our problem. These solutions will contain two constants of integration C_1 and C_2, and will therefore be the most general solutions of Eq. 16.32.

By repeated applications of Eqs. 16.35, we can obtain explicit expressions for the functions $\chi_1^{(m)}$ and $\chi_2^{(m)}$ in the form of an m-fold integral. The structure of these integrals can be seen, for example, from the expressions for $\chi_1^{(4)}$ and $\chi_2^{(4)}$, which have the form

$$\chi_1^{(4)}(z) = C_1 \int_{-\infty}^{z} d\zeta_1 \,\lambda_1(\zeta_1) \int_{\infty}^{\zeta_1} d\zeta_2 \,\lambda_2(\zeta_2) \int_{-\infty}^{\zeta_2} d\zeta_3 \,\lambda_1(\zeta_3) \int_{\infty}^{\zeta_3} d\zeta_4 \,\lambda_2(\zeta_4),$$

$$\chi_2^{(4)}(z) = C_2 \int_{\infty}^{z} d\zeta_1 \,\lambda_2(\zeta_1) \int_{-\infty}^{\zeta_1} d\zeta_2 \,\lambda_1(\zeta_2) \int_{\infty}^{\zeta_2} d\zeta_3 \,\lambda_2(\zeta_3) \int_{-\infty}^{\zeta_3} d\zeta_4 \,\lambda_1(\zeta_4).$$

$$(16.36)$$

Physically, $\chi_1^{(m)}$ and $\chi_2^{(m)}$ must be considered as waves which are obtained as a result of m reflections in the inhomogeneous medium.

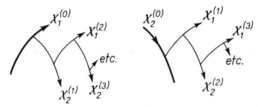

Fig. 85. The successive generation of waves of various orders in an inhomogeneous medium.

Thus, $\chi_1^{(4)}$ and $\chi_2^{(4)}$ are obtained as a result of reflections at the levels ζ_1, ζ_2, ζ_3 and ζ_4. To obtain the complete expressions for $\chi_1^{(4)}$ and $\chi_2^{(4)}$, it is necessary to integrate over all possible levels of reflections.

The successive generation of waves of different orders is shown schematically in Fig. 85.

The solutions we have obtained allow us to estimate the limits of applicability of geometrical optics. Clearly, geometrical optics may be considered a sufficiently good approximation to reality if the "secondary" waves $\chi_1^{(1)}$ and $\chi_2^{(2)}$, etc., which are obtained as a result of reflections at the inhomogeneities of the medium, can be neglected in comparison with the primary waves $\chi_1^{(0)}$ and $\chi_2^{(0)}$.

Using the first of Eqs. 16.35, and taking Eq. 16.31 into account, it is not hard to obtain

$$\epsilon \chi_1^{(1)}(z) = \epsilon C_2 \int_{-\infty}^{z} \lambda_1(\zeta)\,d\zeta,$$

$$(16.37)$$

or, substituting for the function $\epsilon\lambda_1(\zeta)$,

$$\epsilon\chi_1^{(1)} = \tfrac{1}{2}C_2 \int_{-\infty}^{z} \left[\frac{d}{dz}\ln\left(\frac{\cos\vartheta}{n}\right)\right]\exp\left(-2ik_0\int_0^z n\cos\vartheta \, dz\right)dz. \quad (16.38)$$

The exponential under the integral in the last expression is an oscillatory function, with a period equal approximately to

$$\pi/k_0 \, n\cos\vartheta = \lambda(z)/2\cos\vartheta(z).$$

The condition that the integral in Eq. 16.38 be small will be satisfied if the integrals within the limits of each halfperiod in which the exponential retains its sign are small. In view of the assumption that the properties of the medium change only slightly over a wavelength, the expression in the square brackets in Eq. 16.38 can be treated as a constant and taken outside the integral sign in the integration over half a period.

Furthermore, the value of the integral will only increase if the exponential is replaced by its modulus, equal to 1. As a result, the integral over a halfperiod will be no greater than

$$\frac{d}{dz}\left(\ln\frac{\cos\vartheta}{n}\right),$$

multiplied by the magnitude of a halfperiod, i.e. by

$$\lambda/4\cos\vartheta = \lambda_0/4n\cos\vartheta.$$

The smallness requirement applied to this product is

$$\frac{\lambda_0}{4\cos^2\vartheta}\frac{d}{dz}\left(\frac{\cos\vartheta}{n}\right) \ll 1. \quad (16.39)$$

This can be considered the criterion of the applicability of geometrical optics.

This criterion will not be valid if the properties of the medium vary periodically in the z-direction. It could happen that even small amplitude waves reflected from the various layers would add in phase, thereby creating a resultant reflected wave, comparable in amplitude with the incident wave. This is the case of so-called Bragg reflection.

4. *Total reflection of a wave in a layered-inhomogeneous medium*

In § 15 we investigated the total reflection of a wave when the square of the index of refraction is a linear function of z.

In this section we shall investigate the total reflection of a wave in a layered-inhomogeneous medium of arbitrary structure [see below for some limitations imposed on the function $n(z)$]. We shall undertake a detailed investigation of the structure of the field in the region of reflection which was absent in § 15.

Let us note that for certain values of z, the left hand side of inequality (16.39) can become infinite, which indicates that geometrical optics is not applicable at this value of z. This condition occurs both when the index of refraction n becomes zero, and when $\cos \vartheta$ becomes zero. The theory developed below will be valid for both cases. However, for definiteness, we shall first consider the latter case.

Let z_m be the value of z at which $\cos \vartheta$ becomes zero, i.e.

$$\cos \vartheta(z_m) = 0. \tag{16.40}$$

According to Eq. 16.29, this can also be written in the form

$$n(z_m) = \sin \vartheta_0, \quad \text{or} \quad k(z_m) = k_0 \sin \vartheta_0. \tag{16.41}$$

It is clear that geometrical optics is inapplicable not only at the point $z = z_m$, but in the entire region encompassing this point. Therefore, in the layer $z = z_m \pm \Delta$, where the magnitude of Δ will be evaluated later, we must seek a solution without using the methods of geometrical optics.

Equations 16.40 and 16.41 have a very simple physical meaning. It follows from Eq. 16.40 that $\vartheta(z_m) = \pi/2$, that is, the corresponding ray at the point z_m takes a horizontal direction. Thus, $z = z_m$ is a "turning point" of the ray. Although ray representations based on geometrical optics are inapplicable in the vicinity of the point $z = z_m$, we sometimes have recourse to pictures of rays, as was done in Fig. 38, because of the intuitive value of such pictures. We must not forget, of course, the provisional nature of such illustrations (see § 8, Section 7).

In the case we are investigating, we have total reflection of the wave. At points where geometrical optics is applicable, the wave propagates without reflection. Consequently, the entire reflection process must take place in the region $z_m - \Delta < z < z_m + \Delta$.

It is clear from Eq. 16.41 that, in the case of normal incidence on a layered-inhomogeneous medium ($\vartheta_0 = 0$), the condition $\vartheta(z) = \pi/2$ is equivalent to the condition $n(z_m) = 0$. Conditions under which the index of refraction becomes zero are realized, for example, in the ionosphere.[1, 4] The results obtained below will automatically include this case.

We will now proceed to the investigation of the field in the region $z_m - \Delta < z < z_m + \Delta$. Again assuming that the spatial variation of the parameters of the medium is slow, we reduce the problem to the solution of the wave equation (14.3). As shown in § 14, this solution can be represented in the form (14.10), where the function $Z(z)$ satisfies Eq. 14.4. Using Eq. 14.9, we can write Eq. 14.4 as

$$\frac{d^2 Z}{dz^2} + [k^2(z) - k_0^2 \sin^2 \vartheta_0] Z = 0. \tag{16.42}$$

The square brackets become zero at the point $z = z_m$.

The following exposition will be based principally on the work of Gans.[121] We assume that the square brackets can be expanded in a series of powers of $z - z_m$, and that over the entire region $z_m - \Delta < z < z_m + \Delta$ we need consider only the linear term in the expansion.† Then

$$k^2(z) - k_0^2 \sin^2 \vartheta_0 = -ak_0^2(z - z_m), \tag{16.43}$$

where a is a coefficient with the dimensions of inverse length. Since we are assuming that the properties of the medium change only slightly over a wavelength, and in particular, that

$$\frac{1}{k_0^2} \frac{dk^2}{dz} \lambda \ll 1, \tag{16.44}$$

the coefficient a must satisfy the following condition, which is obtained immediately by substituting Eq. 16.43 into Eq. 16.44:

$$a\lambda \ll 1. \tag{16.45}$$

Furthermore, we assume that the retention of only the linear term in expansion (16.43) is justified over the entire region of interest to us if we impose the requirement

$$a\Delta \ll 1. \tag{16.46}$$

Equation 16.42 can now be written

$$\frac{d^2 Z}{dz^2} - k_0^2 a(z_m - z) Z = 0. \tag{16.47}$$

As was indicated in § 15, the solutions of this equation are cylindrical functions of the $\frac{1}{3}$ order.[3] Taking Hankel functions of the first and second kind as linearly independent solutions, we obtain the following

---

† This assumption is invalid only in special cases in which the point $z = z_m$ is near an extremum of the expanded function. This occurs, for example, in the reflection of electromagnetic waves from the ionosphere if the point z_m is near the location of maximum ionization.

general solution of the equation:

$$z < z_m: \quad Z(z) = Av^{\frac{1}{2}} H_{\frac{1}{3}}^{(1)}\left(\frac{2}{3}v^{\frac{3}{2}}\right) + Bv^{\frac{1}{2}} H_{\frac{1}{3}}^{(2)}\left(\frac{2}{3}v^{\frac{3}{2}}\right), \quad v \equiv (ak_0^2)^{\frac{1}{3}}(z_m - z),$$

$$z > z_m: \quad Z(z) = Cu^{\frac{1}{2}} H_{\frac{1}{3}}^{(1)}\left(\frac{2i}{3}u^{\frac{3}{2}}\right) + Du^{\frac{1}{2}} H_{\frac{1}{3}}^{(2)}\left(\frac{2i}{3}u^{\frac{3}{2}}\right), \quad u \equiv (ak_0^2)^{\frac{1}{3}}(z - z_m).$$

$$(16.48)$$

Just as in Eq. 15.7, we must set $D = 0$ here. We shall now show that for sufficiently large $z_m - z$, the term in Eq. 16.48 containing the constant B degenerates into an incident wave, and the term containing the constant A degenerates into a reflected wave, described by the geometrical optics approximation.

Using the arbitrariness available to us in the choice of Δ, we choose it in such a way that the condition

$$\tfrac{2}{3}k_0 a^{\frac{1}{2}} \Delta^{\frac{3}{2}} \gg 1 \qquad (16.49)$$

is satisfied. Conditions (16.49) and (16.45) do not contradict one another, and it is always possible to choose a Δ such that they are satisfied simultaneously. In fact, let $\delta \equiv a\Delta$. According to Eq. 16.46, $\delta \ll 1$. Since $\lambda \ll \Delta$, we can choose Δ such that $a\lambda \sim \delta^2$. Then, according to Eq. 16.49, $k_0 a^{\frac{1}{2}} \Delta^{\frac{3}{2}} \sim \delta^{-\frac{1}{2}} \gg 1$. As a result, all the requirements are compatible.

We will now make use of asymptotic expressions for the Hankel functions[†] for large values of the argument:

$$H_{\frac{1}{3}}^{(1)}(x) \sim \sqrt{\left(\frac{2}{\pi x}\right)} \exp\left[i\left(x - \frac{5}{12}\pi\right)\right],$$

$$H_{\frac{1}{3}}^{(2)}(x) \sim \sqrt{\left(\frac{2}{\pi x}\right)} \exp\left[-i\left(x - \frac{5}{12}\pi\right)\right]. \qquad (16.50)$$

Using these expressions in Eq. 16.48, we obtain for $(z_m - z) \sim \Delta$:

$$Z(z) = \sqrt{\left(\frac{3}{\pi}\right)} v^{-\frac{1}{4}} \left[A \exp\left(\frac{2i}{3}v^{\frac{3}{2}} - \frac{5\pi i}{12}\right) + B \exp\left(-\frac{2i}{3}v^{\frac{3}{2}} + \frac{5\pi i}{12}\right)\right]. \qquad (16.51)$$

This expression can also be written in a form which is typical of geometrical optics. In fact, according to Eq. 16.43, we have

$$v^{-\frac{1}{4}} = \left(\frac{k_0}{a}\right)^{-\frac{1}{6}}(n^2 - \sin^2\vartheta_0)^{-\frac{1}{4}} = [n(z)\cos\vartheta(z)]^{-\frac{1}{2}}\left(\frac{k_0}{a}\right)^{-\frac{1}{6}},$$

---

† See, for example, Ref. 3, pp. 220, 221.

where $n(z) = k(z)/k_0$. Moreover, we also have the identity

$$\tfrac{2}{3} v^{\frac{3}{2}} = \int_0^v v^{\frac{1}{2}} dv = -k_0 \int_{z_m}^z n(z) \cos \vartheta(z) \, dz.$$

As a result, Eq. 16.51 becomes

$$Z(z) = \left(\frac{a}{k_0}\right)^{\frac{1}{4}} \sqrt{\left(\frac{3}{\pi n \cos \vartheta}\right)} \left[B \exp \left(ik_0 \int_{z_m}^z n \cos \vartheta \, dz + \frac{5\pi i}{12} \right) \right.$$
$$\left. + A \exp \left(-ik_0 \int_{z_m}^z n \cos \vartheta \, dz - \frac{5\pi i}{12} \right) \right], \quad (16.52)$$

which has the same structure as Eq. 16.7, except for the factor e^{iax} which we are omitting here.

A rather complete physical interpretation of solutions of the form (16.52) was given above, and it was shown that the first term corresponds to a wave propagating in the positive z-direction, and the second term to a wave in the negative z-direction.

Expression (16.52) which we have obtained from solution (16.48) is valid in the region $z_m - \Delta < z < z_m + \Delta$. However, since geometrical optics is undoubtedly applicable beyond the limits of this region, expression (16.52) will describe the field for any value of $z < z_m - \Delta$. Thus, we see that the constant B represents the amplitude of the incident wave, and consequently, can be prescribed arbitrarily.

The constants A and C can be found from the continuity conditions on Z and dZ/dz at $z = z_m$. To analyze the field at this point and near it, it is best to use the formulas

$$H_{\frac{1}{3}}^{(1)} = \frac{i}{\sin(\pi/3)} \left[\exp \left(-\frac{i\pi}{3} \right) J_{\frac{1}{3}} - J_{-\frac{1}{3}} \right],$$

$$H_{\frac{1}{3}}^{(2)} = -\frac{i}{\sin(\pi/3)} \left[\exp \left(\frac{i\pi}{3} \right) J_{\frac{1}{3}} - J_{-\frac{1}{3}} \right], \quad (16.53)$$

which can be obtained from Eqs. 15.14. It is also expedient to expand the Bessel functions in powers of their arguments:[13]

$$J_{\frac{1}{3}}(x) = \left(\frac{x}{2}\right)^{\frac{1}{3}} \sum_{m=0}^{\infty} \frac{1}{m! \, \Gamma(m + \frac{4}{3})} \left(\frac{ix}{2}\right)^{2m},$$

$$J_{-\frac{1}{3}}(x) = \left(\frac{x}{2}\right)^{-\frac{1}{3}} \sum_{m=0}^{\infty} \frac{1}{m! \, \Gamma(m + \frac{2}{3})} \left(\frac{ix}{2}\right)^{2m}, \quad (16.54)$$

where Γ is the gamma-function. The last formulas give

$$u^{\frac{1}{2}} J_{\frac{1}{3}}\left(\frac{2i}{3} u^{\frac{3}{2}}\right) = \exp\left(i\frac{\pi}{6}\right) \frac{1}{\sqrt[3]{(3)} \, \Gamma(\frac{4}{3})} P(u),$$

$$u^{\frac{1}{2}} J_{-\frac{1}{3}}\left(\frac{2i}{3} u^{\frac{3}{2}}\right) = \exp\left(-i\frac{\pi}{6}\right) \frac{\sqrt[3]{3}}{\Gamma(\frac{2}{3})} Q(u), \qquad (16.55)$$

where

$$P(u) = u\left(1 + \frac{u^3}{12} + \frac{u^6}{504} + \frac{u^9}{45\,360} + \ldots\right),$$

$$Q(u) = 1 + \frac{u^3}{6} + \frac{u^6}{180} + \frac{u^9}{12\,960} + \ldots \qquad (16.56)$$

From Eqs. 16.53 and 16.55, we obtain

$$u^{\frac{1}{2}} H_{\frac{1}{3}}^{(1)}\left(\frac{2i}{3} u^{\frac{3}{2}}\right) = \frac{2}{\sqrt{3}} \exp\left(\pi i/3\right) [C_1 P(u) - C_2 Q(u)], \qquad (16.57)$$

where for brevity we have used the notation

$$C_1 = \frac{1}{\sqrt[3]{(3)} \, \Gamma(\frac{4}{3})} = 0.7762, \quad C_2 = \frac{\sqrt[3]{3}}{\Gamma(\frac{2}{3})} = 1.065. \qquad (16.58)$$

Furthermore, from Eq. 16.54 we find

$$v^{\frac{1}{2}} J_{\frac{1}{3}}(\tfrac{2}{3} v^{\frac{3}{2}}) = C_1 v\left(1 - \frac{v^3}{12} + \frac{v^6}{504} - \frac{v^9}{45\,360} + \ldots\right) = -C_1 P(u),$$

$$v^{\frac{1}{2}} J_{-\frac{1}{3}}(\tfrac{2}{3} v^{\frac{3}{2}}) = C_2\left(1 - \frac{v^3}{6} + \frac{v^6}{180} - \frac{v^9}{12\,960} + \ldots\right) = C_2 Q(u). \qquad (16.59)$$

As a result, according to Eq. 16.53, we obtain

$$v^{\frac{1}{2}} H_{\frac{1}{3}}^{(1)}(\tfrac{2}{3} v^{\frac{3}{2}}) = -\frac{2i}{\sqrt{3}} \{\exp\left[-i(\pi/3)\right] C_1 P(u) + C_2 Q(u)\},$$

$$v^{\frac{1}{2}} H_{\frac{1}{3}}^{(2)}(\tfrac{2}{3} v^{\frac{3}{2}}) = \frac{2i}{\sqrt{3}} \{\exp\left[i(\pi/3)\right] C_1 P(u) + C_2 Q(u)\}. \qquad (16.60)$$

To determine Z and dZ/dz at the point $u = v = 0$, we limit ourselves to the first terms in Eqs. 16.56, which gives $P = u$, $Q = 1$.

Then, from Eqs. 16.48, 16.60 and 16.57, we obtain

$z < z_m$:

$$Z(z) = -\frac{2i}{\sqrt{3}} A\{\exp\left[-i(\pi/3)\right] C_1 u + C_2\} + \frac{2i}{\sqrt{3}} B[\exp\left(i\pi/3\right) C_1 u + C_2];$$

$z > z_m$:
$$Z(z) = \frac{2}{\sqrt{3}} C \exp\left(i\pi/3\right) (C_1 u - C_2), \qquad (16.61)$$

where
$$u = (ak_0^2)^{\frac{1}{3}}(z - z_m).$$

The continuity conditions on Z and dZ/dz at $z = z_m$ yield two equations

$$B - A = iC \exp(i\pi/3),$$

$$B \exp(i\pi/3) - A \exp(-i\pi/3) = -iC \exp(i\pi/3), \qquad (16.62)$$

whence we find

$$A = B \exp(i\pi/3), \quad C = i \exp(i\pi/3) B. \qquad (16.63)$$

Using these relations in Eqs. 16.48, we obtain the solution in the final form

$z_m - \Delta < z < z_m$:

$$Z(z) = B \sqrt{(v)} \left[H_{\frac{1}{3}}^{(2)} (\tfrac{2}{3} v^{\frac{3}{2}}) + \exp(i\pi/3) H_{\frac{1}{3}}^{(1)} (\tfrac{2}{3} v^{\frac{3}{2}}) \right];$$

$z_m < z < z_m + \Delta$:

$$Z(z) = i \exp(i\pi/3) B \sqrt{(u)} H_{\frac{1}{3}}^{(1)} \left(\frac{2i}{3} u^{\frac{3}{2}} \right). \qquad (16.64)$$

The field in the region we have singled out is determined completely by these formulas.†

As we have already seen, at the lower boundary of the region, the solution (16.4) degenerates into the solution given by the geometrical optics approximation (16.52), and continues in this form beyond the boundary of the region. Since $|A| = |B|$ according to Eq. 16.63, the amplitude of the reflected wave is equal to the amplitude of the incident wave at every point. As regards the phase of the waves, we take Eq. 16.63 into account and obtain the following expression for the phase difference between the incident and the reflected waves at an arbitrary point z:

$$\phi = -\frac{\pi}{2} - 2k_0 \int_{z_m}^{z} n(z) \cos \vartheta(z) \, dz = -\frac{\pi}{2} - 2 \int_{z_m}^{z} k_z \, dz. \qquad (16.65)$$

If, in spite of the inapplicability of geometrical optics near the point $z = z_m$, we had used the rules of geometrical optics to calculate the phase change over the entire path from z to z_m, we would have obtained

$$\phi_0 = -2 \int_{z_m}^{z} k_z \, dz,$$

---

† Having prescribed the second of expressions (16.64), valid in the region $z > z_m$, we could have used the so-called continuation relations (see Ref. 3, § 3, Section 62) for cylindrical functions to obtain the first expression, valid for $z < z_m$. It would then not have been necessary for us to join solutions at the point $z = z_m$, which would be of a rather artificial character since the function $n(z)$ is continuous at this point. The continuation relations are particularly useful when absorption is present (n is complex). In this case, the quantity $n^2(z) - \sin^2 \vartheta_0$ becomes zero nowhere, and there is no special plane in which we would have been able to join the solutions.

which differs from Eq. 16.65 only by the term $-\pi/2$. It is often convenient to regard the reflection as taking place at the plane $z = z_m$, in which case the quantity $\pi/2$ must be interpreted as the phase of the reflection coefficient.

For numerical calculations of the field near z_m, it is convenient to use Eqs. 16.64, taking account of Eqs. 16.57 and 16.60. The following tabulation (Table 2) of the functions $P(u)$ and $Q(u)$ is also useful.

TABLE 2

Table of the Functions $P(u)$ and $Q(u)$

u	0.0	0.2	0.4	0.6	0.8	1.0	1.2	1.4	1.6	1.8	2.0
$C_1\,P(u)$	0.0000	0.1553	0.3121	0.4742	0.6479	0.8424	1.071	1.352	1.709	2.177	2.803
$C_2\,Q(u)$	1.065	1.066	1.076	1.104	1.158	1.248	1.390	1.599	1.898	2.319	2.908

5. *Wave leakage through a layer*

We will examine the field in the region $z > z_m$ for values of $z - z_m$ which are sufficiently large to permit the use of the asymptotic representation (16.50) for the function $H_{\frac{1}{3}}^{(1)}[(2i/3)\,u^{\frac{3}{2}}]$ in Eq. 16.64. As a result, we obtain

$$z > z_m, \quad Z(z) = \sqrt{(3/\pi)}\,Bu^{-\frac{1}{4}}\exp\left(\tfrac{1}{6}\pi i - \tfrac{2}{3}u^{\frac{3}{2}}\right),$$
$$u = (ak_0^2)^{\frac{1}{3}}(z - z_m) \gg 1. \tag{16.66}$$

Thus, as $z - z_m$ increases, the field decreases. This is analogous to the exponential decrease of the wave amplitude in a medium from which total internal reflection is taking place.

Equation 16.66 can also be represented in a form similar to geometrical optics. In fact, remembering the definition of u (Eq. 16.48), and using Eq. 16.43, we obtain for $z_m < z < z_m + \Delta$:

$$u^{-\frac{1}{4}} = \left(\frac{a}{k_0}\right)^{\frac{1}{6}}\frac{1}{\sqrt[4]{(\sin^2\vartheta_0 - n^2)}}, \quad \tfrac{2}{3}u^{\frac{3}{2}} = \int_0^u u^{\frac{1}{2}}\,du = k_0\int_{z_m}^z \sqrt{(\sin^2\vartheta_0 - n^2)}\,dz,$$

where

$$\sin^2\vartheta_0 - n^2 > 0.$$

As a result, Eq. 16.66 is written

$$Z(z) = \sqrt{\left(\frac{3}{\pi}\right)}\,B\exp\left(\tfrac{1}{6}\pi i\right)\left(\frac{a}{k_0}\right)^{\frac{1}{6}}\frac{1}{\sqrt[4]{[\sin^2\vartheta_0 - n^2(z)]}}$$
$$\times \exp\left[-k_0\int_{z_m}^z \sqrt{(\sin^2\vartheta_0 - n^2(z)}\,dz\right]. \tag{16.67}$$

In a certain sense, this expression is analogous to the first of the two terms in the square brackets in Eq. 16.52, since $n\cos\vartheta = \sqrt{[n^2(z) - \sin^2\vartheta_0]}$.

However, there is an essential difference between these expressions in that the first and second terms in Eq. 16.52 represent waves running in the positive and negative z-directions, respectively, whereas Eq. 16.67 represents a standing wave with an amplitude decreasing as z increases.

However, Eq. 16.67 (just as Eq. 16.52) is valid for any form of the function $n(z)$,† and can therefore be continued beyond the boundary of the region $z_m < z < z_m + \Delta$.

But if, as z increases beyond the boundaries of this region, we again find some point $z = z'_m$, at which $\sin \vartheta_0 = n(z'_m)$, so that for $z > z'_m$ we have $\sin \vartheta_0 < n(z'_m)$, the wave will become a traveling wave. This wave will carry a definite part of the energy beyond the boundaries of the layer (z_m, z'_m), and the reflection cannot be complete. This is the case of so-called wave "leakage" through the layer and is similar to the penetration of a particle through a potential barrier in quantum mechanics.

Near $z = z'_m$, the solution (16.67) in the geometrical optics approximation will again be inapplicable, since the quantity $1/\sqrt{(\sin^2 \vartheta - n^2)}$ becomes infinite. A rigorous examination of the behavior of a wave in this region can be carried out by analogy with the method used above; the quantity $\sin^2 \vartheta_0 - n^2(z)$ is expanded in a series of powers of $(z - z'_m)$, and cylindrical functions are used.

Without carrying out such a detailed investigation, we shall show that there will be no reflection at the level $z = z'_m$,‡ and the wave leaves

---

† Only if $n(z)$ varies sufficiently slowly with z, and if the inequality $n(z) < \sin \vartheta_0$ is satisfied.

‡ This is clear physically from the following simple considerations. If reflection at the level $z = z_m$ is equivalent to total internal reflection at an optically less dense medium, then the behavior of a wave at the point $z = z'_m$ is equivalent, on the contrary, to that of a wave going from an optically less dense to an optically more dense medium. In this case, reflection could occur only if the properties of the medium changed discontinuously at the point z'_m. However, since these properties are not only continuous, but, according to our assumptions vary sufficiently slowly, there will of course be no reflection.

Mathematically, the absence of reflection arises from the following: as we cross the point $z = z'_m$, the function

$$\sqrt{(u_1)} \, H_{\frac{1}{3}}^{(2)}\left(\frac{2i}{3} u_1^{\frac{3}{2}}\right),$$

where $u_1 = (a_1 k_0^2)^{\frac{1}{3}}(z'_m - z)$, which describes the field near z'_m (but for $z < z'_m$), makes a continuous transition (according to the equation

$$H_{\frac{1}{3}}^{(2)}\left(\frac{2i}{3} u_1^{\frac{3}{2}}\right) = -\exp\left(-i\pi/6\right) H_{\frac{1}{3}}^{(1)}\left(\tfrac{2}{3} v_1^{\frac{3}{2}}\right), \quad v_1 \equiv -u_1),$$

into the function
$$\sqrt{(v_1)} \, H_{\frac{1}{3}}^{(1)}\left(\tfrac{2}{3} v_1^{\frac{3}{2}}\right),$$

which describes the field for $z > z'_m$ and represents a traveling wave.

the layer (z_m, z'_m) with the same amplitude with which it reached the level z'_m. For sufficiently large values of $z - z'_m$, the geometrical solution will be

$$Z(z) = \sqrt{\left(\frac{3}{\pi}\right)} \exp\left(i\pi/6\right) \left(\frac{a}{k_0}\right)^{\frac{1}{4}} \frac{B}{\sqrt{(n \cos \vartheta)}} \exp\left\{-k_0 \int_{z_m}^{z'_m} \sqrt{[\sin^2 \vartheta_0 - n^2(z)]}\, dz\right\}$$

$$\times \exp\left(ik_0 \int_{z'_m}^{z_m} n \cos \vartheta dz\right). \tag{16.68}$$

The amplitude of the transmitted wave will, as a rule, be very small since the factor

$$\exp\left\{-k_0 \int_{z_m}^{z'_m} \sqrt{[\sin^2 \vartheta_0 - n^2(z)]}\, dz\right\}, \tag{16.69}$$

is small.

Having determined the amplitude of the transmitted wave, we can also determine the amplitude of the reflected wave by using the law of conservation of energy (see Ref. 4, § 17).

§ 17. The Reflection Coefficient at a Layer with an Arbitrary Law of Variation of Parameters

1. *The equation for the reflection coefficient*

As has already been indicated, exact solutions of the problem of wave reflection from an inhomogeneous layer have been found only in a few cases. Although the investigation of these cases is of great value and reveals a number of important characteristics, there still remains the problem of wave reflection from layers in which the parameters of the medium can be arbitrary functions of the z-coordinate.

It is also important that under actual conditions, the parameters of the medium do not remain constant, but in the course of time undergo both systematic variations as well as variations of a fluctuational nature. It is necessary to know just how these variations affect the reflection coefficient. Therefore, it is impossible to avoid an investigation of the reflection coefficient from layers in which the parameters vary arbitrarily. In fact, we shall show below (see § 17, Section 5) that even small variations in the spatial dependence of the parameters can affect the reflection coefficient significantly.

The investigation of the coefficient of reflection of a plane wave is also of great value in the study of the field of a point source in a layered-inhomogeneous medium, since a spherical wave can be expanded into plane waves.

A number of results concerning a layered-inhomogeneous medium of a general form have been obtained in Refs. 84, 105, 138, 139 and 182. We shall consider here the method which we have presented in more detail in Ref. 40.

Let the z-dependence of the parameters of the medium be given by the functions $\epsilon = \epsilon(z)$ in electrodynamics, and by the functions $\rho = \rho(z)$, $c = c(z)$ in acoustics, where ϵ is the dielectric constant of the medium (it is complex, in general), and ρ and c are the density of the medium and the sound velocity, respectively. We shall assume that as $z \to -\infty$ and $z \to +\infty$ the parameters of the medium approach constant values, equal respectively to ϵ_0, ρ_0, c_0 and ϵ_1, ρ_1, c_1.

We suppose that at $z = -\infty$ there is a plane wave, propagating in the positive z-direction (incident wave). Its angle of incidence (the angle between the normal to the wave front and the z-axis) will be denoted by ϑ_0.

In the general case, the equations of the electromagnetic and acoustic fields can be satisfied only with the assumption that there is also a reflected wave at $z = -\infty$. Our problem is to find the ratio of the complex amplitudes of the reflected and the incident waves, i.e. the modulus and phase of the reflection coefficient. We shall not follow the usual procedure, according to which it would be necessary to write down the wave equation for the field and to attempt to solve it. We really want to know, not the field, but the reflection coefficient, for which it is possible to obtain a special equation.[40]

We first consider the electromagnetic case, and assume that the electric field vector is perpendicular to the plane of incidence. Furthermore, we align the y-axis with the electric field vector.

For this case, Maxwell's equations (2.1) and (2.2) are written in the form

$$\frac{\partial H_x}{\partial z} - \frac{\partial H_z}{\partial x} = -\frac{i\omega}{c}\epsilon' E_y, \qquad \frac{\partial E_y}{\partial z} = -i\frac{\omega}{c}H_x,$$

$$\frac{\partial H_x}{\partial x} + \frac{\partial H_z}{\partial z} = 0, \qquad \frac{\partial E_y}{\partial x} = i\frac{\omega}{c}H_z. \qquad (17.1)$$

The magnetic permeability μ has been set equal to 1 in these equations.

We define the incident and reflected waves at any point in the following way:

Incident wave:

$$E_y = P(z)\exp[i(k_0 n \sin\vartheta . x - \omega t)],$$

$$H_x = -nP(z)\cos\vartheta \exp[i(k_0 n \sin\vartheta . x - \omega t)],$$

$$H_z = nP(z)\sin\vartheta \exp[i(k_0 n \sin\vartheta . x - \omega t)]. \qquad (17.2)$$

Reflected wave:

$$E_y = R(z) \exp[i(k_0 n \sin \vartheta . x - \omega t)],$$

$$H_x = n R(z) \cos \vartheta \exp[i(k_0 n \sin \vartheta . x - \omega t)],$$

$$H_z = n R(z) \sin \vartheta \exp[i(k_0 n \sin \vartheta . x - \omega t)]. \qquad (17.3)$$

According to the assumptions made above,

$$E_x = E_z = H_y = 0.$$

In Eqs. 17.2 and 17.3, we used the following notation: $k_0 = \omega/c$ is the wave number in a vacuum, $n = \sqrt{[\epsilon'(z)]}$ is the index of refraction, and $P(z)$ and $R(z)$ are for the present arbitrary functions.

Substituting Eqs. 17.2 and 17.3 into Eq. 17.1, it is easily verified that the sum of the incident and reflected waves satisfy Maxwell's equations, if the following conditions are fulfilled:

(1) The angle $\vartheta = \vartheta(z)$ satisfies the condition

$$n(z) \sin \vartheta(z) = \text{const},$$

or since at $z = -\infty$ we have $n = n_0 \equiv \sqrt{(\epsilon_0)}$,

$$n(z) \sin \vartheta(z) = n_0 \sin \vartheta_0. \qquad (17.4)$$

The last equation in Snell's law. When the parameters of the medium vary sufficiently slowly so that the wave may be considered a ray, ϑ will be the angle at an arbitrary point between the ray and the vertical. In this case, Eq. 17.4 is quite well known.

(2) The functions $P(z)$ and $R(z)$ satisfy the equations

$$\frac{dP}{dz} - i\beta P + \gamma(P - R) = 0,$$

$$\frac{dR}{dz} + i\beta R - \gamma(P - R) = 0, \qquad (17.5)$$

where $\qquad \beta = \beta(z) = k_0 n(z) \cos \vartheta(z), \qquad (17.6)$

or, using Eq. 17.4, $\qquad \beta = k_0 \sqrt{[n^2(z) - \sin^2 \vartheta_0]} \qquad (17.6\text{a})$

and $\qquad\qquad\qquad \gamma = \frac{\beta'}{2\beta}. \qquad (17.7)$

Here, and in what follows, the prime denotes differentiation with respect to z.

Multiplying the first of Eqs. 17.5 by R, and the second by P, subtracting one from the other and dividing the result by $P^2(z)$, we obtain the Riccati equation for the reflection coefficient $V(z) = R(z)/P(z)$:

$$\frac{dV}{dz} = -2i\beta V + \gamma(1 - V^2).\tag{17.8}$$

When the electric vector lies in the plane of incidence, Maxwell's equations are written in the form (16.26) and (16.27). In this case, the incident and reflected waves are written in the following way:

Incident wave:

$$E_x = P(z)\frac{\cos\vartheta}{n}\exp[i(k_0 n \sin\vartheta . x - \omega t)],$$

$$E_z = -P(z)\frac{\sin\vartheta}{n}\exp[i(k_0 n \sin\vartheta . x - \omega t)],$$

$$H_y = P(z).\exp[i(k_0 n \sin\vartheta . x - \omega t)].\tag{17.9}$$

Reflected wave:

$$E_x = -R(z)\frac{\cos\vartheta}{n}\exp[i(k_0 n \sin\vartheta . x - \omega t)],$$

$$E_z = -R(z)\frac{\sin\vartheta}{n}\exp[i(k_0 n \sin\vartheta . x - \omega t)],$$

$$H_y = R(z)\exp[i(k_0 n \sin\vartheta . x - \omega t)].\tag{17.10}$$

Just as above, we use the condition that Maxwell's equations be satisfied, and obtain Eqs. 17.5 and 17.8, except that in the latter, γ represents the quantity

$$\gamma = \tfrac{1}{2}\left(\frac{\beta}{n^2}\right)'\frac{n^2}{\beta}.\tag{17.11}$$

We now consider the acoustic case.

The acoustic field in an inhomogeneous medium is described by Eqs. 13.16.

Assuming as above that the plane of incidence is the xz-plane, Eqs. 13.16 yield three scalar equations:

$$\left.\begin{aligned}
\frac{\partial p}{\partial t} + \rho c^2\left(\frac{\partial v_x}{\partial x} + \frac{\partial v_z}{\partial z}\right) &= 0,\\[2mm]
\frac{\partial v_x}{\partial t} = -\frac{1}{\rho}\frac{\partial p}{\partial x}, \quad \frac{\partial v_z}{\partial t} &= -\frac{1}{\rho}\frac{\partial p}{\partial z}.
\end{aligned}\right\}\tag{17.12}$$

We can satisfy these equations in the following way. We first introduce the incident wave:

$$p = P(z) \exp\left[i(k_0 n \sin \vartheta . x - \omega t)\right],$$

$$v_x = \frac{\sin \vartheta}{\rho c} P(z) \exp\left[i(k_0 n \sin \vartheta . x - \omega t)\right],$$

$$v_z = \frac{\cos \vartheta}{\rho c} P(z) \exp\left[i(k_0 n \sin \vartheta . x - \omega t)\right]; \tag{17.13}$$

and the reflected wave:

$$p = R(z) \exp\left[i(k_0 n \sin \vartheta . x - \omega t)\right],$$

$$v_x = \frac{\sin \vartheta}{\rho c} R(z) \exp\left[i(k_0 n \sin \vartheta . x - \omega t)\right],$$

$$v_z = -\frac{\cos \vartheta}{\rho c} R(z) \exp\left[i(k_0 n \sin \vartheta . x - \omega t)\right], \tag{17.14}$$

where $k_0 = \omega/c_0$ is the wave number at $z = -\infty$. It is not difficult to verify that the sum of the incident and reflected waves will satisfy Eqs. 17.12, if $\vartheta(z)$ is subject to the equation

$$n(z) \sin \vartheta(z) = \sin \vartheta_0, \tag{17.15}$$

where

$$n(z) = \frac{c_0}{c(z)}. \tag{17.16}$$

Moreover, proceeding as above, we obtain Eq. 17.8 for the reflection coefficient $V = R(z)/P(z)$, except that now $\gamma = \gamma(z)$ is given by

$$\gamma = \frac{1}{2}\left(\frac{\beta}{\rho}\right)' \frac{\rho}{\beta}. \tag{17.17}$$

Thus, in all cases, the reflection coefficient will be determined by the Riccati equation (17.8), in which γ can be written in the form

$$\gamma(z) = \frac{q'}{2q}, \quad q(z) = \frac{\beta}{m} = \frac{k_0}{m}\sqrt{[n^2(z) - \sin^2 \vartheta_0]}, \tag{17.18}$$

where

$$\left.\begin{array}{l} m = 1 \text{ for a horizontally polarized wave} \\ m = n^2 \text{ for a vertically polarized wave} \\ m = \rho/\rho_0 \text{ for acoustics.} \end{array}\right\} \tag{17.19}$$

As a boundary condition, which is necessary in order to determine the solution of Eq. 17.8 uniquely, we have the condition that the reflection coefficient becomes zero at infinity.

$$V \to 0 \quad \text{as} \quad z \to \infty, \tag{17.20}$$

since as $z \to \infty$ (behind the inhomogeneous layer) there is no reflected wave.

It should be noted that the Riccati equation for the reflection coefficient was also obtained in the acoustic case in Ref. 202.

2. *The first method of successive approximations for the determination of the reflection coefficient*[40, 41]

We present two methods of successive approximations for the determination of the reflection coefficient V. The first method, which is considered in this section, is convenient in the case of thin layers. The second method, which is presented in the next section, yields a rapidly converging series in the case of a weakly reflecting layer, where geometrical optics can be used as a first approximation.

We represent the reflection coefficient $V = V(z)$ in the form

$$V = \frac{qv - q_1 u}{qv + q_1 u}, \tag{17.21}$$

where $u = u(z)$ and $v = v(z)$ are two new unknown functions, and $q_1 = \lim_{z \to \infty} q$.

Substituting Eq. 17.21 into Eq. 17.8 and using the expression (17.18) for γ, we obtain the equation

$$\frac{v'}{v} - \frac{u'}{u} = i\beta \left[\frac{q_1}{q} \frac{u}{v} - \frac{q}{q_1} \frac{v}{u} \right]. \tag{17.22}$$

This equation can be satisfied by subjecting u and v to the two equations

$$u' = i\beta \frac{q}{q_1} v,$$

$$v' = i\beta \frac{q_1}{q} u. \tag{17.23}$$

The solution of these equations, with the boundary conditions

$$u(\infty) = v(\infty) = 1 \tag{17.24}$$

yields Eq. 17.21 for the reflection coefficient, which satisfies Eq. 17.8 and the boundary condition (17.20). This is the solution of our problem.

Letting $q(z_0) = q_0$, we obtain the following expression for the reflection coefficient at the point $z = z_0$, sufficiently far from the layer:

$$V = \frac{q_0 v(z_0) - q_1 u(z_0)}{q_0 v(z_0) + q_1 u(z_0)}. \tag{17.25}$$

The solutions of the system (17.23) can be found by using Picard's method of successive approximations.

We introduce the notation

$$\eta_1 = n \cos \vartheta \, \frac{q}{q_1}, \quad \eta_2 = n \cos \vartheta \, \frac{q_1}{q}. \tag{17.25a}$$

Then system (17.23) is written

$$u' = ik_0 \eta_1 v, \quad v' = ik_0 \eta_2 u. \tag{17.26}$$

Picard's method of successive approximations is applicable to this system.

Replacing the right hand sides of the equations by zero, we obtain $u = \text{const}, v = \text{const}$ in the zeroth approximation and, using the boundary conditions (17.24), we have

$$u = 1, \quad v = 1. \tag{17.27}$$

Substituting these values of u and v into the right hand sides of Eqs. 17.23, and again using the boundary conditions, we obtain

$$u(z) = 1 + ik_0 \int_{\infty}^{z} \eta_1(z) \, dz,$$

$$v(z) = 1 + ik_0 \int_{\infty}^{z} \eta_2(z) \, dz. \tag{17.28}$$

Substituting Eqs. 17.25a into the right hand sides of Eqs. 17.28, and integrating both sides of the equation, we obtain the second approximation for u and v, etc. As a result, $u(z)$ and $v(z)$ will be expressed in the form of infinite series. At an arbitrary, but prescribed, $z = z_0$, we have

$$u(z_0) = 1 + ik_0 \int_{\infty}^{z_0} \eta_1(z) \, dz - k_0^2 \int_{\infty}^{z_0} \eta_1(z) \, dz \int_{\infty}^{z} \eta_2(z) \, dz$$

$$- ik_0^3 \int_{\infty}^{z_0} \eta_1(z) \, dz \int_{\infty}^{z} \eta_2(z) \, dz \int_{\infty}^{z} \eta_1(z) \, dz + \ldots,$$

$$v(z_0) = 1 + ik_0 \int_{\infty}^{z_0} \eta_2(z) \, dz - k_0^2 \int_{\infty}^{z_0} \eta_2(z) \, dz \int_{0}^{z} \eta_1(z) \, dz$$

$$- ik_0^3 \int_{\infty}^{z_0} \eta_2(z) \, dz \int_{\infty}^{z} \eta_1(z) \, dz \int_{\infty}^{z} \eta_2(z) \, dz + \ldots \tag{17.29}$$

In practice, the lower limit of infinity in all the integrals must be replaced by $z = z_1$, corresponding to the boundary of the layer.

It is not difficult to show that the series we have obtained give the expansions of the functions u and v in powers of $k_0 l = 2\pi l/\lambda_0$, where l is the thickness of the layer.

8

To show this, we replace z in Eq. 17.23 by the dimensionless coordinate $\zeta = z/l$, and introduce the notation $\eta_1(z) \equiv \sigma_1(\zeta)$, $\eta_2(z) \equiv \sigma_2(\zeta)$. Then Eq. 17.26 is written

$$\frac{du}{d\zeta} = ik_0 l\sigma_1(\zeta).v, \quad \frac{dv}{d\zeta} = ik_0 l\sigma_2(\zeta).u.$$

Integrating this system by the method of successive approximations, we again obtain a series of the type (17.29), except that, instead of the factor k_0^n, $n = 0, 1, 2, \ldots$, in front of the various terms, we now have the factor $(k_0 l)^n$, which completes the proof.

We note that in the zeroth approximation, in which $u = v = 1$, Eq. (17.21), for the case of polarization perpendicular to the plane of incidence, gives the Fresnel coefficient of reflection from an interface between two media. This is quite natural since this case corresponds to an infinitely thin transitional layer ($l = 0$).

The series (17.29) converge if, over the region we are investigating (z_0, z_1), the function $q(z)$ becomes neither zero nor infinite. This follows from the fact that the right hand sides of Eqs. 17.26 satisfy all the conditions required for the convergence of the method of successive approximations which we have used. In particular, the convergence of these series can be proved without difficulty by using the Cacciopolli-Tikhonov theorem.[12]

3. *The second method of successive approximations for the determination of the reflection coefficient*[40]

Equation 17.8 can be written in the form

$$\frac{d}{dz}\left[V \exp\left(2i \int_{z_0}^{z} \beta dz \right) \right] = \gamma(1 - V^2) \exp\left(2i \int_{z_0}^{z} \beta dz \right), \qquad (17.30)$$

where the lower limit of integration in the exponent is an arbitrary fixed value $z = z_0$.

Equation 17.30 with the boundary conditions (17.20) is equivalent to the integral equation

$$\exp\left(2i \int_{z_0}^{z} \beta dz \right) V(z) = - \int_{z}^{\infty} \gamma(1 - V^2) \exp\left(2i \int_{z_0}^{z} \beta dz \right) dz. \qquad (17.31)$$

We solve this equation by the method of successive approximations. Considering γ small, and neglecting the right hand side in the zeroth approximation, we obtain $V_{(0)}(z) = 0$. This corresponds to the geometrical optics approximation, in which, as we saw in § 16, the wave propagates in the medium without reflection.

Substituting $V = 0$ in the right hand side of Eq. 17.31, and letting

$$s(z) \equiv 2 \int_{z_0}^{z} \beta(z) \, dz, \tag{17.32}$$

we obtain in the first approximation

$$V_{(1)}(z) = -\exp[-is(z)] \int_{z}^{\infty} \gamma(z) \exp[is(z)] \, dz. \tag{17.33}$$

We obtain the second approximation by substituting the last expression into the right hand side of Eq. 17.31, etc. As a result, we obtain the sequence of functions $V_{(0)}, V_{(1)}, V_{(2)}, \ldots$, where $V_{(n+1)}$ is found from $V_{(n)}$ through the relation

$$V_{(n+1)}(z) = -\exp[-is(z)] \int_{z}^{\infty} \gamma(z) [1 - V_{(n)}^2(z)] \exp[is(z)] \, dz. \tag{17.33a}$$

It can be shown that if the function $\gamma(z)$ is bounded (in magnitude) at all points, which (according to Eq. 17.11) means physically that nowhere do the parameters of the medium vary discontinuously, and nowhere does $n(z) = \sin \vartheta_0$, corresponding to reversal of the ray, then as $n \to \infty$, we obtain a convergent sequence for $V(z)$. This follows directly from the general convergence criteria of the Picard method of successive approximations (see Ref. 12, §42).

It is easily seen from what has been presented above, that the smaller the absolute value of the square of the reflection coefficient $|V|^2$, the more rapidly does the sequence of approximations converge.

The various approximations for the reflection coefficient have a simple physical meaning. As has already been indicated, the zeroth approximation $n = 0$, in which $V = 0$, corresponds to the geometrical optics approximation. It can also be shown that the first approximation (17.33) corresponds to taking account of single reflections in the inhomogeneous medium, and the successive approximations correspond to double, triple, etc., reflections (see §16, Section 3).

We will examine Eq. 17.33 in more detail. At $z = z_0$, Eq. 17.33 gives

$$V_{(1)}(z_0) = -\int_{z_0}^{\infty} \gamma(z) \exp\left(2i \int_{z_0}^{z} \beta \, dz\right) dz. \tag{17.34}$$

In order to visualize the physical meaning of this expression more readily, we split the entire inhomogeneous medium from $z = z_0$ to $z = \infty$ into a large number of very thin layers, extending from z_0 to z_1, from z_1 to z_2, from z_2 to z_3, etc. We split the integral (17.34) into a sum of integrals, each of which is taken over a separate layer:

$$V_{(1)}(z_0) = -\int_{z_0}^{z_1} \gamma(z) \exp\left(2i \int_{z_0}^{z} \beta dz\right) dz - \int_{z_1}^{z_2} \gamma(z) \exp\left(2i \int_{z_0}^{z} \beta dz\right) dz - \ldots$$

$$- \int_{z_n}^{z_{n+1}} \gamma(z) \exp\left(2i \int_{z_0}^{z} \beta dz\right) dz - \ldots \qquad (17.35)$$

Let us examine the integral from z_n to z_{n+1}. If z_n and z_{n+1} are sufficiently close to one another, the quantity

$$\exp\left(2i \int_{z_0}^{z} \beta dz\right)$$

will vary only slightly within this layer, and can be taken outside the integral sign at its value say, at $z = z_n$. Taking Eq. 17.12 into account, the remaining integral of $\gamma(z)$ is written

$$\int_{z_n}^{z_{n+1}} \gamma(z) \, dz = \frac{1}{2} \int_{z_n}^{z_{n+1}} \frac{1}{q} \frac{dq}{dz} \, dz.$$

If the layer is sufficiently thin, $q(z)$ will not vary much within the layer; taking $1/q$ outside the integral sign at the value $q = q_n$, we obtain

$$\int_{z_n}^{z_{n+1}} \gamma(z) \, dz = \frac{1}{2q_n} \int_{z_n}^{z_{n+1}} \frac{dq}{dz} \, dz = \frac{q_{n+1} - q_n}{2q_n}.$$

Since q_n and q_{n+1} are very close in value, we can replace $2q_n$ by $q_n + q_{n+1}$.

As a result, we obtain for the nth term in Eq. 17.35:

$$- \int_{z_n}^{z_{n+1}} \gamma(z) \exp\left(2i \int_{z_0}^{z} \beta dz\right) dz = \frac{q_n - q_{n+1}}{q_n + q_{n+1}} \exp\left(2i \int_{z_0}^{z_n} \beta dz\right). \quad (17.36)$$

We thus see that the nth term in Eq. 17.35 is a wave, reflected from the layer (z_n, z_{n+1}). The factor $\exp\left(2i \int_{z_0}^{z} \beta dz\right)$ gives the phase change of the direct wave as it travels from z_0 to z_n, and of the reflected wave as it travels from z_n to z_0. The factor $(q_n - q_{n+1})/(q_n + q_{n+1})$ is the coefficient of reflection from the layer. If we assume that our layers are homogeneous, and that the properties of the medium change at the boundaries between the layers (see § 16, Section 2), the above expression for the reflection coefficient will agree with the well-known expression for the coefficient of reflection from an interface between two media (see §§ 2 and 3).

Thus, Eq. 17.34 represents a sum of waves, reflected in the halfspace (z_0, ∞) and combining at $z = z_0$ according to their respective phases. In this approximation, the direct wave propagating from $z = z_0$ to $z = \infty$ and "generating" reflected waves along the way, is assumed to have a constant amplitude equal to 1, that is, its attenuation is neglected. In the succeeding approximations, both the attenuation of the direct wave, and the multiple reflections of waves in the various layers, are taken into account.

In conclusion, we note that as the coefficient of reflection, we have used the ratio of the complex amplitudes of the direct and reflected waves, taking account of their respective phases. Thus, for example, if a perfectly reflecting plane, for which $V = 1$, were located at some $z = z_1$ in a homogeneous medium, our reflection coefficient at $z = z_0$ would be written in the form

$$V(z_0) = \exp\left[2ik_0(z_1 - z_0)\right],$$

where the quantity $2k_0(z_1 - z_0)$ corresponds to the phase change as the wave travels from z_0 to z_1 and back. If, as is usually done, we had defined the reflection coefficient as the modulus of the ratio of the wave amplitudes, then in the case we are considering, the reflection coefficient would be equal to 1, independently of the point of observation.

It is not hard to see that in the case of reflection from an inhomogeneous layer, if the point at which the reflection coefficient is being determined is moved from $z = z_0$ to $z = z_0 - \Delta$, where both z_0 and $z_0 - \Delta$ lie outside the region in which significant reflection occurs, the following relation between the reflection coefficients at these two points is valid:

$$V(z_0 - \Delta) = V(z_0)\exp\left(-2i\int_{z_0 - \Delta}^{z} \beta dz\right). \qquad (17.37)$$

This can be proved by using Eq. 17.31. The exponential factor on the right hand side gives the phase change of the wave as it traverses the segment Δ twice.

4. *Comparison of the exact and approximate methods of calculation using a concrete example*

As an example, we consider the normal incidence of an electromagnetic wave on a layer, extending from $z = -l/2$ to $z = l/2$, in which the index of refraction varies as

$$n = \frac{A}{A + (z/l + 1/2)}, \qquad (17.38)$$

where
$$A = \frac{n_1}{1 - n_1},$$

and outside this interval is a constant, equal to 1 (for $z < -l/2$) and equal to n_1 (for $z > l/2$). The functions $\beta(z)$ and $\gamma(z)$ introduced earlier and the functions q and q_1 will now have the values

$$\beta = k_0 n, \quad \gamma = \frac{n'}{2n}, \quad q(z) = nk_0, \quad q_1 = n_1 k_0.$$

The case we are considering will also apply to acoustics if the index of refraction n is replaced by the quantity $\rho_0 c_0 / \rho c$, where ρ_0 and c_0 are the density of the medium and the velocity of sound for $z < -l/2$. In the calculations, we shall limit ourselves to the second approximation in both methods. The exact expression for the reflection coefficient in this case is given by Eqs. 14.48, except that now the layer thickness is denoted by l (instead of L).

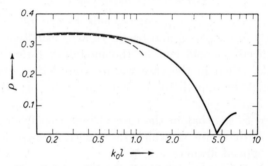

Fig. 86. Comparison of the results of the exact and the approximate methods of calculation of the reflection coefficient.

Calculated values of the modulus of the reflection coefficient $\rho = |V|$ for $n_1 = 2$ and various values of l are shown in Fig. 86. The ordinate represents ρ on a linear scale, while the abscissa represents $k_0 l$ on a logarithmic scale. The dotted curve was calculated by the first method, the solid curve by the second. The two curves coincide for small $k_0 l$. The curve calculated by the exact formula is not shown in the Figure since it would be practically superimposed on the solid curve. It turns out that the second approximation in the second method gives a value of ρ which differs from the exact value by less than 1 per cent for all values of $k_0 l$.

The approximate value of $k_0 l$ at which the reflection coefficient ρ vanishes agrees with the correct value to within 0.02 per cent. It is clear from the Figure that the first method of successive approximations gives a satisfactory result only up to $k_0 l \approx \pi/4$.

Let us note that, in the case under consideration, the function $\gamma = n'/2n$ is discontinuous at the boundaries of the layer. Nevertheless, the methods of successive approximations which we have used are applicable.

5. *One application of the method of successive approximations*

We shall apply the second method of successive approximations to the calculation of the coefficient of reflection from a medium in which the square of the index of refraction varies linearly. We suppose that an inhomogeneous medium of this kind is in contact with a homogeneous medium at the plane $z = 0$. The z-dependence of the index of refraction $n(z) = c_0/c(z)$ will be given by the formulas†

$$z \leqslant 0, \quad n = 1,$$

$$z \geqslant 0, \quad n^2 = 1 + az. \tag{17.39}$$

The wave is assumed to be incident from the homogeneous medium. As previously, the angle of incidence will be denoted by ϑ_0. We also assume that the quantity a is sufficiently small (see inequality 17.47 below). We are thus able to limit ourselves to the first approximation, which is sufficient for many practical applications.

We set $z_0 = 0$ in Eq. 17.34 for the reflection coefficient. This corresponds to the assumption that the point at which the reflection coefficient is calculated lies in the plane $z = 0$, that is, lies on the boundary between the homogeneous and the inhomogeneous media.

For definiteness, we consider the reflection of an electromagnetic wave in which the electric vector is perpendicular to the plane of incidence. According to Eqs. 17.4, 17.6 and 17.7, the functions $\gamma(z)$ and $\beta(z)$ in Eq. 17.34 will now be written

$$\beta(z) = k_0 \sqrt{[n^2(z) - \sin^2 \vartheta_0]} = k_0 \sqrt{(\cos^2 \vartheta_0 + az)},$$

$$\gamma = \frac{\beta'}{2\beta} = \frac{a}{4(\cos^2 \vartheta_0 + az)}. \tag{17.40}$$

Then Eq. 17.34 gives

$$V = -\frac{a}{4} \int_0^\infty \frac{dz}{\cos^2 \vartheta_0 + az} \exp\left[2ik_0 \int_0^z \sqrt{(\cos^2 \vartheta_0 + az)}\, dz \right]. \tag{17.41}$$

Now,
$$2 \int_0^z \sqrt{(\cos^2 \vartheta_0 + az)}\, dz = \frac{4}{3a} [(\cos^2 \vartheta_0 + az)^{\frac{3}{2}}]_0^z$$

$$= \frac{4}{3a} [(\cos^2 \vartheta_0 + az)^{\frac{3}{2}} - \cos^3 \vartheta_0]. \tag{17.42}$$

---

† The problem we are considering here was solved exactly in § 15, using a considerably more complex mathematical apparatus.

In place of z, we will introduce the new variable

$$\frac{4k_0}{3a}[(\cos^2\vartheta_0 + az)^{\frac{3}{2}} - \cos^3\vartheta_0] = i\xi. \qquad (17.43)$$

The limits of integration for this variable will be 0 and $-i\infty$. However, since the integrand has no singularities in the 4th quadrant, the integration along the imaginary axis from 0 to $-i\infty$ can be replaced by integration along the real axis from 0 to ∞. As a result, Eq. 17.41 is written

$$V = -\frac{ia}{8k_0}\int_0^\infty \frac{1}{[\cos^3\vartheta_0 + (3a/4k_0)\,i\xi]}\exp(-\xi)\,d\xi. \qquad (17.44)$$

We expand one factor in the integrand in powers of ξ,

$$\frac{1}{\cos^3\vartheta_0 + (3a/4k_0)\,i\xi} = \frac{1}{\cos^3\vartheta_0}\left(1 - \frac{3a}{4k_0}\frac{i\xi}{\cos^3\vartheta_0} - \frac{9a^2}{16k_0^2}\cdot\frac{\xi^2}{\cos^3\vartheta_0} + \dots\right),$$

and make use of the definite integral

$$\int_0^\infty \exp(-\xi)\,\xi^{p-1}\,d\xi = (p-1)!$$

where p is an integer. Then Eq. 17.44 for the reflection coefficient is written

$$V = \frac{a}{8ik_0\cos^3\vartheta_0}\left[1 - \frac{3i}{4}\frac{a}{k_0\cos^3\vartheta_0} - \frac{9}{8}\left(\frac{a}{k_0\cos^3\vartheta_0}\right)^2 + \dots\right]. \qquad (17.45)$$

In our derivation, we have assumed that $|V|^2 \ll 1$ (only under this condition is Eq. 17.34 valid). Consequently, the quantity $a/(k_0\cos^3\vartheta_0)$ must be small in comparison with unity. Therefore, all the terms in the square brackets, except for 1, can be neglected, and we obtain

$$V = \frac{a}{8ik_0\cos^3\vartheta_0}. \qquad (17.46)$$

This result agrees with Eq. 15.21, which was obtained by a more complicated method. We see that for this result to be valid the following inequality must be satisfied:

$$\frac{a}{k_0\cos^3\vartheta_0} \ll 1. \qquad (17.47)$$

Equation 17.46 was obtained by Gans for the case of normal incidence. However, Gans and succeeding authors treated this expression as the coefficient of reflection from an interface between a homogeneous and an inhomogeneous medium, i.e. from a boundary at which the gradient of the dielectric constant is discontinuous. They assumed that there is a close analogy between the reflection in the present case and reflection from an interface between two homogeneous media (the Fresnel formula), except that in the first case the reflection occurs at a boundary at which the gradient of the dielectric constant is discontinuous, and in the second case, the dielectric constant itself is discontinuous at the boundary. It is clear from our Eqs. 17.34 and 17.41 that actually in the first case, the reflection occurs not at the boundary $z = 0$, but in the entire inhomogeneous halfspace.

Thus, in particular, according to Gans, there would be no reflection if the function $n(z)$ were smoothed in such a way that dn/dz was a continuous function of z.

Using our results, it can easily be shown that smoothing the discontinuity of dn/dz over an interval Δz which is small compared with the effective wavelength in the z-direction, equal to $\lambda_0/\cos \vartheta_0$, has no effect on the value of the reflection coefficient.

6. *Concerning the nonuniqueness of the splitting of the total field into a direct and a return wave*

In order to avoid misunderstandings, we must keep in mind that the field equations from which we start (Maxwell's equations (17.1) or the equations of the acoustic field (17.12)) determine only the total field. However, splitting the total field into a sum of incident and reflected waves, as was done in § 17, Section 1, involves a certain degree of arbitrariness. The only exceptions are homogeneous media or media with slowly varying properties; only in these cases can the field be split uniquely into waves propagating in two directions, allowing the reflection coefficient to be determined.

In an inhomogeneous medium, an expression of the form $A(z)\,e^{i\Phi(z)}$, where A is the wave amplitude and $\Phi(z)$ is the phase, is usually called a traveling wave. However, if $A(z)$ is not a constant or a slowly varying function, this expression can also represent a standing wave. To show this, we consider the following example, due to Schelkunoff.[189]

Consider the function

$$E(z) = \cos \beta z + \epsilon\, e^{i\beta z}. \qquad (17.48)$$

When $\epsilon \ll 1$, this function describes an essentially standing wave, since the first term will be dominant. However, this function can be written in the form

$$E(z) = A(z)\,e^{i\Phi(z)}, \qquad (17.49)$$

where

$$A(z) = \sqrt{[(1+\epsilon)^2 \cos^2 \beta z + \epsilon^2 \sin^2 \beta z]},$$

$$\Phi(z) = \arctan \frac{\epsilon \sin \beta z}{(1+\epsilon)\cos \beta z}.$$

Thus, this expression appears to take the form of a traveling wave.

In general, the field in an inhomogeneous medium can be written in the form $A(z)\,e^{i\Phi(z)}$, but it is not possible to split this expression uniquely into the sum of "direct" and "reflected" waves. Moreover, such a splitting would in general have no physical meaning. An understanding of this circumstance is sometimes lacking even in very reliable works on wave propagation. In particular, this explains one important error in the above-mentioned work of Gans.[121] In studying the total reflection of a wave in an inhomogeneous medium, he obtained an entirely correct expression for the field in the reflection region (see our § 16).

$$w^{\frac{1}{3}}\{\exp\left[i(\pi/3)\right] H_{\frac{1}{3}}^{(1)}(w) + H_{\frac{1}{3}}^{(2)}(w)\}.$$

He considered the term $w^{\frac{1}{3}} H_{\frac{1}{3}}^{(1)}(w)$ to represent the incident wave, and the other term—the reflected wave. The only basis he had for this interpretation was that, far from the point of reflection (w large), where geometrical optics is applicable, the one term becomes the incident wave, and the other term becomes the reflected wave.

However, there are many ways in which the above expression can be split into "incident" and "reflected" waves, each of which leads to the same result as $w \to \infty$, yet gives completely different values for the two waves when $w \leqslant 1$.

As a result of his erroneous assumption, Gans was led to a conclusion which is devoid of meaning, namely, that at the so-called "point of reflection" (the point corresponding to $w = 0$), there is a finite angle between the incident and the reflected ray (concerning this point, see also §§ 8 and 7).

7. A nonreflecting layer can always be found for a wave of a given frequency

As a conclusion to this Paragraph, we will show, following Kofink,[146] that, at a given frequency, there is always some distribution of parameters over the thickness of the layer such that reflection will be absent. For simplicity, we consider the case of normal incidence.

It was shown in § 16 that the expressions

$$F_+(z) = \frac{C_1}{\sqrt{p}} \exp\left[ik \int_{z_0}^{z} p(z)\,dz\right] \quad \text{and} \quad F_-(z) = \frac{C_2}{\sqrt{p}} \exp\left[-ik \int_{z_0}^{z} p(z)\,dz\right] \tag{17.50}$$

are approximate solutions of the equation

$$F'' + k^2 p^2 F = 0, \tag{17.51}$$

if the function $p(z)$ varies sufficiently slowly. Here, F_+ and F_- are waves propagating in opposite directions. However, the coefficient of F in Eq. 17.51 can be modified in such a way that Eqs. 17.50 will be no longer approximate, but *exact* solutions. It is easily verified by direct substitution that the modified equation will be

$$F'' + k^2 \epsilon F = 0, \tag{17.52}$$

where
$$\epsilon = p^2 + \frac{1}{2k^2} \left[\frac{p''}{p} - \frac{3}{2} \left(\frac{p'}{p} \right)^2 \right]. \tag{17.53}$$

In comparison with Eq. 17.51, the quantity ϵ contains the additional term

$$P = \frac{1}{2k^2} \left[\frac{p''}{p} - \frac{3}{2} \left(\frac{p'}{p} \right)^2 \right]. \tag{17.54}$$

If we assume that at $z = -\infty$ and $z = \infty$ the quantity p becomes constant, equal respectively to $\sqrt{\epsilon_1}$ and $\sqrt{\epsilon_2}$, the additional term $P(z)$ will vanish at $z \to \pm 1$. If the dielectric constant in the layer is given by Eq. 17.53, the layer will be nonreflecting, because expressions (17.50), which satisfy the wave equation (17.52), represent noninteracting waves, each propagating in its own direction. At those points of space at which p varies sufficiently slowly and the additional term can be neglected, expressions (17.50) represent the geometrical optics approximation studied earlier. The constants C_1 and C_2 are independent. If, for example, we set $C_1 \neq 0$, $C_2 = 0$, we obtain only an incident wave, and no reflected wave.

As an example, we will consider the function $p^2(z)$ as given by Epstein, namely

$$p^2(z) = \frac{\epsilon_2 + \epsilon_1}{2} + \frac{\epsilon_2 - \epsilon_1}{2} \tanh \frac{mz}{2} + \frac{\epsilon_3}{4} \frac{1}{\cosh^2 (mz/2)}, \tag{17.55}$$

which coincides with Eq. 14.20 if we set

$$\epsilon_1 = 1, \quad \epsilon_2 = 1 - N, \quad \epsilon_3 = -4M, \quad p^2 = \frac{k^2(z)}{k_0^2}.$$

In this case, the additional term (17.54), which reduces the reflection to zero, will be

$$P(z) = -\frac{1}{S^2} \frac{a(z)}{[(\epsilon_1 + \epsilon_2 + \epsilon_3) + \epsilon_2 e^{mz} + \epsilon_1 e^{-mz}]^2}, \tag{17.56}$$

where

$$a(z) = \tfrac{1}{4}[(\epsilon_2 - \epsilon_1)^2 + 2\epsilon_3^2] + 2\epsilon_3(\epsilon_2 + \epsilon_1) + [\epsilon_2(\epsilon_2 - \epsilon_1) - \epsilon_2 \epsilon_3] e^{mz}$$
$$- [\epsilon_1(\epsilon_2 - \epsilon_1) + \epsilon_1 \epsilon_3] e^{-mz} + \tfrac{3}{2}\epsilon_3(\epsilon_2 - \epsilon_1) . \tanh (mz/2)$$
$$- (\epsilon_3^2/4) \tanh^2 (mz/2). \tag{17.57}$$

The quantity S, which is proportional to the thickness of the layer, has the same meaning as in § 14 (see Eqs. 14.21 and 14.22).

We see from Eq. 17.56 that the additional term is inversely proportional to S^2. Thus, when the layer is very thick, even a small addition to ϵ is sufficient to reduce the reflection to zero.

We now consider in detail the two Epstein cases investigated in § 14:

a. A transitional layer ($\epsilon_3 = 0$). In this case, the reflection coefficient is given by Eq. 14.46, where $n_\infty = \sqrt{(\epsilon_2/\epsilon_1)}$. The additional term, which reduces the reflection to zero, will be

$$P(z) = -\frac{\epsilon_2}{S^2} \cdot \frac{(\epsilon_2 + \epsilon_1) \cdot e^{mz} - \epsilon_1 e^{-mz} + \frac{1}{4}\epsilon_2}{[\epsilon_2(1 + e^{mz}) + \epsilon_1(2 + e^{mz} + e^{-mz})]^2}. \qquad (17.58)$$

The additional term $P(z)$ is shown graphically in Fig. 87, with $S = 2$, $\epsilon_2 = 2$, $\epsilon_1 = 1$ (Ref. 146).

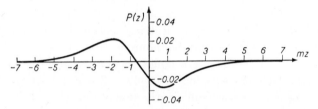

Fig. 87. Graph of the additional term which reduces the reflection at a transitional layer to zero.

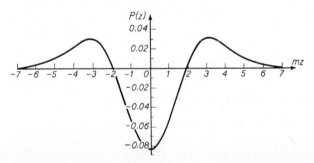

Fig. 88. The same as Fig. 83, but for a symmetrical layer.

The maximum positive value of the additional term is 0.0232, and occurs at $mz \approx -2$. The modulus of the maximum negative value (at $mz \approx 0.73$) is -0.027. According to Eq. 14.22, the value $S = 2$ signifies that the effective thickness of the layer is approximately $\lambda/2$. Since the additional term is inversely proportional to S^2, and is shown in Fig. 87 for $S = 2$, it can be obtained from the Figure for all other values of S by multiplying the ordinate by $(2/S)^2$. Thus, when $S = 20$, and the effective thickness of the layer is approximately 10λ, the addition to ϵ does not exceed 0.00232.

b. A symmetrical layer ($\epsilon_2 = \epsilon_1$). The coefficient of reflection from the layer is given by Eqs. 14.50 and 14.51. The additional term, which reduces the reflection to zero will be

$$P(z) = -\frac{1}{S^2} \cdot \frac{\frac{1}{4}(\epsilon_1^2 + 2\epsilon_3^2) + 6\epsilon_3\,\epsilon_1 + 2\epsilon_1(\epsilon_1 - \epsilon_3)\,\mathrm{e}^{mz} - \epsilon_1(\epsilon_1 + \epsilon_3)\,\mathrm{e}^{-mz} + \frac{3}{2}\epsilon_1\,\epsilon_3 \cdot \tanh\,(mz/2) - (\epsilon_3^2/2)\tanh^2\,(mz/2)}{[(3\epsilon_1 + \epsilon_3) + 2\epsilon_1\,\mathrm{e}^{mz} + \epsilon_1\,\mathrm{e}^{-mz}]^2}.$$

This term is shown graphically in Fig. 88 for $S = 2$, $\epsilon = \epsilon_2 = 1$, $\epsilon_3 = 8$. We see that, just as in the case of the transitional layer, the addition to $\epsilon(z)$ is small, even for $S = 2$, and will decrease as S increases.

CHAPTER IV

REFLECTION AND REFRACTION OF SPHERICAL WAVES

THE theory of the reflection and refraction of spherical waves has become of immediate importance as a result of the intensive development of radiophysics and acoustics during the last few decades. All of the fundamental problems which arose earlier in optics were solved through the use of plane wave representations. Even in optical phenomena involving diverging waves, it was always possible to consider the field only at distances at which the radius of curvature of the wave front was large compared with the wavelength. Under these conditions, a spherical wave could be considered plane, over a limited portion of the wave front. Only in radiophysics and acoustics do we find conditions under which the distances between the reflecting objects (in particular, interfaces between media) and the radiator are comparable with the wavelength.

These circumstances explain the fact that a complete theory of the reflection and refraction of spherical waves was created only during the last two decades, whereas the theory of the reflection and refraction of plane waves had already been developed by Fresnel.

The problem we are considering was first studied by Sommerfeld,[195] who investigated the field of a variable electric dipole located on a plane interface between two media. Later, fundamental investigations were performed by Weil,[207] Fok (see Ref. 24, Chap. 23, edited by Fok) and Leontovich.[57]

In the presentation below, I will follow principally my own work.[32, 33] Using the method developed in these articles, we shall be able to apply a unified point of view to problems which until now have been studied by various methods (for example, a source located at an interface, a raised source, moderate and high conductivity of one of the media, acoustical case, etc.). We shall use the same method to solve the problem of the refraction of spherical waves.

The basis of the method is the utilization and further development of the expansion of the spherical wave into plane waves, as suggested by Weil.[207]

In view of the abundance of literature on the problem we are considering in this chapter, we shall not be able to dwell on all of the work which has some bearing on the material we shall present. Thus, for example, we shall not touch upon some articles containing calculations and experimental results with respect to the reflection of acoustic waves.[142, 153, 185]

§ 18. SPHERICAL WAVES

1. *The field of a simple source of electromagnetic waves*

The wave field of any concentrated radiating system can be represented as the field of a set of simple sources, each of which radiates a spherical wave.

Consider a simple source of electromagnetic waves in the form of a variable electric dipole. We recall that the vector potential of an arbitrary system of currents is expressed by the formula[20]

$$\mathbf{A} = \frac{1}{c} \int \frac{\mathbf{j}[t - (r/c)]\,dv}{r}, \tag{18.1}$$

where the integration is carried over the volume of all bodies in which the current density $\mathbf{j}$ is not zero, and r is the distance from an arbitrary volume element dv to the point of observation. We are using the Gaussian system of units.

In the simplest case, the variable electric dipole, or, as it is sometimes called, the electric vibrator, can be represented in the form of a rectilinear segment of conductor of length l ($l \ll \lambda$) with a variable voltage applied to its ends. Assuming that the current excited in the conductor is constant over its length l, Eq. 18.1 can be written in the form

$$\mathbf{A} = \frac{\mathbf{I}[t - (R/c)]\,l}{cR}, \tag{18.2}$$

where R is the distance from the center of the dipole to the point of observation.

If $\mathbf{A}$ is known, the scalar potential ϕ can be found from the relation

$$\operatorname{div}\mathbf{A} = -\frac{\epsilon\mu}{c}\frac{\partial\phi}{\partial t}, \tag{18.3}$$

after which the electric and magnetic fields are given by

$$\mathbf{H} = \operatorname{curl}\mathbf{A},$$

$$\mathbf{E} = -\operatorname{grad}\phi - \frac{1}{c}\frac{\partial\mathbf{A}}{\partial t}. \tag{18.4}$$

In the investigation of the field of an elementary dipole, it is convenient to replace the two functions $\mathbf{A}$ and ϕ by a single function, the Hertz vector $\mathbf{\Pi}$, through which $\mathbf{A}$ and ϕ are expressed as follows:

$$\mathbf{A} = \frac{\epsilon \cdot \mu}{c} \frac{\partial \mathbf{\Pi}}{\partial t},$$

$$\phi = -\operatorname{div} \mathbf{\Pi}. \tag{18.5}$$

Equation (18.3) is then automatically satisfied. As is easily seen from Eqs. (18.4) and (18.5), the vectors $\mathbf{E}$ and $\mathbf{H}$ are expressed in terms of the Hertz vector as follows:

$$\mathbf{E} = -\frac{\epsilon \mu}{c^2} \frac{\partial^2 \mathbf{\Pi}}{\partial t^2} + \operatorname{grad} \operatorname{div} \mathbf{\Pi},$$

$$\mathbf{H} = \frac{\epsilon \mu}{c} \frac{\partial}{\partial t} \operatorname{curl} \mathbf{\Pi}. \tag{18.6}$$

In Eq. (18.2), we let $l\mathbf{I} = d\mathbf{p}/dt$, where $\mathbf{p}$ is the dipole moment. Comparing the expression thus obtained with the first of Eqs. (18.5) we obtain an explicit expression for the Hertz vector:

$$\mathbf{\Pi} = \frac{\mathbf{p}[t - (R/c)]}{\mu \epsilon R}. \tag{18.7}$$

In what follows, we shall consider almost exclusively radiators with a sinusoidal time variation. As above, we prescribe the time variation by the factor $e^{-i\omega t}$. We then obtain

$$\mathbf{\Pi} = \frac{\mathbf{p}_0}{\mu \epsilon} \frac{e^{i(kR-\omega t)}}{R}, \tag{18.8}$$

where $\mathbf{p}_0$ is the amplitude of the dipole moment. Thus, we have shown that the Hertz vector of the elementary radiator is expressed in the form of a spherical wave. As a result of the influence of inhomogeneities in the medium, the primary spherical wave may be strongly modified. However, in all cases, having obtained the Hertz vector of the field, we can always find the vectors $\mathbf{E}$ and $\mathbf{H}$ by using Eqs. 18.6. Since the transition to the vectors $\mathbf{E}$ and $\mathbf{H}$ never involves any fundamental difficulties, we shall consider our problem solved if we have found the vector $\mathbf{\Pi}$.

Let us note that, with a sinusoidal time variation, $\partial/\partial t = -i\omega$, and Eqs. 18.6 become

$$\mathbf{E} = k^2 \mathbf{\Pi} + \operatorname{grad} \operatorname{div} \mathbf{\Pi},$$

$$\mathbf{H} = -\frac{i\omega \epsilon \mu}{c} \operatorname{curl} \mathbf{\Pi}. \tag{18.9}$$

It is not hard to show that these expressions for $\mathbf{E}$ and $\mathbf{H}$ satisfy Maxwell's equations (2.2), if the vector $\mathbf{\Pi}$ satisfies the wave equation.

The reasoning presented above permits an important generalization. We note that Eq. 2.2 has a definite symmetry with respect to the vectors $\mathbf{E}$ and $\mathbf{H}$. Therefore, Maxwell's equations are satisfied not only by Eqs. 18.9, but also by the expressions

$$\mathbf{H} = -k^2\,\mathbf{\Pi}_M + \operatorname{grad}\operatorname{div}\mathbf{\Pi}_M,$$

$$\mathbf{E} = \frac{i\omega\mu}{c}\operatorname{curl}\mathbf{\Pi}_M, \qquad (18.10)$$

where, compared with Eqs. 18.9, the roles of $\mathbf{E}$ and $\mathbf{H}$ are interchanged (to within the sign).

The solution (18.10) is called the magnetic solution of Maxwell's equations, and the vector $\mathbf{\Pi}_M$ in Eqs. 18.10 is called the magnetic Hertz vector. We shall prescribe the vector $\mathbf{\Pi}_M$ in the form of a spherical wave

$$\mathbf{\Pi}_M = \mathbf{M}\,\frac{e^{ikR}}{R}.$$

It is not hard to show that the $\mathbf{E}$ and $\mathbf{H}$ vectors which are obtained in this way from Eqs. 18.10 comprise the electromagnetic field of a variable magnetic dipole of moment $\mathbf{M}$. In fact, when we prescribed $\mathbf{\Pi}$ in the form (18.8) and used Eqs. 18.9 to make the transition from $\mathbf{\Pi}$ to $\mathbf{E}$, we obtained the field of a variable *electric* dipole, in which the electric lines of force lay in the meridian planes, passing through the axis of the dipole, and the magnetic lines of force coincided with circles of latitude.

Now, since $\mathbf{E}$ and $\mathbf{H}$ are interchanged, we obtain a field in which the magnetic lines of force are directed along the meridians, and the electric lines of force along latitudinal circles. This is the field of a variable magnetic dipole.

A practical example of this type of dipole is a conducting loop, the dimensions of which are small compared with a wavelength, and which carries a variable current.

2. *The elementary source in acoustics*

The simplest source in acoustics is a pulsating sphere of small radius. The acoustic pressure or the acoustic potential of this source will also be expressed as a spherical wave of the form indicated above. If we again limit ourselves to a sinusoidal time variation, and assume that the radius of the sphere is small compared to the wavelength, the acoustic potential at a distance R from the sphere will be given by the formula[25]

$$\psi = \frac{V_0}{4\pi}\,\frac{e^{i(kR-\omega t)}}{R},$$

where $V_0 = 4\pi r_0^2 v_0$ is the so-called volume velocity of the source, equal to the product of the area of the sphere and the amplitude of its surface velocity.

In what follows, we study the reflection and refraction of spherical waves without reference to whether they represent electromagnetic or acoustic fields.

3. The expansion of a spherical wave into plane waves

The difficulty of the problem of the reflection and refraction of a spherical wave at a plane interface between two media is due to the difference between the symmetry of the wave and the form of the boundary. Whereas the wave has spherical symmetry, the boundary is plane. It is therefore natural to solve the problem by expanding the spherical wave into plane waves, especially since the theory of the reflection and refraction of plane waves is well known.

Dropping the factor $e^{-i\omega t}$, and also the factor characterizing the strength of the source, the spherical wave is written in the form e^{ikR}/R. Moreover, assuming temporarily that the source is at the origin, we have

$$R = \sqrt{(x^2 + y^2 + z^2)}.$$

In the plane $z = 0$, the field of the spherical wave will have the form e^{ikr}/r, where $r = \sqrt{(x^2 + y^2)}$. We expand this field in a double Fourier integral in the variables x and y

$$\frac{e^{ikr}}{r} = \iint_{-\infty}^{\infty} A(k_x, k_y) \exp[i(k_x x + k_y y)]\, dk_x\, dk_y. \qquad (18.11)$$

As is well known, $A(k_x, k_y)$ is given by

$$(2\pi)^2 A(k_x, k_y) = \iint_{-\infty}^{\infty} \frac{e^{ikr}}{r} \exp[-i(k_x x + k_y y)]\, dx\, dy. \qquad (18.12)$$

We transform to polar coordinates and use the notation

$$k_x = q\cos\psi, \quad k_y = q\sin\psi, \quad q = \sqrt{(k_x^2 + k_y^2)},$$

$$x = r\cos\phi, \quad y = r\sin\phi, \quad dx\, dy = r\, dr\, d\phi. \qquad (18.13)$$

We then obtain

$$(2\pi)^2 A(k_x, k_y) = \int_0^{2\pi} d\phi \int_0^{\infty} \exp\{ir[k - q\cos(\psi - \phi)]\}\, dr.$$

The integral over r is elementary. Moreover, if we assume that the medium is absorbing, even if ever so slightly, i.e. that k has a positive imaginary part, substitution of the upper limit yields zero, and we obtain

$$(2\pi)^2 A(k_x, k_y) = i \int_0^{2\pi} \frac{d\phi}{k - q \cos(\psi - \phi)} = \frac{i}{k} \int_0^{2\pi} \frac{d\delta}{1 - (q/k) \cos \delta}.$$

Using the Table of Integrals, we find

$$A(k_x, k_y) = \frac{i}{2\pi} \frac{1}{\sqrt{(k^2 - q^2)}} = \frac{i}{2\pi \sqrt{(k^2 - k_x^2 - k_y^2)}}. \tag{18.14}$$

Thus,

$$\frac{e^{ikr}}{r} = \frac{i}{2\pi} \iint_{-\infty}^{+\infty} \frac{\exp[i(k_x x + k_y y)]}{\sqrt{(k^2 - k_x^2 - k_y^2)}} \, dk_x \, dk_y. \tag{18.15}$$

Just as was done in § 8 for a bounded beam, the last expression, describing the field in the xy-plane, can easily be "continued" into space. As is well known, each Fourier component will then correspond to a plane wave in space. Formally, it is sufficient for this "continuation" to add the term

$$\pm i k_z \cdot z$$

to the exponent in the integrand; we have used the notation

$$k_z = \sqrt{(k^2 - k_x^2 - k_y^2)}. \tag{18.16}$$

The plus sign corresponds to points lying in the halfspace $z > 0$, and to waves propagating in the positive z-direction. The minus sign corresponds to points for which $z < 0$. Thus

$$z > 0 \quad \frac{e^{ikR}}{R} = \frac{i}{2\pi} \iint_{-\infty}^{+\infty} \exp[i(k_x x + k_y y + k_z z)] \frac{dk_x \, dk_y}{k_z},$$

$$z < 0 \quad \frac{e^{ikR}}{R} = \frac{i}{2\pi} \iint_{-\infty}^{+\infty} \exp[i(k_x x + k_y y - k_z z)] \frac{dk_x \, dk_y}{k_z}. \tag{18.17}$$

The validity of the present "continuation" is based on the fact that the right hand side of the last expression satisfies the wave equation (since it is satisfied by the integrand) and gives the correct value for the field at $z = 0$.

Equations 18.17 are the formulas for the expansion of a spherical wave into plane waves. The exponent in the integrand represents a plane wave, propagating in the direction given by the components k_x, k_y, k_z of the wave vector.

The integration over the horizontal components k_x and k_y of the wave vector $\mathbf{k}$ can be replaced, in Eq. 18.17, by integration over the angles ϑ and ϕ, characterizing the propagation direction of each of the plane waves (see Fig. 89).

The components of the wave vector are expressed in terms of these angles as follows:

$$k_x = k \sin \vartheta \cos \phi, \quad k_y = k \sin \vartheta \sin \phi, \quad k_z = k \cos \vartheta. \qquad (18.18)$$

The integration with respect to ϕ will be performed between the limits $(0, 2\pi)$. The integration with respect to ϑ cannot be restricted to real values of this angle. According to Eq. 18.16, k_z varies from the value $k_z = k$ when $k_x = k_y = 0$ to $k_z \to i \infty$ when $k_x \to \pm \infty$ or $k_y \to \pm \infty$.

Since, according to Eq. 18.18, we have $\cos \vartheta = k_z/k$, ϑ will vary from $\vartheta = 0$ to $\vartheta = (\pi/2) - i \infty$. We choose the path of integration over ϑ in the form of the contour Γ_0, shown in Fig. 90.

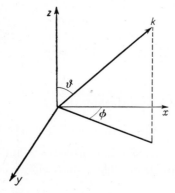

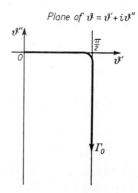

Fig. 89. The spatial position of the vector **k**, and the orientation of the angles ϑ and ϕ.

Fig. 90. The path of integration in the complex plane.

Using the usual formulas for the transformation of variables, we obtain

$$\frac{dk_x \, dk_y}{k_z} = k \sin \vartheta . d\vartheta \, d\phi.$$

As a result, the expansion (18.17) can also be written in the form

$$z \geqslant 0 \quad \frac{e^{ikR}}{R} = \frac{ik}{2\pi} \int_0^{\frac{1}{2}\pi - i\infty} \int_0^{2\pi} \exp[i(k_x x + k_y y + k_z z)] \sin \vartheta \, d\vartheta \, d\phi,$$

$$z \leqslant 0 \quad \frac{e^{ikR}}{R} = \frac{ik}{2\pi} \int_0^{\frac{1}{2}\pi - i\infty} \int_0^{2\pi} \exp[i(k_x x + k_y y - k_z z)] \sin \vartheta \, d\vartheta \, d\phi, \qquad (18.19)$$

where k_x, k_y and k_z are expressed in terms of ϑ and ϕ through Eqs. 18.18.

Thus, we see that in the expansion of a spherical wave, in addition to the usual waves in all possible directions within the limits $0 \leqslant \phi \leqslant 2\pi$, $0 \leqslant \vartheta \leqslant \pi/2$, there will also be waves corresponding to complex values of ϑ. Waves of this type, also called inhomogeneous waves, were considered in §1. At the points $\vartheta = (\pi/2) - ia$, corresponding to the contour of integration Γ_0 (see Fig. 90), where a is real and positive, these waves are propagated with a shortened wavelength along some direction in the xy-plane (given by the angle ϕ) and with an exponentially decreasing amplitude in the z-direction.

This type of wave is necessary in the expansion of a spherical wave since a superposition of only the usual plane waves could not yield a field which would have the necessary singularity at $R \to 0$, and would remain bounded at all other points. However, on the basis of some easily visualized arguments, it is not hard to understand how such a singularity can be obtained through the use of inhomogeneous waves.

Setting $\vartheta = (\pi/2) - ia$, Eqs. 18.18 yield the following values for the components of the wave vector of the inhomogeneous waves:

$$k_x = k \cos \phi \cdot \cosh a, \quad k_y = k \sin \phi \cdot \cosh a, \quad k_z = i \sinh a. \tag{18.20}$$

As $a \to \infty$, we obtain

$$k_x \to \infty \cos \phi, \qquad k_y \to \infty \sin \phi, \qquad k_z \to i \infty.$$

This signifies that we have waves propagating in the horizontal plane (the xy-plane) with a wavelength tending toward zero, and simultaneously attenuating in the vertical direction with an attenuation coefficient tending toward infinity.

At $x = y = z = 0$, a superposition of an infinite number of these waves [integral (18.19)] gives an infinite value for the field. As we depart from this point we obtain finite values, either because of attenuation (for $z \neq 0$) or because of phase interference (for $x \neq 0$ or $y \neq 0$), since all the waves have the same phase only at $x = y = 0$.

It is appropriate to remark that in the expansion of a spherical wave in plane waves, the direction of the coordinate axes can be chosen arbitrarily. Therefore, a spherical wave can be expanded into plane waves in such a way that the inhomogeneous waves entering into the expansion are attenuated not in the z-direction, but in any other direction desired.

We note, finally, that in what follows, instead of the angle of incidence ϑ, we shall frequently use the grazing angle α, which is the complement of ϑ. Replacing ϑ by $\vartheta = (\pi/2) - \alpha$ in Eqs. 18.19, we obtain (for $z \geqslant 0$)

$$\frac{e^{ikR}}{R} = -\frac{ik}{2\pi} \int_{\pi/2}^{i\infty} \int_0^{2\pi} \exp\left[i(k_x x + k_y y + k_z z)\right] \cos \alpha \, d\alpha \, d\phi, \tag{18.21}$$

where $k_x = k \cos \alpha \cos \phi$, $k_y = k \cos \alpha \sin \phi$, $k_z = k \sin \alpha$.

The path of integration over α passes from $\alpha = \pi/2$ along the real axis to $\alpha = 0$, and then along the imaginary axis to $\alpha = i\infty$.

Here, we have considered the plane wave expansion of a harmonic spherical wave, for which the time dependence is given by the factor $e^{-i\omega t}$. An analogous expansion for a spherical wave of the form $(1/R) F(ct - R)$, with an arbitrary function F, was given in Ref. 180.

§ 19. The Reflection of a Spherical Wave at a Plane Interface

1. *The reflected wave as a superposition of plane waves*

Let a spherical wave be radiated at the point O, at a distance z_0 from the interface (Fig. 91). Taking the reflection at the boundary into account, the total field will be

$$\psi = \frac{e^{ikR}}{R} + \psi_{\text{refl}}, \qquad (19.1)$$

where ψ_{refl} is the reflected wave. Our problem is the analysis of ψ_{refl}.

In what follows, we assume that the origin of a rectangular system of coordinates is located on the boundary between the media (Fig. 91).

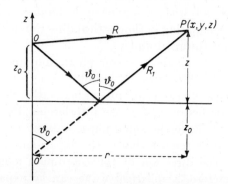

Fig. 91. The positions of the source O and the point of observation P with respect to the interface.

The plane wave expansion of the spherical wave incident on the boundary will be written in the form of Eqs. 18.19, where in place of z we now have $z - z_0$.

Clearly, the reflected wave ψ_{refl} can be represented as a superposition of plane waves, resulting from the reflection of the plane waves into which the original spherical wave was expanded. Upon reflection, the amplitude of each plane wave must be multiplied by the reflection coefficient $V(\vartheta)$, where ϑ is the angle of incidence. Furthermore, we must take into account the phase change as the wave travels from the source to the boundary and then to the point of observation.

Since the projections of the path traversed by the wave are x, y and $z + z_0$, the expression for the reflected plane wave will be

$$V(\alpha) \exp \{i[k_x x + k_y y + k_z(z + z_0)]\}, \tag{19.2}$$

where, as previously, the projections k_x, k_y and k_z of the wave vector are given by Eqs. 18.18.

Integrating over all the plane waves, we obtain an expression for ψ_{refl} in a form similar to Eq. 18.19,

$$\psi_{\text{refl}} = \frac{ik}{2\pi} \int_0^{\frac{1}{2}\pi - i\infty} \int_0^{2\pi} \exp\{ik[x \sin \vartheta \cos \phi + y \sin \vartheta \sin \phi + (z + z_0) \cos \vartheta]\}$$
$$\times V(\vartheta) \sin \vartheta \, d\vartheta \, d\phi. \tag{19.3}$$

The limits of integration over ϑ and ϕ are the same as in Eq. 18.19 for the original spherical wave.

The last expression can be written in a different form by recognizing that the integral over ϕ reduces to a Bessel function of zero order. In fact, we introduce the notation

$$x = r \cos \phi_1, \quad y = r \sin \phi_1.$$

Then in Eq. 19.3 we have

$$\int_0^{2\pi} \exp[ik(x \cos \phi + y \sin \phi) \sin \vartheta] d\phi$$
$$= \int_0^{2\pi} \exp[ikr \sin \vartheta . \cos(\phi - \phi_1)] d\phi = 2\pi J_0(u), \tag{19.4}$$

where $u \equiv kr \sin \vartheta$. As a result, Eq. 19.3 becomes

$$\psi_{\text{refl}} = ik \int_0^{\frac{1}{2}\pi - i\infty} J_0(u) \exp[ik(z + z_0) \cos \vartheta] V(\vartheta) \sin \vartheta \, d\vartheta. \tag{19.5}$$

It is convenient to transform the last expression by changing the limits of integration and replacing the Bessel function by Hankel functions. For this purpose, we note that

$$J_0(u) = \tfrac{1}{2}[H_0^{(1)}(u) + H_0^{(2)}(u)],$$

where $H_0^{(1)}$ and $H_0^{(2)}$ are Hankel functions of the first and second kinds.

Substituting the last expression into Eq. 19.5 we obtain a sum of two integrals. In the integral containing $H_0^{(2)}(u)$, we replace ϑ by $-\vartheta$, and make use of the fact that (see Ref. 3, p. 89) $H_0^{(2)}(e^{-\pi i} u) = -H_0^{(1)}(u)$

and also that $V(-\vartheta) = V(\vartheta)$. We then obtain two integrals with exactly the same integrands, the limits of one integral being 0 to $(\pi/2) - i\infty$, and the limits of the other being $(-\pi/2) + i\infty$ to 0. Combining both integrals into one, with the limits $(-\pi/2) + i\infty$ to $(\pi/2) - i\infty$ over the path Γ_1 (see Fig. 93, below), we obtain

$$\psi_{\text{refl}} = \frac{ik}{2} \int_{-\frac{1}{2}\pi + i\infty}^{\frac{1}{2}\pi - i\infty} H_0^{(1)}(u) \exp\left[ik(z + z_0)\cos\vartheta\right] V(\vartheta)\sin\vartheta\, d\vartheta. \qquad (19.6)$$

By using the appropriate reflection coefficient $V(\vartheta)$† we can use Eq. 19.6, not only for a wave reflected from an interface between two homogeneous media, but also for a wave reflected from an arbitrary inhomogeneous layer.

The reflection coefficients for electromagnetic and acoustic waves were obtained in §§ 2 and 3, respectively, for the simplest case, in which the reflection takes place at an interface between two homogeneous media. In particular, the reflection coefficient for an acoustic wave is given by Eq. 3.16.

Some additional remarks are required with regard to the electromagnetic waves. For simplicity, we consider a plane electromagnetic wave, with the xz-plane as the plane of incidence. In this case, the only nonzero component of the Hertz vector is the component Π_z, and Eq. 18.6 yields

$$E_x = \frac{\partial^2 \Pi}{\partial x \partial z} = -k_x k_z \Pi, \quad E_z = -k^2 \Pi + \frac{\partial^2 \Pi}{\partial z^2} = k_x^2 \Pi,$$

$$H_y = \frac{i\omega\epsilon}{c} \frac{\partial \Pi}{\partial x} = -\frac{\omega\epsilon k_x}{c} \Pi, \quad E_y = H_x = H_z = 0.$$

Thus, we have the case of a wave polarized in the plane of incidence. The reflection coefficient for this case, defined as the ratio of the values of the component E_x in the reflected and incident waves, is given by Eq. 2.18. However, as is clear from the last equation, E_x is proportional to Π, with the proportionality constant $-k_x k_z$. Whereas k_x has the same value for the incident and reflected waves, the values of k_z for the two waves differ in sign. Therefore, the ratio of the values of Π in the incident and reflected waves will have a sign opposite to that of the ratio of the values of E_x. As a result, the reflection coefficient for the Hertz vector will be the same as that given by Eq. 2.18 (to

---

† Using the method presented here, S. S. Voit[45] investigated the reflection of a spherical wave as it passed from a stationary to a moving medium.

within the sign), that is,

$$V = \frac{(\mu/\mu_1)\, n^2 \cos \vartheta - \sqrt{(n^2 - \sin^2 \vartheta)}}{(\mu/\mu_1)\, n^2 \cos \vartheta + \sqrt{(n^2 - \sin^2 \vartheta)}}. \tag{19.7}$$

On the other hand, it can be shown that when the wave is "horizontally" polarized (the vector **E** is perpendicular to the plane of incidence), it can be described by the magnetic Hertz vector, for which the reflection coefficient is given by Eq. 2.16.

2. The method of steepest descents

We shall analyze integral (19.6) in the wave zone, i.e. at distances from the source which are great compared to the wavelength. It is then possible to represent the integral in such a way that an important rôle is played only by the plane waves traveling in directions close to that of the ray $O'P$ (Fig. 91), corresponding to reflection according to the laws of geometrical optics.

It will be convenient for us to use the so-called method of steepest descents (or saddle point method). This method is used to estimate the values of integrals of the form

$$I = \int_C \exp\left[\rho f(\zeta)\right] F(\zeta)\, d\zeta \tag{19.8}$$

for large values of the parameter ρ. The functions $f(\zeta)$ and $F(\zeta)$ are arbitrary analytic functions of the complex variable ζ, and C is the path of integration in the ζ-plane, which in a special case may involve only real values of ζ.

The reader is referred to Refs. 7, 18, 30 (p. 437), 95, 115 and 144 for the general theory of the method of steepest descents. The method reduces essentially to the following. Within certain limits, the path of integration in the complex plane can be deformed without changing the value of the integral. Using this fact, we try to select a path such that the entire value of the integral is determined by a comparatively short portion of the path. Then, as we shall see, the integrand can be replaced by another, simpler function which coincides sufficiently closely with the integrand over the essential portion of the path. The behavior of this function over the remaining, nonessential portions of the path is of no concern.

Without limiting the generality, ρ in Eq. 19.8 can be considered real and positive. We separate the real and imaginary parts of $f(\zeta)$:

$$f(\zeta) = f_1(\zeta) + i f_2(\zeta).$$

Then the exponential in the integrand in Eq. 19.8 becomes

$$\exp\left(i\rho f_2 + \rho f_1\right). \tag{19.9}$$

The path of integration will satisfy the above requirement if it is chosen in such a way that the function f_1 has a maximum at some point and decreases as rapidly as possible with distance from this point. But the imaginary and real parts of an analytic function—in our case f_1 and f_2—have the property that, in the plane of the complex variable, the lines of most rapid decrease of one part are lines of constant values of the other.

Since the path of integration must pass along the line of most rapid decrease of f_1, it must coincide with the line $f_2 = $ const, a line of constant phase. The point of the path at which f_1 is a maximum is called the saddle point.† The derivative of f_1 must be zero at this point. Since $f_2 = $ const on this path its derivative will also be zero. Thus, the saddle point can be found from the equation

$$\frac{df}{d\zeta} = 0. \tag{19.10}$$

Hence, the most advantageous path of integration must go through the saddle point determined by Eq. 19.10, and must leave this point along the line of the most rapid decrease of the function f_1, which coincides with the line $f_2 = $ const. For brevity, we will call this path the path of constant phase. If ρ is large, the exponential in Eq. 19.9 will decrease rapidly with distance from the saddle point, and only a small portion of the path of integration, including the saddle point, will play an essential rôle.

Let us suppose that we have solved Eq. 19.10, and obtained $\zeta = \zeta_0$ for the saddle point. It is then easy to see that the path of integration will be the locus of the points determined by the equation

$$f(\zeta) = f(\zeta_0) - s^2, \tag{19.11}$$

where s varies over all real values from $-\infty$ to $+\infty$. The saddle point corresponds to the value $s = 0$. In fact, taking the imaginary and real parts of Eq. 19.11 we obtain

$$f_2(\zeta) = f_2(\zeta_0), \quad f_1(\zeta) = f_1(\zeta_0) - s^2.$$

Hence, it is clear that the imaginary part of the function $f(\zeta)$ remains constant, while the real part is maximum at $s = 0$, and decreases as s becomes different from zero on either side.

---

† Translator's note: In Russian, the method of steepest descents (as described here) is known as the *method of passage*; the saddle point is then known as the *passage point*. The term *method of steepest descent* is applied in Russian to integration from the saddle point outward. See page 356.

Using Eq. 19.11, we transform from the variable ζ to the new complex variable s in Eq. 19.8. As we have just seen, the path of integration will coincide with the real axis in the s-plane. However, it is of interest to investigate Eq. 19.11 not only for real values of s, but also for arbitrary complex values. Setting $s = s' + is''$, and separating the imaginary and real parts, we obtain

$$f_1 = f_1(\zeta_0) - (s'^2 - s''^2), \quad f_2 = f_2(\zeta_0) - 2s' s''.$$

Hence, in the s-plane, the lines $f_1 = $ const and $f_2 = $ const form two mutually orthogonal systems of hyperbolas (Fig. 92).

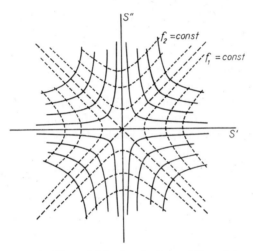

Fig. 92. The neighborhood of a saddle point.

As already indicated, the saddle point is $s = 0$, and the path of integration in the method of steepest descents is the real axis ($s'' = 0$). We see from Fig. 92 that the real axis is one of the lines $f_2 = $ const and is perpendicular to the lines $f_1 = $ const, i.e. it goes along the line of most rapid decrease of f_1. We note that one other line of $f_2 = $ const passes through the saddle point $s = 0$, namely, the imaginary axis ($s' = 0$). However, along this line, the function f_1 does not decrease with distance from the point $s = 0$, but rather, increases. Thus, the path $s' = 0$ is not a path of most rapid decrease of f_1, but is rather a path of the *most rapid increase* of f_1.

Let us represent the relief of the function f_1 on the s-plane as follows: at each point (s', s'') we erect a segment perpendicular to the s-plane, the length of the segment being proportional to the value of the function $f_1(s', s'')$, and then pass a surface through the ends of the segments. In the neighborhood of the point $s = 0$, this relief will have the form of a saddle, since on both sides of $s = 0$ along the real axis it decreases, and in the perpendicular direction, along the imaginary axis, it increases. Going along the real axis from left to right, we at first go upwards along the relief as we approach the point $s = 0$, and then, passing this point, and having gone, as it were, through a pass, we go downward. This is why the point $s = 0$ is sometimes called the *point of passage*. Since the relief of the surface $f(s)$ has the form of a saddle in the neighborhood of the point $s = 0$, this point is called a *saddle point*.†

The integral (19.8), taken over some arbitrary path C, can always be replaced by the integral of the same integrand, taken over the new path of integration. It may be necessary to add some terms obtained in going around singular points, if such points are encountered as the path C is deformed. In particular, if we go around a pole, we must add the residue at the pole. If the function is multiple-valued, and we encounter a branch point on the path, we must add the integral over the edges of an appropriate cut made at this point. Without investigating additions of this sort, we consider the evaluation of the integral over the deformed path.‡

In the change of variable from ζ to s, we use the notation

$$F(\zeta)\frac{d\zeta}{ds} \equiv \Phi(s). \tag{19.12}$$

Using Eq. 19.11, the integral (19.8) over the "passage path" can be written in the form

$$I = \exp\left[\rho f(\zeta_0)\right]\int_{-\infty}^{+\infty} \exp\left(-\rho s^2\right)\Phi(s)\,ds. \tag{19.13}$$

Since ρ is assumed to be large, only small values of s will contribute significantly to the integral. It is therefore convenient to expand the function $\Phi(s)$ in a series of powers of s

$$\Phi(s) = \Phi(0) + \Phi'(0)s + \tfrac{1}{2}\Phi''(0)s^2 + \dots \tag{19.14}$$

Substituting this series into the integral, and using the values of the definite integrals

$$\int_{-\infty}^{+\infty} \exp\left(-\rho s^2\right)ds = \sqrt{\left(\frac{\pi}{\rho}\right)}, \quad \int_{-\infty}^{+\infty} \exp\left(-\rho s^2\right)s^2\,ds = \frac{1}{2}\sqrt{\left(\frac{\pi}{\rho^3}\right)}, \tag{19.15}$$

---

† Translator's note: The sense of the original text is modified here to conform to common English usage. See the previous footnote.

‡ Translator's note: In Russian, the "passage path".

we obtain
$$I = \exp\left[\rho f(\zeta_0)\right]\sqrt{\left(\frac{\pi}{\rho}\right)}\left[\Phi(0) + \frac{1}{4\rho}\Phi''(0) + \ldots\right]. \tag{19.16}$$

Thus, the method of steepest descents allows us to represent the integral in the form of a series of inverse powers of the large parameter ρ. If the function $\Phi(s)$ varies sufficiently slowly compared with the exponential $\exp(-\rho s^2)$, that is, if its derivatives are sufficiently small, we can limit ourselves to the first term or to the first few terms in Eq. 19.16.

Let us note that frequently the function $\Phi(s)$, defined by Eq. 19.12, must be sought directly in the form of a power series in s, since it would be too difficult to find an explicit expression in closed form (see, for example, the next Section).

Limiting ourselves to the terms written out in Eqs. 19.14 and 19.16, we can find this series in a general form. Differentiating Eq. 19.11, we obtain

$$f'(\zeta)\frac{d\zeta}{ds} = -2s.$$

Solving for $d\zeta/ds$ and substituting into Eq. 19.12, we find

$$\Phi(s) = -2s\frac{F(\zeta)}{f'(\zeta)}. \tag{19.17}$$

We now represent the functions $f(\zeta)$ and $F(\zeta)$ as series of powers of $x \equiv \zeta - \zeta_0$,
$$f(\zeta) = f(\zeta_0) - Ax^2 + Bx^3 + Cx^4 + \ldots,$$
$$A \equiv -\tfrac{1}{2}f''(\zeta_0), \quad B \equiv \tfrac{1}{6}f'''(\zeta_0), \quad C \equiv \tfrac{1}{24}f^{IV}(\zeta_0), \tag{19.18}$$
$$F(\zeta) = F(\zeta_0)[1 + Px + Qx^2 + \ldots],$$
$$P \equiv \frac{F'(\zeta_0)}{F(\zeta_0)}, \quad Q \equiv \frac{1}{2}\frac{F''(\zeta_0)}{F(\zeta_0)}. \tag{19.19}$$

Substituting $f(\zeta)$ from Eq. 19.18 into Eq. 19.11, we obtain
$$-Ax^2 + Bx^3 + Cx^4 + \ldots = -s^2. \tag{19.20}$$

Inverting this series, we can represent x as a power series in s. For this purpose, we set
$$x = \frac{s}{\sqrt{A}}(1 + a_1 s + a_2 s^2 + \ldots). \tag{19.21}$$

Substituting this expression into Eq. 19.20, and equating the coefficients of like powers of s, we obtain
$$a_1 = \frac{B}{2A^{\frac{3}{2}}}, \quad a_2 = \frac{C}{2A^2} + \frac{5}{8}\frac{B^2}{A^3}. \tag{19.22}$$

From Eq. 19.17, taking into account Eqs. 19.18 and 19.19, we have

$$\Phi(s) = -2sF(\zeta_0)\frac{1+Px+Qx^2}{-2Ax+3Bx^2+4Cx^3}.$$

Substituting for x by using Eq. 19.21, it is easy to represent $\Phi(s)$ in the form of a power series in s. As a result, we obtain

$$\Phi(s) = \frac{F(\zeta_0)}{\sqrt{A}}\left[1+\left(\frac{P}{A^{\frac{1}{2}}}+\frac{B}{A^{\frac{3}{2}}}\right)s\right.$$
$$\left.+\left(\frac{Q}{A}+\frac{15}{8}\frac{B^2}{A^3}+\frac{3}{2}\frac{C}{A^2}+\frac{3}{2}\frac{BP}{A^2}\right)s^2+\ldots\right]. \qquad (19.23)$$

Using the values of A, B, C, P and Q as given by Eqs. 19.18 and 19.19, we obtain

$$\Phi(0) = \sqrt{\left[-\frac{2}{f''(\zeta_0)}\right]}F(\zeta_0),$$

$$\tfrac{1}{2}\Phi''(0) = \Phi(0)\left[\frac{f'''}{(f'')^2}\frac{F'}{F}+\frac{1}{4}\frac{f^{\mathrm{IV}}}{(f'')^2}-\frac{5}{12}\frac{(f''')^2}{(f'')^3}-\frac{F''}{Ff''}\right]. \qquad (19.24)$$

3. *Analysis of the field of the reflected wave*[32, 33, 167]

We shall apply the method of steepest descents to analyze the field of the reflected wave (19.6) at distances from the source that are large compared to the wavelength (the wave zone). Under these conditions, it is convenient to use an asymptotic representation of the Hankel function

$$H_0^{(1)}(u) \approx \sqrt{\left(\frac{2}{\pi u}\right)}e^{i(u-\frac{1}{4}\pi)}\left(1+\frac{1}{8iu}+\ldots\right). \qquad (19.25)$$

Setting $u \equiv kr\sin\vartheta$, we obtain the exponential

$$\exp\{ik[(z+z_0)\cos\vartheta+r\sin\vartheta]\}$$

in the integrand in Eq. 19.6.

Remembering that $z+z_0 = R_1\cos\vartheta_0$, $r = R_1\sin\vartheta_0$, where R_1 is the distance from the image O' of the source to the point of observation P (Fig. 91), this exponential can be written in the form

$$\exp[ikR_1\cos(\vartheta-\vartheta_0)].$$

As a result, the reflected wave is written

$$\phi_{\mathrm{refl}} = e^{i\frac{1}{4}\pi}\sqrt{\left(\frac{k}{2\pi r}\right)}\int_{-\frac{1}{2}\pi+i\infty}^{\frac{1}{2}\pi-i\infty}\exp[ikR_1\cos(\vartheta-\vartheta_0)]$$
$$\times\left(1+\frac{1}{8ikr\sin\vartheta}\right)V(\vartheta)\sqrt{(\sin\vartheta)}\,d\vartheta. \qquad (19.26)$$

Since, according to our assumptions, kR_1 is large compared with unity, this integral is readily treated by the method of steepest descents. We let Γ_1 (Fig. 93) denote the path of integration going from the point $(-\pi/2)+i\infty$ to the point $(\pi/2)-i\infty$, and will use the rules indicated above to find the passage path Γ.†

Comparing the integral in Eq. 19.26 with Eq. 19.8, we set

$$\rho = kR_1, \quad f(\vartheta) = i\cos(\vartheta - \vartheta_0),$$

$$F(\vartheta) = e^{i\frac{1}{4}\pi}\sqrt{\left(\frac{k}{2\pi r}\right)}\left(1 + \frac{1}{8kr\sin\vartheta}\right)V(\vartheta)\sqrt{(\sin\vartheta)}. \qquad (19.27)$$

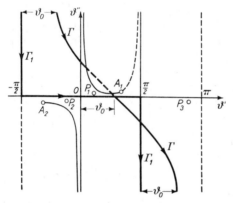

Fig. 93. Transformation of the path of integration in the complex plane.

The saddle point, defined by $df/d\vartheta = 0$, will be $\vartheta = \vartheta_0$. In the present case, Eq. 19.11 for the transformation to the variable s will be

$$\cos(\vartheta - \vartheta_0) = 1 + is^2. \qquad (19.28)$$

The passage path of integration corresponds to real values of s. Taking the real parts of both sides of Eq 19.28, and remembering that for $\vartheta = \vartheta' + i\vartheta''$ we have

$$\sin\vartheta = \sin\vartheta'\cosh\vartheta'' + i\cos\vartheta'\sinh\vartheta'',$$

$$\cos\vartheta = \cos\vartheta'\cosh\vartheta'' - i\sin\vartheta'\sinh\vartheta'', \qquad (19.29)$$

---

† There is one obvious weakness in our arguments, namely, that there are points on the path at which u is not large and even $u = 0$, so that the asymptotic representation of the Hankel function cannot be used over the entire path. However, the arguments will be completely rigorous if the asymptotic representation is used only after the path of integration has been transformed to the path Γ, defined below. The result will then be the same.

we obtain the equation for the passage path in the complex plane

$$\cos(\vartheta' - \vartheta_0)\cosh\vartheta'' = 1. \qquad (19.30)$$

It is not difficult to show that this path will intersect the real axis at the point $\vartheta = \vartheta_0$ at an angle of 45°, and goes toward $-(\pi/2) + \vartheta_0 + i\infty$ on one side, and towards $(\pi/2) + \vartheta_0 - i\infty$ on the other (path Γ in Fig. 93).

It now remains for us to evaluate the integral over the path Γ. However, before proceeding with the evaluation, we examine in detail the possibility of replacing the path of integration Γ_1 by Γ. For this purpose, we must find out what kind of singularities of the integrand can be encountered in the ϑ-plane during the continuous deformation of Γ_1 into Γ.

First of all, since $\sqrt{(\sin\vartheta)}$ enters into the integrand, we have the branch point $\vartheta = 0$. However, this point is of no great significance, since we can go around it from above on the contour Γ_1, and consequently, we shall not pass through it as Γ_1 is deformed into Γ.

The singularities introduced into the integrand by the function $V(\vartheta)$ are extremely important. In both the electromagnetic case (Eqs. 19.7 and 2.16) and the acoustic case (Eq. 3.14), this function contains the radical $\sqrt{(n^2 - \sin^2\vartheta)}$, as a result of which the points $\vartheta = \pm \arcsin n$ will be branch points. At each point ϑ, the reflection coefficient $V(\vartheta)$ can take two values, depending on which sign we choose for the radical. As is always done, it is convenient to talk about the two sheets of the ϑ-plane (formed by a two-sheeted Riemann surface), on each of which the function $V(\vartheta)$ will be single-valued. On one of the sheets (which we shall call the "upper" sheet) we assume that $\mathrm{Im}\sqrt{(n^2 - \sin^2\vartheta)} > 0$, and on the other sheet, the "lower", $\sqrt{(\mathrm{Im}\, n^2 - \sin^2\vartheta)} < 0$. According to the footnote on p. 9, the path Γ_1 must pass over the upper sheet. The two sheets will be joined along the lines

$$\mathrm{Im}\sqrt{(n^2 - \sin^2\vartheta)} = 0,$$

starting at the branch points.

For convenience in what follows, we shall make "cuts" in the complex plane along these lines. Clearly, the equation of these cuts will be

$$n^2 - \sin^2\vartheta = x^2,$$

where x^2 is real and positive, and varies between the limits $(0, \infty)$. The value $x = 0$ gives $\sin\vartheta = n$ (a branch point). As $x^2 \to \infty$, we have $\sin\vartheta \to \pm i\infty$, whence, using the first of Eqs. 19.29, we obtain $\vartheta' \to 0$, $\vartheta'' \to \pm\infty$. Two branch points A_1 and A_2 and the cuts issuing from them are shown in Fig. 93 for the case $|n| < 1$.

The function $V(\vartheta)$ entering into the integrand in Eq. 19.26 also has poles as singularities. They must be examined separately for electrodynamics and for acoustics.

   a. Acoustics. According to Eq. 3.14, the reflection coefficient $V(\vartheta)$ will have a pole at $\vartheta = \vartheta_p$, where ϑ_p is found from the equation

$$m \cos \vartheta_p + \sqrt{(n^2 - \sin^2 \vartheta_p)} = 0, \tag{19.31}$$

where $m = \rho_1/\rho$ is the ratio of the densities of the two adjoining media. Hence, we obtain

$$\sin \vartheta_p = \pm \sqrt{\left(\frac{m^2 - n^2}{m^2 - 1}\right)}. \tag{19.32}$$

   Now, substituting into Eq. 19.31, we can find $\sqrt{(n^2 - \sin^2 \vartheta_p)}$ and from the sign of its imaginary part, we can determine on which sheet the given pole is located. We first consider the case $n < 1$, $m > 1$. We see from Eq. 19.32 that in this case $\sin \vartheta_p$ is real, and moreover $|\sin \vartheta_p| > 1$. This gives four values for $\vartheta_p = \vartheta'_p + i\vartheta''_p$:

$$(1) \quad \vartheta'_p = \tfrac{1}{2}\pi, \quad \vartheta''_p > 0,$$

$$(2) \quad \vartheta'_p = -\tfrac{1}{2}\pi, \quad \vartheta''_p < 0,$$

$$(3) \quad \vartheta'_p = \tfrac{1}{2}\pi, \quad \vartheta''_p < 0,$$

$$(4) \quad \vartheta'_p = -\tfrac{1}{2}\pi, \quad \vartheta''_p > 0.$$

   Poles 2 and 1 are symmetrical with respect to the origin; poles 4 and 3 are also.

   When Eq. 19.31 is used to determine the sign of $\mathrm{Im} \sqrt{(n^2 - \sin^2 \vartheta_p)}$ for each of the poles, it becomes apparent that poles 1 and 2 lie on the upper sheet, and poles 3 and 4 lie on the lower sheet. From this, and also from Fig. 93, we see that none of the poles are crossed in the deformation of Γ_1 into Γ; poles 1 and 2 are not crossed because they do not lie between Γ_1 and Γ, and poles 3 and 4 are not crossed because they lie on the lower sheet while Γ_1 and Γ lie on the upper sheet.

   It is not hard to prove, similarly, that the poles are not crossed for other values of m and n.

   b. Electrodynamics. For simplicity, we suppose that the magnetic permeabilities of the media are equal, that is, $\mu = \mu_1$. Then, according to Eq. 19.7, the equation for the poles in the case of a wave polarized in the plane of incidence (vertical electric dipole) will coincide with Eq. 19.31, if we assume that $m = n^2$. We thus obtain

$$\sin \vartheta_p = \pm \frac{n}{\sqrt{(n^2 + 1)}}. \tag{19.33}$$

We set $n = |n|e^{i\phi}$. In all practical cases we can assume that $0 \leqslant \phi \leqslant \pi/4$. Therefore,

$$\sqrt{(n^2+1)} = |\sqrt{(n^2+1)}|e^{i\psi}, \quad \text{where} \quad \psi \leqslant \phi,$$

and consequently,

$$\sin \vartheta_p = \pm \left| \frac{n}{\sqrt{(n^2+1)}} \right| e^{i(\phi - \psi)}. \tag{19.34}$$

Hence, we obtain four values for $\vartheta_p = \vartheta'_p + i\vartheta''_p$:

$$P_1 : 0 \leqslant \vartheta'_p \leqslant \tfrac{1}{2}\pi, \quad \vartheta''_p \geqslant 0,$$

$$P_2 \text{—symmetrical to } P_1,$$

$$P_3 : \tfrac{1}{2}\pi \leqslant \vartheta'_p \leqslant \pi, \quad \vartheta''_p \leqslant 0,$$

$$P_4 \text{—symmetrical to } P_3.$$

In order to find on which sheet each of these poles is located, we note that the square root

$$\sqrt{(n^2 - \sin^2 \vartheta_p)} = \pm \frac{n^2}{\sqrt{(n^2+1)}} = \pm \left| \frac{n^2}{\sqrt{(n^2+1)}} \right| e^{i(2\phi - \psi)} \tag{19.35}$$

must have a positive imaginary part on the upper sheet. Consequently, we must choose the positive sign. Then the expression

$$(1/n^2) \sqrt{(n^2 - \sin^2 \vartheta_p)} = -\cos \vartheta_p$$

must be proportional to $e^{-i\phi}$, that is, it must lie in the fourth quadrant, and $\cos \vartheta_p$ must lie in the second quadrant. Hence, it is easily found that poles P_1 and P_2 lie on the lower sheet, and poles P_3 and P_4 lie on the upper sheet.

The poles P_1, P_2 and P_3 are shown in Fig. 93 for the case $|n| < 1$. Again, none of these poles will be crossed as the contour Γ_1 is deformed into the contour Γ.

It will be necessary in what follows for us to determine the position of the pole P_1 more accurately. In particular, we shall show that it lies below (and not above) the cut issuing from the branch point A_1. For this purpose, in addition to the cut represented by the line $\text{Im} \sqrt{(n^2 - \sin^2 \vartheta)} = 0$, we draw the line $\text{Re} \sqrt{(n^2 - \sin^2 \vartheta)} = 0$. It is shown by the dotted line in Fig. 93. The signs of $\text{Im} \sqrt{(n^2 - \sin^2 \vartheta)}$ and $\text{Re} \sqrt{(n^2 - \sin^2 \vartheta)}$ change only when one of these lines is crossed. As is clear from Eq. 19.35, both signs are negative on the lower sheet (the negative sign in Eq. 19.35). They are negative at the point $\vartheta = 0$, at which, on the lower sheet, $\sqrt{(n^2 - \sin^2 \vartheta)} = -n = -|n|e^{i\phi}$. The same holds, as is easily established, at the pole P_1. Therefore, the pole P_1 is on the same side of the cut as the origin.

And so, we see that the poles of the integrand need not be taken into account in the analysis of the field of the reflected wave. It can be shown that the same holds true in the case of horizontal polarization of E. In specific cases, as Γ_1 is deformed into Γ, an additional integral along the edges of the cut may appear; this integral yields peculiar ("lateral") waves, the analysis of which we will put off until § 21. We now return to the evaluation of the integral in Eq. 19.26, over the path of steepest descent (the passage path). By use of Eqs. 19.16, 19.24 and 19.27, the evaluation becomes elementary. In this case, the angle ϑ will play the rôle of the variable. As a result, neglecting quantities of the order $1/(kR_1)^2$ compared with unity, we obtain

$$\psi_{\text{refl}} = \frac{e^{ikR_1}}{R_1}\left[V(\vartheta_0) - \frac{iN}{kR_1}\right], \tag{19.36}$$

where
$$N = \tfrac{1}{2}[V''(\vartheta_0) + V'(\vartheta_0)\cot\vartheta_0]. \tag{19.37}$$

Here, $V'(\vartheta_0)$ and $V''(\vartheta_0)$ are the derivatives with respect to ϑ of the reflection coefficient, at the point $\vartheta = \vartheta_0$. It is frequently convenient to express V in terms of the quantity $\gamma = \cos\vartheta$. Then, both in acoustics as well as in the electromagnetic case we are presently considering (vertical polarization), V is given, according to Eqs. 3.14 and 19.7, by the formula

$$V = \frac{m\gamma - \sqrt{(n^2 - 1 + \gamma^2)}}{m\gamma + \sqrt{(n^2 - 1 + \gamma^2)}}, \tag{19.38}$$

where $m \equiv \rho_1/\rho$ in acoustics, and $m \equiv n^2$ in electrodynamics.

Equation 19.37 can also be written in a somewhat different form:

$$N = \tfrac{1}{2}(1 - \gamma_0^2)\,V''(\gamma_0) - \gamma_0\,V'(\gamma_0), \quad \gamma_0 = \cos\vartheta_0, \tag{19.39}$$

where the derivatives of V with respect to γ appear.

Taking Eq. 19.38 into account, and calculating the derivatives of the function $V(\gamma)$, we obtain

$$N = \frac{m(1 - n^2)}{q_0^3(m\gamma_0 + q_0)^3}\,[2m(n^2 - 1) + 3m\gamma_0^2$$
$$+ q_0\gamma_0(2n^2 + 1 - \gamma_0^2) - m\gamma_0^4], \tag{19.40}$$

where
$$q_0 \equiv \sqrt{(n^2 - 1 + \gamma_0^2)} = \sqrt{(n^2 - \sin^2\vartheta_0)}. \tag{19.41}$$

4. Some conclusions. Limits of applicability of geometrical optics

It is easy to give a physical interpretation of the mathematical operations used to obtain Eq. 19.36 for the reflected wave. The deformation of the original path of integration Γ_1 into the passage path Γ signifies that one and the same field is represented by sets of plane waves chosen in different ways. The choice of the path Γ corresponds to the fact that the field is composed of plane waves which, at the point of observation, have the *same phase*, in particular, the same phase as the wave reflecting from the boundary at the angle ϑ_0. According to the general properties of analytic functions, the path over which the phase is constant is such that the integrand decreases most rapidly with distance from the saddle point.† It therefore turns out that in the analysis of the integral, the only essential portions of the passage path are those near the saddle point, corresponding to angles ϑ close to the angle ϑ_0. This signifies that the field at the point of observation is composed principally of the plane waves reflected from the boundary at angles close to ϑ_0, the angle of reflection (and consequently, of incidence) of the ray constructed according to the rules of geometrical optics.

In accordance with this, the principal term in Eq. 19.36 is the first term in the brackets, i.e. the term giving the reflected wave in the geometrical optics approximation. In this approximation, the spherical wave is reflected with the same reflection coefficient as a plane wave. The second term can be treated as a correction term (vanishing as $k \to 0$), which, however, plays an extremely important rôle in a number of cases, in particular, as we shall see below, when the distances of the source and the receiver from the boundary are small compared with the wavelength.

According to Eq. 19.37, the correction term becomes identically zero when $V = \text{const}$, i.e. when the reflection coefficient is independent of the angle of incidence. This occurs, in particular, for a perfectly reflecting surface ($V = \pm 1$).‡ The applicability of geometrical optics in this case is well known, the reflected wave being treated as if it were emitted by the fictitious ("image") source O', shown in Fig. 91. In acoustics, the reflection coefficient is independent of angle under conditions of total reflection, and also when the acoustic velocity in both media is the same. In fact, setting $n = 1$ in Eq. 19.38 we obtain,

$$V = \frac{\rho_1 - \rho}{\rho_1 + \rho}.$$

In this case, geometrical acoustics is rigorously valid.

---

† A line of constant phase can also be a path over which the integrand *increases* most rapidly. However, these two paths are easily distinguished by the sign of $f''(\zeta_0)$.

‡ The case $V = -1$ is realized approximately, for example, in the reflection of a sound wave, incident from water onto a water–air boundary.

In many practical cases, the correction term in Eq. 19.36, although not identically zero, contributes so little to the total field of the direct and reflected waves, that it can be neglected. We shall derive criteria indicating when this can be done.

We first consider the case in which the source or the receiver (for definiteness, let us say the source) is situated on the interface. Then $R = R_1$ (see Fig. 91), and the condition that the correction term in Eq. 19.36 be small compared with the sum of the direct wave e^{ikR}/R and the reflected wave $V(\vartheta_0)e^{ikR}/R$ is written, in the geometrical optics approximation,

$$kR\,|\,1 + V(\vartheta_0)| \gg |\,N\,|. \tag{19.42}$$

Using Eq. 19.38, we obtain

$$1 + V(\vartheta_0) = \frac{2m\gamma_0}{m\gamma_0 + \sqrt{(n^2 - 1 + \gamma_0^2)}}. \tag{19.43}$$

Since kR is assumed to be large, condition (19.42) will not be satisfied only for small values of $\gamma_0 = \cos\vartheta_0$, which corresponds to angles of incidence close to grazing ($\vartheta_0 \approx \pi/2$). Taking into account that $\cos\vartheta_0 = z/R$, and that γ_0 is small, condition (19.42) can be written in the form

$$kz \gg \left| \frac{m\sqrt{(n^2 - 1)}}{[(m\gamma_0 + \sqrt{(n^2 - 1)}]^2} \right|. \tag{19.44}$$

If γ_0 is so small that $m\gamma_0 \ll \sqrt{(n^2 - 1)}$, the last equation can also be written in the form

$$kz \gg \frac{m}{\sqrt{(n^2 - 1)}}. \tag{19.45}$$

Thus, it is necessary for the applicability of geometrical optics that the elevation of the receiver above the interface be sufficiently great compared with the wavelength.

As $|\,m\,| \to \infty$, we have the transition to a perfectly reflecting boundary, since under this condition, according to Eq. 19.38, $V \to 1$. The right hand side of Eq. 19.44 then approaches zero, whence it follows that, in this case, geometrical optics (acoustics) is valid for any value of z.

In the case of the reflection of radio waves from the surface of the earth, the modulus of $m \equiv n^2 \equiv \epsilon$ is usually large compared to unity. Then Eq. 19.45 is written quite simply as

$$kz \gg |\,n\,|. \tag{19.46}$$

It is not hard to understand why the geometrical optics approximation becomes less and less applicable as the receiver approaches the interface. We see that under these conditions, $\gamma_0 = \cos\vartheta_0 = z/R$

decreases, since the angle of incidence approaches $\pi/2$, and the incident wave becomes closer and closer to grazing. Then, as is seen from Eq. 19.38, the reflection coefficient approaches -1. As a result, the incident and reflected waves begin more and more to cancel one another, and the correction term plays an all the more important rôle. If the receiver is situated on the boundary itself ($\gamma_0 = 0$), the field will be identically zero in the geometrical optics approximation (the direct and the reflected waves will completely cancel one another). In this case, the term which we call the correction term will completely determine the field, which will decrease with distance between the point of observation and the source, as $1/R^2$.

The reader will find many important results concerning the reflection of spherical sound waves from an interface between two media, in Refs. 142, 153 and 185.

The problem of the total internal reflection of a sound pulse, radiated by a point source, was considered in Ref. 204. For the reflection of a spherical sound pulse at an infinite absorbing plane, see Ref. 103.

Finally, we note that the order of magnitude of the correction term could also have been evaluated by starting with Huygens' principle and the construction of Fresnel zones on the reflecting plane.[34]

5. *Generalization to the case of other sources: the magnetic dipole and horizontal electric dipole*

Everything presented above can be generalized without difficulty to the case of other sources. In particular, when the source is a vertical magnetic dipole, Eqs. 19.36 and 19.37 remain valid for the reflected wave; however, when the magnetic permeabilities of the media are the same, Eq. 19.38 for the reflection coefficient will be replaced by the formula (see Eq. 2.16).

$$V = \frac{\gamma - \sqrt{(n^2 - 1 + \gamma^2)}}{\gamma + \sqrt{(n^2 - 1 + \gamma^2)}}. \qquad (19.47)$$

The polarization in this case corresponds to the vector $\mathbf{E}$ being perpendicular to the plane of incidence. This is clear if only from the following: in the transition from an electric to a magnetic dipole, the rôles of $\mathbf{E}$ and $\mathbf{H}$ change, and consequently the nature of the polarization of the plane waves into which the spherical wave is expanded also changes.

The case of the horizontal electric dipole is considerably more complicated. We shall assume that the dipole is directed along the x-axis. The spherical wave which it radiates will be described by the component Π_x of the Hertz vector. Each of the plane waves into which the spherical wave is expanded will also contain only the component Π_x. However, it turns out that, in addition to Π_x, the reflected and the refracted waves will also contain the component Π_z, since otherwise, the four boundary conditions, expressing the continuity of the field components E_x, E_y, H_x and H_y across the interface, could not be satisfied. The amplitudes of all four waves (two reflected and two refracted) are easily obtained from these conditions. If the amplitude of the incident plane wave for Π_x is taken as unity, the amplitudes of these waves will be the corresponding reflection and transmission coefficients. The complex amplitude of Π_x in the reflected wave will then be

$$V_{xx} = \frac{\gamma - \sqrt{(n^2 - 1 + \gamma^2)}}{\gamma + \sqrt{(n^2 - 1 + \gamma^2)}}, \tag{19.48}$$

and the amplitude of Π_z in the reflected wave will be

$$V_{xz} = \frac{2\gamma(1 - n^2)\sqrt{(1 - \gamma^2)}}{[\gamma + \sqrt{(n^2 - 1 + \gamma^2)}][n^2\gamma + \sqrt{(n^2 - 1 + \gamma^2)}]}. \tag{19.49}$$

We see that V_{xx} coincides with Eq. 19.47 for V, the reflection coefficient for horizontally polarized waves.

Having determined V_{xx} and V_{xz}, the calculations are continued in accordance with the rule presented above. As a result, we obtain the following expressions for the total field of the reflected spherical wave:

$$\Pi_x^{\text{refl}} = \frac{e^{ikR_1}}{R_1}\left[V_{xx}(\gamma_0) - \frac{iN_{xx}}{kR_1}\right], \tag{19.50}$$

$$\Pi_z^{\text{refl}} = \frac{e^{ikR_1}}{R_1}\left[V_{xz}(\gamma_0) - \frac{iN_{xz}}{kR_1}\right], \tag{19.51}$$

where N_{xx} and N_{xz} will be given by Eq. 19.39, in which V_{xx} and V_{xz}, respectively, must be substituted for $V(\gamma_0)$.

Having obtained the Hertz vector of the reflected wave $\mathbf{\Pi}^{\text{refl}}$ with the components Π_x^{refl} and Π_z^{refl}, we can find $\mathbf{E}$ and $\mathbf{H}$ in the reflected wave by using Eqs. 18.6.

6. *The relation between the field of a source located on the boundary and the field of an elevated source*

There is a very simple relation between the field of an arbitrarily located source and the field of a source situated on the boundary. As previously, we shall use a vertical dipole as the radiator.†

---

† All that will be presented below will apply equally well to the field of a vertical magnetic dipole, and to the field of an acoustic source consisting of a pulsating sphere of small radius.

In the latter case, ψ will denote the acoustic potential.

We let $\psi(r, z, z_0)$ denote the vertical component of the Hertz vector at the point (r, z) for a source located at an elevation of z_0 above the boundary. Similarly, $\psi(r, z + z_0, 0)$ will be the value of the Hertz vector at the point $(r, z + z_0)$ for a source situated on the boundary. We shall show that the following identity holds:

$$\psi(r, z, z_0) = \frac{e^{ikR}}{R} - \frac{e^{ikR_1}}{R_1} + \psi(r, z + z_0, 0), \qquad (19.52)$$

where, as above, R and R_1 are the distances from the point of reception to the source and to the "image" source.

Thus, at an arbitrary point (r, z), we obtain the vertical component of the Hertz vector of the dipole located at an elevation of z_0, by taking the direct wave e^{ikR}/R, and adding to it:

(a) the wave $-e^{ikR_1}/R_1$, issuing from the image source, and taken with a minus sign.

(b) the vertical component of the Hertz vector of a dipole situated on the boundary, but calculated not at the point $P(r, z)$, but at the auxiliary point $P(r, z + z_0)$ (Fig. 94).

This very useful prescription for the calculation of the field of an elevated source was first obtained, apparently, by Hörschelmann in 1911, and later, using another method, by Niessen.[163] It has a somewhat mysterious appearance, and the derivations given by both of the above-mentioned authors are very complicated. But essentially, Eq. 19.52 and the prescription which follows from it, can be obtained from very simple and graphic considerations.[33]

The vertical component of the Hertz vector at an arbitrary point consists of a direct and a reflected wave

$$\psi(r, z, z_0) = \frac{e^{ikR}}{R} + \psi_{\text{refl}}(r, z, z_0). \qquad (19.53)$$

As is clear from the initial expression (19.3) for the reflected wave, the quantities z and z_0 enter only as the sum $z + z_0$, and consequently, the value of ψ_{refl} remains unchanged if the source is lowered and the receiver is raised by the same amount. Thus,

$$\psi_{\text{refl}}(r, z, z_0) = \psi_{\text{refl}}(r, z + z_0, 0). \qquad (19.54)$$

This relation is also understandable from graphical considerations: when the elevations of the source and receiver are varied as described above, neither the total phase change nor the angle of incidence at the boundary change for any of the plane waves comprising the field of the reflected wave. This is shown in Fig. 94 for a wave, incident on the boundary at the angle ϑ. Equation 19.52 follows immediately from this property of the reflected wave.

In fact, when the source is situated on the boundary, the reflected wave, that is $\psi_{\text{refl}}(r, z + z_0, 0)$, can be expressed in terms of the total field, since the total field is the sum of the direct and the reflected waves

$$\psi(r, z + z_0, 0) = \frac{e^{ikR_1}}{R_1} + \psi_{\text{refl}}(r, z + z_0, 0), \qquad (19.55)$$

where the first term represents the direct wave, and the distance from the source to the receiver is $CP' = R_1$ (Fig. 94). Expressing

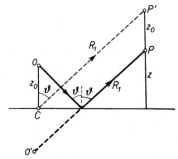

Fig. 94. Regarding the relation between the field of an elevated source and the field of a source situated on the interface.

$\psi_{\text{refl}}(r, z + z_0, 0)$ in this way, and substituting into Eq. 19.54, we obtain

$$\psi_{\text{refl}}(r, z, z_0) = \psi(r, z + z_0, 0) - \frac{e^{ikR_1}}{R_1}.$$

Finally, substituting this expression into Eq. 19.53 we obtain Eq. 19.52.

Thus, if the field of a dipole situated on the boundary between two media has been calculated, the transition to an elevated dipole presents no essential difficulties.

§ 20. THE WEYL–VAN DER POL FORMULA

1. *The limits of applicability of Eq. 19.36. The concept of numerical distance*

The applicability of Eq. 19.36 obtained above for the reflected wave has an important limitation, arising from the fact that in the case of a vertical electric dipole (only this case will be examined in what follows), one of the poles of the reflection coefficient can approach the saddle point, and the usual method of steepest descents becomes inapplicable. We shall investigate this problem in detail in this Paragraph, and will generalize the method of steepest descents to include the case in which a pole is located arbitrarily close to the saddle point. As a result, a new formula will be obtained for the reflected wave. This formula is of great importance in the theory of radiowave propagation along the surface of the earth, and we therefore limit ourselves to the case of electrodynamics.

The main steps in the derivation of Eq. 19.36 were the replacement of the variable ϑ by s according to Eq. 19.28, and the expansion of the entire integrand in Eq. 19.26 (other than the exponential), in a power series in s. It is of course necessary that this series be convergent. However, the radius of convergence of the integrand is bounded, since $V(\vartheta)$ becomes infinite (it has a pole of the first order) at the point $\vartheta = \vartheta_p$, where ϑ_p is found from the equation

$$n^2 \cos \vartheta_p + \sqrt{(n^2 - \sin^2 \vartheta_p)} = 0, \tag{20.1}$$

which gives
$$\sin \vartheta_p = \pm \frac{n}{\sqrt{(n^2 + 1)}}. \tag{20.2}$$

The location of the pole in the s-plane will be denoted by s_0. According to Eq. 19.28, s_0 and ϑ_p are connected by the relation

$$\cos (\vartheta_p - \vartheta_0) = 1 + i s_0^2. \tag{20.3}$$

The power series in s will converge inside a circle of radius s_0, on whose boundary the pole is located. The smaller s_0, the greater the extent to which the circle of convergence will shrink, up to the saddle point $s = 0$.

For the method of steepest descents (which we used above) to be applicable, it is necessary that the entire region of "significant" values of s, for which the integrand is not yet very small, occupy a small portion of the circle of convergence near its center. The upper limit of the significant values of s will be denoted by s_1. Since the integrand contains the exponential $e^{-kR_1 s^2}$, which decreases rapidly with increasing s, it can be shown that $s_1 \sim 1/\sqrt{(kR_1)}$. In expanding the integrand in a power series in s, we limited ourselves to the term in s^2; therefore, the condition $|s_1/s_0|^2 \ll 1$, or

$$kR_1 |s_0|^2 \gg 1, \tag{20.4}$$

must be satisfied. The quantity $w = \sqrt{(kR_1)} s_0$ is called the numerical distance. Consequently, in order that Eq. 19.36 be applicable, the numerical distance must be large.

According to Eq. 20.3, we have

$$w^2 = ikR_1[1 - \cos(\vartheta_p - \vartheta_0)]$$
$$= ikR_1(1 - \cos\vartheta_p \cos\vartheta_0 - \sin\vartheta_p \sin\vartheta_0). \tag{20.5}$$

We are interested in the case in which ϑ_p is close to ϑ_0 (the pole is situated close to the saddle point). This can occur in the case of grazing propagation, $\vartheta_0 \approx \pi/2$, $\sin\vartheta_0 \approx 1$, and sufficiently large n, so that with the positive sign in Eq. 20.2, we also obtain $\sin\vartheta_p \approx 1$. Thus, for the pole of interest to us, we obtain

$$\sin\vartheta_p = \frac{n}{\sqrt{(n^2+1)}}. \tag{20.6}$$

For the same pole we have

$$\cos\vartheta_p = -\frac{1}{\sqrt{(n^2+1)}}. \tag{20.7}$$

The positive sign in front of the square root is inadmissible here, since Eq. 20.1 would then give

$$\mathrm{Im}\sqrt{(n^2 - \sin^2\vartheta_p)} = -\mathrm{Im}\, n^2/\sqrt{(n^2+1)} < 0,$$

that is, instead of being on the upper sheet of the Riemann surface, we would be on the lower sheet, over which the path of integration does not pass.

Substituting Eqs. 20.6 and 20.7 into Eq. 20.5, we obtain for the numerical distance

$$w^2 = ikR_1\left[1 - \frac{n\sin\vartheta_0 - \cos\vartheta_0}{\sqrt{(n^2+1)}}\right]. \tag{20.8}$$

In the most interesting case, when $|n|^2$ is large, we expand $\sqrt{(n^2+1)}$ in powers of $1/n^2$, and obtain

$$w^2 = ikR_1\left(1 - \sin\vartheta_0 + \frac{\cos\vartheta_0}{n} + \frac{\sin\vartheta_0}{2n^2}\right). \tag{20.9}$$

Finally, remembering that the angle of incidence is close to $\pi/2$, and using the notation $\chi_0 = (\pi/2) - \vartheta_0$, we obtain

$$w^2 = \frac{ikR_1}{2n^2}(1 + n\chi_0)^2, \tag{20.10}$$

where there is now no limitation to the magnitude of the product $n\chi_0$ (if $n\chi_0 > 1$, we must require that $n\chi_0^3 \ll 1$ in order that Eq. 20.10 be valid).

When $|n|$ is large, the modulus of the numerical distance w can be of the order of unity or less, in spite of the fact that kR_1 is large. In this case, the method of analysis of the reflected wave, presented in the preceding Section, is inapplicable.

For example, let us examine the case of sea water ($\epsilon = 81$, $\sigma = 4 \times 10^{10}$). For a receiver and a radiating antenna situated on the ground ($\chi_0 = 0$), we obtain

$$| w |^2 = \frac{kR_1}{2 | n |^2} = \frac{kR_1}{2} \frac{1}{\sqrt{[\epsilon^2 + (4\sigma^2/f^2)]}}.$$

For the wavelength $\lambda = 300$ m, we obtain $| w |^2 = 1.3 \times 10^{-4} kR_1$. It is thus evident that the condition $| w |^2 \gg 1$ can turn out to be considerably more rigid than the condition $kR_1 \gg 1$.

2. *The method of steepest descents when a pole is located near the saddle point*

From the mathematical point of view, the inapplicability, for small w, of the method of steepest descents presented in the preceding Paragraph is due to the presence of a pole near the saddle point. However, the method of steepest descents can be modified in such a way that the presence of poles is taken into account from the very beginning. We thus obtain a result which is valid for any value of w, so long as kR_1 is large.[33, 68] We suppose that we have an integral of the form

$$I = \int_{-\infty}^{+\infty} e^{-kRs^2} \Phi(s) \, ds, \qquad (20.11)$$

where kR is large. Furthermore, let the function $\Phi(s)$ have a pole of the first order at the point $s = s_0$, where in general s_0 is complex. We multiply and divide the integrand by $s^2 - s_0^2$:

$$I = \int_{-\infty}^{+\infty} e^{-kRs^2} \Phi(s) \, (s^2 - s_0^2) \frac{ds}{s^2 - s_0^2}.$$

The function $\Phi(s) \, (s^2 - s_0^2)$ has no singularities at $s = s_0$. We expand this function in a power series in s

$$\Phi(s) \, (s^2 - s_0^2) = \sum_{m=0}^{\infty} \frac{c_m}{m!} s^m. \qquad (20.12)$$

It can be shown that

$$\int_{-\infty}^{+\infty} \frac{e^{-kRs^2} s^{2m} \, ds}{s^2 - s_0^2} = \frac{1 . 3 . 5 \ldots (2m-1) \sqrt{(\pi)} \, w^{2m-1}}{2^{m-1}(kR)^{m-\frac{1}{2}}} e^{-w^2} \int_w^{i\infty} \frac{e^{u^2}}{u^{2m}} du, \qquad (20.13)$$

where $w \equiv \pm s_0 \sqrt{(kR)}$, and we choose the sign in such a way that w lies in the upper halfplane. We will prove this, basing our analysis on the work of Ott.[168]

As a preliminary, we investigate the integral $\int_{-\infty}^{+\infty} e^{-\xi s^2} s^{2m} \, ds$, which is convergent for all nonzero $\xi = |\xi|e^{2i\beta}$, if $|2\beta| \leqslant \pi/2$. We introduce the new variable $t = s\sqrt{\xi}$, choosing the sign of the square root in such a way that $\mathrm{Re}\sqrt{\xi} > 0$. In the t-plane, the path of integration goes along the straight line which passes through the origin and has the slope β with respect to the real axis. However, without changing the value of the integral, this path can be transposed to the real axis. As a result, we obtain the integral

$$\frac{1}{\xi^{m+\frac{1}{2}}} \int_{-\infty}^{+\infty} e^{-t^2} t^{2m}\, dt,$$

which is in the Tables of Integrals. Using this integral, we obtain

$$\int_{-\infty}^{+\infty} e^{-\xi s^2} s^{2m}\, ds = \frac{\Gamma(m+\frac{1}{2})}{\xi^{m+\frac{1}{2}}} = \frac{\sqrt{(\pi)}\,(2m)!}{m!\,2^{2m}}\frac{1}{\xi^{m+\frac{1}{2}}}. \qquad (20.14)$$

We now multiply both sides of the last equation by $\exp(\xi s_0^2)$ and integrate with respect to ξ. The path of integration goes from the point $\xi = kR$ to infinity in such a way that the exponential $\exp(\xi s_0^2)$ decreases along the path. Using the Table of Integrals for the integral with respect to ξ, we finally obtain

$$\exp(kRs_0^2)\int_{-\infty}^{+\infty} \frac{\exp(-kRs^2)\, s^{2m}\, ds}{s^2 - s_0^2} = \Gamma(m+\tfrac{1}{2})\int_{kR} \frac{\exp(\xi s_0^2)}{\xi^{m+\frac{1}{2}}}\, d\xi. \qquad (20.15)$$

The requirement that $\exp(\xi s_0^2)$ vanish at infinity is compatible with the requirement that $\mathrm{Re}\sqrt{(\xi)} > 0$. In fact, in whatever part of the complex plane the initial value of the quantity $\xi s_0^2 = kRs_0^2$ does not appear [making use of the fact that in $\xi = |\xi|e^{2i\beta}$, the quantity $|2\beta|$ can vary arbitrarily within the limits 0, $\pi/2$] we can choose a path of integration in such a way that ξs_0^2 goes to infinity in the left half-plane, where the exponential $\exp(\xi s_0^2)$ approaches zero. We note that under these conditions, there is no need for us to cross the positive real semiaxis of the ξs_0^2-plane at any point.

We introduce a new variable $u = \sqrt{(\xi s_0^2)}$ in the last integral. Since this leads to a double-valued quantity, we construct a two-sheeted Riemann surface with a cut going along the real axis from 0 to ∞, and, as we see, not intersecting the path of integration. One sheet of the Riemann surface will be mapped onto the upper halfplane of the variable u, and the other sheet onto the lower halfplane. Which of the sheets is mapped onto the upper, and which onto the lower halfplane, depends on the choice of the sign of the square root $\sqrt{(\xi s_0^2)}$ in the change of variable. We will choose this sign in such a way that the sheet containing the path of integration will be mapped onto the upper halfplane. The path of integration will then go from $u = w$, where

$$w \equiv \pm s_0 \sqrt{(kR)}, \tag{20.16}$$

to $u = i\infty$. In order that the value $u = w$ lie in the upper halfplane, we must take the positive sign if s_0 lies in the upper halfplane and the negative sign if s_0 lies in the lower halfplane. Thus, the change of variable must be

$$u = \pm s_0 \sqrt{\xi}$$

with the above-indicated choice of sign. Now, taking Eq. 20.16 into account, we easily obtain from Eq. 20.15,

$$\int_{-\infty}^{+\infty} \frac{e^{-kRs^2} s^{2m}\, ds}{s^2 - s_0^2} = \frac{2w^{2m-1}}{(kR)^{m-\frac{1}{2}}} e^{-w^2} \Gamma(m+\tfrac{1}{2}) \int_w^{i\infty} \frac{e^{u^2}}{u^{2m}}\, du. \tag{20.17}$$

Remembering that

$$\Gamma(m+\tfrac{1}{2}) = \sqrt{\pi}\, \frac{1 \cdot 3 \cdot 5 \dots (2m-1)}{2^m}, \tag{20.18}$$

we obtain Eq. 20.13, which completes the proof.

We note that as the pole $s = s_0$ crosses the real axis, the value of the integral (20.13) changes discontinuously, since we must change the sign in Eq. 20.16, which defines w. However, this is quite natural, since the path of integration goes along the real axis, and as the pole crosses this axis, we must add the residue of the function at the pole to the integral.

Let us again return to the original integral (20.11). Using expansion (20.12) and Eq. 20.13, we obtain

$$I = \sum_{m=0}^{\infty} \frac{2\sqrt{(\pi)}\, c_{2m}\, w^{2m-1}}{4^m \cdot m! \, (kR)^{m-\frac{1}{2}}} e^{-w^2} \int_w^{i\infty} \frac{e^{u^2}\, du}{u^{2m}}. \tag{20.19}$$

In integrating over s, the terms containing odd powers of s vanish.

Thus, the problem is solved for any distribution of poles, no matter how closely some of them may approach the saddle point.

In distinction from the usual method of steepest descents the final result in Eq. 20.19 is not a power series in $1/kR$, because kR also enters into w. However, if the numerical distance is large, and we represent the integral in Eq. 20.19 in the form of a semiconvergent power series in $1/w$ (using integration by parts), it is not hard to show that we immediately obtain the same power series in $1/kR$ that is obtained if the usual method of steepest descents is applied to Eq. 20.11 (that is, we expand $\Phi(s)$ in a power series in s, and integrate term by term).

3. *The Weyl–Van der Pol formula*

We now apply the method developed above to our problem. We rewrite Eq. 19.3 for the reflected wave in a somewhat different form, adding and subtracting unity from the function $V(\vartheta)$ in the integrand. We then obtain a result which can be written in the form

$$
\psi_{\text{refl}} = \frac{ik}{2\pi} \int_0^{\frac{1}{2}\pi - i\infty} \int_0^{2\pi} \exp\{ik[x \sin\vartheta \cos\phi + y \sin\vartheta \sin\phi
$$
$$
+ (z + z_0)\cos\vartheta]\}\sin\vartheta \, d\vartheta \, d\phi
$$
$$
+ \frac{ik}{2\pi} \int_0^{\frac{1}{2}\pi - i\infty} \int_0^{2\pi} \exp\{ik[x \sin\vartheta \cos\phi + y \sin\vartheta \sin\phi
$$
$$
+ (z + z_0)\cos\vartheta]\}[V(\vartheta) - 1]\sin\vartheta \, d\vartheta \, d\phi.
$$

In this expression, the first term represents the spherical wave e^{ikR_1}/R_1 {where $R_1 = \sqrt{[x^2 + y^2 + (z + z_0)^2]}\}$, issuing from the "image" source O' (see Fig. 91). We can easily convince ourselves of this by comparing this term with the first of Eqs. 18.19. We transform the second term in the same way that we transformed Eq. 19.3. We then arrive at a formula, completely analogous to Eq. 19.26, the only difference being that in place of $V(\vartheta)$, we now have $V(\vartheta) - 1$. Furthermore, we will neglect the term $1/(8kr\sin\vartheta)$ compared with unity. As a result, we obtain, for the reflected wave

$$
\psi_{\text{refl}} = \frac{e^{kR_1}}{R_1} + \sqrt{\left(\frac{k}{2\pi r}\right)}\exp\left(i\frac{\pi}{4}\right)\int_{-\frac{1}{2}\pi + i\infty}^{\frac{1}{2}\pi - i\infty} \exp[ikR_1 \cos(\vartheta - \vartheta_0)]
$$
$$
\times [V(\vartheta) - 1]\sqrt{(\sin\vartheta)}\, d\vartheta : \tag{20.20}
$$

and from Eq. 19.7, we find

$$
V(\vartheta) - 1 = -\frac{2q(\vartheta)}{M(\vartheta)}, \tag{20.21}
$$

where $\qquad q(\vartheta) \equiv \sqrt{(n^2 - \sin^2\vartheta)}, \quad M(\vartheta) = n^2\cos\vartheta + q.$

As in §19, we change the variable of integration in accordance with relation (19.28), and change the path of integration in such a way that the integration is carried over real values of s (the passage path of integration). It is not hard to show that under these conditions

$$
d\vartheta = \sqrt{(2)}\exp\left(-i\frac{\pi}{4}\right)\frac{ds}{\sqrt{[1 + (is^2/2)]}}. \tag{20.22}
$$

As a result, we reduce the integral in Eq. 20.20 to the form of

Eq. 20.11, after which, according to Eq. 20.19, we have

$$\psi_{\text{refl}} = \frac{e^{ikR_1}}{R_1} + \sqrt{\left(\frac{2k}{r}\right)} \exp\left[ikR_1 + i(\pi/4)\right]$$

$$\sum_{m=0}^{\infty} \frac{c_{2m}}{4^m\,m!\,(kR_1)^{m-\frac{1}{2}}} w^{2m-1} \exp(-w^2) \int_w^{i\infty} \frac{e^{u^2}}{u^{2m}} du, \quad (20.23)$$

where w is determined from Eq. 20.8; in extracting the square root, the sign is chosen such that the imaginary part of w is positive; c_{2m} are the coefficients of the expansion

$$-\frac{2q(\vartheta)}{M(\vartheta)} \sqrt{(\sin\vartheta)} \frac{d\vartheta}{ds} (s^2 - s_0^2) = \sum_{m=0}^{\infty} \frac{c_m}{m!} s^m. \quad (20.24)$$

To determine these coefficients, we must substitute $d\vartheta/ds$ from Eq. 20.22 into Eq. 20.24; in addition, we must express ϑ in terms of s in the expressions $q(\vartheta)\sqrt{(\sin\vartheta)}$ and $M(\vartheta)$. For this purpose, Eq. 19.28 gives, to within s^2,

$$\vartheta - \vartheta_0 \approx \sqrt{(2)} \exp\left[-i(\pi/4)\right] \cdot s\left(1 + \frac{s^2}{12i}\right).$$

As a result, grouping the terms on the left hand side of Eq. 20.24 which do not contain s, and the terms which are proportional to s^2, we obtain for the most interesting case $\vartheta_0 \approx \pi/2$, $|n| \gg 1$:

$$c_0 = \frac{2s_0}{n}, \quad c_2 = \frac{2\cos\vartheta_0}{s_0}, \quad s_0 = \frac{\exp\left[i(\pi/4)\right]}{\sqrt{(2)}\,n} (1 + n\cos\vartheta_0). \quad (20.25)$$

Substituting these values of c_0 and c_2 into Eq. 20.23, we note that the second term can be completely neglected (particularly if we remember that ϑ_0 is close to $\pi/2$, and consequently, $\cos\vartheta$ is small), and as a result, we obtain

$$\psi_{\text{refl}} = \frac{e^{ikR_1}}{R_1}\left[1 + \frac{1}{n}\sqrt{(8kR_1)} \exp\left(i\frac{\pi}{4} - w^2\right) \int_w^{i\infty} e^{u^2} du\right]. \quad (20.26)$$

If to this expression we add the expression for the direct wave e^{ikR}/R, we obtain the total field.

Equations 19.36 and 20.26 are mutually complementary. The first is suitable for calculations with arbitrary n (arbitrary characteristics of the medium), but only for large values of the numerical distance. The second is suitable only for large values of $|n|$, but with any value of the numerical distance. For large $|n|$ and large numerical distances w^2, both formulas are applicable, and must give the same result, as will be shown below.

In the special case in which the source is situated on the boundary, we have $R = R_1$ (see Fig. 91), and for the total field in the upper medium, we obtain

$$\psi = \frac{2\,e^{ikR}}{R}\left[1 + \frac{1}{n}\sqrt{(2kR)}\exp\left(i\frac{\pi}{4} - w^2\right)\int_w^{i\infty} e^{u^2}\,du\right]. \qquad (20.27)$$

The last expression is usually called the Weyl–Van der Pol formula. As we see, our method made it possible to obtain Eq. 20.26, which is also valid for an elevated dipole.

The quantity $2\,e^{ikR}$ in Eq. 20.27 corresponds to the Hertz vector in the case of a perfectly reflecting boundary. The factor in the parentheses characterizes the additional attenuation of the field due to the flow of electromagnetic energy into the lower medium, and is called the amplitude function. We note that the amplitude function becomes unity for small values of the numerical distance w^2.

When the numerical distance is large, Eq. 20.26 must agree with Eq. 19.36, obtained by the usual method of steepest descents. We shall verify this. When w^2 is large, integration by parts gives

$$\int_w^{i\infty} e^{u^2}\,du = -\frac{1}{2w}e^{w^2} + \frac{1}{2}\int_w^{i\infty}\frac{e^{u^2}}{u^2}\,du = -\frac{1}{2w}e^{w^2} - \frac{1}{4w^3}e^{w^2} - \cdots$$

Limiting ourselves to the two terms written above, we substitute into Eq. 20.26, and obtain

$$\psi_{\text{refl}} = \frac{e^{ikR_1}}{R_1}\left[1 - \frac{1}{nw}\sqrt{(2kR_1)}\,e^{i(\pi/4)}\left(1 + \frac{1}{2w^2}\right)\right],$$

or, using Eq. 20.10 for w^2,

$$\psi_{\text{refl}} = \frac{e^{ikR_1}}{R_1}\left\{1 - \frac{2}{1 + n\chi_0}\left[1 + \frac{n^2}{ikR_1(1 + n\chi_0)^2}\right]\right\}.$$

The last equation is easily transformed into

$$\psi_{\text{refl}} = \frac{e^{ikR_1}}{R_1}\left[\frac{n\chi_0 - 1}{n\chi_0 + 1} + \frac{2in^2}{kR_1(1 + n\chi_0)^3}\right]. \qquad (20.28)$$

This expression is in complete agreement with Eq. 19.36, since with our assumptions $[\,|n| \gg 1, \chi_0 = (\pi/2) - \vartheta_0 \ll 1\,]$ we obtain therein, according to Eqs. 19.38 and 19.40:

$$\gamma_0 \approx \chi_0, \qquad V \approx \frac{n\chi_0 - 1}{n\chi_0 + 1}, \qquad N \approx -\frac{2n^2}{(1 + n\chi_0)^3}. \qquad (20.29)$$

The generalized method of steepest descents which we have used in this Section permits us to elucidate satisfactorily the question of the existence of the Zenneck surface wave in the field of a vertical dipole situated on an interface between two media.[169]

§ 21. LATERAL WAVES

1. *Expressions for the lateral wave*

The analysis of the reflected wave carried out in §§ 19 and 20 is incomplete because the fact that the function $V(\vartheta)$ in the integrand in Eq. 19.26 is double-valued was not taken into account. Taking this circumstance into account requires us to add one more term to Eq. 19.36, obtained above for the reflected wave. This term corresponds to the so-called lateral wave. First, we shall present formally the mathematical basis for the appearance of this term, and then we shall discuss its physical significance.

In § 19, we replaced the integration over the original path Γ_1 by integration over the path Γ (Fig. 93). We now introduce some corrections to this transformation.

As has already been indicated in § 19 (page 252), the integrand in Eq. 19.6 is double-valued, since the square root $\sqrt{(n^2 - \sin^2\vartheta)}$ enters into the expression for $V(\vartheta)$. Our original path of integration Γ_1 passes over the upper sheet of the corresponding Riemann surface, and can be deformed into the passage path of integration Γ only when at least the beginning and the end of the latter lie on this sheet.

Let us consider, for example, the case in which the imaginary part of n is vanishingly small, and the real part is less than unity. If, under these conditions, the angle of incidence ϑ_0 satisfies the condition

$$\vartheta_0 < \delta, \quad \delta = \arcsin n, \tag{21.1}$$

that is, it does not exceed the angle of total internal reflection δ, then, in the complex plane, we have the picture shown in Fig. 93 (the point A_1 corresponds to $\vartheta = \delta$). Here, the passage path intersects the cut twice. The transition from the path Γ_1 to the passage path Γ is accomplished without any complications. Except for a part, shown dotted in the Figure, the path Γ will lie in the upper sheet.

In this case, the analysis carried out in § 19 is complete, and there will be no lateral wave. The situation will be different in the opposite case, in which the angle ϑ_0 is greater than *the angle of total internal reflection*, i.e. when

$$\vartheta_0 > \arcsin n. \tag{21.2}$$

The relative positions of the branch point and the passage path of integration for this case are shown in Fig. 95. Here, the passage path of integration intersects the cut once, crosses from the upper sheet to the lower sheet, and goes to infinity along the lower sheet, without connecting anywhere with the path Γ_1, which is inadmissible. However, we can construct a more complicated path of integration, supplementing the passage path by a contour, encompassing the cut in such a way that the beginning and the end of the more complicated path will again lie in the upper sheet.

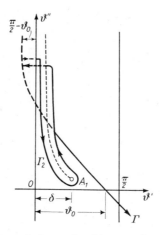

Fig. 95. Transformation of the path of integration in the complex plane, with a circuit around a cut.

The complete path of integration is shown in Fig. 95. This path goes from the point $-(\pi/2)+\vartheta_0+i\infty$ to the point $i\infty$, and then goes around the cut along contour Γ_2. It then intersects the cut and goes along the lower sheet to the initial point $-(\pi/2)+\vartheta_0+i\infty$. From this point, the integration is carried along the passage path Γ, the initial part of which, shown by the dotted line, now lies in the lower sheet.

As a result, the complete expression for the reflected wave will consist of two parts

$$\psi = \psi_{\text{refl}} + \psi_{\text{lat}}, \tag{21.3}$$

where ψ_{refl}, the reflected wave proper, is expressed as an integral over the passage path and was studied in § 19, and ψ_{lat} is the "lateral" wave, the expression for which is given by the integral over the contour Γ_2, encompassing the cut, or, as is sometimes said, by the integral around the edges of the cut.

Using Eq. 19.26, and neglecting $1/8kr \sin \vartheta$ compared with unity, we obtain

$$\psi_{\text{lat}} = \sqrt{\left(\frac{k}{2\pi r}\right)} \exp\left[i(\pi/4)\right] \left\{ \int_{i\infty}^{A_1} \exp\left[ikR_1 \cos(\vartheta - \vartheta_0)\right] V(\vartheta) \sqrt{(\sin \vartheta)}\, d\vartheta \right.$$

$$\left. + \int_{A_1}^{i\infty} \exp\left[ikR_1 \cos(\vartheta - \vartheta_0)\right] V^+(\vartheta) \sqrt{(\sin \vartheta)}\, d\vartheta \right\},$$

where $V(\vartheta)$ is the value of the reflection coefficient on the left border of the cut, and $V^+(\vartheta)$ is the value on the right. The imaginary part of the square root $\sqrt{(n^2 - \sin^2 \vartheta)}$ entering into the expression for the reflection coefficient is zero on the cut, and the real part has different signs on the two borders of the cut. Therefore, $V(\vartheta)$ and $V^+(\vartheta)$ will differ from one another only in the signs of the square roots. As was shown in § 19, Section 3, the real part of the square root is positive on the left border of the cut (and at $\vartheta = 0$).

Interchanging the limits of integration in the first integral, we can combine the two integrals into one, taken from A_1 to $i\infty$ along the cut. We then obtain

$$\psi_{\text{lat}} = \sqrt{\left(\frac{k}{2\pi r}\right)} \exp\left[i(\pi/4)\right] \int_{A_1}^{i\infty} \exp\left[ikR_1 \cos(\vartheta - \vartheta_0)\right] \Phi(\vartheta) \sqrt{(\sin \vartheta)}\, d\vartheta.$$

$$(21.4)$$

Using Eq. 19.38, and remembering that $V^+(\vartheta)$ is obtained from $V(\vartheta)$ by changing the sign of the square root $\sqrt{(n^2 - \sin^2 \vartheta)}$, we obtain

$$\Phi(\vartheta) = V^+(\vartheta) - V(\vartheta) = \frac{4m \cos \vartheta \sqrt{(n^2 - \sin^2 \vartheta)}}{(m \cos \vartheta)^2 - n^2 + \sin^2 \vartheta}. \qquad (21.5)$$

It will be convenient in what follows to use another form of the method of steepest descents.† For this purpose, the path of integration is deformed in such a way that it goes from the branch point A_1 along the line on which the exponential in the integrand decreases most rapidly. This will be the line on which the real part of $kR_1 \cos(\vartheta - \vartheta_0)$ is constant:

$$\text{Re} \cos(\vartheta - \vartheta_0) = \text{const.} \qquad (21.6)$$

It is then necessary that

$$\text{Im} \cos(\vartheta - \vartheta_0) > 0. \qquad (21.7)$$

As previously, we use the notation

$$\delta = \arcsin n \quad (\text{at the point } A_1, \vartheta = \delta)$$

---

† Translator's note: This particular form is actually called the *method of steepest descents* in Russian, and is thus distinguished from the so-called *method of passage* previously outlined.

and assume that n is real. Then, setting $\vartheta = \delta$ in Eq. 21.6, we obtain

$$\text{const} = \cos(\delta - \vartheta_0).$$

Now, remembering that

$$\cos(\vartheta - \vartheta_0) = \cos(\vartheta' + i\vartheta'' - \vartheta_0)$$

$$= \cos(\vartheta' - \vartheta_0)\cosh\vartheta'' - i\sin(\vartheta' - \vartheta_0)\sinh\vartheta'', \qquad (21.8)$$

we obtain the equation of the path of integration from Eq. 21.6:

$$\cos(\vartheta_0 - \vartheta')\cosh\vartheta'' = \cos(\delta - \vartheta_0). \qquad (21.9)$$

This path is shown as a solid line (path Γ_2') in Fig. 96. Condition (21.7)

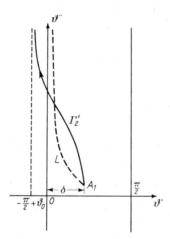

Fig. 96. The transformation of the integral over the cut L into an integral over the path of steepest descent Γ_2.

is also satisfied on this path. The deformation of the path of integration from the cut L to the contour Γ_2' is performed without difficulty, because there are no singularities of the integrand in the part of the ϑ-plane lying between them. In particular, as was shown in § 19, the pole P_1 (see Fig. 93) lies outside this region.

As a result, if we take Eq. 21.8 into account, Eq. 21.4 gives

$$\psi_{\text{lat}} = \sqrt{(k/2\pi r)}\exp\left[ikR_1\cos(\delta - \vartheta_0) + i(\pi/4)\right]$$

$$\times \int_{\Gamma_2'}\Phi(\vartheta)\exp\left[-kR_1\sin(\vartheta_0 - \vartheta')\sinh\vartheta''\right]\sqrt{(\sin\vartheta)}\,d\vartheta. \qquad (21.10)$$

Now, using Eq. 21.9, we express the entire integrand in terms of ϑ'', which we treat as a new variable of integration. As is clear from Fig. 96, the limits of integration over ϑ'' will be 0 and ∞. Since kR_1 is assumed to be large, and we have chosen the path of most rapid decrease, only small values of ϑ'' will play an important rôle in the integrand. Moreover, as is seen in Fig. 96, we can set $\vartheta' \approx \delta$ on the initial part of the path of integration. Therefore, we set

$$\sin(\vartheta_0 - \vartheta') \sinh \vartheta'' \approx \sin(\vartheta_0 - \delta)\,\vartheta'',$$

$$d\vartheta \approx i\,d\vartheta''.$$

In Eq. 21.5 for $\Phi(\vartheta)$, we can set $\vartheta \approx \delta$ everywhere except in the square root $\sqrt{(n^2 - \sin^2 \vartheta)}$, which becomes zero here, since $\sin \delta = n$. We therefore calculate it to a higher approximation. We have

$$\sqrt{(n^2 - \sin^2 \vartheta)} = \sqrt{[n^2 - (\sin \vartheta' \cosh \vartheta'' + i \cos \vartheta' \sinh \vartheta'')^2]}.$$

Setting $\vartheta' \approx \delta$, $\sinh \vartheta'' \approx \vartheta''$, $\cosh \vartheta'' \approx 1$, we obtain

$$\sqrt{(n^2 - \sin^2 \vartheta)} = \sqrt{(-2in \cos \delta . \vartheta'')} = -\sqrt{(2)} \exp[-i(\pi/4)] \sqrt{(n \cos \delta . \vartheta'')}.$$

Of the two possible values of $\sqrt{-i}$, we have chosen the one which

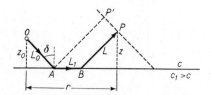

Fig. 97. The ray representation of a lateral wave.

guarantees that the imaginary part of $\sqrt{(n^2 - \sin^2 \vartheta)}$ will be positive. Equation 21.10 is now written

$$\psi_{\text{lat}} = \frac{-4i\sqrt{(k)} . n}{\sqrt{(\pi r)}\, m\, \sqrt{(\cos \delta)}} \exp[ikR_1 \cos(\delta - \vartheta_0)]$$

$$\times \int_0^\infty \exp[-kR_1 \sin(\vartheta_0 - \delta)\,\vartheta'']\sqrt{(\vartheta'')}\,d\vartheta''. \quad (21.11)$$

Introducing the new variable $\sqrt{\vartheta''} = x$, and using the second integral in Eqs. 19.15, we obtain

$$\psi_{\text{lat}} = \frac{-2in \exp[ikR_1 \cos(\delta - \vartheta_0)]}{km\sqrt{(r \cos \delta)}\,[R_1 \sin(\delta - \vartheta_0)]^{\frac{3}{2}}}. \quad (21.12)$$

This is the expression we are seeking for the lateral wave. It can also be represented in a somewhat different form. We have

$$kR_1 \cos(\delta - \vartheta_0) = kR_1 (\cos \delta \cos \vartheta_0 + \sin \delta \sin \vartheta_0).$$

We will take into account that (Fig. 91) $R_1 \sin \vartheta_0 = r$, $R_1 \cos \vartheta_0 = z + z_0$, and that according to Fig. 97,

$$r = L_1 + (z_0 + z) \tan \delta.$$

Furthermore, we have $k \sin \delta = kn = k_1$. Using all of these relations, we easily obtain

$$kR_1 \cos (\delta - \vartheta_0) = k [nr + \sqrt{(1 - n^2)} (z + z_0)] = k(L_0 + L) + k_1 L_1,$$

$$R_1 \sin (\vartheta_0 - \delta) = R_1 (\sin \vartheta_0 \cos \delta - \cos \vartheta_0 \sin \delta)$$

$$= \cos \delta [r - (z + z_0) \tan \delta] = L_1 \cos \delta. \qquad (21.13)$$

As a result, Eq. 21.12 for the lateral wave can be written in the form

$$\psi_{\text{lat}} = \frac{2in}{km(n^2 - 1) \sqrt{(r)} L_1^{\frac{3}{2}}} \exp [ik(L_0 + L) + ik_1 L_1]. \qquad (21.14)$$

The expression $k(L_0 + L) + k_1 L_1$ can be treated as the phase change along the ray $OABP$ (Fig. 97), joining the source and the point of observation. This ray consists of the segments L_0 and L along which the wave propagates in the upper medium at the angle of total internal reflection with respect to the normal to the boundary, and the segment L_1, in which the wave propagates along the boundary with a velocity

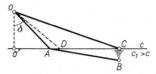

Fig. 98. Elucidation of the nature of a lateral wave.

equal to the velocity in the lower medium. At large distances, for which $r \gg (z + z_0)$, we have $L_1 \approx r$, whence it is clear that the amplitude of the lateral wave will decrease with distance as $1/r^2$. As we see, the expression we have obtained for the lateral wave loses meaning when $n \to 1$ and when $L_1 \to 0$ (when $\vartheta_0 \to \delta$). We shall remove the latter limitation in § 22. The case in which n is very close to unity was considered by the author in Ref. 43.

2. Physical meaning of the lateral wave

It is clear from what has been presented above that the lateral wave is associated with propagation along L_1 (Fig. 97) in the lower medium. Therefore, we cannot carry out a complete analysis of the nature of the lateral wave without considering wave processes in the lower medium. These processes will be analyzed in more detail in § 23. For the present, we shall dwell only on that which is necessary for an understanding of the nature of the lateral wave.

We shall assume that the source is situated at the point O in Fig. 98. At the point B, sufficiently far from the source O, but situated near the boundary in the lower medium, the wave is incident along the two paths corresponding to the rays OAB and OC. The ray OC is incident on the boundary at an angle greater than the angle of total internal reflection, and being totally reflected, creates a wave in the lower medium which is attenuated exponentially with depth. The ray OA undergoes the usual refraction, and travels along AB in the lower medium to the point B. The closer the ray OA approaches the dotted line OD, corresponding to the angle of total internal reflection, the closer will the ray AB approach the boundary. The wave represented by the ray AB is the cause of the lateral wave. In fact, it propagates along the boundary with the velocity c_1, creating a corresponding disturbance on the boundary. This gives rise to a new wave in the upper medium. Since the spatial period of this disturbance along the boundary is equal to λ_1, the wavelength in the lower medium, the wave in the upper medium can "fit" this period only if the angle δ between its propagation direction and the normal to the boundary is such that $\lambda = \lambda_1 \sin \delta$ (Fig. 99). This is just the direction of the lateral wave.

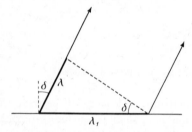 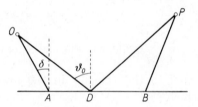

Fig. 99. Relation between the wavelengths in the upper and lower media and the angle of inclination of the lateral wave front.

Fig. 100. Paths of rays arriving at the point of observation P.

As an additional basis for the above considerations, it can be shown (see, also, § 24) that the entire wave process propagating along the boundary can be separated into two groups of waves. The first group contains the usual incident wave (ray OC, Fig. 98), the corresponding reflected wave, and the exponentially attenuating refracted wave CB. The second group contains the wave traveling along the path OAB, and the lateral wave. Each group propagates along the boundary with its own velocity, and satisfies the boundary conditions separately.

It is important to note that over the portion $O'D$ of the boundary (Fig. 98), the only waves present are the incident, reflected and refracted waves, contained in the first of the two groups mentioned above. Therefore, this portion of the boundary will not radiate a lateral wave.

Thus, if the angle ϑ_0 is greater than δ (Fig. 100), two waves will arrive at the point P—a reflected wave corresponding to the ray ODP, and a lateral wave, the phase of which is given by the optical length of the ray $OABP$.

In the case of a pulse, the lateral wave arrives at the point of observation first. Very effective methods of seismic exploration are based on the recording and analysis of this lateral wave. The theory of the lateral wave for some simple pulse forms is considered in Ref. 49.

As is seen from Eqs. 21.13 and 21.14, the wave front of the lateral wave is given by the equation

$$k(L_0+L)+k_1 L_1 = k[nr + \sqrt{(1-n^2)}\,(z+z_0)] = \text{const.} \qquad (21.15)$$

This is a straight line in the xz-plane. In space, the wave front will be a conic as a result of the cylindrical symmetry of the problem.

The wave fronts of the direct, reflected and lateral waves, and of the wave in the lower medium giving rise to the lateral wave, are shown in Fig. 101. According to Eq. 21.15, the normal to the wave front of the

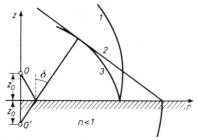

Fig. 101. Wave fronts of various types. 1. Direct wave. 2. Lateral wave. 3. Reflected wave.

lateral wave makes an angle of $\delta = \arcsin n$ with respect to the normal to the boundary. The lower edge of the front of the lateral wave coincides with the edge of the front of the wave propagating in the lower medium with the velocity c_1 ($c_1 > c$). The upper edge of the wave front of the lateral wave merges with the front of the reflected wave, which can be represented as a wave arriving from the image source O'. The amplitude of the lateral wave increases along its front from the interface to the point at which it merges with the reflected wave. This is clear from Eq. 21.14 since under these conditions the magnitude of L_1 decreases (see Fig. 97, in which the front of the lateral wave is shown by the dotted line PP').

We note that all the calculations presented in the preceding Section could also have been performed for the case $n > 1$ ($c_1 < c$).

We would then have obtained an expression for the lateral wave, agreeing completely with Eq. 21.12, except that $\delta = \arcsin n$ would be complex. Using Eq. 21.13 for the phase of the lateral wave, and setting $n > 1$, we obtain the factor $\exp[-k\sqrt{(n^2-1)}(z+z_0)]$, in Eq. 21.12, which gives an exponentially decreasing wave amplitude for increase in z as well as in z_0.

It is interesting to examine the nature of the lateral wave in this case. Since for $n > 1$, the usual type of wave refraction occurs (without total internal reflection), and upon refraction, the normal to the wave front approaches the normal to the boundary of separation, it would seem at first glance that in this case there could not be a wave in the lower medium, propagating along the interface.

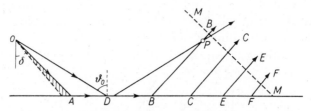

Fig. 102. The relation between the ray displacement upon reflection and the lateral wave.

However, we must remember that the point source also radiates inhomogeneous plane waves. Upon refraction into the lower medium, one of these inhomogeneous waves, with the wave vector components $k_z = ik\sqrt{(n^2-1)}$, $k_x = kn$, is transformed into an *ordinary plane wave* propagating along the boundary with the wave vector components $k_x = kn$, $k_z = 0$. This plane wave will also radiate a lateral wave into the upper medium, and the amplitude of the lateral wave will be attenuated exponentially.

We see that in all cases, the lateral wave has the nature of an offshoot of a wave propagating in the lower medium. It propagates sideways, as it were, from the main path (lying in the upper medium), which justifies its name.

3. *Ray displacement upon reflection and the lateral wave*

The excitation of the lateral wave can also be easily understood by considering the ray displacement upon reflection, which we examined in § 8. Strictly speaking, it is true that in § 8 we considered the displacement only of bounded beams. However, it will be shown in the next Paragraph that the displacement obtained in § 8 must be ascribed to every ray issuing from a point source and undergoing total internal reflection.

The closer the angle of incidence is to the angle of total internal reflection, the greater will be the ray displacement along the boundary upon reflection. As a result, the shaded beam of rays shown issuing from O in Fig. 102, will, after reflection, be broken into a set of rays BB, CC, EE, FF, etc., traveling almost parallel to one another. These rays form the lateral wave with the front MM. Thus, at an arbitrary point P there will be two waves: a lateral wave represented by the ray $OABP$, and an ordinary reflected wave represented by the ray ODP, which suffers very little displacement if its angle of incidence is not too close to δ.

These considerations allow us to obtain not only the phase of the lateral wave, and consequently, the location of its front, but also the dependence of its amplitude on distance.

In fact, it was shown in § 8 that a ray, incident on the boundary at the angle ϑ $(\vartheta > \delta)$, is, upon reflection, displaced along the boundary by the amount

$$\Delta \sim \frac{1}{\sqrt{(\sin^2 \vartheta - n^2)}}.$$

As a result, after reflection, the beam of rays issuing from O and having an angular divergence of $d\vartheta$ in the rz-plane, will be broken into a beam extending along the boundary through a distance proportional to

$$\frac{d\Delta}{d\vartheta} d\vartheta \sim \frac{\sin \vartheta \cos \vartheta \, d\vartheta}{(\sin^2 \vartheta - n^2)^{\frac{3}{2}}} \sim d\vartheta \, \Delta^3.$$

For a given $d\vartheta$, the intensity of the wave is inversely proportional to the cross sectional area of the ray tube. Remembering that the divergence of the ray tube in the plane perpendicular to the plane of Fig. 102 is proportional to r, we obtain

$$I \sim \frac{1}{r\Delta^3}$$

for the intensity, or

$$A \sim \frac{1}{\sqrt{(r)}\,\Delta^{\frac{3}{2}}}$$

for the amplitude. The same dependence of the amplitude on distance is given by Eq. 21.14, in which the displacement Δ is denoted by L_1. We can obtain Eq. 21.14 in a complete form through arguments based on ray representations, if they are carried through properly, taking account of all factors of proportionality.

The connection between the lateral wave and the ray displacement upon reflection is also considered in Ref. 158.

4. *The case of a source and receiver located in a strongly absorbing medium*

We have already mentioned that the lateral wave plays an essential rôle in seismic exploration, because the pulse corresponding to it arrives first. Another case in which the lateral wave is of the greatest importance is the case of radio communication between two points, situated in a strongly absorbing medium, such as, for example, in the earth or in the sea. In this case, the direct and the reflected waves will propagate in the absorbing medium (for brevity, we say in the earth), and consequently, will be very rapidly attenuated. This leaves the lateral wave, which penetrates from the earth into the air, propagates through the air, and then returns into the earth.

If the source is a vertical dipole,† the corresponding component of the Hertz vector of the lateral wave will be given by Eq. 21.14. However, this expression must be transformed somewhat. In Eq. 21.14 we have $n = c/c_1$, where c is the velocity in the medium in which the source and receiver are situated (the earth in our case), and c_1 is the velocity in air. If, as usual, we let n_1 denote the index of refraction of the earth with respect to air, and not the opposite, then $n_1^2 = \epsilon + 4\pi i\sigma/\omega$, where ϵ and σ are the dielectric constant and conductivity, respectively, and $n = 1/n_1$. We also make this substitution in Eq. 21.14. We must also remember that in this expression $k = \omega/c_1 = k_0 n_1$, $m = n^2 = 1/n_1^2$, where k_0 is the wave number in air. We use Eq. 21.13 for the phase, and limit ourselves to the case in which the distances of the source and receiver from the boundary are small compared with the horizontal distance between them. In this case, we can set $L_1 \approx r$ (see Fig. 97). As a result, Eq. 21.14 gives

$$\psi_{\text{lat}} = \frac{2in_1^2}{k_0 r^2(1 - n_1^2)} \exp\{ik_0[r + \sqrt{(n_1^2 - 1)}\,(z + z_0)]\}. \qquad (21.16)$$

As the depth of the source as well as that of the receiver increases, ψ_{lat} will decrease rapidly. The amplitude will decrease according to the law $\exp[-k_0\kappa(z + z_0)]$, where $\kappa = \text{Im}\sqrt{(n_1^2 - 1)}$, and z and z_0 are the depths of the receiver and the radiating antenna.

---

† If the dipole is located in a cylindrical or spherical cavity, a supplementary investigation of the radiation from the system consisting of the dipole and the cavity is required. This is beyond the scope of the present work.

§ 22. The Field in the Region Close to the Angle of Total Internal Reflection

1. Qualitative considerations. The region of applicability of the formulas obtained above

As was noted in § 21, the expression for the lateral wave (21.14) loses meaning as the point of observation approaches the point at which the front of the lateral wave merges with the reflected wave. In this region, the angle between the reflected ray and the normal to the boundary is close to the angle of total internal reflection (Fig. 103). The point P approaches the point P' (Fig. 97), and L_1 approaches zero. Applying Eq. 21.14, we obtain $\psi_{\text{lat}} \to \infty$, which has no meaning. Letting $\vartheta_0 \to \delta$ in Eq. 21.12 leads to the same result.

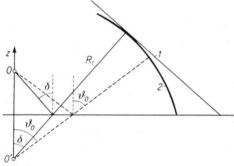

Fig. 103. The region of angles close to the angle of total internal reflection.
1. The lateral wave front. 2. The reflected wave front.

In this region of angles, Eq. 19.36, obtained above for the reflected wave, also loses meaning. In fact, according to Eq. 19.40, the quantity N (in Eq. 19.36) contains q_0 in its denominator, and q_0 approaches zero as $\vartheta_0 \to \delta$ ($\sin \vartheta_0 \to n$). As a result, the expression for the reflected wave becomes infinite.

Our formulas become inapplicable as $\vartheta_0 \to \delta$ for the following reason. In the derivations, we used the method of steepest descents, and treated the reflection coefficient $V(\vartheta)$ as a slowly varying function. But near the angle of total internal reflection this is not so. In particular, the derivative $(dV/d\vartheta)_{\vartheta \to \delta}$ not only is not small, but, on the contrary, becomes infinite. In fact, we have

$$V(\vartheta) = \frac{m \cos \vartheta - \sqrt{(n^2 - \sin^2 \vartheta)}}{m \cos \vartheta + \sqrt{(n^2 - \sin^2 \vartheta)}},$$

$$\left(\frac{dV}{d\vartheta}\right) = \frac{2m \sin \vartheta (1 - n^2)}{\sqrt{(n^2 - \sin^2 \vartheta)} \, [m \cos \vartheta + \sqrt{(n^2 - \sin^2 \vartheta)}]^2}; \tag{22.1}$$

and as $\vartheta \to \delta$ we have $\sin \vartheta \to n$, whence $(dV/d\vartheta)_\delta \to \infty$. Hence, it is clear that at these angles, the function $V(\vartheta)$ cannot be treated as slowly varying and, consequently, the method of steepest descents cannot be applied.

In this Section, we carry out a special analysis of the field in the region of angles close to the angle of total internal reflection. It will be shown that the results obtained above are valid only under the condition

$$kR_1(\vartheta_0 - \delta)^2 \gg 1. \tag{22.2}$$

With the exact theory, we can see how the front of the lateral wave branches off from the front of the reflected wave in the transition from the region $\vartheta_0 < \delta$ to the region $\vartheta_0 > \delta$.

It is also interesting to note that at $\vartheta_0 = \delta$, the reflected wave can be calculated to the first approximation by geometrical optics, but the correction in the next approximation will not be of the order $1/kR_1$ as, for example, in Eq. 19.36, but will be of the order $1/(kR_1)^{\frac{1}{4}}$.

2. Complete expressions for the field

We shall begin with the analysis of the lateral wave, again starting from integral (21.4), in which $\Phi(\vartheta)$ is given by Eq. 21.5. Since we are interested in the angles ϑ close to δ, it is convenient to introduce the new variable of integration $\beta = \delta - \vartheta$. The exponential in the integrand is then

$$
\begin{aligned}
\exp\left[ikR_1 \cos(\vartheta - \vartheta_0)\right] &= \exp\{ikR_1[\cos(\delta - \vartheta_0)\cos\beta + \sin(\delta - \vartheta_0)\sin\beta]\} \\
&= \exp\left[ikR_1\cos(\delta - \vartheta_0)\right] \\
&\quad \times \exp\{-2ikR_1[\cos(\delta - \vartheta_0)\sin^2(\beta/2) \\
&\quad\quad - \sin(\delta - \vartheta_0)\sin(\beta/2)\cos(\beta/2)]\}. \tag{22.3}
\end{aligned}
$$

Since kR_1 is large, we can choose the path of integration over β in such a way that only small values of β, not exceeding

$$|\beta|_{\max} \sim \frac{1}{\sqrt{(kR_1)}} \tag{22.4}$$

in order of magnitude, will play an essential rôle in the integrand. Using this, we transform the remaining part of the integrand in Eq. 21.4. The function $\Phi(\vartheta)$ in this expression is given by Eq. 21.5, in which

$$
\begin{aligned}
\sqrt{(n^2 - \sin^2 \vartheta)} &= \sqrt{(\sin^2 \delta - \sin^2 \vartheta)} = \sqrt{[(\sin\delta - \sin\vartheta)(\sin\delta + \sin\vartheta)]} \\
&= \sqrt{[\sin(\delta + \vartheta)\sin(\delta - \vartheta)]}.
\end{aligned}
$$

Expressing ϑ in terms of β, we obtain

$$\sqrt{(n^2 - \sin^2\vartheta)} = \sqrt{[\sin(2\delta - \beta)\sin\beta]}.$$

The sign of the square root must be chosen in such a way that the condition $\operatorname{Im}\sqrt{(n^2 - \sin^2\vartheta)} > 0$ is satisfied.

Remembering that β is small, we set

$$\sin(2\delta - \beta) \approx \sin 2\delta$$

and as a result, $\sqrt{(n^2 - \sin^2\vartheta)} \approx \sqrt{(\sin 2\delta \sin \beta)}.$ (22.5)

This approximation can always be made, except when δ is close to $\pi/2$, and consequently, n is close to 1 (since $n = \sin\delta$). Setting $\delta = (\pi/2) - \chi$ in this case, we obtain

$$\cos\chi = n, \quad \sin\chi = \chi = \sqrt{(1 - n^2)}$$

and moreover, $\sin(2\delta - \beta) = \sin(2\chi - \beta).$

We now see that β can be neglected if $\beta \ll \chi$. Remembering the meaning of χ, and using Eq. 22.4, we obtain the necessary condition

$$|\sqrt{[kR_1(1 - n^2)]}| \gg 1.$$ (22.6)

Thus, n must not be too close to 1[†]. Keeping all the above remarks in mind, and using Eq. 22.5 we can without difficulty transform Eq. 21.4 into the form

$$\psi_{\text{lat}} = \frac{8\sqrt{(k)}\, n \exp[i(\pi/4)]\,(-\sqrt{2})}{m\sqrt{(\pi r \cos\delta)}}$$

$$\times \int_0^{\delta - i\infty} \exp\{ikR_1[\cos(\vartheta_0 - \delta)\cos\beta - \sin(\vartheta_0 - \delta)\sin\beta]\}$$

$$\times \left(\tan\frac{\beta}{2}\right)^{\frac{3}{2}} \frac{d\beta}{2\sin(\beta/2)}.$$ (22.7)

We have also used the fact that β is small. An integral in this form can be expressed immediately in terms of Weber functions (parabolic cylinder functions). In fact, using the integral representation of these functions, it can be shown that the following identity holds:[‡]

$$\frac{1}{\Gamma(-n)}\int_0^{\frac{1}{2}\pi + \beta_0 - i\infty} \exp[(i/2)(\xi^2 - \eta^2)\cos\beta - i\xi\eta\sin\beta]\left(i\tan\frac{\beta}{2}\right)^{-n}\frac{d\beta}{2\sin(\beta/2)}$$

$$= D_n(\xi - i\xi)\,D_n(\eta + i\eta),$$ (22.8)

---

† We have investigated the case in which n is close to unity in Ref. 43.

‡ The existence of this identity was pointed out to the author by V. A. Fok. We note that integral (22.7) can be expressed in terms of Weber functions without having recourse to Eq. 22.8. In fact, if in the integrand, we take the exponential in the form (22.3), and let $\sin(\beta/2) = t$, with $\cos(\beta/2) \sim 1$, we obtain an integral which is the same as the one in the integral representation of the Weber functions (Ref. 22, p. 157). However, it is more convenient for us to use Eq. 22.8.

where β_0 is an arbitrary real quantity, within the limits $-\pi/2 < \beta_0 < \pi/2$. If we set

$$\xi = \sqrt{(2kR_1)} \cos \frac{\vartheta_0 - \delta}{2}, \quad \eta = \sqrt{(2kR_1)} \sin \frac{\vartheta_0 - \delta}{2}, \quad (22.9)$$

our integral (22.7) is reduced to the form (22.8); as a result, we obtain

$$\psi_{\text{lat}} = \frac{4in}{m} \sqrt{\left(\frac{2k}{r \cos \delta}\right)} D_{-\frac{3}{2}}(\xi - i\xi)\, D_{-\frac{1}{2}}(\eta + i\eta). \quad (22.10)$$

Since $\xi \gg 1$, we can replace the function $D_{-\frac{3}{2}}(\xi - i\xi)$ by its asymptotic representation (Ref. 22, p. 154)

$$D_{-\frac{3}{2}}(\xi - i\xi) \approx [\sqrt{(2)}\xi]^{-\frac{3}{2}} \exp\left[(3\pi i/8) + i(\xi^2/2)\right].$$

Remembering, in addition, that

$$\tfrac{1}{2}(\xi^2 - \eta^2) = kR_1 \cos(\vartheta_0 - \delta), \quad \zeta\eta = kR_1 \sin(\vartheta_0 - \delta),$$

and using Eq. 21.13, we obtain

$$\psi_{\text{lat}} = \frac{2in}{km\,(n^2 - 1)\sqrt{(r)}\,L_1^{\frac{3}{2}}} \exp\left[ik(L_0 + L) + ik_1 L_1\right] F(\eta), \quad (22.11)$$

where $\quad F(\eta) = \exp\left[-(5\pi i/8)\right] (2\eta^2)^{\frac{3}{4}} \exp(i\eta^2/2)\, D_{-\frac{3}{2}}(\eta + i\eta). \quad (22.12)$

Thus, in the more exact theory, we obtain Eq. 21.14 for the lateral wave, but with the additional factor $F(\eta)$. The asymptotic expansion of $F(\eta)$ in powers of $1/\eta^2$ is

$$F(\eta) = 1 - \frac{3.5.7.9}{2.4(8\eta^2)^2} + \frac{3.5\ldots 15.17}{2.4.6.8(8\eta^2)^4} - \cdots$$

$$+ \frac{15i}{16\eta^2} \left(1 - \frac{7.9.11.13}{4.6(8\eta^2)^2} + \frac{7.9\ldots 19.21}{4.6.8.10(8\eta^2)^2} - \cdots\right). \quad (22.13)$$

When the angle of incidence ϑ_0 is sufficiently far from the angle of total internal reflection, we have $\eta \gg 1$, $F(\eta) \approx 1$, and we obtain the result of the preceding Paragraph, namely, Eq. 21.14.

The expansion of $F(\eta)$ in powers of η is

$$F(\eta) = \sqrt{(\pi)} \exp(3\pi i/8)\, \eta^{\frac{3}{2}} \left\{ \frac{4}{\Gamma(\frac{1}{4})} \left[1 + \frac{3\eta^2}{2} i - \frac{3.7}{2!\,1.3} \left(\frac{\eta^2}{2}\right)^2 \right.\right.$$

$$\left. - \frac{3.7.11}{3!\,.1.3.5} \left(\frac{\eta^2}{2}\right)^3 + \cdots\right] - \frac{\sqrt{(2)}\,(1+i)\,\eta}{\Gamma(\frac{3}{4})} \left[1 + \frac{5}{3} \left(\frac{\eta^2}{2}\right) i\right.$$

$$\left.\left. - \frac{5.9}{2!\,3.5} \left(\frac{\eta^2}{2}\right)^2 - \frac{5.9.13}{3!\,3.5.7} \left(\frac{\eta^2}{2}\right)^3 i + \cdots\right]\right\}. \quad (22.13a)$$

Using Eqs. 22.13 and 22.13a, we can tabulate the function $F(\eta)$ (Table 3).

Thus, Eq. 22.10 gives the field of the lateral wave *over its entire region of existence* $\vartheta_0 \leqslant \delta$ (Fig. 103).

It is interesting to note that as $\vartheta_0 \to \delta$ ($\eta \to 0$), i.e. as the boundary of the region in which the lateral wave is present is approached, the amplitude of the lateral wave does not become zero, as might be expected at first glance. Actually, remembering that, according to Eqs. 21.13 and 22.9, $kL_1 \cos \delta = \sqrt{(2kR_1)}\,\eta$, Eq. 22.11 gives, for $\vartheta_0 = \delta$ ($\eta = 0$),

$$\psi_{\text{lat}} = \frac{4in\,\sqrt{(2\pi)}\exp\left[-(\pi i/8)\right]}{m\,\Gamma(\tfrac{1}{4})\,\sqrt{(\sin 2\delta)}\,(kR_1/2)^{\frac{1}{4}}}\frac{\exp\left(ikR_1\right)}{R_1}. \tag{22.14}$$

For the reflected wave, we limit ourselves to the analysis of the integral (19.26), with ϑ_0 very close to the angle of total internal reflection δ. We neglect the term $1/(8kr\sin\vartheta)$ compared with unity. Moreover, the

TABLE 3

Values of the Function $F(\eta) = |F|\,e^{i\phi}$

η^2	$|F|$	$\dfrac{4}{\pi}\phi$	η^2	$|F|$	$\dfrac{4}{\pi}\phi$
0.1	0.25	1.12	1.6	0.82	0.48
0.2	0.36	1.00	1.9	0.84	0.43
0.4	0.51	0.85	2.2	0.86	0.39
0.7	0.65	0.71	3.0	0.89	0.30
1.0	0.72	0.61	4.0	0.93	0.25
1.3	0.78	0.53	5.0	0.94	0.22

square root $\sqrt{(n^2 - \sin^2\vartheta)}$ in Eq. 22.1 can be considered small, since we are assuming that the passage point ϑ_0 is close to δ. Therefore, we can set

$$V(\vartheta) \approx 1 - \frac{2}{m\cos\vartheta}\sqrt{(n^2 - \sin^2\vartheta)}.$$

Again replacing ϑ by the variable of integration $\beta = \delta - \vartheta$, and using Eq. 22.5, we obtain

$$V(\vartheta) = 1 - \frac{2}{m\cos\vartheta}\sqrt{(\sin 2\delta \sin\beta)}. \tag{22.15}$$

In applying the method of steepest descents here, ϑ can be replaced by ϑ_0, or, assuming that

$$\left|\frac{\vartheta_0 - \delta}{\vartheta_0}\right| \ll 1, \tag{22.16}$$

ϑ_0 in turn can be replaced by δ. As a result

$$V(\vartheta) \approx 1 - \frac{2\sqrt{(2n)}}{m} \bigg/ \sqrt{\left(\frac{\sin \beta}{\cos \delta}\right)}.$$

Let us note, incidentally, that Eq. 22.16 guarantees that the omitted terms in the last expansion are small.

Now, using the above results, we can without difficulty express integral (19.26) in terms of two integrals of the type (22.8). Taking into account the rule indicated above for choosing the sign of the square root $\sqrt{(\sin \beta)}$ (page 283), we obtain

$$\psi_{\text{refl}} = \frac{e^{ikR_1}}{R_1} - \frac{2in}{m} \sqrt{\left(\frac{2k}{r \cos \delta}\right)} D_{-\frac{3}{2}}(\xi - i\xi) [\pm D_{-\frac{3}{2}}(\eta + i\eta)$$
$$- i D_{-\frac{3}{2}}(-\eta - i\eta)], \quad (22.17)$$

where the plus sign refers to the case $\vartheta_0 < \delta$, and the minus sign refers to the case $\vartheta_0 > \delta$.

When $\vartheta_0 < \delta$, Eq. 22.17 is the complete expression for the field in the upper medium (excluding, of course, the incident wave), since there is no lateral wave in this region. Under these conditions, the linear combination of Weber functions of the $-\frac{3}{2}$ order in the square brackets can be combined into a Weber function of the $\frac{1}{2}$ order (see Ref. 22, p. 155). As a result, we obtain

$$\psi_{\text{refl}} = \frac{e^{ikR_1}}{R_1} + \frac{8n}{m} \sqrt{\left(\frac{k}{r \cos \delta}\right)} e^{3\pi i/4} D_{-\frac{3}{2}}(\xi - i\xi) D_{\frac{1}{2}}(\eta - i\eta). \quad (22.18)$$

When $\vartheta > \delta$, we must use the minus sign in Eq. 22.17, and must add Eq. 22.10 for the lateral wave. Then we again obtain Eq. 22.18. Thus, the latter is an expression for the reflected wave, suitable for any ϑ_0, close to δ. We study this expression for various values of η, that is, for various values of $\vartheta_0 - \delta$.

Using the expansion of the Weber function in powers of its argument, and taking Eq. 22.10 into account, we obtain

$$\psi_{\text{refl}} = \frac{e^{ikR_1}}{R_1} \Bigg\{ 1 - \frac{4n\, e^{i(\pi/8)}}{m\Gamma(\frac{1}{4})\,(kR_1/2)^{\frac{1}{4}}} \sqrt{\left(\frac{\pi}{\sin 2\delta}\right)} \Bigg[1 + \frac{i\eta^2}{2} + \frac{3}{2!\,.1.3}\left(\frac{\eta^2}{2}\right)^2$$
$$- i\frac{3.7}{3!\,1.3.5}\left(\frac{\eta^2}{2}\right)^3 - \ldots + \frac{(1-i)\,\Gamma^2(\frac{1}{4})\eta}{4\sqrt{\pi}}\left(1 - \frac{i}{3}\cdot\frac{\eta^2}{2}\right.$$
$$\left. - \frac{1.5}{2!\,3.5}\left(\frac{\eta^2}{2}\right)^2 + i\frac{1.5.9}{3!\,3.5.7}\left(\frac{\eta^2}{2}\right)^3 + \ldots\right) \Bigg] \Bigg\}. \quad (22.19)$$

When $\vartheta_0 = \delta$, that is when $\eta = 0$, only two terms remain in the curly brackets. The first of these terms (and the principal one), unity, is equal to the coefficient of reflection of a plane wave for the case $\vartheta_0 = \delta$. Thus, geometrical optics is suitable as a first approximation in this case. However, as kR_1 increases, the second term, which is a correction to geometrical optics, decreases considerably more slowly than the correction term in Eq. 19.36, namely, it decreases as $(kR_1)^{-\frac{1}{4}}$. Part of this term comprises the lateral wave, which we analyzed separately earlier (see Eq. 22.14). The expansion in powers of $1/\eta$ depends on the sign of η. For $\eta < 0$ ($\vartheta_0 < \delta$), we have

$$\psi_{\text{refl}} = \frac{e^{ikR_1}}{R_1} \left\{ 1 - \frac{2\sqrt{[\sin 2\delta \sin (\delta - \vartheta_0)]}}{m \cos \delta} \left[1 - \frac{i}{2(8\eta^2)} \right. \right.$$
$$\left. \left. + \frac{1.3.5}{2.4(8\eta^2)^2} + i \frac{1.3.5.7}{2.4.6(8\eta^2)^3} - \ldots \right] \right\}. \qquad (22.20)$$

When $\eta^2 \gg 1$, all the terms in the square brackets, except unity, can be neglected. The two remaining terms in the curly brackets are the first two terms of the expansion of the reflection coefficient in a power series in the square root $\sqrt{(n^2 - \sin^2 \vartheta_0)} \equiv \sqrt{[\sin 2\delta \sin (\delta - \vartheta_0)]}$, on the assumption of Eq. 22.16, that is, we obtain the geometrical optics approximation. Remembering the definition of η, we note that the condition $\eta^2 \gg 1$ can be written

$$kR_1(\vartheta_0 - \delta)^2 \gg 1. \qquad (22.21)$$

When $\eta > 0$, the asymptotic value for $D_{\frac{1}{2}}(\eta - i\eta)$ has a different form, and contains two terms. One of these terms agrees exactly with Eq. 22.20, and the other gives a wave with a conical wave front (22.11).

Thus, we have elucidated all the details of the wave behavior at the angle of total internal reflection, and in particular, we have clarified the way in which the lateral and reflected waves merge in this region. Moreover, the analysis we have just presented makes it possible for us to describe the reflected wave ψ_{refl} for all angles of incidence ϑ_0.

Actually, when condition (22.21) is fulfilled, the analysis presented in § 19 is valid. On the other hand, if this condition is not fulfilled, it means that $kR_1(\vartheta_0 - \delta)^2 \lesssim 1$. We can write the latter in the form $(\vartheta_0 - \delta)/\vartheta_0 \lesssim 1/[\vartheta_0 \sqrt{(kR_1)}]$ or, since $kR_1 \gg 1$, we have $(\vartheta_0 - \delta)/\vartheta_0 \ll 1$. However, the fulfilment of this condition permits us to perform the expansion (22.15), as a result of which we obtain Eq. 22.17.

3. Ray representations. The caustic

The peculiarities arising at angles close to the angle of total internal reflection, can be easily explained through the use of ray representations.

We shall show, first of all, that the expression obtained in §8 for the displacement of a bounded beam of rays upon reflection is also valid for each of the rays issuing from the point source O (see Fig. 102).

For this purpose, we again use Eq. 19.26 for the reflected wave, neglecting the term $1/(8kr\sin\vartheta)$ compared with unity. Setting $V(\vartheta) = e^{i\phi(\vartheta)}$, where the phase $\phi(\vartheta)$ of the reflection can be a complex function, we obtain

$$\psi_{\text{refl}} = \exp[i(\pi/4)]\sqrt{\left(\frac{k}{2\pi r}\right)}\int_{-(\pi/2)+i\infty}^{(\pi/2)-i\infty}\exp[if(\vartheta)]\sqrt{(\sin\vartheta)}\,d\vartheta, \quad (22.22)$$

where

$$f(\vartheta) = kR_1\cos(\vartheta-\vartheta_0)+\phi(\vartheta). \quad (22.23)$$

We will treat integral (22.22) by the method of steepest descents. To

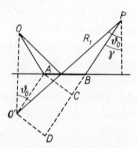

Fig. 104. The relation between the ray displacement and the angle at which it arrives at the point of observation.

determine the saddle point $\vartheta = \gamma$, we have the equation $f'(\gamma) = 0$; that is,

$$\sin(\gamma-\vartheta_0) = \frac{\phi'(\gamma)}{kR_1}. \quad (22.24)$$

Remembering that kR_1 is large, the last equation can be solved by the method of successive approximations. Setting the right hand side equal to zero in the zeroth approximation, we find

$$\gamma \approx \vartheta_0.$$

Substituting this into the right hand side of the equation, we find, in the first approximation,

$$\gamma = \vartheta_0 + \arcsin\frac{\phi'(\vartheta_0)}{kR_1}; \quad (22.25)$$

γ is the angle at which the ray arrives at the receiver. In the absence of ray displacement, the angle of arrival would have been ϑ_0 (Fig. 104).

Expanding $f(\vartheta)$ in Eq. 22.22 into a series in ϑ around the passage point γ, and taking $\sqrt{(\sin \vartheta)}$ outside the integral sign at the value $\vartheta = \gamma$, since it is being treated as a slowly varying function, we obtain

$$\psi_{\text{refl}} = \sqrt{\left(\frac{k \sin \gamma}{2\pi r}\right)} \exp\left[if(\gamma) + i(\pi/4)\right] \int_{-(\pi/2)+i\infty}^{(\pi/2)-i\infty} \exp\left[(i/2)f''(\gamma)(\vartheta-\gamma)^2\right.$$
$$\left. + (i/6)f'''(\gamma)(\vartheta-\gamma)^3 + \ldots\right] d\vartheta, \qquad (22.26)$$

where $$f''(\gamma) = -kR_1 \cos(\gamma - \vartheta_0) + \phi''(\gamma). \qquad (22.27)$$

For the present, we neglect the cubic and higher order terms in the exponent. Then, introducing the new variable of integration $\gamma - \vartheta = is$, and using the definite integral

$$\int_{-\infty}^{+\infty} \exp\left[-(i/2)f''(\gamma)s^2\right] ds = \exp\left[i(\pi/4)\right] \sqrt{\left[\frac{2\pi}{-f''(\gamma)}\right]}, \qquad (22.28)$$

we obtain $$\psi_{\text{refl}} = \sqrt{\left[\frac{k \sin \gamma}{-rf''(\gamma)}\right]} \exp\left[if(\gamma)\right]. \qquad (22.29)$$

Knowing the angle γ at which the ray arrives at the receiver, we can determine the ray displacement $\Delta = AB$ (Fig. 104). Upon reflection we have
$$O'D = O'P \sin(\vartheta_0 - \gamma) = R_1 \sin(\vartheta_0 - \gamma),$$

where $R_1 = O'P$ is, as usual, the distance from the image source to the point of observation. On the other hand,

$$O'D = AC = \Delta \cos \gamma.$$

Equating the last two expressions, and using Eq. 22.24, we obtain

$$\Delta = -\frac{1}{k \cos \gamma} \left(\frac{d\phi}{d\vartheta}\right)_{\vartheta=\gamma}. \qquad (22.30)$$

This expression is identical to Eq. 8.25 for the displacement in the case of the reflection of a bounded beam.

The field has an interesting singularity at points for which the second derivative of the function f becomes zero, simultaneously with the first derivative, i.e. the following two equations are satisfied simultaneously:

$$kR_1 \cos(\gamma - \vartheta_0) = \phi''(\gamma), \quad kR_1 \sin(\gamma - \vartheta_0) = \phi'(\gamma). \qquad (22.31)$$

It is clear from Eq. 22.29 that at these points, ψ_{refl} approaches ∞ in the ray approximation [since $f''(\gamma) \to 0$]. These points are distributed over a surface called the *caustic*. The equation of the caustic in polar coordinates (R_1, ϑ_0) with center at the point O' (see Fig. 103) is obtained by eliminating the variable γ from Eqs. 22.31.

Thus, due to the ray displacement upon reflection, the reflecting plane has its own sort of focusing action over some region of angles of incidence. The reason for the occurrence of the caustic is easily understood from ray representations. For this purpose, we consider a beam of rays incident on the boundary at angles close to the angle of total internal reflection δ, but somewhat greater than this angle. Upon reflection, each ray is displaced along the boundary, and the closer the angle of incidence of the ray is to δ, the greater will be the displacement. As a result, the reflected beam of rays behaves as shown in Fig. 105. The envelope of the family of rays is the caustic.

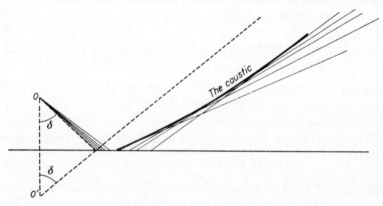

Fig. 105. The formation of a caustic as a result of ray displacement.

To determine the field near the caustic, where Eq. 22.29 is invalid, we must start from integral (22.26), taking account of the terms of third order in the exponent. As a result, we obtain the Airy integral,† which has been thoroughly investigated. It is known that the field has a maximum on the caustic, and decreases in an oscillatory manner as we go away from the caustic in the direction that leads us into the region where at every point, two rays intersect (in our case, to the right of the caustic). The oscillation is due precisely to the interference between these rays. On the other side of the caustic, the field decreases monotonically.

We shall examine the form of the caustic in detail. In the case $\sin \vartheta > n$ (total internal reflection), Eq. 22.1 for the reflection coefficient takes the form

$$V = \frac{m \cos \vartheta - i \sqrt{(\sin^2 \vartheta - n^2)}}{m \cos \vartheta + i \sqrt{(\sin^2 \vartheta - n^2)}}. \tag{22.32}$$

† Regarding the Airy integral, see Ref. 23. Regarding the calculation of the field near the caustic, see Ref. 95.

If we write V in the form $V = \exp[i\phi(\vartheta)]$, then

$$\phi(\vartheta) = -2\arctan\frac{\sqrt{(\sin^2\vartheta - n^2)}}{m\cos\vartheta}. \qquad (22.33)$$

Limiting ourselves to angles close to the angle of total internal reflection δ ($\sin\vartheta$ close to n), and using the formulas obtained in § 22, Section 2, we obtain

$$\sqrt{(\sin^2\vartheta - n^2)} = \sqrt{[\sin(\delta + \vartheta)\sin(\vartheta - \delta)]} \approx \sqrt{[(\vartheta - \delta)\sin 2\delta]}.$$

As a result, Eq. 22.33 can be written

$$\phi(\vartheta) = -a(\vartheta - \delta)^{\frac{1}{2}}, \qquad a \equiv \frac{2(2n)^{\frac{1}{2}}}{m(1 - n^2)^{\frac{1}{4}}}. \qquad (22.34)$$

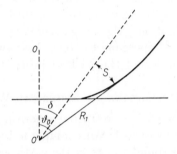

Fig. 106. Regarding the equation of the caustic in polar coordinates.

We will transform Eq. 22.31, taking Eq. 22.34 into account. Since $\gamma - \vartheta_0$ is small, we can set

$$\sin(\gamma - \vartheta_0) \approx \gamma - \vartheta_0, \qquad \cos(\gamma - \vartheta_0) \approx 1.$$

As a result, Eq. 22.31 is written

$$a(\gamma - \delta)^{-\frac{3}{2}} = 4kR_1, \qquad a(\gamma - \delta)^{-\frac{1}{2}} = 2kR_1[(\vartheta_0 - \delta) - (\gamma - \delta)].$$

Eliminating $\gamma - \delta$ from these two equations, we find

$$\vartheta_0 - \delta = \frac{3}{4}\left(\frac{2a}{kR_1}\right)^{\frac{2}{3}}. \qquad (22.35)$$

This is the equation of the caustic in a polar coordinate system with its center at the image source O' (Fig. 106). As $R_1 \to \infty$, we have $\vartheta_0 \to \delta$. However, the distance from the caustic to the straight line $\vartheta_0 = \delta$ (the dotted straight line in Fig. 106) increases continuously, because

$$S \equiv R_1(\vartheta_0 - \delta) = \frac{3}{4}\left(\frac{2a}{k}\right)^{\frac{2}{3}}R_1^{\frac{1}{3}}. \qquad (22.36)$$

In conclusion, and on the basis of ray representations, we shall make some remarks regarding the interconnection between the reflected and the lateral waves. In the preceding Paragraph, we saw that the lateral wave could be explained by the ray displacement upon reflection. If the angle of incidence ϑ_0 is far from the angle δ, the rays corresponding to the reflected and to the lateral waves will be separated in a natural way (the rays ODP and $OABP$ in Fig. 102), and will arrive at the point of observation at different angles. As ϑ_0 approaches δ, the two waves cannot be separated in this way, and the difference between them becomes less distinct. Both kinds of waves correspond to rays which are strongly displaced upon reflection, and have directions close to δ. The envelope of these rays is the caustic.†

From the purely mathematical point of view, the impossibility of separating the two waves when ϑ_0 is close to δ stems from the fact that under these conditions, the saddle point approaches the branch point, and it becomes impossible to treat these two points separately during the transformation of the path of integration in the complex plane.

§ 23. Refraction of Spherical Waves

The calculation of the field of refracted spherical waves is of practical interest in a number of cases.‡ An example is the calculation of the field of radiowaves or sound waves in the earth or in water, when the radiating antenna is situated in the air. Just as in the case of the reflected wave, we shall obtain geometrical optics as the first approximation, and the successive approximations will be corrections (sometimes quite significant) to geometrical optics. We shall start by analyzing the refracted wave on the basis of ray representations.

1. *The field of the refracted wave in the geometrical optics approximation*

Continuing with the nomenclature used above, the medium in which the source is situated will be called the "upper" medium. Consequently, the problem of the present Paragraph is the analysis of the field in the "lower" medium.

One of the rays which left the source O and was refracted at the boundary, in accordance with the laws of refraction of geometrical optics, will pass through some point S. In Fig. 107, which refers to the case $n < 1$ ($c_1 > c$), this is the ray OTS. As is well known, the angle

---

† It can be shown that the rays BB, CC, EE, etc., in Fig. 102, strictly speaking, are not parallel, but intersect, forming a caustic.

‡ We must not confuse the "refracted" wave in seismology, which is actually the lateral wave, with the refracted wave used here in the usual sense.

of incidence ϑ and the angle of refraction ϑ_1 are related by the equation

$$n \sin \vartheta_1 = \sin \vartheta, \quad n \equiv \frac{k_1}{k} \equiv \frac{c}{c_1}. \tag{23.1}$$

The wave amplitude at the point S can be determined, in the geometrical optics approximation, from the condition that there be conservation of energy in a ray tube. The phase is determined from the optical length of the ray. We first calculate the phase. It is equal to

$$2\pi \left(\frac{OT}{\lambda} + \frac{TS}{\lambda_1} \right) = k \cdot OT + k_1 TS,$$

where k and k_1 are wave numbers, and OT and TS are the lengths of the segments composing the ray. According to Fig. 107, these lengths

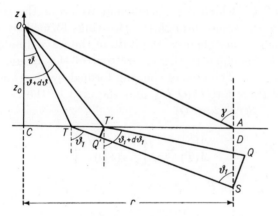

Fig. 107.   On the calculation of the field in the lower medium by the ray method.

are given by the expression

$$OT = \frac{z_0}{\cos \vartheta}, \quad TS = \frac{D}{\cos \vartheta_1},$$

where $D = -z$ is the distance from the point of observation S to the interface.

Using the law of refraction (23.1), we obtain for the phase change

$$k \left[\frac{z}{\cos \vartheta} + \frac{n^2 D}{\sqrt{(n^2 - \sin^2 \vartheta)}} \right]. \tag{23.2}$$

For a given arrangement of the source O and the receiver S, the angle ϑ

is found by eliminating ϑ_1 from Eq. 23.1 and the equation

$$z_0 \tan \vartheta + D \tan \vartheta_1 = r, \tag{23.3}$$

the geometrical meaning of which is evident.

To determine the amplitude of the wave at the point S, we construct the ray $OT'Q$ (Fig. 107), lying in the plane OTS and incident on the boundary at the angle $\vartheta + \Delta\vartheta$, where $\Delta\vartheta$ is vanishingly small. We now examine all the rays contained between the surfaces formed by rotating the rays OTS and $OT'Q$ around the z-axis. Part of the energy carried by these rays is reflected, and part enters the lower medium and propagates between TS and $T'Q$. Let the wave amplitude in the lower medium at the point T be $A(T)$. To determine $A(S)$, the wave amplitude at S, we must establish the relation between the length of the segment SQ, which is perpendicular to TS at S, and the length of the segment $T'Q'$, which is perpendicular to TS at the point Q'. The energy flux is the same through the two rings formed by the rotation of $Q'T'$ and QS. Consequently, the ratio of the wave amplitudes at the points T and S is equal to the square root of the inverse ratio of the areas of the rings. The area of the ring formed by the rotation of $Q'T'$ is $2\pi(CT')(Q'T') \approx 2\pi(CT)(Q'T')$, and the area of the ring formed by the rotation of SQ is $2\pi r(SQ)$.

Consequently, $\qquad \dfrac{A(S)}{A(T)} = \sqrt{\left[\dfrac{(CT)(Q'T')}{r(SQ)}\right]}. \tag{23.4}$

But $\qquad\qquad CT = z_0 \tan \vartheta,$

$$TT' = \frac{d(CT)}{d\vartheta}\, d\vartheta = \frac{z_0\, d\vartheta}{\cos^2 \vartheta},$$

$$Q'T' = TT' \cos \vartheta_1 = \frac{z_0 \cos \vartheta_1\, d\vartheta}{\cos^2 \vartheta}. \tag{23.5}$$

Now, SQ is equal to the length of the segment $Q'T'$ at the boundary plus an additional segment resulting from the rotation of the ray in the plane of incidence:

$$SQ = Q'T' + TS\, d\vartheta_1 = Q'T' + D\frac{d\vartheta_1}{\cos \vartheta_1}. \tag{23.6}$$

Differentiating Eq. 23.1, we obtain an expression for $d\vartheta_1$ in terms of $d\vartheta$:

$$\cos \vartheta\, d\vartheta = n \cos \vartheta_1\, d\vartheta_1.$$

The field ψ_1 at the point T directly below the interface is connected with the field ψ at the same point above the boundary by the relation†

$$\psi_1 = \frac{1}{m}\psi, \qquad (23.7)$$

where, as above, $m = \rho_1/\rho$ in acoustics, $m = n^2$ in electrodynamics in the case of vertical polarization, and $m = 1$ in electrodynamics in the case of horizontal polarization.

Above the boundary of separation, where the direct and the reflected waves add, the amplitude of the combined wave is

$$|\psi| = \frac{1}{R}|1 + V(\vartheta)|.$$

Now, using Eq. 23.7, it is not hard to find the amplitude $A(T) = |\psi_1|$, and using Eq. 23.4, we find $A(S)$, the wave amplitude at the point S. With the use of Eq. 23.2 for the phase change, and Eq. 3.16 for the reflection coefficient $V(\vartheta)$, we obtain, after some straightforward transformations, the total value of the vertical component of the Hertz vector or the acoustic potential in the geometrical optics approximation:

$$\psi_1(S) = \frac{2\sqrt{(\sin\vartheta)}\exp\{ik[(z_0/\cos\vartheta) + (nD/\cos\vartheta_1)]\}}{\sqrt{\{r[z_0\cos^{-3}\vartheta + (1/n)D\cos^{-3}\vartheta_1]\}(m\cos\vartheta + n\cos\vartheta_1)}}, \qquad (23.8)$$

where, according to Eq. 23.1,

$$\cos\vartheta_1 = \frac{1}{n}\sqrt{(n^2 - \sin^2\vartheta)},$$

and the angle ϑ is found from Eqs. 23.3 and 23.1.

Equation 23.8 is valid if the distances from the source and receiver to the interface are sufficiently large compared with the wavelength. This will be clear from the more exact solution presented in the next Section.

2. *Corrections to geometrical optics for the refracted wave*

An exact expression for the field of the refracted wave (the field in the "lower" medium) can be obtained in an integral form, analogous to expression (19.3) for the reflected wave. For this purpose, we must again expand the spherical wave in plane waves. As each of the plane

---

† In acoustics, Eq. (23.7) follows from the equality of the acoustic pressures on both sides of the boundary: $p = p_1$. Hence, we obtain $\rho\psi = \rho_1\psi_1$, or $m\psi_1 = \psi$. In the electrodynamic case, with vertical polarization, Eq. 18.6 gives $H_y = -(\omega\epsilon k_x/c)\Pi$. Since H_y must be continuous across the boundary, and k_x does not change, we obtain the condition $\epsilon\Pi = \epsilon_1\Pi_1$, or $\Pi = m\Pi_1$.

incident waves crosses the boundary between the media, its amplitude is multiplied by the transmission coefficient, which we denote by $W(\vartheta)$. If the amplitude of the incident wave is set equal to unity, the amplitude of the reflected wave will be V, and that of the transmitted wave will be W. Remembering that the total field near the boundary in the upper medium is $\psi = 1 + V$, and in the lower medium, $\psi_1 = W$,† we obtain the relation between the reflection and transmission coefficients from Eq. 23.7:

$$W = \frac{1}{m}(1 + V).$$

Now, taking the corresponding phase change into account, we obtain the following expression for the field in the lower medium $(z < 0)$:

$$\psi_1 = \frac{ik}{2\pi m} \int_0^{(\pi/2)-i\infty} \int_0^{2\pi} \exp\left[ik(x\cos\phi + y\sin\phi)\sin\vartheta\right.$$
$$\left. + ikz_0\cos\vartheta - ik_1 z\cos\vartheta_1\right](1 + V)\sin\vartheta\, d\vartheta\, d\phi.$$

At distances from the source which are large compared to the wave-length, it is convenient to analyze this expression by the method of steepest descents, as was done in § 19 for the reflected wave. We shall not go into the calculations here, because they are identical to those in § 19. Instead, we will go immediately to the results of the calculations and will discuss their physical significance. In this connection, the corrections to the above results of geometrical optics are of principal interest.

Let us consider Fig. 108, in which the two cases of refraction $n > 1$ and $n < 1$ are illustrated. In both cases, OTS is the ray constructed according to the laws of geometrical optics. The expression for the Hertz vector or the acoustic potential corresponding to this ray is given by Eq. 23.8.

As in the case of the reflected wave, the improvement of geometrical optics proceeds along two lines:

1. Taking account of the second approximation in the method of steepest descents. This leads to the addition of one more term of the next order in $1/kR$ (which is a small quantity) to Eq. 23.8.

2. The addition of a new type of wave, obtained from the integration over the borders of the cut, and analogous to the lateral wave in the case of reflected waves. The energy transferred from the source to the receiver by this wave travels along a path very different from that in geometrical optics. In Fig. 108, this is the path OMS. In case (b), this

---

† To within a phase factor common to all three waves.

wave has a very simple significance. It is the well-known wave, attenuating exponentially with depth in the lower medium, which is obtained upon reflection of the ray OM, incident on the boundary at an angle greater than the angle of total internal reflection $(\sin \beta > n)$. In case (a), this wave results from the fact that the inhomogeneous plane waves, attenuating exponentially along z, which are present in the expansion of the spherical wave issuing from O (these waves are shown by the horizontal shading in Fig. 108), when incident on the boundary excite the usual plane waves in the lower medium, which are propagated at all angles of incidence β, satisfying the condition

$$\sin \beta > \frac{1}{n}.$$

We note that case (a) is obtained from case (b) by interchanging the source and the receiver in the latter.

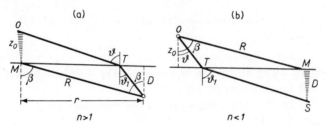

Fig. 108. Paths of wave penetration into the lower medium for the cases $n > 1$ and $n < 1$.

As we stated above, we shall not go into a detailed investigation in the complex plane, since it is just like the case of the reflected waves, but shall write out the final results for the vertical component of the Hertz vector and the acoustic potential. We consider the case of small grazing angles, since it is only under these conditions that the corrections to geometrical optics can be significant. Accordingly, we assume that $(\pi/2) - \vartheta \ll 1$, and $m[(\pi/2) - \vartheta] \ll \sqrt{(n^2 - 1)}$. As a result, we have:

(a) $n > 1$. Eq. 23.8 in the improved form is written

$$\psi_1(s) = \frac{2}{r} \exp \{ik[r \sin \vartheta + z_0 \cos \vartheta + D \sqrt{(n^2 - \sin^2 \vartheta)}]\}$$

$$\times \left[\sqrt{\left(\frac{r}{z_0}\right)} \frac{\cos^{\frac{3}{2}} \vartheta}{m \cos \vartheta + \sqrt{(n^2 - \sin^2 \vartheta)}} + \frac{im}{(n^2 - 1)} \cdot \frac{1}{kr} \right], \qquad (23.9)$$

where, as mentioned above, the angle ϑ is found by eliminating ϑ_1 from Eqs. 23.1 and 23.3. If we now use the condition that $(\pi/2) - \vartheta$ be

small, we obtain
$$\frac{\pi}{2} - \vartheta \approx \frac{z_0}{r - [D/\sqrt{(n^2 - 1)}]}. \tag{23.10}$$

If we neglect $D \cos^{-3} \vartheta_1$ compared with $z_0 \cos^{-3} \vartheta$ in Eq. 23.8 (which can be done in view of our assumption that $(\pi/2) - \vartheta$ is small), and transform the exponential in a straightforward manner, this equation turns into Eq. 23.9 with the second term in the parentheses thrown away, as it must be, since this term gives the correction to geometrical optics. When $\vartheta \to \pi/2$, and the first term, corresponding to geometrical optics, vanishes, the second (correction) term becomes the principal term.†

We must still add a wave of the lateral wave type, which is obtained from the integration over the borders of the cut in the complex plane, to Eq. 23.9. This occurs only at sufficiently great angles β (Fig. 108), satisfying the condition $\sin \beta > 1/n$. Its analytic expression has the form

$$\psi_{1\,\text{lat}} = \frac{2n}{R} \exp [ik_1 R - kz_0 \sqrt{(n^2 \sin^2 \beta - 1)}]$$
$$\times \left[\frac{\cos \beta}{n \cos \beta + im \sqrt{(n^2 \sin^2 \beta - 1)}} + \frac{i}{m(1 - n^2) kR} \right], \tag{23.11}$$

were $R = \sqrt{(r^2 + D^2)}$.

As we see, the amplitude of this wave decreases exponentially as the source moves away from the boundary. The total field in the lower medium will be the sum $\psi_1 + \psi_{1\,\text{lat}}$.

(b) $n < 1$. In this case, the appropriate formulas are:

$$\psi_1 = \frac{2n}{r} \exp [i(kr \sin \vartheta + kz_0 \cos \vartheta + k_1 D \cos \vartheta_1)]$$
$$\times \left[\sqrt{\left(\frac{r}{D}\right)} \frac{\cos^{\frac{3}{2}} \vartheta_1}{n \cos \vartheta_1 + m \sqrt{(1 - n^2 \sin^2 \vartheta_1)}} + \frac{i}{m^2(1 - n^2) kr} \right]. \tag{23.12}$$

It is assumed here that $(\pi/2) - \vartheta_1 \ll 1$ and $n[(\pi/2) - \vartheta_1] \ll m(1 - n^2)^{\frac{3}{2}}$. With these assumptions, Eqs. 23.1 and 23.3 give

$$\frac{\pi}{2} - \vartheta_1 \approx \frac{D}{r - [nz_0/\sqrt{(1 - n^2)}]}. \tag{23.13}$$

The additional wave of the lateral wave type, which occurs when $\sin \beta > n$, will be

$$\psi_{1\,\text{lat}} = \frac{2}{R} \exp [ikR - kD \sqrt{(\sin^2 \beta - n^2)}]$$
$$\times \left[\frac{\cos \beta}{m \cos \beta + i \sqrt{(\sin^2 \beta - n^2)}} + \frac{im}{kR(n^2 - 1)} \right], \tag{23.14}$$

where $R = \sqrt{(r^2 + z_0^2)}$.

---

† To simplify the formulas, this term was written for the case $z \cos \vartheta \ll z_0 \sqrt{(n^2 - 1)}$.

It was assumed here that $m[(\pi/2) - \beta] \ll \sqrt{(1 - n^2)}$. If, on the contrary, we had $m[(\pi/2) - \beta] \gg \sqrt{(1 - n^2)}$, we would have obtained

$$\frac{2}{kRm^2} \sqrt{(1 - n^2)} \cos^3 \beta$$

for the second term in the brackets.

It is clear from Fig. 108 that one case is transformed into the other if we make the replacements: $z_0 \gtrless D$, $n \to 1/n$, $m \to 1/m$, $\vartheta \gtrless \vartheta_1$. Thus, with the above replacements, Eqs. 23.9 and 23.11 must be transformed into Eqs. 23.12 and 23.14 respectively, and vice versa. It is not hard to convince ourselves that this actually occurs. In this connection, we must remember that it is necessary to divide each of the formulas by m when we make the replacements. This arises in acoustics, for example, from the fact that we are basing ourselves here on the reciprocity principle which is valid for the acoustic pressure, whereas our ψ denotes the acoustic potential, which differs from the acoustic pressure by the factor ρ in the upper medium, and by the factor ρ_1 in the lower medium.

On the basis of the formulas we have obtained, we shall find the limits of applicability of the geometrical optics approximation. In the case $n > 1$, the condition that the second term in the parentheses in Eq. 23.9 be negligible, as in § 19, gives

$$kz_0 \gg \frac{m}{\sqrt{(n^2 - 1)}}. \tag{23.15}$$

Thus, for geometrical optics to be applicable, it is necessary that the elevation of the source above the boundary be sufficiently great compared with the wavelength. We note that, as z_0 increases, the amplitude of the wave (23.11) approaches zero, as it should, since this wave does not occur in geometrical optics.

Similarly, in the case $n < 1$, the condition that the second term in Eq. 23.12 be small compared with the first gives

$$kz \gg \frac{1}{m \sqrt{(1 - n^2)}}. \tag{23.16}$$

Thus, for the applicability of geometrical optics when $n > 1$, it is necessary that the source be sufficiently far from the boundary, while the position of the receiver does not play an essential rôle (however, we must keep in mind that we are considering the case $z \ll r$, $z_0 \ll r$). When $n < 1$, on the other hand, it is necessary that the receiver be sufficiently far from the boundary.

3. *The case in which one of the media has marked absorption*

We now consider the case in which the receiver is situated in an absorbing medium, and the source is in a nonabsorbing medium. In electrodynamics, this case is realized when the receiving antenna is situated in the earth or in the sea. In hydro-acoustics, a situation similar to this arises when the receiver is located in water saturated with bubbles.

When the source and the receiver are sufficiently far from the boundary, the problem can be solved in an elementary manner by using geometrical optics (see Eq. 23.8 in which n must be considered complex, which, as is well known, corresponds to the presence of absorption). We examine the case in which the distances are not large compared with the wavelength.

When absorption is present, the only wave (of the two considered above) which will have significant amplitude at the receiver will be the one whose path lies principally in the upper, nonabsorbing medium. This will be the wave *OTS* in case (a), and the wave *OMS* in case (b) (Fig. 108).

Thus, in these two cases, the Hertz vector or the acoustic potential will be given by Eqs. 23.9 and 23.14 respectively. These equations differ from one another only in that ϑ and r enter into one, and β and R enter into the other. However, if we neglect D compared with r in Eq. 23.10 for ϑ, we obtain $\vartheta \approx \beta \approx (\pi/2) - z_0/r$, $r \approx R$.

Thus, for our case,

$$\psi_1(r, D) = \frac{2}{R} \exp\left[ikR + ikD\sqrt{(n^2 - \sin^2\vartheta)}\right]$$
$$\times \left[\frac{\cos\vartheta}{m\cos\vartheta + \sqrt{(n^2 - \sin^2\vartheta)}} + \frac{im}{(n^2 - 1)kR}\right], \quad (23.17)$$

where $\cos\vartheta \approx z_0/r$.

It is interesting to compare this expression for the field in the absorbing medium with the expression for the field in the upper medium directly above the boundary, at the same distance r from the source. We obtain the latter by adding the direct wave $\exp(ikR)/R$ to the reflected wave given by Eq. 19.36. It should be noted that the angle ϑ_0 in Eq. 19.36 is identical with the angle ϑ of the present Paragraph. Taking Eq. 19.40 into account, and again limiting ourselves to small grazing angles ($\gamma_0 = \cos\vartheta_0$ is small), we obtain the following expression for the total field in the upper medium near the interface ($z = 0$):

$$\psi(r, 0) = \frac{2m}{R} e^{ikr} \left[\frac{\cos\vartheta}{m\cos\vartheta + \sqrt{(n^2 - \sin^2\vartheta)}} + \frac{im}{kR(n^2 - 1)}\right].$$

Comparing the last expression with Eq. 23.17, we obtain

$$m\psi_1(r, D) = \psi(r, 0) \exp\left[ikD\sqrt{(n^2 - \sin^2\vartheta)}\right].$$

This formula, in a very simple way, connects the field in the lower medium at the point S (Fig. 107) with the field in the upper medium at the very boundary of separation (point A). It shows that the only effect of lowering the receiver is the appearance of the factor $\exp[ikD\sqrt{(n^2 - \sin^2\vartheta)}]$. Taking the modulus of both sides of the last equation, we obtain the expression describing the decrease of the wave amplitude with depth:

$$m|\psi_1(r, D)| = |\psi(r, 0)|\, e^{-\kappa D}, \tag{23.18}$$

where
$$\kappa = \frac{2\pi}{\lambda} \operatorname{Re}\sqrt{(\sin^2\vartheta - n^2)}.$$

In acoustics, remembering the connection between the potential and the acoustic pressure, Eq. 23.18 can be rewritten in the form

$$|p_1(r, D)| = |p(r, 0)|\, e^{-\kappa D}, \tag{23.19}$$

where p and p_1 are the sound pressures in the upper and the lower media, respectively. Similarly, in electrodynamics, instead of Eq. 23.18, we can use formulas which connect the field components in the upper and lower media, directly.[35] These formulas have the form

$$|n|^2|\,E_z(r, D)| = |E_z(r, 0)|\, e^{-\kappa D} \tag{23.20}$$

$$|E_{1z}(r, D)| = |E_r(r, 0)|\, e^{-\kappa D} \tag{23.21}$$

and the formula for H_ϕ is similar to Eq. 23.21.

It is interesting to note that the formulas which we used in the derivation of Eq. 23.18, in particular, Eq. 19.36, are valid only under the condition that a pole does not lie too close to the passage point (see § 20). This signifies that for a given distance r and a given wavelength, the conductivity of the ground must not be too great. However, using the generalized method of passage, developed in § 20, it can be shown[35] that Eq. 23.18 is valid in the much more general case, in which the only conditions to be satisfied are $kR \gg 1$, $(\pi/2) - \vartheta \ll 1$.

4. *The acoustic field in water from a source located in air*

The acoustic field in water due to a source located in air, is made up of two parts, given by Eqs. 23.12 and 23.14. Moreover, for this special case, we must set $m \approx 800$, $n \approx 2/9$. Since m is very large, the second term in the brackets in Eqs. 23.12 and 23.14 can be neglected (see what was said following Eq. 23.14). Thus, the first component of the acoustic field, given by Eq. 23.12, corresponds to the usual geometrical acoustics. If we also take Eq. 23.13 into account, we obtain the following expression for the amplitude of this component:

$$|\psi_1(r, D)| = \frac{2nD}{m\sqrt{[r(1-n^2)]}\{r - [nz_0/\sqrt{(1-n^2)}]\}^{\frac{3}{2}}}. \tag{23.22}$$

The second part of the acoustic potential differs from zero only when the condition

$$\sin\beta = \frac{r}{\sqrt{(r^2+z_0^2)}} > n, \quad \text{i.e. } \beta > 77°$$

is satisfied. The amplitude of this part of the field is obtained from Eq. 23.14 under the assumption that $m\beta \gg 1$:

$$|\psi_{1\,\text{lat}}| = \frac{2\exp[-kD\sqrt{(\sin^2\beta - n^2)}]}{m\sqrt{(r^2+z_0^2)}}. \tag{23.23}$$

We see that this part of the acoustic potential is damped exponentially with depth in the water. However, for small D, the amplitude of this wave can be many times greater than the amplitude of the wave (23.22), corresponding to geometrical optics, because the latter is proportional to D. The refraction of spherical waves in the case in which the "lower" medium has a greater propagation velocity, is also considered in Ref. 123.

§ 24. REFLECTION AND REFRACTION OF SPHERICAL WAVES AT AN INTERFACE BETWEEN TWO ELASTIC MEDIA

In the present Paragraph, we shall study the reflection and refraction of a spherical wave at an interface between two elastic media with finite values of the Lamé constants λ and μ. Since the material presented here and in some of the following Paragraphs is of fundamental value in seismology and seismic exploration, it is necessary to make some preliminary remarks.

The theoretical problems of seismology have attracted the attention of many important mathematicians. The most significant results were obtained in the works of V. I. Smirnov and S. L. Sobolev, who developed new methods and approaches to this intricate complex of problems. However, it is characteristic of these works, in our opinion, that, along with the extremely high level of the mathematical methods used in working out the problem, the solutions are given in a form which makes comparison with experiment difficult. In particular, with this theory, it is difficult to predict the kind of process that will be recorded by an instrument with one or another "real" frequency characteristic. It is true that the theoretical results regarding the times of arrival of the various kinds of waves can be verified experimentally, but these results could have been obtained by the simpler method of geometrical seismology.

It seems to us that the theoretical method best suited to the experimental conditions is the method using spectral representations of the processes (the Fourier Integral method). In this approach to the problem, the initial pulse as well as the pulse arriving at the detector is represented as a super-position of waves harmonic in time. The correction for the frequency characteristic of the instrument can then be made without difficulty, and the investigation of the asymptotic behavior of the wave as the frequency approaches infinity permits us to determine the character of the onset or the order of the corresponding discontinuities in the solution.

In what follows, we shall use the spectral method. The work of G. I. Petrashen' and his collaborators seems to be evolving along the lines indicated above.

1. *Source of a spherical wave in a solid medium*

In a liquid medium, the simplest source of elastic waves is a pulsating sphere of small radius. However, any other source of zero order (a source with a finite volume velocity) can also radiate a spherical wave, as long as its dimensions are small compared with the wavelength. Thus, the radiator was assumed to have a spherical form only to simplify the discussion. The situation becomes more complicated in the case of a radiator in a solid medium. Here, the character of the wave will be quite dependent on the form of the source. We shall assume that the source has cylindrical symmetry. The deformations excited by the source will also have cylindrical symmetry. Therefore, the field of the elastic strains and stresses can be described with the aid of the three auxiliary functions ("potentials") ϕ_0, ψ_0 and χ_0, satisfying the wave equations

$$\nabla^2 \phi_0 = \frac{1}{c^2} \cdot \frac{\partial^2 \phi_0}{\partial t^2},$$

$$\nabla^2 \psi_0 = \frac{1}{b^2} \cdot \frac{\partial^2 \psi_0}{\partial t^2},$$

$$\nabla^2 \chi_0 = \frac{1}{b^2} \cdot \frac{\partial^2 \chi_0}{\partial t^2}, \tag{24.1}$$

where $c = (\lambda + 2\mu)^{\frac{1}{2}}/\rho^{\frac{1}{2}}$ and $b = \mu^{\frac{1}{2}}/\rho^{\frac{1}{2}}$ are the compressional and shear wave velocities, respectively, in the solid.

We shall use the cylindrical coordinate system r, ϑ and z. The displacement components u, v and w in the directions r, ϑ and z, respectively, will be expressed in terms of the potentials by the formulas†

$$u = \frac{\partial \phi_0}{\partial r} - \frac{\partial^2 \psi_0}{\partial r\, \partial z},$$

$$v = \frac{\partial \chi_0}{\partial r},$$

$$w = \frac{\partial \phi_0}{\partial z} + \frac{1}{r} \cdot \frac{\partial}{\partial r}\left(r\, \frac{\partial \psi_0}{\partial r}\right). \tag{24.2}$$

As we see, ϕ_0 is the potential of compressional waves, and ψ_0 and χ_0 are the potentials of the shear waves (isometric), polarized in the plane of incidence and perpendicular to it, respectively. The stresses are expressed in terms of the potentials as follows:

$$\widehat{rr} = \lambda \nabla^2 \phi_0 + 2\mu \frac{\partial}{\partial r}\left(\frac{\partial \phi_0}{\partial r} - \frac{\partial^2 \psi_0}{\partial r\, \partial z}\right),$$

$$\widehat{rz} = \mu \frac{\partial}{\partial r}\left(2\frac{\partial \phi_0}{\partial z} + \nabla^2 \psi_0 - 2\frac{\partial^2 \psi_0}{\partial z^2}\right),$$

$$\widehat{zz} = \lambda \nabla^2 \phi_0 + 2\mu \frac{\partial}{\partial z}\left(\frac{\partial \phi_0}{\partial z} + \nabla^2 \psi_0 - \frac{\partial^2 \phi_0}{\partial z^2}\right),$$

$$\widehat{r\vartheta} = \mu \left(\frac{\partial^2 \chi_0}{\partial r^2} - \frac{1}{r} \cdot \frac{\partial \chi_0}{\partial r}\right),$$

$$\widehat{z\vartheta} = \mu \frac{\partial^2 \chi_0}{\partial r\, \partial z}. \tag{24.3}$$

As in §§ 18 and 19, the potentials can be expanded into plane waves or into cylindrical functions. We shall use the latter expansion. Moreover, assuming that in the general case we shall be dealing with a pulsed source, we represent the potentials in the form of Fourier integrals with respect to time.

We now assume, as previously, that the source is situated on the z-axis, at a distance z_0 from the origin.

---

† Instead of the velocity potentials (as in § 4), we are now using displacement potentials.

As a result, we can set

$$\phi_0 = \operatorname{Re} \int_0^\infty \exp\left(-ikct\right) dk \int_{\Gamma_1} f_0(k, \vartheta) \, H_0^{(1)}(kr \sin \vartheta)$$
$$\times \exp\left[ik(z_0 - z) \cos \vartheta\right] \sin \vartheta \, d\vartheta,$$

$$\psi_0 = \operatorname{Re} \int_0^\infty \exp\left(-ikct\right) dk \int_{\Gamma_1} g_0(k, \vartheta) \, H_0^{(1)}(kr \sin \vartheta)$$
$$\times \exp\left[i\kappa(z_0 - z) \cos \gamma\right] \sin \vartheta \, d\vartheta,$$

$$\chi_0 = \operatorname{Re} \int_0^\infty \exp\left(-ikct\right) dk \int_{\Gamma_1} h_0(k, \vartheta) \, H_0^{(1)}(kr \sin \vartheta)$$
$$\times \exp\left[i\kappa(z_0 - z) \cos \gamma\right] \sin \vartheta \, d\vartheta, \tag{24.4}$$

where κ and γ are connected with k and ϑ by the relations

$$\kappa b = kc, \quad k \sin \vartheta = \kappa \sin \gamma. \tag{24.5}$$

The path of integration Γ_1 in the complex plane is shown in Fig. 93. In the further analysis of these integrals using the method of steepest descents, the functions f_0, g_0 and h_0 will belong to the class of slowly varying functions with respect to both variables k and ϑ.

A case of interest in seismology is that in which the source of the wave is an impulsive stress applied to the walls of a cylindrical cavity (for example, an explosion in the cavity). If the radius and length of the cavity are small in comparison with the wavelength, and the axis of the cylinder is vertical, the functions f_0, g_0 and h_0 at great distances, in our notation, are[135]

$$f_0 = ip_1(k) \left(2 \frac{b^2}{c^2} \cos^2 \vartheta - 1\right) \frac{k\Delta}{8\pi\mu},$$

$$g_0 = p_1(k) \cos \vartheta \, \frac{\Delta b^2}{4\pi\mu c^2},$$

$$h_0 = s_1(k) \frac{\cos \vartheta}{\cos \gamma} \frac{\Delta b k}{4\pi i \mu c}, \tag{24.6}$$

where Δ is the volume of the cavity, and $p_1(k)$ and $s_1(k)$ are the Fourier components of the normal and tangential stresses applied to the walls of the cavity. Moreover,

$$p(t) = \operatorname{Re} \int_0^\infty p_1(k) \, \mathrm{e}^{-ikct} \, dk,$$

$$s(t) = \operatorname{Re} \int_0^\infty s_1(k) \, \mathrm{e}^{-ikct} \, dk, \tag{24.7}$$

where $p(t)$ and $s(t)$ are the corresponding stresses as functions of time.

Substituting Eq. 24.6 into Eq. 24.4, and then substituting the expressions for ϕ_0, ψ_0 and χ_0 into Eq. 24.2, it is not hard to obtain the expressions for the displacements.

As an example, we obtain the expression for the vertical displacement in a compression wave. We shall use the subscript p for the compressional wave, the subscript SV for the shear wave polarized in the vertical plane, and the subscript SH for the shear wave polarized in the horizontal plane. We see from the expression for u in Eqs. 24.2 that

$$u = u_p + u_{SV},$$

i.e. the vertical component of the displacement is composed of the displacement in the compressional wave, and the displacement in the shear wave (SV). In this connection,

$$u_p = \frac{\partial \phi_0}{\partial r}, \qquad u_{SV} = -\frac{\partial^2 \psi_0}{\partial r\, \partial z}.$$

Using Eqs. 24.4 and 24.6, and differentiating under the integral sign, we obtain

$$u_p = \mathrm{Re} \int_0^\infty \mathrm{e}^{-ikct} U_p(k, r, z)\, dk, \tag{24.8}$$

where

$$U_p(k, r, z) = \frac{ip_1(k)\, k\Delta}{8\pi\mu} \int_{\Gamma_1} \left(2\frac{b^2}{c^2} \cos^2\vartheta - 1 \right) \frac{\partial}{\partial r}$$
$$\times H_0^{(1)}(kr\sin\vartheta) \exp\left[ik(z_0 - z)\cos\vartheta \right] \sin\vartheta\, d\vartheta. \tag{24.9}$$

On the same basis as in § 19, we will use the asymptotic representation of the Hankel function (19.25), but here, we need not take into account the correction term $(1/8)\,iu$ in the parentheses. Moreover, we keep in mind that, according to Fig. 109,

$$(z_0 - z) = \mathrm{R}\cos\vartheta_0, \qquad r = \mathrm{R}\sin\vartheta_0. \tag{24.10}$$

Then Eq. 24.9 can be written in the form

$$U_p(k, r, z) = \frac{ikp_1(k)\,\Delta}{4\pi\mu} \sqrt{\left(\frac{k}{2\pi r} \right)} \exp\left[i(\pi/4) \right] \int_{\Gamma_1} \left(2\frac{b^2}{c^2}\cos^2\vartheta - 1 \right)$$
$$\exp\left[ikR\cos(\vartheta - \vartheta_0) \right] \sin^{\frac{3}{2}}\vartheta\, d\vartheta. \tag{24.11}$$

Making a change of variable in accordance with the equation

$$\cos(\vartheta - \vartheta_0) = 1 + is^2, \tag{24.12}$$

and transforming to the path of integration over which s takes on all real values from $-\infty$ to $+\infty$, we can apply the method of steepest descents presented in § 19, Section 2. However, since the situation is simplified considerably by the fact that in the present case we are interested only in the principal term, we need not turn to the formulas of §19. Remembering that only small values of s, and consequently, only values of ϑ close to ϑ_0, play an essential rôle in the integrand, we expand $\cos(\vartheta - \vartheta_0)$ in powers of $(\vartheta - \vartheta_0)$, and obtain from Eq. 24.12

$$(\vartheta - \vartheta_0)^2 = -2is^2,$$

$$(\vartheta - \vartheta_0) = \pm\sqrt{(2)}\exp[-i(\pi/4)]s,$$

$$d\vartheta = \pm\sqrt{(2)}\exp[-i(\pi/4)]ds. \qquad (24.13)$$

Of the two signs in Eq. 24.13, we choose the positive sign, which, as is not hard to show, corresponds to the following: as the variable ϑ goes along the path Γ_1 from $-\frac{1}{2}\pi + i\infty$ to $\frac{1}{2}\pi - i\infty$ (Fig. 93), s varies from $-\infty$ to $+\infty$, and not the contrary. Substituting this value of $d\vartheta$ into

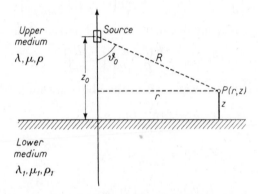

Fig. 109. On the calculation of the field of the direct wave.

Eq. 24.11, taking the entire integrand with the exception of the exponential e^{-kRs^2} outside the integral sign at the value $\vartheta = \vartheta_0$ as a slowly varying function, and remembering the value of the first of integrals (19.15), we obtain

$$U_p = \frac{ikp_1(k)\Delta}{4\pi\mu}\left(2\frac{b^2}{c^2}\cos^2\vartheta_0 - 1\right)\sin\vartheta_0\frac{e^{ikR}}{R}. \qquad (24.14)$$

Now, substituting U_p into Eq. 24.8, we first of all note that if we ignore the time factor, which is independent of k, the following integral with respect to k is obtained:

$$\text{Re}\int_0^\infty ikp_1(k)\exp\left[-ikc\left(t - \frac{R}{c}\right)\right]dk = -\frac{1}{c}\frac{\partial}{\partial t}\text{Re}\int_0^\infty p_1(k)\exp\left[-ikc\left(t - \frac{R}{c}\right)\right]$$

$$= -\frac{1}{c}\frac{\partial p[t - (R/c)]}{\partial t}.$$

Then, taking the remaining factors into account, we obtain the following expression for the vertical displacement in the compressional wave[†]

$$u_p = \frac{\Delta}{4\pi\mu c R}\left(1 - 2\frac{b^2}{c^2}\cos^2\vartheta_0\right)\sin\vartheta_0\frac{d}{dt}\left[p\left(t-\frac{R}{c}\right)\right].$$

The other displacement components can be obtained in exactly the same way. As a result, we will have for the P-wave

$$\begin{bmatrix}u_p\\w_p\end{bmatrix} = \left[\frac{F_1(\vartheta_0)}{R}\cdot\frac{d}{dt}\left\{p\left(t-\frac{R}{c}\right)\right\}\right]\begin{bmatrix}\sin\vartheta_0\\-\cos\vartheta_0\end{bmatrix}. \tag{24.15}$$

We see that the displacement given by the last equation is directed along the radius-vector from the source O, that is, the P-wave is a longitudinal wave. For the shear SV-wave, we will have

$$\begin{bmatrix}u_{SV}\\w_{SV}\end{bmatrix} = \left[\frac{F_2(\vartheta_0)}{R}\cdot\frac{d}{dt}\left\{p\left(t-\frac{R}{b}\right)\right\}\right]\begin{bmatrix}\cos\vartheta_0\\\sin\vartheta_0\end{bmatrix}. \tag{24.16}$$

As we see, this is a longitudinal wave. For the shear SH-wave, we obtain

$$v_{SH} = \frac{K(\vartheta_0)}{R}\frac{d}{dt}\left\{s\left(t-\frac{R}{b}\right)\right\}. \tag{24.17}$$

The following notation was used in the above formulas:

$$F_1(\vartheta_0) = \frac{\Delta}{4\pi\mu c}\left(1 - 2\frac{b^2}{c^2}\cos^2\vartheta_0\right),$$

$$F_2(\vartheta_0) = \frac{\Delta}{4\pi\mu b}\sin 2\vartheta_0,$$

$$K(\vartheta_0) = \frac{\Delta}{4\pi\mu b}\sin\vartheta_0. \tag{24.18}$$

2. Formal solution for the reflected and the refracted waves

We will denote the potentials of the reflected wave by ϕ, ψ and χ, and the potentials of the refracted wave (in the lower medium) by ϕ', ψ' and χ'. The six new functions introduced in this way must satisfy the wave equations

$$\frac{\partial^2\phi}{\partial t^2} - c^2\nabla^2\phi = 0, \qquad \frac{\partial^2\phi'}{\partial t^2} - c'^2\nabla^2\phi' = 0,$$

$$\frac{\partial^2\psi}{\partial t^2} - b^2\nabla^2\psi = 0, \qquad \frac{\partial^2\psi'}{\partial t^2} - b'^2\nabla^2\psi' = 0,$$

$$\frac{\partial^2\chi}{\partial t^2} - b^2\nabla^2\chi = 0, \qquad \frac{\partial^2\chi'}{\partial t^2} - b'^2\nabla^2\chi' = 0,$$

[†] This displacement will be picked up and recorded by an instrument with an "ideal" frequency characteristic, i.e. an instrument with constant sensitivity for all k from 0 to ∞. To obtain the result given by an actual instrument, the integrand in Eq. 24.8 must be multiplied by the frequency characteristic of the instrument. This remark continues to apply for the remainder of the exposition.

where† $\quad c^2 = \dfrac{\lambda+2\mu}{\rho}, \quad b^2 = \dfrac{\mu}{\rho}, \quad c'^2 = \dfrac{\lambda'+2\mu'}{\rho'}, \quad b'^2 = \dfrac{\mu'}{\rho'}.$

It is natural to assume that the functions $\phi, \ldots, \chi'$ can be represented in the form of the integrals

$$\phi = \operatorname{Re} \int_0^\infty e^{-ikct}\,dk \int_{\Gamma_1} f_1\,H_0^{(1)}(\sigma r)\,e^{-\alpha z}\sin\vartheta\,d\vartheta,$$

$$\psi = \operatorname{Re} \int_0^\infty e^{-ikct}\,dk \int_{\Gamma_1} g_1\,H_0^{(1)}(\sigma r)\,e^{-\beta z}\sin\vartheta\,d\vartheta,$$

$$\chi = \operatorname{Re} \int_0^\infty e^{-ikct}\,dk \int_{\Gamma_1} h_1\,H_0^{(1)}(\sigma r)\,e^{-\beta z}\sin\vartheta\,d\vartheta,$$

$$\phi' = \operatorname{Re} \int_0^\infty e^{-ikct}\,dk \int_{\Gamma_1} f'\,H_0^{(1)}(\sigma r)\,e^{-\alpha' z}\sin\vartheta\,d\vartheta,$$

$$\psi' = \operatorname{Re} \int_0^\infty e^{-ikct}\,dk \int_{\Gamma_1} g'\,H_0^{(1)}(\sigma r)\,e^{-\beta' z}\sin\vartheta\,d\vartheta.$$

$$\chi' = \operatorname{Re} \int_0^\infty e^{-ikct}\,dk \int_{\Gamma_1} h'\,H_0^{(1)}(\sigma r)\,e^{-\beta' z}\sin\vartheta\,d\vartheta, \qquad (24.19)$$

where
$$\alpha = -ik\cos\vartheta, \quad \beta = -i\kappa\cos\gamma,$$
$$\alpha' = ik'\cos\vartheta', \quad \beta' = i\kappa\cos\gamma',$$
$$\sigma = k\sin\vartheta = k'\sin\vartheta' = \kappa\sin\gamma = \kappa'\sin\gamma', \qquad (24.20)$$

and the relation $kc = \kappa b = k'c' = \kappa'b'$ holds. The contour of integration Γ_1 is the same as previously.

At the interface $z = 0$, we have six boundary conditions, expressing the continuity of the displacements and the stresses. Substituting the expressions for the incident, reflected and refracted system of waves into the boundary conditions, we obtain a system of six equations for the six unknown functions $f_1, \ldots, h'$. Solving this system, we find

$$f_1 = f_0 V_{11}\,e^{-\alpha z_0} + g_0 V_{21}\,e^{-\beta z_0}, \quad f' = f_0 D_{11}\,e^{-\alpha z_0} + g_0 D_{21}\,e^{-\beta z_0},$$

$$g_1 = f_0 V_{12}\,e^{-\alpha z_0} + g_0 V_{22}\,e^{-\beta z_0}, \quad g' = f_0 D_{12}\,e^{-\alpha z_0} + g_0 D_{22}\,e^{-\beta z_0},$$

$$h_1 = h_0 V_{33}\,e^{-\beta z_0}, \qquad\qquad h' = h_0 D_{33}\,e^{-\beta z_0}. \qquad (24.21)$$

In the last equation, $V_{11}, \ldots$ denote the reflection coefficients of plane waves at the boundary, and $D_{11}, \ldots$ denote the transmission coefficients at the boundary. The subscripts 1, 2 and 3 refer, respectively, to

---

† In the present paragraph, quantities referring to the lower medium will be denoted by a prime, to avoid confusion in the subscripts.

the compressional wave (the P-wave), the shear wave polarized in the vertical plane (the SV-wave), and the shear wave polarized in the horizontal plane (the SH-wave). Accordingly, V_{11} characterizes the amplitude of the reflected compressional wave when a compressional wave is incident on the boundary at the angle ϑ, V_{12} characterizes the amplitude of the SV-wave when a compressional wave is incident on the boundary at the same angle (the order of the subscripts in V_{12} signifies that, upon reflection, wave 1 is transformed into wave 2), V_{21} characterizes the amplitude of the compressional wave which arises when a shear SV-wave is incident on the boundary at the angle γ ($\kappa \sin \gamma = k \sin \vartheta$), and so forth. Finally, V_{33} denotes the reflection coefficient of SH-waves. The transmission coefficients $D_{11}, \dots$ have analogous meanings. We see that upon reflection at the boundary, the P- and SV-waves interact with one another, one partially transforming into the other. However, the SH-wave is reflected and is transmitted into the lower medium independently of the other two waves, "not mixing" with them. In addition to the notation (24.20), we set

$$\xi = -\kappa^2 \cos 2\gamma,$$

$$\xi' = -\kappa'^2 \cos 2\gamma'. \tag{24.22}$$

Then the explicit expressions for the reflection and transmission coefficients $V_{11}, \dots, D_{33}$ are† (here and below in the present paragraph, we will use the results of Ref. 136)

$$V_{11} = \frac{\alpha' P_1 + Q_1}{\alpha' P + Q}, \tag{24.23}$$

where
$$P = 4\alpha\beta\beta'\sigma^2(\mu'-\mu)^2 - \beta'(\mu\xi - 2\mu'\sigma^2)^2 - \mu\mu' \kappa^2 \kappa'^2 \beta,$$

$$Q = -\alpha\beta(\mu'\xi' - 2\mu\sigma^2)^2 - \mu\mu' \kappa^2 \kappa'^2 \alpha\beta' + \sigma^2(\mu'\xi' - \mu\xi)^2,$$

$$P_1 = 4\alpha\beta\beta'\sigma^2(\mu-\mu')^2 + \beta'(\mu\xi - 2\mu'\sigma^2)^2 + \mu\mu' \kappa^2 \kappa'^2 \beta,$$

$$Q_1 = -\alpha\beta(\mu'\xi' - 2\mu\sigma^2)^2 - \mu\mu' \kappa^2 \kappa'^2 \alpha\beta' - \sigma^2(\mu'\xi' - \mu\xi)^2. \tag{24.24}$$

Furthermore,
$$V_{21} = \frac{\beta' P_2 + Q_2}{\alpha' P + Q}, \tag{24.25}$$

where
$$P_2 = 4\alpha'\beta\sigma^2(\mu'-\mu)(\mu\xi - 2\mu'\sigma^2),$$

$$Q_2 = 2\beta\sigma^2(\mu'\xi' - \mu\xi)(\mu'\xi' - 2\mu\sigma^2), \tag{24.26}$$

$$V_{12} = \frac{\alpha' P_3 + Q_3}{\alpha' P + Q}, \tag{24.27}$$

---

† See Ref. 162 for a detailed analysis of the system of reflection coefficients at the boundary between two solid bodies.

where
$$P_3 = 4\alpha\beta'(\mu'-\mu)(\mu\xi - 2\mu'\sigma^2),$$

$$Q_3 = 2\alpha(\mu'\,\xi'-\mu\xi)(\mu'\,\xi - 2\mu\sigma^2);\qquad(24.28)$$

and
$$V_{22} = \frac{\beta'\,P_4 + Q_4}{\alpha'\,P + Q},\qquad(24.29)$$

where

$$P_4 = 4\alpha\alpha'\,\beta\sigma^2(\mu'-\mu)^2 + \alpha'(\mu\xi - 2\mu'\sigma^2)^2 + \mu\mu'\,\kappa^2\,\kappa'^2\,\alpha,$$

$$Q_4 = -\,\mu\mu'\,\kappa^2\kappa'^2\,\alpha'\,\beta - \alpha\beta(\mu'\,\xi'-2\mu\sigma^2)^2 - \sigma^2(\mu'\,\xi'-\mu\xi)^2,\qquad(24.30)$$

$$V_{33} = \frac{\mu\beta - \mu'\,\beta'}{\mu\beta + \mu'\,\beta'}.\qquad(24.31)$$

Finally,

$$(\alpha'\,P + Q)\,D_{11} = 2\mu\kappa^2\,\alpha\beta(\mu'\,\xi'-2\mu\sigma^2) + 2\mu\kappa^2\,\alpha\beta'(\mu\xi - 2\mu'\sigma^2),$$

$$(\alpha'\,P + Q)\,D_{21} = 4\mu\kappa^2\,\alpha\beta\beta'\sigma^2(\mu'-\mu) - 2\mu\kappa^2\sigma^2\,\beta(\mu'\,\xi'-\mu\xi),$$

$$(\alpha'\,P + Q)\,D_{12} = 4\mu\kappa^2\,\alpha\alpha'\,\beta(\mu'-\mu) - 2\mu\kappa^2\,\alpha(\mu'\,\xi'-\mu\xi),$$

$$(\alpha'\,P + Q)\,D_{22} = 2\mu\kappa^2\,\alpha\beta(\mu'\,\xi'-2\mu\sigma^2) + 2\mu\kappa^2\,\alpha'\,\beta(\mu\xi - 2\mu'\sigma^2),\qquad(24.32)$$

$$D_{33} = \frac{2\mu\beta}{\mu\beta + \mu'\,\beta'}.$$

Using Eqs. 24.15 and 24.16, we can obtain expressions for the displacements which are analogous to Eq. 24.19 for the potentials. For brevity in writing, we omit the operation

$$\int_0^\infty e^{-ikct}\,dk$$

in our formulas.

The displacement in the reflected compressional wave will consist of the two parts (u_1, w_1) and (u_2, w_2).

The first part, which we will call the PP-part, is due to the incident compressional wave; the second part, called the SP-part, is due to the incident shear wave, which, as we have seen, is partially transformed into a compressional wave upon reflection.

Using Eqs. 24.19 for the potentials, and also using the formulas for the transition from the potentials to the displacements, analogous to Eqs. 24.2, we obtain

$$u_1 = -\int_{\Gamma_1} \sigma f_0 V_{11} H_1^{(1)}(\sigma r) \exp\left[-\alpha(z+z_0)\right] \sin \vartheta \, d\vartheta,$$

$$w_1 = -\int_{\Gamma_1} \alpha f_0 V_{11} H_0^{(1)}(\sigma r) \exp\left[-\alpha(z+z_0)\right] \sin \vartheta \, d\vartheta; \qquad (24.33)$$

$$u_2 = -\int_{\Gamma_1} \sigma g_0 V_{21} H_1^{(1)}(\sigma r) \exp\left(-\alpha z - \beta z_0\right) \sin \vartheta \, d\vartheta,$$

$$w_2 = -\int_{\Gamma_1} \alpha g_0 V_{21} H_0^{(1)}(\sigma r) \exp\left(-\alpha z - \beta z_0\right) \sin \vartheta \, d\vartheta. \qquad (24.34)$$

The displacement in the shear SV-wave also consists of two parts: (u_3, w_3) corresponding to the PS-wave, and (u_4, w_4) corresponding to the SS-wave,

$$u_3 = -\int_{\Gamma_1} \sigma \beta f_0 V_{12} H_1^{(1)}(\sigma r) \exp\left(-\beta z - \alpha z_0\right) \sin \vartheta \, d\vartheta,$$

$$w_3 = -\int_{\Gamma_1} \sigma^2 f_0 V_{12} H_0^{(1)}(\sigma r) \exp\left(-\beta z - \alpha z_0\right) \sin \vartheta \, d\vartheta; \qquad (24.35)$$

$$u_4 = -\int_{\Gamma_1} \sigma \beta g_0 V_{22} H_1^{(1)}(\sigma r) \exp\left[-\beta(z+z_0)\right] \sin \vartheta \, d\vartheta,$$

$$w_4 = -\int_{\Gamma_1} \sigma^2 g_0 V_{22} H_0^{(1)}(\sigma r) \exp\left[-\beta(z+z_0)\right] \sin \vartheta \, d\vartheta. \qquad (24.36)$$

Finally, the shear SH (horizontally polarized) displacement consists of one term,

$$v_1 = -\int_{\Gamma_1} \sigma h_0 V_{33} H_1^{(1)}(\sigma r) \exp\left[-\beta(z+z_0)\right] \sin \vartheta \, d\vartheta. \qquad (24.37)$$

The displacements in the lower medium (in the transmitted wave) are written similarly. In particular, the displacement in the compressional wave will consist of two parts: the displacements u_1', w_1' of the type $P\underline{P}$†, and the displacements u_2', w_2' of the type $S\underline{P}$. The expressions are:

$$u_1' = -\int_{\Gamma_1} \sigma f_0 D_{11} H_1^{(1)}(\sigma r) \exp\left(\alpha' z - \alpha z_0\right) \sin \vartheta \, d\vartheta,$$

$$w_1' = \int_{\Gamma_1} \alpha' f_0 D_{11} H_0^{(1)}(\sigma r) \exp\left(\alpha' z - \alpha z_0\right) \sin \vartheta \, d\vartheta; \qquad (24.38)$$

$$u_2' = -\int_{\Gamma_1} \sigma g_0 D_{21} H_1^{(1)}(\sigma r) \exp\left(\alpha' z - \beta z_0\right) \sin \vartheta \, d\vartheta,$$

$$w_2' = \int_{\Gamma_1} \alpha' g_0 D_{21} H_0^{(1)}(\sigma r) \exp\left(\alpha' z - \beta z_0\right) \sin \vartheta \, d\vartheta. \qquad (24.39)$$

---

† The bar under a symbol signifies that the wave corresponding to this symbol is traveling in the lower medium.

The shear displacement will also consist of two parts: (u_3', w_3') corresponding to a wave of the $P\underline{S}$ type, and (u_4', w_4') corresponding to a wave of the $S\underline{S}$ type:

$$u_3' = -\int_{\Gamma_1} \sigma\beta' f_0 D_{12} H_1^{(1)}(\sigma r) \exp(\beta' z - \alpha z_0) \sin\vartheta\, d\vartheta,$$

$$w_3' = -\int_{\Gamma_1} \sigma^2 f_0 D_{12} H_0^{(1)}(\sigma r) \exp(\beta' z - \alpha z_0) \sin\vartheta\, d\vartheta; \qquad (24.40)$$

$$u_4' = \int_{\Gamma_1} \sigma\beta' g_0 D_{22} H_1^{(1)}(\sigma r) \exp(\beta' z - \beta z_0) \sin\vartheta\, d\vartheta,$$

$$w_4' = -\int_{\Gamma_1} \sigma^2 g_0 D_{22} H_0^{(1)}(\sigma r) \exp(-\beta' z - \beta z_0) \sin\vartheta\, d\vartheta. \quad (24.41)$$

Finally, the shear SH (horizontally polarized) displacement will be

$$v' = -\int_{\Gamma_1} \sigma h_0 D_{33} H_1^{(1)}(\sigma r) \exp(\beta' z - \beta z_0) \sin\vartheta\, d\vartheta. \qquad (24.42)$$

Equations 24.33–24.42, taken together, are the complete solution of the problem of the reflection of a spherical wave at an interface between two elastic media.

3. Analysis of the solution. The reflected and lateral waves

The integral formulas obtained above for the displacements are quite complicated, and cannot be used to arrive at any far-reaching conclusions of a general nature regarding the character of the wave processes accompanying the reflection and refraction of waves. As in the preceding paragraphs, we limit ourselves to the analysis of the displacements at sufficiently great distances, at which the asymptotic representations become valid.† As above, the method of steepest descents is a suitable mathematical method for this purpose.

As an example, we examine the expression for u_1 as given by Eq. 24.33:

$$u_1 = -\int_{\Gamma_1} \sigma f_0 V_{11} H_1^{(1)}(\sigma r) \exp[-\alpha(z + z_0)] \sin\vartheta\, d\vartheta. \qquad (24.43)$$

---

† If we let λ_{max} denote the maximum wavelength of any of the elastic waves, corresponding to the lowest frequency which the experimental apparatus can detect, our results will be valid at distances R, satisfying the condition $R \gg \lambda_{max}$. But if we assume that the process is recorded by an "ideal" instrument with an almost uniform frequency characteristic from $\omega = 0$ to $\omega = \infty$, our asymptotic formulas will still be valid at instants of time, close to the times of arrival of the waves, where the process varies comparatively rapidly with time, and consequently, the high frequency part of the process is the principal part.

Following a procedure similar to that in § 19, the path of integration Γ_1 can be deformed in such a way that on the new path, at sufficiently great r, the condition $\sigma r \gg 1$ will be fulfilled everywhere. Then the Hankel function can be replaced by its asymptotic representation

$$H_1^{(1)}(\sigma r) \approx \sqrt{\left(\frac{2k}{\pi \sigma r}\right)} \exp\left(i\sigma r - \frac{3\pi i}{4}\right).$$

Moreover, we recall that (see Fig. 91)

$$r = R_1 \sin\vartheta_0, \quad z + z_0 = R_1 \cos\vartheta_0,$$

and also keep in mind the notation

$$\alpha = -ik\cos\vartheta, \quad \sigma = k\sin\vartheta.$$

As a result, the expression for u_1 is written

$$u_1 = \sqrt{\left(\frac{2k}{\pi r}\right)} \exp\left(\frac{\pi i}{4}\right) \int_\Gamma f_0 V_{11} \exp\left[ikR_1 \cos(\vartheta_0 - \vartheta)\right] \sin^{\frac{3}{2}}\vartheta \, d\vartheta. \quad (24.44)$$

It is convenient to choose the new path of integration in such a way that it passes through the saddle point, at which the exponential has an extremum, and leaves this point along the path on which the absolute value of the exponential decreases most rapidly.[†]

As in § 19, this is the path (see Fig. 93) on which the variable s, introduced through the relation $\cos(\vartheta - \vartheta_0) = 1 + is^2$, takes on all real values from $-\infty$ to $+\infty$. The saddle point corresponds to $s = 0$.

The integral over the path Γ is easily calculated by the method of steepest descents presented in § 19. We limit ourselves here to the first approximation, in which the entire integrand, except for the exponential, can be taken outside the integral sign at the value $\vartheta = \vartheta_0$. As a result, we obtain[‡]

$$u_1 = 2f_0(\vartheta_0) \sin\vartheta_0 \, V_{11}(\vartheta_0) \frac{e^{ikR_1}}{R}. \quad (24.45)$$

[†] It is assumed that the reader is familiar with the contents of §§ 19 and 21, where similar, but simpler, problems are examined.

[‡] If, experimentally, the reflected wave is recorded separately from the direct wave, improving the result by finding successive approximations has no meaning, since, compared with Eq. 24.45, the successive corrections will be of the order $1/kR_1$, $1/(kR_1)^2$, and so forth. But if the direct and reflected waves are superimposed, then at small grazing angles $(\vartheta_0 \to \pi/2)$, at which $V_{11} \to -1$, the term written out in Eq. 24.45 can be canceled to a considerable extent by the direct wave. The remainder may turn out to be of the same order as the first correction term. However, it can be shown that even in this case, the correction terms will play no rôle, if the combined distance of the source and the receiver from the interface is large compared with the wavelength.

We have obtained the expression for the reflected PP-wave. The factor $2f_0(\vartheta_0)\sin\vartheta_0$ characterizes the amplitude of the incident compressional wave, and V_{11} is the reflection coefficient. We will write the latter in the form

$$V(\vartheta_0) = A(\vartheta_0) + iB(\vartheta_0), \tag{24.46}$$

where it can be shown that if $\vartheta_0 < \arcsin(c/c_1)$, then $B \equiv 0$.

If we now apply the operation $\mathrm{Re}\int_0^\infty \exp(-ikct)\,dk$ to Eq. 24.45, we finally obtain (see Eq. 24.15)

$$u_1 = \frac{F_1(\vartheta_0)}{R_1}\left\{A(\vartheta_0)\frac{d}{dt}\left[P_{\mathrm{Re}}\left(t-\frac{R_1}{c}\right)\right] - B(\vartheta_0)\frac{d}{dt}\left[P_{\mathrm{Im}}\left(t-\frac{R_1}{c}\right)\right]\right\}\sin\vartheta_0,$$

where $\tag{24.47}$

$$\frac{d}{dt}\left[P_{\mathrm{Re}}\left(t-\frac{R_1}{c}\right)\right] = \mathrm{Re}\int_0^\infty(-ikc)\,P_1(k)\exp\left[ikc\left(t-\frac{R_1}{c}\right)\right]dk,$$

$$\frac{d}{dt}\left[P_{\mathrm{Im}}\left(t-\frac{R_1}{c}\right)\right] = \mathrm{Im}\int_0^\infty(-ikc)\,P_1(k)\exp\left[-ikc\left(t-\frac{R_1}{c}\right)\right]dk.$$

We obtain all the other displacement components similarly. In particular, the expression for w_1 will differ from Eq. 24.47 only in that $\sin\vartheta_0$ will be replaced by $\cos\vartheta_0$. This shows that in the first approximation, the compressional wave PP is a longitudinal wave.

In transforming the path of integration Γ_1 into the passage path Γ, it may become necessary to go around singularities of the integrand in the ϑ-plane. The singularities can be of two types: poles and branch points. The poles are found from the equation $\alpha'P = -Q$. The circuit of the poles adds the residues at these poles to Eq. 24.47. As a result, we obtain the expressions for the surface waves (the Stonely and Rayleigh waves), the amplitude of which decreases exponentially with distance on both sides of the boundary, and decreases as $1/\sqrt{r}$ with distance from the source along the interface. Since the character and properties of these waves have been thoroughly investigated (see, for example, Refs. 47 and 55), we will spend no time on them here.†

The presence of branch points (which arise because the integrand is not single-valued) makes it necessary to supplement our solution with integrals over paths encompassing the cuts (drawn from the branch points) in an appropriate manner. As a result, we obtain a system of

---

† It would have been of great interest to determine the excitation coefficients of these waves as a function of the source distribution with respect to the boundary. However, as far as we know, no one has as yet carried out such calculations.

"lateral" waves with conical wave fronts. These waves are sometimes called head waves or Mintrop waves. A wave of this type was investigated in detail in § 21 for a simple case. Since the analysis of these waves is a repetition of the exposition in § 21, we shall go through it here in an abbreviated form, concentrating our attention on the physical picture of the phenomena.

For definiteness, we will again have Eq. 24.44 for u_1 in mind. In this equation, the reflection coefficient V_{11} is not a single-valued function of the variable of integration ϑ. According to Eq. 24.23, V_{11} contains the quantities α', β and β', which, using Eqs. 24.20, can be expressed in terms of ϑ in the following way:

$$\alpha' = \sqrt{(k^2 \sin^2 \vartheta - k'^2)}, \quad \beta = \sqrt{(k^2 \sin^2 \vartheta - \kappa^2)},$$

$$\beta' = \sqrt{(k^2 \sin^2 \vartheta - \kappa'^2)}; \tag{24.48}$$

V_{11} can take on various values, depending on the choice of the signs of these radicals. For the branch points, at which the radicals become zero, we have

$$\vartheta_{k'} = \arcsin\frac{k'}{k}, \quad \vartheta_{\kappa} = \arcsin\frac{\kappa}{k}, \quad \vartheta_{\kappa'} = \arcsin\frac{\kappa'}{k}. \tag{24.49}$$

We consider the lateral wave associated with the branch point $\vartheta_{k'}$. The course of the arguments will be exactly the same here as in § 21. During the calculations, expressions of the type (21.4) and (21.10) again appear, and $\Phi(\vartheta)$, in analogy with Eq. 21.5, will be determined by the expression

$$\Phi(\vartheta) = V_{11}^+(\vartheta) - V_{11}(\vartheta),$$

where V_{11}^+ and V_{11} differ from one another in the sign of the square root $\alpha' = ik\sqrt{(n^2 - \sin^2 \vartheta)}$. According to Eqs. 24.23 and 24.24 we have, for $\Phi(\vartheta)$,

$$\Phi(\vartheta) = \frac{2\alpha'(P_1 Q - Q_1 P)}{Q^2 - \alpha'^2 P^2}.$$

As a result, to within the same accuracy as in § 21, the displacement u_1 in the lateral wave is

$$u_1 = \left[f_0(\delta) \frac{P_1 Q - Q_1 P}{Q^2 \cos\delta} \right]_{\vartheta = \delta} \cdot \frac{n \exp(ikR_{\text{lat}})}{r^{\frac{1}{2}} L_1^{\frac{3}{2}}} \sin\delta. \tag{24.50}$$

For convenience in comparing the present results with those of § 21, we used the notation $\vartheta_{k'} \equiv \delta$, $k'/k \equiv n$. The quantity R_{lat} in Eq. 24.50 has the value

$$R_{\text{lat}} = \frac{z + z_0}{\cos\delta} + nL_1 = L_0 + L + nL_1.$$

The meanings of the quantities L_0, L and L_1, which have the dimensions of length, are clear from Fig. 97.

We now apply the operation $\mathrm{Re} \int_0^\infty \exp(-ikct)\,dk$ to u_1 and take Eqs. 24.6, 24.7, 24.18 and 24.19 into account. We finally obtain

$$u_1 = -ik'c \left[\frac{QP_1 - PQ_1}{Q^2 \cos \delta} \right] \frac{F_1(\delta)\,P(t_1)}{r^{\frac{1}{2}} L_1^{\frac{3}{2}}} \sin \delta, \qquad (24.51)$$

where

$$t_1 = t - \frac{L_0 + L}{c} - \frac{L_1}{c_1}. \qquad (24.52)$$

The expression for the displacement w_1 will be the same as Eq. 24.51, except that $\sin \delta$ will be replaced by $\cos \delta$. Thus, the displacement in the lateral wave is along the ray BP (Fig. 97), i.e. the wave is longitudinal.

A pulse, corresponding to the lateral wave, travels from the source to the point of observation during the time

$$t - t_1 = \frac{L_0 + L}{c} + \frac{L_1}{c_1},$$

i.e. travels along the segments L_0 and L in Fig. 97 with the velocity c, and along the segment L_1 with the velocity c_1. It is convenient to denote a lateral wave of this type by the symbol $P_1 P_2 P_1$, where the subscript 1 signifies that over the corresponding portion of the path, the wave travels in the upper medium, and the subscript 2 signifies the same for the lower medium.

Let us note some important properties of the lateral wave in the case under consideration:

1. The factor $F_1(\delta)$ in Eq. 24.51 indicates that the amplitude of the lateral wave is proportional to the amplitude of the primary ray, incident on the boundary at the angle of total internal reflection δ.

2. The time dependence of the displacement u_1 is given by the factor $p(t)$. If the source is a cylindrical cavity of small dimensions, the time dependence of the force applied to the lateral walls of the cavity is characterized by this same function. We note that the time dependence of the displacements in the direct and the reflected waves was characterized by the function $(d/dt)\,p(t - R/c)$.

3. Just as in § 21, Eq. 24.51 for the lateral wave loses meaning on the ray issuing from the image source O' at the angle δ. On this ray, $L_1 = 0$. When necessary, the value of the displacements on this ray and in its vicinity can be obtained by the method presented in § 22.

11

4. *Analysis of the complete system of waves arising upon the reflection and refraction of a spherical wave*

Above, we analyzed the reflected wave PP and the lateral wave $P_1P_2P_1$, due to the presence of the branch point $\vartheta_{k'}$. In what follows, the reflected and refracted waves, as well as the surface waves given by residues at poles, will be called waves of the first order. The lateral waves, and also the surface waves into which the lateral waves degenerate when there are certain relations between the propagation velocities of the waves in the adjoining media (see, for example, the case $n > 1$ in § 21), will be called waves of the second order.

The waves of the second order are distinguished by the fact that until the application of the operation $\int_0^\infty \exp(-ikct)\,dk$, their amplitude is a factor of kR less than the amplitude of the waves of the first order. Accordingly, in the final expressions for the displacements, after integration over k, this leads to the following two significant features.

1. The time dependence of waves of the second order is given by the function $p(t)$, while for waves of the first order, it is given by the function dp/dt. Consequently, if we have a discontinuity in the function $p(t)$ itself, or in its derivative, the discontinuity in the waves of the second order will be weaker.

2. The amplitude of the waves of the second order decreases approximately as $1/r^2$ with distance along the boundary, while the waves of the first order decrease approximately as $1/r$ (for the reflected and refracted wave) or as $1/\sqrt{r}$ (for the surface waves of the Rayleigh wave type). The expressions for all of the waves of the second order are obtained by integration over the borders of the cuts, drawn from the corresponding branch points.

Above, we obtained the expression for the lateral wave corresponding to the branch point $\vartheta_{k'}$. We obtain another wave when we integrate over the borders of the cut made at the branch point $\vartheta_{\kappa'} = \arcsin(\kappa'/k)$. This point will lie on the real axis if the velocity of transverse waves in the lower medium is greater than the velocity of longitudinal waves in the upper medium ($\kappa' < k$). In this case, the wave will have the character of a lateral wave, and on the segment corresponding to propagation in the lower medium (L_1 in Fig. 97), the wave will travel as a transverse wave. We will therefore denote this wave by the symbol $P_1S_2P_1$. The angle of incidence at the boundary of the corresponding ray (the angle δ in Fig. 97) will be equal to $\vartheta_{\kappa'}$. But if the relation between the wave velocities is the inverse, i.e. $\kappa' > k$, then $\vartheta_{\kappa'}$ will be a complex angle, and we will obtain an inhomogeneous surface wave. We will denote the latter by the symbol $(S_2)_1$. Here, and in what follows, parentheses will denote waves which are attenuated with distance from the boundary.

Still one more wave is obtained as a result of the integration over the borders of the cut made at the branch point $\vartheta_\kappa = \arcsin(\kappa/k)$. However, since we always have $\kappa > k$, we always obtain the surface wave in this case. We will denote this wave by $(S_1)_1$. Thus, we have obtained three waves of the second order for the displacement u_1 (and for the analogous component w_1 along the z-axis).

TABLE 4
The System of Waves in the Upper Medium

Compressional waves (P-waves) and shear SV-waves, polarized in the vertical plane

	Waves of first order		Waves of second order				
	Saddle point	Roots of the equation $\alpha' P = -Q$	k'	k	κ'	κ	Type of wave
u_1, w_1	$P_1 P_1$	Surface wave of	$P_1 P_2 P_1$	—	(a)-$P_1 S_2 P_1$ (b) $(S_2)_1$	$(S_1)_1$	Compressional wave
u_2, w_2	$S_1 P_1$	the Rayleigh	$S_1 P_2 P_1$	—	(a)-$S_1 S_2 P_1$ (b) $S_1 (S_2)$	$(S_1)_2$	Compressional wave
u_3, w_3	$P_1 S_1$	type	$P_1 P_2 S_1$	—	(a)-$P_1 S_2 S_1$ (b) $S_2 S_1$	$(S_1)_3$	Shear wave
u_4, w_4	$S_1 S_1$		$S_1 P_2 S_1$	$S_1 P_1 S_1$	$S_1 S_2 S_1$	—	Shear wave

Shear SH-waves, polarized in the horizontal plane

v_1	$S_1 S_1$				$S_1 S_2 S_1$	—	Shear wave

Case (a) in the Table corresponds to $b' > c$, that is, $k > \kappa'$.

Case (b) corresponds to $b' < c$, that is, $k < \kappa'$.

In the same way, we will obtain three waves for the components of the longitudinal displacement u_2, w_2, due to a *transverse wave* incident on the boundary.

There will also be three waves for the transverse displacements (u_3, w_3) and (u_4, w_4). As a result, we obtain a system of 12 waves of the second order in the upper medium. Some of these waves will have the character of lateral waves, and some will have the character of surface waves.

A complete summary of all the waves is given in Table 4. The second row from the top indicates the character of the singularity in the complex plane associated with the given wave (passage point, pole), and for waves of the second order, this row gives the wave number with which the wave propagates along the boundary. The wave number immediately determines the location (in the complex plane) of the branch point corresponding to the wave. Thus, in particular, the wave number k' corresponds to the branch point

$$\vartheta_{k'} = \arcsin(k'/k)$$

in the ϑ-plane.

It is worthwhile to note that in obtaining the expressions for the various lateral waves, it is not always convenient to perform the integration in the plane of the complex variable ϑ. In particular, in obtaining

TABLE 5

The Regions of Existence of the Waves indicated in Table 4

Compressional waves (P-waves) and shear SV-waves

	Saddle point	Roots of the equation $\alpha' P = -Q$	k'	k	κ'	κ
u_1, w_1	$z > 0$	Regions of existence are determined by the position of the roots	$\epsilon > i_1$	—	(a) $\epsilon > i'_3$ (b) $\epsilon > i_3$	$\epsilon > i_5$
u_2, w_2	$z > 0$					
			$\epsilon_{21} > i_2$	—	(a) $\epsilon_{21} > i_4$ $\dfrac{r}{b'} >$	$\dfrac{r}{b} > \dfrac{z_0}{c \cos \epsilon_{21}} + \dfrac{z}{b \cos \epsilon'_{21}}$
u_3, w_3	$z > 0$					
			$\epsilon_{12} > i_1$	—	(a) $\epsilon_{12} > i'_3$ $\dfrac{r}{b'} >$	$\dfrac{r}{b} > \dfrac{z_0}{c \cos \epsilon_{12}} + \dfrac{z}{b \cos \epsilon'_{12}}$
u_4, w_4	$z > 0$					
			$\epsilon > i_2$	$\epsilon > i_5$	$\epsilon > i_4$	—

Shear SH-waves

v_1	$z > 0$				$\epsilon > i_4$	—

the expression for the wave $S_1 P_1 S_1$, all of the arguments become considerably simpler if we use the variable γ, connected with ϑ by the relation $\cos \vartheta = \sqrt{[1 - (\kappa^2/k^2) \sin^2 \gamma]}$. In this plane, we have the branch point $\gamma_k = \arcsin(k/\kappa)$, which gives rise to the lateral wave $S_1 P_1 S_1$. In this wave, the transverse oscillations at the boundary are transformed into longitudinal oscillations which propagate along the boundary with the velocity of longitudinal waves and are radiated back into the upper medium in the form of transverse waves.

It should also be noted that in the column corresponding to κ', we observe in a number of cases, (a) a lateral wave or (b) a surface wave, depending on the relation between the pertinent velocities. Case (a) is realized when $b' > c$, that is, when $k > \kappa'$. Case (b) is realized when $b' < c$, that is when $k < \kappa'$.

The boundaries of the regions within which the waves in Table 4 exist, are indicated in Table 5. For an arbitrary location of the point of observation, the angles ϵ, ϵ_{12}, ϵ_{21}, ϵ'_{12} and ϵ'_{21} in this table are given by the relations

$$r = (z + z_0) \tan \epsilon,$$

$$r = z \tan \epsilon'_{12} + z_0 \tan \epsilon_{12}, \qquad \frac{c}{\sin \epsilon_{12}} = \frac{b}{\sin \epsilon'_{12}},$$

$$r = z \tan \epsilon'_{21} + z_0 \tan \epsilon_{21}, \qquad \frac{b}{\sin \epsilon_{21}} = \frac{c}{\sin \epsilon'_{21}}. \qquad (24.53)$$

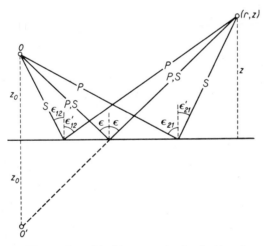

Fig. 110. The angles of incidence and of reflection for various types of waves.

As is clear from Fig. 110, ϵ, ϵ_{12} and ϵ_{21} are the angles of incidence, and $\epsilon' = \epsilon$, ϵ'_{12} and ϵ'_{21} are the corresponding angles of reflection of the four waves arriving at the arbitrary point (r, z). Previously, the angle ϵ was also denoted by ϑ_0.

The "critical" angles $i_1, i_2, \ldots$ are given by the relations

$$i_1 = \arcsin \frac{c}{c'}, \qquad i_2 = \arcsin \frac{b}{c'}, \qquad i_3 = \arcsin \frac{b'}{c} \quad (\text{for } c > b'),$$

$$i_3' = \arcsin \frac{c}{b'} \quad (\text{for } c < b'), \qquad i_4 = \arcsin \frac{b}{b'},$$

$$i_5 = \arcsin \frac{b}{c}, \qquad i_6 = \arcsin \frac{b'}{c'}. \tag{24.54}$$

Thus, for example, it is clear from Table 5, that the wave $P_1 P_2 P_1$ exists only in the region of space in which $\epsilon > i_1$, that is, when the angle of incidence of the wave is greater than the angle of total internal reflection of the longitudinal waves.

TABLE 6

The System of Waves in the Lower Medium

Compressional waves (P-waves) and shear SV-waves

		Waves of first order		Waves of second order				
	Saddle point	Roots of the equation $\alpha' P = -Q$	k'	k	κ'	κ	Type of wave
u_1', w_1'	$P_1\underline{P_2}$	Surface	$-$	$(P_1)_1$	(a) $P_1(S_2)$	$(S_1)_1$	Compressional wave
		waves of			(b) $\underline{(S_2)}$		
u_2', w_2'	$S_1\underline{P_2}$	the	$-$	$S_1(P_1)_2$	$\overline{S_1(S_2)}$	$(S_1)^2$	,, ,,
u_3', w_3'	$P_1\underline{S_2}$	Rayleigh	$P_1P_2\underline{S_2}$	(a) $\underline{(P_1)_3}$	(b) $\underline{(S_2)_3}$	$\underline{(S_1)_3}$	Shear wave
u_4', w_4'	$S_1\underline{S_2}$	type	$S_1P_2\underline{S_2}$	(a) $\overline{S_1(P_1)_4}$	(b) $\overline{S_1P_1\underline{S_2}}$	$\overline{(S_1)_4}$	,, ,,

				Shear SH-wave				
v'	$S_1\underline{S_2}$					(S_1)	Shear wave

(a) $b' > c$, that is $k > \kappa'$.

(b) $b' < c$, that is $k < \kappa'$.

A summary of the waves arising in the lower medium is given in Table 6. In the present notation, the bar below a symbol signifies that the corresponding path of the wave passes through the lower medium, and the parentheses indicate that the amplitude of the wave decreases exponentially with distance from the boundary. The regions within which these waves exist are shown in Table 7. This whole complicated picture of waves of the first and second order will become clear to the reader if he will carefully read § 23, where exactly the same effect was examined for the simple example of the refraction of a spherical wave at an interface between two liquid media.

TABLE 7

The Regions of Existence of the Waves indicated in Table 6

Compressional waves (P-waves) and shear SV-waves

Saddle point	Roots of the equation $\alpha'P=-Q$	k'	k	κ'	κ	
u_1', w_1' $z<0$	Regions of existence are deter-	—	$\dfrac{r}{c}>$	$\dfrac{r}{b'}>$	$\dfrac{r}{b}>$	$\dfrac{z_0}{c\cos\eta_{11}} - \dfrac{z}{c'\cos\eta_{11}'}$
u_2', w_2' $z<0$	mined by the position of the roots	—	$\dfrac{r}{c}>$	$\dfrac{r}{b'}>$	$\dfrac{r}{b}>$	$\dfrac{z_0}{b\cos\eta_{21}} - \dfrac{z}{c'\cos\eta_{21}'}$
u_3', w_3' $z<0$		$\eta_{12}>i_1$	(a) $\dfrac{r}{c}>$	(b) $\dfrac{r}{b'}>$	$\dfrac{r}{b}>$	$\dfrac{z_0}{c\cos\eta_{12}} - \dfrac{z}{b'\cos\eta_{12}'}$
u_4', w_4' $z<0$		$\eta_{22}>i_2$	(a) $\dfrac{r}{c}>$	(b) $\eta_{22}>i_5$	$\dfrac{r}{b}>$	$\dfrac{z_0}{b\cos\eta_{22}} - \dfrac{z}{b'\cos\eta_{22}'}$

Shear SH-waves

v' $z<0$					$\dfrac{r}{b}>$	$\dfrac{z_0}{b\cos\eta_{22}} - \dfrac{z}{b'\cos\eta_{22}'}$

The angles $\eta_{11}, \ldots, \eta_{11}', \ldots$ in Table 4 are determined, for an arbitrary point of observation (r,z), where $z<0$, by the following system of equations:

$$r = z_0 \tan\eta_{11} - z\tan\eta_{11}', \quad \frac{c}{\sin\eta_{11}} = \frac{c'}{\sin\eta_{11}'},$$

$$r = z_0 \tan\eta_{12} - z\tan\eta_{12}', \quad \frac{c}{\sin\eta_{12}} = \frac{b'}{\sin\eta_{12}'},$$

$$r = z_0 \tan\eta_{22} - z\tan\eta_{22}', \quad \frac{b}{\sin\eta_{22}} = \frac{b'}{\sin\eta_{22}'},$$

$$r = z_0 \tan\eta_{21} - z\tan\eta_{21}', \quad \frac{b}{\sin\eta_{21}} = \frac{c'}{\sin\eta_{21}'}. \tag{24.55}$$

As is clear from Fig. 111, the angles $\eta_{11}, \ldots$ are the angles of incidence, and the angles $\eta_{11}', \ldots$ are the angles of refraction of the various rays, connecting the receiver with the source.

Finally, it is interesting to note that the various waves of the second order can be combined into dynamically independent groups. Waves belonging to one and the same group correspond to one and the same critical angle of incidence at the boundary, propagate along the boundary with one and the same phase velocity, and, as a group, satisfy the boundary conditions independently of the other groups. As an example, we will consider one such group, consisting of the lateral waves $P_1 P_2 P_1$, $P_1 P_2 S_1$, $P_1 P_2 \underline{S_2}$ and the refracted wave $P_1 \underline{P_2}$. All three lateral waves arise at the "critical" angle of incidence $i_1 = \arcsin(c/c')$, and propagate along the boundary with the velocity c', which corresponds to the wavelength $\Lambda' = 2\pi/k'$. At $z = 0$, the term of the first order in the wave $P_1 P_2$ becomes identically zero. But the term of the second order, which can be found by the method of steepest

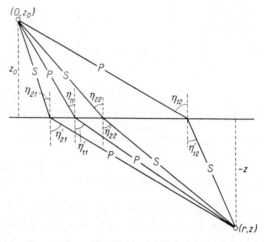

Fig. 111. The angles of incidence and refraction for various types of waves.

descents, has the same order of magnitude as the amplitude of the lateral waves. It can be shown that the collective system of waves $P_1 P_2 P_1$, $P_1 P_2 S_1$, $P_1 P_2 \underline{S_2}$ and $P_1 \underline{P_2}$ satisfies all the conditions at the boundary, independently of all other waves. The energy relations between these waves is also interesting. As is evident from the form of the wave fronts of the lateral waves, they all carry energy away from the boundary. It can be shown that the refracted wave $P_1 \underline{P_2}$ carries just the right amount of energy to the boundary for the waves $\overline{P_1 P_2 P_1}$, $P_1 P_2 S_1$ and $P_1 P_2 \underline{S_2}$. The oscillations in this wave are not purely longitudinal, and have a component normal to the boundary. It is somewhat difficult to give graphic representations here because geometrical optics is inapplicable (the term of the first order becomes identically zero).

Chapter V

WAVE PROPAGATION IN LAYERS

In the present chapter, we shall develop the theory of wave propagation in a homogeneous layer, bounded on both sides by plane-parallel boundaries. The investigation of this phenomenon is important in various branches of physics. Thus, radiowave propagation through great distances in the atmosphere is essentially propagation in a layer bounded on one side by the earth, and on the other side by the ionosphere. In the case of very long waves (wavelengths of the order of tens of kilometers and greater), this layer exhibits typical waveguide properties.[2, 72]

The study of sound propagation in the sea is of great importance. In this case, the layer of water is bounded on one side by the ocean bottom, and on the other side by the surface of the water. However, the theory of wave propagation in layers is probably of greatest importance in seismic exploration.

A detailed theory of the propagation of elastic waves in the crust of the earth, taking its layered structure into account, was developed by Petrashen' and collaborators.[64] In this chapter, we shall consider only some of the fundamental problems of the theory, which are of equally great importance both for elastic wave propagation as well as for various cases of electromagnetic wave propagation in layers. We shall also try to analyze some physical representations connected with wave propagation in layers.

§ 25. A Layer with Perfectly Reflecting Boundaries

1. *The picture of image sources*

We consider a layer with the boundaries $z = 0$ and $z = h$ (Fig. 112). The source is assumed to be situated on the z-axis, at the point $z = z_0$. The wave propagation velocity in the layer will be denoted by c.

Our discussion will refer simultaneously to sound wave propagation in the layer, and to electromagnetic wave propagation in the layer. In the first case, the source, as usual, will be assumed to be a pulsating sphere of small radius; we shall also assume that the medium comprising the layer has no shear resistance (a liquid or a gas). In the second case, as usual, the radiator will be a vertical dipole.

The function ψ, which in the electromagnetic case is the vertical component of the Hertz vector, and in the acoustic case is the acoustic potential, satisfies the wave equation (within the layer)

$$\nabla^2 \psi + k^2 \psi = 0, \quad k = \frac{\omega}{c}. \tag{25.1}$$

The boundaries of the layer are assumed to be perfectly reflecting. This is expressed by the condition

$$\frac{\partial \psi}{\partial z} = 0 \begin{cases} z = 0, \\ z = h. \end{cases} \tag{25.2}$$

These conditions are fulfilled in acoustics if the boundaries of the layer are rigid walls. In electrodynamics, these conditions indicate the presence of perfectly conducting surfaces at $z = 0$ and $z = h$.

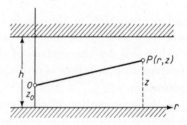

Fig. 112. Locations of the source and the receiver in the layer.

We now show that the field at an arbitrary point (r, z) can be represented as the sum of the direct wave and waves radiated by an infinite network of "image" sources, which are obtained as a result of the successive specular reflections of the source at the boundaries of the layer.

In fact, let us add to the field of our source, the field of the image source, obtained by specular reflection at the lower boundary of the layer. It will be convenient to denote the source by O_{01}, and the image source, obtained by reflection at the lower boundary, by O_{02} (Fig. 113). The total field will be

$$\psi = \psi_{01} + \psi_{02} = \frac{\exp(ikR_{01})}{R_{01}} + \frac{\exp(ikR_{02})}{R_{02}}, \tag{25.3}$$

where $\quad R_{01} = \sqrt{[r^2 + (z - z_0)^2]}, \quad R_{02} = \sqrt{[r^2 + (z + z_0)^2]} \tag{25.4}$

are the distances from O_{01} and O_{02} to the receiver at the point $P(r, z)$.

Equation 25.3 for ψ will satisfy the wave equation (25.1), because each of the two terms satisfy this equation. It will also satisfy the boundary condition at the boundary $z = 0$, as is clear from the fact that the system consisting of the sources O_{01} and O_{02} is symmetrical with respect to the boundary $z = 0$, and it follows immediately from the symmetry condition that the equation $\partial\psi/\partial z = 0$ is also satisfied at this boundary. However, the sum (25.3) cannot be the solution of our problem because it does not satisfy condition (25.2) at the boundary $z = h$.

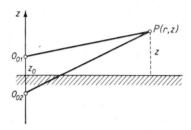

Fig. 113. The source and its mirror image in the lower boundary.

Now, to the sources O_{01} and O_{02}, we add the pair of image sources O_{03} and O_{04}, which are obtained by specular reflection of the first two sources at the upper boundary of the layer (Fig. 114). This solution will satisfy the wave equation and the boundary condition at the upper boundary (because the system of four sources is symmetrical with respect to this boundary), but will not satisfy the boundary condition at the lower boundary.

We next add to our solution the waves radiated by the image sources O_{11} and O_{12}. Since the addition of these sources symmetrizes the picture of the sources with respect to the lower boundary, the boundary conditions will now be satisfied at the lower boundary, but not at the upper boundary, and so forth. Continuing with this construction of a network of image sources, we alternately satisfy the boundary conditions at one of the boundaries and not at the other. However, since each additional pair of sources is farther away, the addition to the field at the point $P(r, z)$ will be less each time, and in the limit of an infinite network of sources, the boundary conditions will be satisfied at both boundaries of the layer.

As a result, the total field can be written in the form

$$\psi = \sum_{l=0}^{\infty} \left(\frac{\exp{(ikR_{l1})}}{R_{l1}} + \frac{\exp{(ikR_{l2})}}{R_{l2}} + \frac{\exp{(ikR_{l3})}}{R_{l3}} + \frac{\exp{(ikR_{l4})}}{R_{l4}} \right), \qquad (25.5)$$

where
$$R_{l1} = \sqrt{[r^2 + (2lh + z - z_0)^2]},$$
$$R_{l2} = \sqrt{[r^2 + (2lh + z + z_0)^2]},$$
$$R_{l3} = \sqrt{\{r^2 + [2(l+1)h - z - z_0]^2\}},$$
$$R_{l4} = \sqrt{\{r^2 + [2(l+1)h + z_0 - z]^2\}}. \tag{25.6}$$

The solution (25.5) satisfies the wave equation, because it is composed of spherical waves, each of which satisfies the wave equation. Furthermore it satisfies the condition at the source, since as the point P approaches the principal source O_{01}, the only important term in the

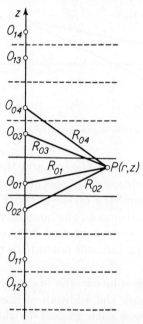

Fig. 114. The network of image sources obtained as a result of multiple reflections at the boundaries of the layer.

sum (25.5) will be the term $e^{ikR_{01}}/R_{01}$, which yields the required singularity. We saw above that the solution must also satisfy boundary conditions (25.2). However, we shall give a special proof of this. To save space, we shall limit the proof to the upper boundary $(z = h)$.

According to Eq. 25.5, the derivative $\partial\psi/\partial z$ can be written

$$\frac{\partial\psi}{\partial z} = \sum_{l=0}^{\infty} \left[\frac{\partial}{\partial R_{l1}} \left(\frac{\exp(ikR_{l1})}{R_{l1}} \right) \frac{\partial R_{l1}}{\partial z} + \frac{\partial}{\partial R_{l3}} \left(\frac{\exp(ikR_{l3})}{R_{l3}} \right) \frac{\partial R_{l3}}{\partial z} \right]$$
$$+ \sum_{l=0}^{\infty} \left[\frac{\partial}{\partial R_{l2}} \left(\frac{\exp(ikR_{l2})}{R_{l2}} \right) \frac{\partial R_{l2}}{\partial z} + \frac{\partial}{\partial R_{l4}} \left(\frac{\exp(ikR_{l4})}{R_{l4}} \right) \frac{\partial R_{l4}}{\partial z} \right]. \tag{25.7}$$

From Eq. 25.6, we have

$$\frac{\partial R_{l1}}{\partial z} = \frac{2lh + z - z_0}{R_{l1}}, \qquad \frac{\partial R_{l2}}{\partial z} = \frac{2lh + z + z_0}{R_{l2}},$$

$$\frac{\partial R_{l3}}{\partial z} = -\frac{2(l+1)h - z - z_0}{R_{l3}}, \quad \frac{\partial R_{l4}}{\partial z} = -\frac{2(l+1)h - z + z_0}{R_{l4}}. \qquad (25.8)$$

Moreover, $\quad \dfrac{\partial}{\partial R_{l1}}\left(\dfrac{\exp{(ikR_{l1})}}{R_{l1}}\right) = \dfrac{ikR_{l1} - 1}{R_{l1}}\dfrac{\exp{(ikR_{l1})}}{R_{l1}}, \quad$ etc. $\qquad (25.9)$

From Eqs. 25.6, 25.8 and 25.9, we obtain at $z = h$:

$$R_{l1} = R_{l3}, \quad R_{l2} = R_{l4},$$

$$\frac{\partial}{\partial R_{l1}}\left(\frac{\exp{(ikR_{l1})}}{R_{l1}}\right) = \frac{\partial}{\partial R_{l3}}\left(\frac{\exp{(ikR_{l3})}}{R_{l3}}\right),$$

$$\frac{\partial}{\partial R_{l2}}\left(\frac{\exp{(ikR_{l2})}}{R_{l2}}\right) = \frac{\partial}{\partial R_{l4}}\left(\frac{\exp{(ikR_{l4})}}{R_{l4}}\right),$$

$$\frac{\partial R_{l1}}{\partial z} = -\frac{\partial R_{l3}}{\partial z}, \quad \frac{\partial R_{l2}}{\partial z} = -\frac{\partial R_{l4}}{\partial z}. \qquad (25.10)$$

Using the above relations, Eq. 25.7 gives

$$\left(\frac{\partial \psi}{\partial z}\right)_{z=h} = 0,$$

which completes the proof.

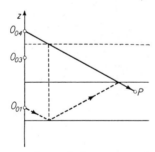

Fig. 115. The image source O_{04} and the ray associated with it, propagating along the layer.

The proof for the lower boundary ($z = 0$) is performed similarly, except that, in grouping the terms, it is convenient to pair terms corresponding to sources which are situated symmetrically with respect to the lower boundary.

It is useful to note that to each image source there corresponds a ray propagating along the layer and undergoing a definite number of reflections from the boundaries as it travels from the source to the receiver. This equivalence is seen graphically for the image source O_{04}, in Fig. 115.

2. *The case of a negative reflection coefficient*

Instead of the condition $\partial\psi/\partial z = 0$, the condition $\psi = 0$ can be satisfied at a perfectly reflecting boundary. This occurs, for example, in the electromagnetic case, when a horizontally polarized wave is incident on a perfectly conducting boundary. In the propagation of a sound wave in a layer of liquid with a completely rigid bottom and a

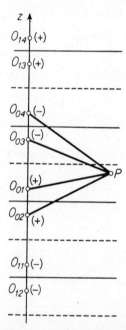

Fig. 116. The network of image sources for the case in which the reflection coefficient at the upper boundary is equal to -1.

free surface, the condition $\partial\psi/\partial z = 0$ (zero normal velocity) is fulfilled at the bottom, and the condition $\psi = 0$ (zero pressure) is fulfilled at the free surface.

At the surface fulfilling the condition $\psi = 0$, the reflection coefficient is equal to -1, while at the surface fulfilling the condition $\partial\psi/\partial z = 0$, it is equal to $+1$. This means that the phase of the wave changes discontinuously by half a period upon reflection from the boundary. The network of image sources constructed for a layer, in which one or both boundaries have a reflection coefficient equal to -1, contains image sources which are 180° out of phase with the principal source. These are the sources which correspond to rays which are reflected an odd number of times from the boundary at which $V = -1$. The picture of the image sources for the case in which the lower boundary corresponds to $V = 1$, and the upper to $V = -1$ (the above-mentioned acoustic case) is constructed in Fig. 116. The principal source and those in phase with it are marked $+$, and the sources which are 180° out of phase with the principal source are marked $-$. It is not hard to see that the lines joining the latter sources with the point of observation P, have an odd number of intersections with the dotted horizontal lines, which represent the upper boundary of the layer and its successive reflections.

In this case, the expression for the field analogous to Eq. 25.5 is written

$$\psi = \sum_{l=0}^{\infty} (-1)^l \left[\frac{\exp(ikR_{l1})}{R_{l1}} + \frac{\exp(ikR_{l2})}{R_{l2}} - \frac{\exp(ikR_{l3})}{R_{l3}} - \frac{\exp(ikR_{l4})}{R_{l4}} \right].$$

(25.11)

It is useful to generalize the discussion to some extent. The reflection coefficient at the lower boundary will be denoted by V_1, and at the upper boundary by V_2. Either one can take the value ± 1. Tracing through the number of reflections undergone at the boundaries by the rays corresponding to each of the image sources, it is not hard to show that in this general case the expression for the field will be

$$\psi = \sum_{l=0}^{\infty} (V_1 V_2)^l \left[\frac{\exp(ikR_{l1})}{R_{l1}} + V_1 \frac{\exp(ikR_{l2})}{R_{l2}} \right.$$
$$\left. + V_2 \frac{\exp(ikR_{l3})}{R_{l3}} + V_1 V_2 \frac{\exp(ikR_{l4})}{R_{l4}} \right]. \quad (25.12)$$

§ 26. A Layer with Perfectly Reflecting Boundaries (Continued)

1. *Integral representation of the field in the layer*

The field in the layer can also be represented as a set of "normal modes", each of which individually satisfies the wave equation and the boundary conditions, and propagates along the layer with its own velocity.

We shall use the integral representation of the field of a spherical wave to transform the field of the set of image sources into the field of the normal modes.

Using Eq. 18.19, and taking into account the transformations (19.3) to (19.6), we obtain

$$\frac{\exp(ikR)}{R} = \frac{ik}{2}\int_{\Gamma_1} \exp(\pm ikz\cos\vartheta)\, H_0^{(1)}(kr\sin\vartheta)\sin\vartheta\, d\vartheta,$$

$$R = \sqrt{(r^2 + z^2)}, \tag{26.1}$$

where the signs $+$ and $-$ refer, respectively, to the cases $z > 0$ and $z < 0$. We apply the last relation to each of the terms of the sum (25.12). The quantity z will take on all of the values $z = z_{lj}$, $l = 0, 1, \ldots$, $j = 1, 2, 3, 4$, where, according to Eq. 25.6,

$$z_{l1} = 2lh + z - z_0,$$

$$z_{l2} = 2lh + z + z_0,$$

$$z_{l3} = 2(l+1)h - z - z_0, \tag{26.2}$$

$$z_{l4} = 2(l+1)h - z + z_0,$$

$$R_{lj} = \sqrt{(r^2 + z_{lj}^2)}.$$

Under these conditions, we have $z_{lj} > 0$ for all l and j, with the exception $z_{01} = z - z_0$. The sign of the last difference depends on the relation between the magnitudes of z and z_0.

For definiteness, we shall first take the case $z > z_0$. We use Eq. 26.1 to transform the expression $\exp(ikR_{lj}/R_{lj})$, and then substitute the result into Eq. 25.12. Interchanging the order of summation and integration, and carrying out some straightforward transformations, we are finally able to write Eq. 25.12 in the form

$$\psi = \frac{ik}{2}\int_{\Gamma_1} H_0^{(1)}(kr\sin\vartheta) \times \{\exp[b(z-z_0)] + V_1\exp[b(z+z_0)]$$

$$+ V_2\exp[b(2h-z-z_0)] + V_1V_2\exp[b(2h-z+z_0)]\}$$

$$\times \sum_{l=0}^{\infty} (V_1V_2)^l \exp(2lbh)\sin\vartheta\, d\vartheta; \tag{26.3}$$

where we have used the notation

$$b \equiv ik\cos\vartheta. \tag{26.4}$$

We keep in mind that the identity

$$\exp[b(z-z_0)] + V_1\exp[b(z+z_0)] + V_2\exp[b(2h-z-z_0)]$$

$$+ V_1V_2\exp[b(2h-z+z_0)]$$

$$\equiv [\exp(-bz_0) + V_1\exp(bz_0)]\{\exp(bz) + V_2\exp[b(2h-z)]\} \tag{26.5}$$

holds, and that, moreover,

$$\sum_{l=0}^{\infty} (V_1 V_2)^l \exp{(2lbh)} = \frac{1}{1 - V_1 V_2 \exp{(2bh)}}. \tag{26.6}$$

There is no doubt as to the convergence of the series on the left hand side of the last equation, if we take account of a small imaginary part in the wave number k (absorption in the medium) or of the difference (however small) between V_1 or V_2 and unity.

As a result, we obtain from Eq. 26.3, for $z > z_0$,

$$\psi = \frac{ik}{2} \int_{\Gamma_1} \frac{[\exp{(-bz_0)} + V_1 \exp{(bz_0)}] \{\exp{[-b(h-z)]} + V_2 \exp{[b(h-z)]}\}}{\exp{(-bh)} [1 - V_1 V_2 \exp{(2bh)}]}$$
$$\times H_0^{(1)}(kr \sin{\vartheta}) \sin{\vartheta} \, d\vartheta. \tag{26.7}$$

It can be shown in the same way that, in the case $z < z_0$, the field is obtained from Eq. 26.7 by replacing z by z_0 and vice versa. This also follows from the reciprocity principle.

It will be shown below that Eq. 26.7 has a very general significance. However, at this point we shall consider only some special cases. We begin, as in § 25, by setting $V_1 = V_2 = 1$, which corresponds to the normal derivative $\partial \psi / \partial z$ equal to zero on both boundaries.

Then Eq. 26.7 gives

$$z > z_0, \quad \psi = 2ik \int_{\Gamma_1} \frac{\cosh{bz_0} \cosh{b(h-z)}}{e^{-bh} - e^{bh}} H_0^{(1)}(kr \sin{\vartheta}) \sin{\vartheta} \, d\vartheta, \tag{26.8a}$$

$$z < z_0, \quad \psi = 2ik \int_{\Gamma_1} \frac{\cosh{bz} \cosh{b(h-z_0)}}{e^{-bh} - e^{bh}} H_0^{(1)}(kr \sin{\vartheta}) \sin{\vartheta} \, d\vartheta. \tag{26.8b}$$

We shall use these expressions in what follows.

We shall show that the integral (26.8) converges everywhere, except at the point $z = z_0$, $r = 0$, at which it yields the required singularity of the form $1/R$.

On infinitely remote portions of the path of integration Γ_1 (see Fig. 93), that is, when $\vartheta = \pm (\frac{1}{2}\pi - ia)$, where $a \to \infty$, we have

$$b = ik \cos{\vartheta} = -k \sinh{a} \to \infty,$$

$$\sin{\vartheta} = \pm \cosh{a} \to \pm \infty. \tag{26.9}$$

On these portions, the integrand in Eq. 26.8a can be written in the form

$$\exp{[-k(z-z_0) \sinh{a}]} H_0^{(1)}(\pm kr \cosh{a}) \cosh{a}.$$

Hence, it is clear that integral (26.8a) converges. Even at $z = z_0$, when the exponentially decreasing factor $\exp{[-k(z-z_0) \sinh{a}]}$ degenerates to unity, the integral converges because of the oscillatory character of the Hankel function at large values of its argument.

As $r \to 0$, our arguments remain valid if $z \neq z_0$, since the exponential $\exp[-k(z-z_0)\sinh a]$ guarantees the convergence of the integral (as before). The case $z \to z_0$, $r \to 0$, simultaneously, requires special consideration. Since in this case the function $H_0^{(1)}(kr\sin\vartheta)$ is in the integrand, the portions of the path of integration far from the origin, where $\sin\vartheta$ is large, will play the principal rôle. Taking Eq. 26.9 into account, Eq. 26.8a can be written in the form

$$\psi = \frac{ik}{2}\int_{\Gamma_1} \exp\left[-ik(z_0-z)\cos\vartheta\right]H_0^{(1)}(kr\sin\vartheta)\sin\vartheta\,d\vartheta.$$

Comparison with Eq. 26.1 shows that, in this case, ψ is a spherical wave e^{ikR}/R, which completes the proof.

2. Normal modes

We shall now transform the integral expression (26.8) for ψ into a sum of normal modes. For this purpose, it is convenient to introduce a new variable of integration ξ, connected with ϑ by the relation

$$\xi = k\sin\vartheta.$$

The integration over this variable will go from $-\infty$ to $+\infty$. As a result, Eqs. 26.8 become

$$z > z_0, \quad \psi = \int_{-\infty}^{+\infty}\frac{\cosh bz_0 \cosh(h-z)b}{b\sinh bh}H_0^{(1)}(\xi r)\,\xi\,d\xi, \qquad (26.10a)$$

$$z < z_0, \quad \psi = \int_{-\infty}^{+\infty}\frac{\cosh bz \cosh b(h-z_0)}{b\sinh bh}H_0^{(1)}(\xi r)\,\xi\,d\xi,$$

$$b = \sqrt{(\xi^2-k^2)}. \qquad (26.10b)$$

These integrals can be reduced to the sum of the residues at the poles of the integrand.

The poles are located at the points determined by the roots of the equation

$$\sinh bh = 0, \qquad (26.11)$$

the solution of which is

$$bh = il\pi, \qquad (26.12)$$

where l is zero or an integer.

The integrand becomes infinite at the poles. In particular, near the root corresponding to $l = 0$, we have

$$\frac{1}{b\sinh bh} \approx \frac{1}{hb^2} = \frac{1}{h(\xi^2-k^2)} \qquad (26.13)$$

in Eqs. 26.10.

The last expression has poles of the first order at the points $\xi_0 = \pm k$. Generally, when l in Eq. 26.12 is arbitrary, the poles are given by

$$\xi_l = \pm \sqrt{\left[k^2 - \left(\frac{l\pi}{h}\right)^2\right]}. \qquad (26.14)$$

When $l\pi < kh$, i.e. when $\qquad\qquad l\lambda < 2h, \qquad\qquad\qquad (26.15)$

they lie on the real axis in the ξ-plane; when l is large, they lie on the imaginary axis.

The analysis of the integrals (26.10) in the complex plane is simplified if we take into account that there is always some wave absorption in the medium comprising the layer. Therefore, we shall assume that k

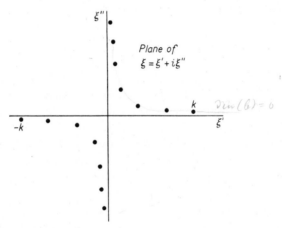

Fig. 117. Schematic illustration of the location of the poles in the complex plane.

has a small (positive) imaginary part, which can be allowed to approach zero in the final results. In this case, the poles will not be situated on the coordinate axes, but will be distributed as shown schematically in Fig. 117.

We now pull the path of integration away from the real axis to infinity in the upper halfplane. As long as $r \neq 0$, the function $H_0^{(1)}(\xi r)$ will approach zero at infinity. It can be shown that the integral over the infinitely distant portions of the path of integration will vanish. As a result, the expression for ψ is reduced to the sum of the residues at the poles in the first quadrant. Using the symbol Res for the residues, we obtain for any l, except $l = 0$:

$$\mathrm{Res}\left[\frac{\xi \cosh bz_0 \cosh b(h-z)}{b \sinh bh}\right]_{\xi=\xi_l} = \left[\frac{\xi \cosh bz_0 \cosh b(h-z)}{b(d/d\xi)(\sinh bh)}\right]_{\xi=\xi_l}.$$

But at $\xi = \xi_l$ we have $\sinh bh = 0$. Using this, we obtain

$$[\cosh b(h - z)]_{\xi=\xi_l} = \cosh bh \cosh bz.$$

Moreover, we have $\qquad \dfrac{d}{d\xi}(\sinh bh) = \dfrac{\xi h}{b} \cosh bh;$

and as a result, $\qquad \operatorname{Res}\left[\qquad\right] = \dfrac{1}{h} \cosh bz \cosh bz_0.$

For the pole $l = 0$ we have, according to Eq. 26.13,

$$\operatorname{Res}\left[\frac{1}{b \sinh bh}\right]_{\xi=k} = \operatorname{Res}\frac{1}{h(\xi - k)(\xi + k)} = \frac{1}{2kh}.$$

As a result, still using Eq. 26.12, the field in the layer is

$$\psi = \frac{2\pi i}{h}\left\{\tfrac{1}{2} H_0^{(1)}(kr) + \sum_{l=1}^{\infty} \cos\frac{l\pi z}{h} \cos\frac{l\pi z_0}{h} H_0^{(1)}(\xi_l r)\right\}, \qquad (26.16)$$

where ξ_l is given by Eq. 26.14.

The last expression for ψ does not change when z is replaced by z_0 and vice versa, and consequently, for any z $(0 \leqslant z \leqslant h)$.

Solution (26.16) satisfies all the conditions of the problem. In particular, it is easy to verify by substitution, that it satisfies the wave equation (25.1) and the boundary conditions (25.2), term by term.

Each term in Eq. 26.16 is a so-called normal mode. We will examine its physical significance.

At large (compared with the wavelength) distances from the source, the Hankel function can be replaced by its asymptotic representation. As a result, we obtain

$$\psi = \frac{2\exp(i\tfrac{1}{4}\pi)}{h}\sqrt{\left(\frac{2\pi}{r}\right)}\left\{\frac{1}{2\sqrt{k}}\exp(ikr) + \sum_{l=1}^{\infty}\cos\frac{l\pi z}{h}\cos\frac{l\pi z_0}{h}\frac{1}{\sqrt{\xi_l}}\exp(i\xi_l r)\right\}.$$
$$(26.17)$$

We see that each normal mode propagates along the layer with the velocity

$$c_l = \frac{\omega}{\xi_l} = \frac{c}{\sqrt{[1 - (l\lambda/2h)^2]}}, \qquad (26.18)$$

where c is the velocity in the medium comprising the layer. The amplitude of each mode decreases along the layer as $1/\sqrt{r}$.

We see from the last formula that in the general case (with the exception of the case $l = 0$), the velocity c_l is greater than the propagation velocity in free space. If $l\lambda > 2h$, the velocity c_l becomes imaginary. The corresponding normal waves are inhomogeneous waves with amplitudes decreasing exponentially with increasing r. Just as in § 18, these waves are necessary for the singularity required at the source.

When $r \geqslant h$, it is necessary in practice to take into account in the sum (26.17) only the finite number of terms satisfying the condition $l\lambda < 2h$. The number of such terms is equal to the thickness of the layer in half wavelengths.

Each of the terms in Eq. 26.17 represents a wave, traveling in the r-direction and standing in the z-direction. The z-dependence of the amplitude of each of the waves is given by the factor $\cosh b_l z$. This dependence is shown in Fig. 118 for $l = 0$, 1, 2 and 3. The derivative $\partial \psi / \partial z$ is equal to zero at each of the boundaries. When $l = 0$, we have a plane wave.

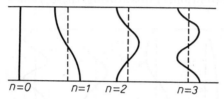

Fig. 118. The amplitude variations of the first four normal modes over the layer. The boundaries of the layer are assumed to be perfectly reflecting, with a reflection coefficient equal to 1.

So far, we have excluded the case in which the layer is an integral number of half wavelengths thick. If this occurs, one of the ξ_l becomes zero, and the amplitude of the corresponding wave in Eq. 26.16 becomes infinite because of the unbounded increase of the Hankel function. This is a case of a rather peculiar resonance, in which, if some kind of loss is not taken into account, it is not possible to establish a steady state.†

It is also useful to note that each normal mode can be represented as the superposition of two traveling plane waves with their fronts at definite angles of inclination. In fact, from Eq. 26.17, we have

$$\cosh b_l z \exp (i \xi_l r) = \tfrac{1}{2} \{\exp [i(\xi_l r + \kappa_l z)] + \exp [i(\xi_l r - \kappa_l z)]\},$$

where (see Eq. 26.12) $\kappa_l = -ib_l = l\pi/h$,

$$\kappa_l^2 + \xi_l^2 = k^2. \tag{26.19}$$

Equation 26.19 is a superposition of two plane waves, propagating in different directions with respect to z, and in the same direction with respect to r.

_____

† Under these conditions, the radiation resistance of the source becomes infinite.

The angle between the normal to the front of each wave and the z-axis is

$$\tan \vartheta_l = \frac{\xi_l}{\kappa_l} = \frac{\sqrt{[(kh)^2 - (l\pi)^2]}}{l\pi},$$

or

$$\cos \vartheta_l = \frac{l\pi}{kh}. \qquad (26.20)$$

As the number l of the wave increases (assuming, however, that $l\pi < kh$), the angle ϑ_l decreases, i.e. the inclination of the plane waves with respect to the boundaries of the layer increases.

The case $V_1 = 1, V_2 = -1$, corresponding to the propagation of acoustic waves in a liquid layer with a free surface and a perfectly rigid bottom, deserves special attention. Substituting these values of the reflection

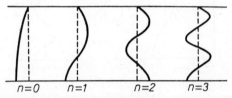

Fig. 119. The same as Fig. 118, but with the assumption that the reflection coefficient at the upper boundary is equal to -1.

coefficients into the integral expression (26.7), and following the same procedure as in the case considered above, we obtain, similar to Eq. 26.16,

$$\psi = \frac{2\pi i}{h} \sum_{l=0}^{\infty} \cosh b_l z_0 \cosh b_l z H_0^{(1)}(\xi_l r), \qquad (26.21)$$

where

$$b_l = \frac{i(l + \frac{1}{2})\pi}{h},$$

$$h\xi_1 = h\sqrt{(b_l^2 + k^2)} = \sqrt{[(kh)^2 - (l + \frac{1}{2})^2 \pi^2]}, \quad l = 0, 1, 2, \ldots \quad (26.22)$$

The normal modes can propagate in the layer without attenuation only if either $kh > \pi/2$ or $h > \lambda/4$, that is, when the depth of the water is greater than a quarter wavelength. In the opposite case, as is clear from Eq. 26.22, there is no l for which ξ_l would be real. The frequency $\omega_{cr} = \pi c/2h$, corresponding to $h = \lambda/4$, is called the critical frequency. Waves with frequencies below the critical frequency propagate in the layer with very great attenuation.

The distribution of the acoustic pressure amplitude with respect to the z-axis is shown in Fig. 119 for the first four normal modes. In all cases, the acoustic pressure at the surface of the liquid (the upper boundary of the layer) is zero. In contradistinction to the preceding case, plane waves cannot exist here, since, as a result of the requirement that the acoustic pressure be zero at the liquid surface, such pressure would have to be zero throughout the layer.

3. *The relation between the picture of the image sources and the normal modes*

The infinitely extended network of image sources can be considered as a self-luminous diffraction grating. The field of such a grating can

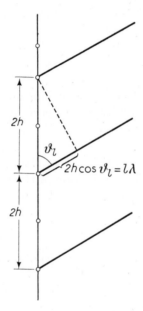

Fig. 120.   On the equivalence of the normal modes and the spectra of various orders.

be represented as a set of spectra, each of which propagates at its own angle with respect to the axis of the grating. It can be shown that each of the normal modes can also be represented as a set of a pair of spectra, the directions of which are symmetrical with respect to the plane perpendicular to the axis of the grating. For definiteness, we take the case $V_1 = V_2 = 1$. We showed earlier that each of the normal modes is equivalent to a pair of waves, the directions of the normals to the wavefronts being given by Eq. 26.20. We now show that this same equation will also determine the inclination of the spectra, if they are determined by using the elementary theory of the diffraction grating.

The angle determining the direction of the lth spectrum is found, as is well known, from the condition that the path difference between two rays, traveling parallel to one another, and starting from two identical elements in adjacent periods of the grating, must be equal to an integral number of wavelengths. Figure 120 shows several elements of the network of image sources corresponding to Fig. 114. Hence, we obtain the condition

$$2h \cos \vartheta_{\bar{l}} = l\lambda,$$

which is in agreement with condition (26.20), with $k = 2\pi/\lambda$.

From the point of view of the picture of the image sources, it is interesting to explain the fact that sound waves propagating in a liquid layer with a rigid bottom are subject to extremely great attenuation when the depth of the water is less than a quarter of a wavelength (see above). In this case, because of the small "period" of the grating, neighboring image sources, working in opposite phases, short circuit one another as it were. In other words, instead of acoustic radiation, we have periodic transfer of liquid between neighboring image sources (when one of the sources forces liquid outward, the other sucks it inward, and vice versa).

In the first Sections of the present Paragraph, we formally carried out the mathematical transformation of the fields of the image sources into a set of normal waves. It is also of interest to carry out the reverse transformation.[178]

For this purpose, we use the Poisson formula:[21]

$$\beta^{\frac{1}{2}}\left[\tfrac{1}{2}F(0) + \sum_{l=1}^{\infty} F(l\beta)\right] = \alpha^{\frac{1}{2}}\left[\tfrac{1}{2}f(0) + \sum_{l=1}^{\infty} f(l\alpha)\right], \tag{26.23}$$

where

$$F(x) = \sqrt{\left(\frac{2}{\pi}\right)} \int_0^\infty f(t) \cos(xt)\, dt, \quad \alpha\beta = 2\pi, \quad \alpha > 0. \tag{26.24}$$

As we see, the Poisson formula permits us to transform one series into another, where the terms of the new series are obtained from the terms of the old series by a Fourier transformation.

In the present case, in accordance with Eq. 26.16,

$$f = \frac{2\pi i}{h} H_0^{(1)}\left[r\left(k^2 - \frac{l^2 \pi^2}{h^2}\right)^{\frac{1}{2}}\right] \cos\frac{l\pi z}{h} \cos\frac{l\pi z_0}{h}.$$

We introduce the new variable $u = l\pi/h$. Then, using Eq. 26.24, we obtain

$$F(x) = 2i\sqrt{\left(\frac{2}{\pi}\right)} \int_0^\infty \cos\left(\frac{hxu}{\pi}\right) H_0^{(1)}[r(k^2 - u^2)^{\frac{1}{2}}] \cos uz \cos uz_0\, du. \tag{26.25}$$

We note that the identity

$$\int_0^\infty H_0^{(1)}[r(k^2-u^2)^{\frac{1}{2}}]\cos zu\, du = -\frac{i}{R}e^{ikR},$$

$$R = \sqrt{(z^2+r^2)} \qquad\qquad (26.26)$$

holds. Moreover,

$$\cos\left(\frac{hxu}{\pi}\right)\cos uz\cos uz_0 = \tfrac{1}{4}[\cos(v+z+z_0)+\cos(v-z+z_0)$$
$$+\cos(v+z-z_0)+\cos(v-z-z_0)],$$

where $v = hx/\pi$. As a result, we obtain from Eq. 26.25

$$F(x) = \frac{1}{\sqrt{(2\pi)}}\{g[r^2+(v+z+z_0)^2]+g[r^2+(v+z-z_0)^2]$$
$$+g[r^2+(v-z+z_0)^2]+g[r^2+(v-z-z_0)^2]\},$$

where, for brevity, we have introduced the symbolic notation

$$g(R^2) \equiv \frac{e^{ikR}}{R}.$$

Now, setting $\alpha = 1$, $\beta = 2\pi$, $x = l\beta$, the left hand side of Eq. 26.23 becomes an infinite series corresponding to the field of the infinite network of image sources. It is not hard to verify, if only by term-by-term comparison, that this series will agree with series (25.5), and will differ from it only in the numbering of the terms and their grouping into separate sets of four.

§ 27. A Layer with Arbitrary Boundaries

1. *Generalization of the results of the preceding Section to include the case of arbitrary boundaries*

Equation 26.7, obtained for the case of perfectly reflecting boundaries, turns out to be valid for the case of arbitrary boundaries, in which the reflection coefficients are no longer constants, but are functions of the angle of incidence. We shall now prove this by again obtaining Eq. 26.7 for the more general case in which the picture of the image sources, from which we started previously, may be invalid. Here, we start from the representation of a spherical wave as a superposition of plane waves, and will generalize the discussion presented in § 19.

If the boundaries of the layer in Fig. 112 are infinitely far from the source and receiver, only a spherical wave will be observed at the point $P(r, z)$. As we have seen, this spherical wave can be represented as a superposition of plane waves of the type

$$\exp\{i[k_x x + k_y y + k_z(z-z_0)]\},$$

where k_x, k_y and k_z are given by Eqs. 18.18. Here, x, y and $z - z_0$ are the projections on the coordinate axes of the segments joining the source with the point of observation.

If the lower boundary is at a finite distance from the source and receiver, then a reflected wave will also be observed, in addition to the direct wave. This reflected wave can be represented as a superposition of plane waves of the form $V_1(\vartheta) \exp\{i[k_x x + k_y y + k_z(z + z_0)]\}$, where k_x, k_y and k_z are the same as above, and $V_1(\vartheta)$ is the reflection coefficient at the lower boundary.

By generalizing these considerations, we can obtain an expression for the field of a point source situated in a layer. In this case, in addition to the direct wave at the receiver, we have an infinite series of waves which have been reflected at the boundaries of the layer various numbers of times. The phase of each of these plane waves at the point of observation will be given by the expression

$$k_x x + k_y y + k_z z_{lj},$$

where z_{lj} is the projection on the z-axis of the path traversed by the wave. The subscripts lj characterize the number of reflections undergone by the wave at the boundaries of the layer, and run through the values $l = 0, 1, 2, \ldots, \infty$, $j = 1, 2, 3, 4$.

In particular:

1. For the direct wave ($l = 0, j = 1$)

$$z_{01} = z - z_0.$$

2. For the wave reflected from the lower boundary ($l = 0, j = 2$),

$$z_{02} = z + z_0.$$

3. For the wave reflected from the upper boundary ($l = 0, j = 3$),

$$z_{03} = 2h - z - z_0.$$

4. For the wave reflected first from the lower and then from the upper boundary ($l = 0, j = 4$),

$$z_{04} = 2h - z + z_0.$$

The values of z_{lj} for arbitrary l and j are given by Eqs. 26.2.

We obtain the total field by summing over all the plane waves with a given set of direction cosines but different numbers of reflections from the boundaries, and then integrating the sum over all values of the direction cosines. Taking account of the original expansion (18.19) of a spherical wave into plane waves, and also remembering that, at each reflection from the boundary, the amplitude of the wave is multiplied by the appropriate reflection coefficient, we obtain the following expression for the total field in the layer for $z > z_0$:

$$\psi = \frac{ik}{2\pi} \int_0^{\frac{1}{2}\pi - i\infty} \int_0^{2\pi} \exp\left[i(k_x x + k_y y)\right] \sum_{l=0}^{\infty} \{\exp\left[ik_z(z - z_0)\right]$$
$$+ V_1 \exp\left[ik_z(z_0 + z)\right] + V_2 \exp\left[ik_z(2h - z - z_0)\right]$$
$$+ V_1 V_2 \exp\left[ik_z(2h - z + z_0)\right]\} (V_1 V_2)^l \exp\left(2ik_z lh\right) \sin\vartheta \, d\vartheta \, d\phi. \quad (27.1)$$

We shall spend some more time on this expression in the next Section, but for now we shall carry out some further transformations.

Integrating the last expression with respect to ϕ, as was done in Eqs. 19.3–19.6, and taking Eqs. 26.5 and 26.6 into account, we obtain for $z > z_0$:

$$\psi = \frac{ik}{2} \int_{\Gamma_1} \frac{\left[(\exp\left(-bz_0\right) + V_1 \exp\left(bz_0\right)\right]\{\exp\left[-b(h-z)\right] + V_2 \exp\left[b(h-z)\right]\}}{\exp\left(-bh\right)\left[1 - V_1 V_2 \exp\left(2bh\right)\right]}$$
$$\times H_0^{(1)}(kr \sin\vartheta) \sin\vartheta \, d\vartheta, \quad (27.2a)$$

where $b = ik_z = ik \cos\vartheta$.

This expression agrees with Eq. 26.7, the only difference being that now V_1 and V_2 are not constants, but are functions of the angle ϑ. Replacing z by z_0 and vice versa, we obtain for $z < z_0$:

$$\psi = \frac{ik}{2} \int_{\Gamma_1} \frac{\left[\exp\left(-bz\right) + V_1 \exp\left(bz\right)\right]\{\exp\left[-b(h-z_0)\right] + V_2 \exp\left[b(h-z_0)\right]\}}{\exp\left(-bh\right)\left[1 - V_1 V_2 \exp\left(2bh\right)\right]}$$
$$\times H_0^{(1)}(kr \sin\vartheta) \sin\vartheta \, d\vartheta. \quad (27.2b)$$

It can be shown that the solution (27.2) satisfies all the conditions that our solution is required to fulfil, namely:

1. It satisfies the wave equation.

2. It satisfies the boundary conditions.

3. The integral (26.7) converges everywhere, and has the necessary singularity as the point of observation approaches the source.

The verification of the first of these three statements is elementary since the integrand itself satisfies the wave equation. The second and third statements are proven in Ref. 41, and also below, in § 34, for a more general case than that considered here. Therefore, we shall only prove the second statement here; in the process, we shall obtain an expression for the field outside the layer, which will be useful in the future.

Let us consider the conditions at the lower boundary of the layer ($z = 0$). As a preliminary, we obtain an expression for the field ψ_1 in the lower medium ($z < 0$). We follow the same path as in the analysis of the field in the layer, namely, we sum over the successive reflections of the plane waves at the lower boundary, taking into account that each reflection is accompanied by penetration of the wave into the lower medium.

Figure 121 illustrates four of the simplest paths by which the wave can travel from the source O to the point P_1 in the lower medium.

We shall first consider one of the plane waves into which the spherical wave is expanded. We suppose that by traveling directly from O to P_1, and also by partial reflection, the plane wave creates the field

$$f(z, \vartheta) \exp[i(k_x x + k_y y + k_z z_0)], \qquad (27.3)$$

at the point P_1, where as above, k_x, k_y and k_z are the components of the wave vector of the plane wave under consideration as it propagates in the layer.

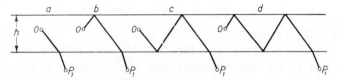

Fig. 121. Regarding the expression for the field in the lower medium.

Let us consider the special case in which the lower medium is homogeneous. We then have, clearly,

$$f(z, \vartheta) = W_1 \exp(-ik)_{1z} z,$$

$$k_{1z} = \sqrt{(k_1^2 - k_x^2 - k_y^2)} = k\sqrt{(n_1^2 - \sin^2 \vartheta)}, \qquad n_1 = \frac{k_1}{k}, \qquad (27.4)$$

as follows directly from the laws of refraction of plane waves. Here, W_1 is the transmission coefficient of the boundary (see §§ 2 and 3). But, in general, if the form of the function $f(z, \vartheta)$ is not as yet prescribed, our arguments will also be valid for the general case in which the medium $z < 0$ is layered-inhomogeneous.

When the wave penetrates into the lower medium after one reflection at the upper boundary (Fig. 121b), we have for the field in the lower medium

$$f(z, \vartheta) \exp\{i[k_x x + k_y y + kz(2h - z_0)]\} V_2(\vartheta).$$

The fields created by the waves undergoing a greater number of reflections, as for example, cases (c) and (d) in Fig. 121, are obtained similarly.

Adding all the components of the field in the lower medium with different numbers of reflections from the boundaries, we obtain, as above, the series

$$f(z, \vartheta) \exp[i(k_x x + k_y y)] \sum_{l=1}^{\infty} \{\exp(ik_z z_0)$$

$$+ V_2(\vartheta) \exp[ik_z(2h - z_0)]\} (V_1 V_2)^l \exp(2ik_z lh).$$

Taking into account the value of the sum entering into this expression (see Eq. 26.6), and integrating over all direction cosines of the plane waves, we obtain an expression, similar to Eq. 26.7,

$$\psi_1 = \frac{ik}{2} \int_{\Gamma_1} \frac{\exp[-b(h - z_0)] + V_2 \exp[b(h - z_0)]}{\exp(-bh)[1 - V_1 V_2 \exp(2bh)]}$$

$$\times f(z, \vartheta) H_0^{(1)}(kr \sin \vartheta) \sin \vartheta \, d\vartheta, \qquad (27.5)$$

where again, $b \equiv ik_z$.

Let us verify that Eqs. 27.2b and 27.5 for ψ and ψ_1 satisfy the conditions at the lower boundary. We examine the case of a homogeneous lower medium. In this case, the boundary conditions for the acoustic potential and the vertical component of the Hertz vector are written

$$\frac{\partial \psi}{\partial z} = \frac{\partial \psi_1}{\partial z}, \qquad (27.6)$$

$$\psi = m\psi_1,$$

where
$$m = \frac{\rho_1}{\rho} \text{—in acoustics,}$$

$$\left. \begin{array}{c} \\ \\ \\ \\ \end{array} \right\} \qquad (27.7)$$

$$m = \left(\frac{k_1}{k}\right)^2 \text{—in electrodynamics.}$$

Substituting Eqs. 27.2b and 27.5 into Eq. 27.6, taking Eq. 27.4 into account, differentiating under the integral sign and equating the integrands, we obtain

$$k_{1z} W_1 = k_z(1 - V_1),$$

$$m W_1 = 1 + V_1. \qquad (27.8)$$

However, we obtain the same equations for V_1 and W_1 by solving the problem of the reflection of a plane wave at an interface between two media. This is easily verified by prescribing the field in one medium as the sum of the incident and the reflected wave

$$\psi = \exp[i(k_x x + k_y y)][\exp(-ik_z z) + V_1 \exp(ik_z z)],$$

and the field in the other medium as the refracted wave

$$\psi_1 = W_1 \exp[i(k_x x + k_y y - k_{1z} z)].$$

Substituting these values of ψ and ψ_1 into the boundary conditions (27.6) and (27.7), we obtain relations identical to (27.8). We must assume, of course, that these relations are fulfilled.

The upper boundary of the layer is treated similarly, but we use Eq. 27.2a instead of Eq. 27.2b.

2. Concerning the picture of image sources

We proved above that the picture of the image sources is rigorously applicable to the case of perfectly reflecting boundaries. In the case of boundaries for which the reflection coefficient differs from unity and depends on the angle of incidence, the sum of the fields of the image sources will not, in general, be a correct solution of the problem, and in certain cases will be only an approximation to reality.

Let us return to the exact expression (27.1) for the field in the layer, and write it in the form

$$\psi = \sum_{l=0}^{\infty} \sum_{j=1}^{4} \psi_{lj}, \tag{27.9}$$

where

$$\psi_{lj} \equiv \frac{ik}{2} \int_0^{\frac{1}{2}\pi - i\infty} \int_0^{2\pi} \exp\left[i(k_x x + k_y y + k_z z_l)\right] V_{lj}(\vartheta) \sin\vartheta \, d\vartheta \, d\phi. \tag{27.10}$$

Here, z_{lj} is given by Eqs. 26.2, and V_{lj} is determined by the following equations:

$$V_{l1} = (V_1 V_2)^l, \quad V_{l2} = V_1^{l+1} V_2^l, \quad V_{l3} = V_1^l V_2^{l+1}, \quad V_{l4} = (V_1 V_2)^{l+1}. \tag{27.11}$$

Each of the ψ_{lj} can be associated with a definite image source. In particular, if V_1 and V_2 are constants, i.e. are independent of the angle of incidence, then ψ_{lj} will be the exact expression for the field of the image source. In fact, taking V_{lj} outside the integral sign, and using Eq. 26.1, we obtain

$$\psi_{lj} = V_{lj} \frac{\exp(ikR_{lj})}{R_{lj}}, \tag{27.12}$$

where R_{lj} is given by Eqs. 25.6, and is the distance from the image source (lj) to the point of observation.

In the general case, in which V_1 and V_2 are functions of ϑ, the integral (27.10) can be evaluated by the method of steepest descents in exactly the same way as was done in § 19 in the analysis of a wave undergoing a single reflection at an interface. We can use the formulas of § 19 if we associate the reflection coefficient $V(\vartheta)$ therein with the quantity V_{lj}.

In the second approximation of the method of passage, we obtain for ψ_{lj}, in analogy with Eqs. 19.36 and 19.37,

$$\psi_{lj} = \frac{\exp(ikR_{lj})}{R_{lj}} \left[V_{lj}(\vartheta_{lj}) - \frac{iN_{lj}}{kR_{lj}} \right], \tag{27.13}$$

where $$N_{lj} \equiv \tfrac{1}{2}\left[V_{lj}''(\vartheta_{lj}) + V_{lj}'(\vartheta_{lj})\cot\vartheta_{lj}\right], \tag{27.14}$$

$$\vartheta_{lj} = \arctan\frac{r}{z_{lj}}. \tag{27.15}$$

In the special case in which the upper boundary of the layer is completely reflecting $(V_2 = 1)$, we have $V_{l1} = V_{l3} = V_1^l$; $V_{l2} = V_{l4} = V_1^{l+1}$. Setting

$$V_1(\vartheta_{lj}) = \frac{m\gamma - q}{m\gamma + q}, \quad \gamma = \cos\vartheta_{lj}, \quad q = \sqrt{(n_1^2 - \sin^2\vartheta_{lj})}, \tag{27.16}$$

for brevity, and calculating the derivatives, we obtain

$$N_{l1} = N_{l3} = \frac{2m(n_1^2 - 1)}{q^3}\frac{(m\gamma - q)^{l-1}}{(m\gamma + q)^{l+2}}\left\{\frac{l(l-1)\,m(n_1^2 - 1)\,(1 - \gamma^2)\,q}{m\gamma - q}\right.$$
$$\left. - \frac{l}{2}\left[-m\gamma^4 + \gamma(2n_1^2 + 1 - \gamma^2)q + 2m(n_1^2 - 1) + 3m\gamma^2\right]\right\}. \tag{27.17}$$

We then obtain N_{l2} and N_{l4} by increasing l by unity.

The evaluation of the rôle of the correction term, containing N_{lj}, in Eq. 27.13, is carried out just as in §19. It turns out here that as a rule, we can limit ourselves to the first approximation of the method of steepest descents (i.e. we take into account only the first term in the square brackets in Eq. 27.13), if z_{lj} is large compared with the wavelength. Thus, in layers which are thick compared with the wavelength, we can, in practice, always use the image source representation. In the case of thin layers (with thicknesses comparable with the wavelength or smaller) the image source picture is valid only when the number of reflections is sufficiently great.

3. Normal modes

We will transform Eq. 27.2 by changing the path of integration in the complex plane. As before, we will consider homogeneous lower and upper media.

As is easily seen from Fig. 122, the integral over the path Γ_1 is equivalent to the sum of the integrals over the contours Γ_2 and Γ_3. It can be shown that the integral over Γ_2 is identically zero. It is convenient for this purpose to introduce the grazing angle $\alpha = \tfrac{1}{2}\pi - \vartheta$ as the variable of integration instead of ϑ. In the transition from ϑ to α, the path Γ_2 is transformed into a contour coinciding with the imaginary axis.

For definiteness, we consider the case $z > z_0$. Then, according to Eq. 27.2a, the integral over Γ_2 can be written in the form

$$\int_{-i\infty}^{+i\infty} \Phi H_0^{(1)}(kr\cos\alpha)\cos\alpha\,d\alpha, \tag{27.18}$$

where $\qquad \Phi \equiv \dfrac{ik}{2}\dfrac{(e^{-bz_0}+V_1 e^{bz_0})(e^{-b(h-z)}+V_2 e^{b(h-z)})}{e^{-bh}(1-V_1 V_2 e^{2bh})}.$ (27.19)

However, it is not hard to see that the integral (27.18) is identically zero, because the function $\Phi(\alpha)$ is odd. In fact, let us replace α by $-\alpha$ in Eq. 27.19. Then $b = ik \sin\alpha$ also changes sign. For the reflection coefficients V_1 and V_2, we have the relations

$$V_1(-\alpha) = \frac{1}{V_1(\alpha)}, \qquad V_2(-\alpha) = \frac{1}{V_2(\alpha)}.$$ (27.20)

The validity of these expressions for the acoustic and the electromagnetic cases follows directly from Eq. 19.38, with $\gamma = \sin\alpha$.

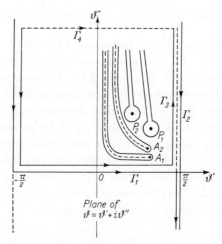

Fig. 122.   Regarding the transformation of the path of integration in the complex plane.

Now, we obtain without difficulty

$$\Phi(-\alpha) = -\Phi(\alpha).$$

It is interesting to note that the validity of relations (27.20), which guarantee the vanishing of integral (27.18), can be proved for a much broader class of cases than those examined here. In particular, it can be proved, with no special difficulty, for the reflection coefficients at transitional layers, and at an interface between a liquid and a solid.

In the first case, we can use Eq. 17.25 for the reflection coefficient.†

† See, also, Ref. 41, pp. 513 and 516.

Of all the quantities entering into this equation, only q_0 changes sign when α changes sign. Either the angle of refraction ϑ_1 or the angle of incidence of the wave in the layer ϑ enter into all the remaining quantities. They are expressed in terms of the angle α as follows:

$$\cos \vartheta_1 = \sqrt{(n_1^2 - \cos^2 \alpha)}, \quad \cos \vartheta = \sqrt{[n^2(z) - \cos^2 \alpha]}$$

and do not change when the sign of α is changed. It is then simple to show that $V(\alpha) = 1/V(-\alpha)$.

For the same purpose in the second case, we must use Eq. 4.25 for the reflection coefficient at the boundary between a liquid and a solid.

Thus, it remains for us to calculate the integral over the path Γ_3. We deform the path Γ_3 toward infinity in the positive imaginary direction, so that it passes into the infinitely remote path Γ_4 (Fig. 122).

It can be shown that the integral over the path Γ_4 is zero. As a result, the path of integration "hangs" over the singular points of the integrand which are circuited during the deformation of the path. We then obtain for the field in the layer

$$\psi = \int_{\Gamma_1} \Phi(\vartheta) H_0^{(1)}(kr \sin \vartheta) \sin \vartheta \, d\vartheta = 2\pi i \sum \operatorname{Res} [\Phi(\vartheta_l)] H_0^{(1)}(kr \sin \vartheta_l) \sin \vartheta_l$$

$$+ \int_{i\infty}^{A_1^+} \Phi(\vartheta) H_0^{(1)}(kr \sin \vartheta) \sin \vartheta \, d\vartheta + \int_{i\infty}^{A_2^+} \Phi(\vartheta) H_0^{(1)}(kr \sin \vartheta) \sin \vartheta \, d\vartheta.$$

$$(27.21)$$

The first term is the sum of the residues of the integrand at the poles situated in the region encompassed during the deformation of the path. These poles are denoted by $P_1, P_2, \ldots$ in Fig. 122. Their positions are determined by the roots of the equation

$$1 - V_1(\vartheta) V_2(\vartheta) \exp (2ikh \cos \vartheta) = 0. \qquad (27.22)$$

The last two terms in Eq. 27.21 are the integrals over the borders of the cuts, beginning at the points $\vartheta = \arcsin n_1$ and $\vartheta = \arcsin n_2$ (A_1 and A_2 in Fig. 122), and going along the lines

$$\operatorname{Im} \sqrt{(n_1^2 - \sin^2 \vartheta)} = 0 \quad \text{and} \quad \operatorname{Im} \sqrt{(n_2^2 - \sin^2 \vartheta)} = 0,$$

respectively. The symbol $\int_{i\infty}^{A_1^+}$ in Eq. 27.21 denotes the integral over the path which goes along the borders of the cut from $i\infty$ and goes around the point A_1 in the positive direction (counter-clockwise). These integrals appear because the functions $V_1(\vartheta)$ and $V_2(\vartheta)$, and, consequently, also the function $\Phi(\vartheta)$ are not single-valued, since they contain the square roots $\sqrt{(n_1^2 - \sin^2 \vartheta)}$ and $\sqrt{(n_2^2 - \sin^2 \vartheta)}$, where n_1 and

12

n_2 are the indices of refraction of the media bounding the layer from below and from above, respectively. The sign of each of the square roots can be chosen in two ways. As a result, we will have four combinations of signs $(++, +-, -+, --)$, that is the integrand will be quadruple-valued. We can construct a four-sheeted Riemann surface for this function, and the function will be single-valued on each of the sheets. The paths of integration Γ_1, Γ_2 and Γ_3 lie on the sheet (we will call it the upper sheet) on which[†]

$$\mathrm{Im}\,\sqrt{(n_1^2 - \sin^2\vartheta)} > 0, \quad \mathrm{Im}\,\sqrt{(n_2^2 - \sin^2\vartheta)} > 0. \qquad (27.23)$$

The path Γ_4 will lie on the upper sheet only if, in deforming the path of integration, we go around the cuts which extend from the branch points A_1 and A_2. If this is not done, the path Γ_4 will intersect both cuts, and one of its ends will lie on the sheet on which the imaginary parts of both square roots are negative. For the same reason, it will not join with the appropriate end of Γ_3, lying on the upper sheet.

As a result, the path of integration hangs over the poles and cuts as shown in Fig. 122.

Using the terminology of P. E. Krasnushkin[54] the field given by the first term in Eq. 27.21 can be called the discrete spectrum or the normal modes, and the field given by the two remaining terms can be called the continuous spectrum. These terms yield lateral waves, propagating in the media bounding the layer from below and from above. We shall first study the normal modes.

Qualitatively, we have the same picture here as in the case of the perfectly reflecting boundaries considered above. The greatest difficulty in the investigation of the field of the normal modes is finding the solutions of the equation for the poles (27.22). However, this problem is simplified considerably when the thickness of the layer is large compared with the wavelength. We shall examine this case first.

We set[‡]

$$V_1(\vartheta) = -\exp[i\phi_1(\vartheta)], \quad V_2(\vartheta) = -\exp[i\phi_2(\vartheta)], \qquad (27.24)$$

[†] In fact, according to Eqs. 27.4 and 27.5, the field of the refracted waves in the lower medium will contain the factor

$$\exp[-ik\,\sqrt{(n_1^2 - \sin^2\vartheta)z}].$$

Therefore, as $z \to -\infty$, the field will remain bounded only if the first of conditions (27.23) is fulfilled. The second of conditions (27.23) has a similar basis.

[‡] The minus sign in front of the exponentials is chosen for convenience in the series expansion of V_1 and V_2 near $\vartheta = \pi/2$ (see below).

in Eq. 27.22, where in the general case, ϕ_1 and ϕ_2 are complex. Equation 27.22 then becomes

$$\exp\left[i(2kh\cos\vartheta + \phi_1 + \phi_2)\right] = 1$$

or, if we use the notation $\phi \equiv \phi_1 + \phi_2$, we obtain

$$2kh\cos\vartheta + \phi = 2\pi l, \tag{27.25}$$

where l is an integer.

As we shall see below, when there is considerable absorption at the boundaries of the layer, those normal waves for which the angle ϑ is close to $\pi/2$ will play the principal rôle at great distances. The absorption at the boundaries is least for these waves. Therefore, instead of the angle of incidence, it is more convenient to use the grazing angles $\alpha = (\pi/2) - \vartheta$, which in the present case will be small. Then Eq. 27.25 can be written

$$2kh\sin\alpha + \phi(\alpha) = 2\pi l. \tag{27.26}$$

We first examine the normal waves with comparatively high values of l ($l \gg 1$). Since $|\phi| \leqslant 2\pi$, we can neglect ϕ compared with $2\pi l$. As a result, to the first approximation, we obtain the following solutions of Eq. 27.26:

$$\alpha_l^{(1)} = \arcsin\frac{\pi l}{kh}. \tag{27.27}$$

Substituting this solution into $\phi(\alpha)$ in Eq. 27.26, we obtain the next approximation, which is sufficient for our purposes:

$$\alpha_l = \arcsin\left[\frac{\pi l}{kh} - \frac{\phi(\alpha_l^{(1)})}{2kh}\right] \approx \arcsin\frac{\pi l}{kh} - \frac{\phi(\alpha_l^{(1)})}{2\sqrt{[(kh)^2 - (\pi l)^2]}}. \tag{27.28}$$

From Eq. 27.24, we find

$$\phi_1(\vartheta) = -i\ln(-V_1)$$

and similarly for ϕ_2. Therefore, $\phi = \phi_1 + \phi_2 = -i\ln V_1 V_2$. As a result, Eq. 27.28 is written

$$\alpha_l = \arcsin\frac{\pi l}{kh} + i\frac{\ln[V_1(\alpha_l^{(1)})V_2(\alpha_l^{(1)})]}{2\sqrt{[(kh)^2 - (\pi l)^2]}}. \tag{27.29}$$

As we shall see later, the attenuation of each of the normal waves will be determined by the imaginary part. Using the notation

$$\Delta_l = -\operatorname{Im}\alpha_l, \tag{27.30}$$

in Eq. 27.29, we obtain

$$\Delta_l = -\frac{\ln|V_1(\alpha_l^{(1)})V_2(\alpha_l^{(1)})|}{2\sqrt{[(kh)^2 - (\pi l)^2]}}, \tag{27.31}$$

where we have assumed that $kh > \pi l$.

In the case of a completely reflecting boundary, we have $V_1 V_2 = 1$, $\Delta_l = 0$, and consequently, there will be no attenuation of the wave. But generally, since $|V_1 V_2| < 1$, Δ_l will be positive.

The method of solution used above is not suitable for small l, which (for large kh) according to Eq. 27.27, corresponds to small grazing angles α_l. In this case, we use another method, based on the expansion of ϕ_1 and ϕ_2 in a power series in α.

Using the grazing angle $\alpha = (\pi/2) - \vartheta$ instead of the angle of incidence ϑ in Eq. 22.1, the reflection coefficient at the lower boundary is

$$V_1(\alpha) = \frac{m_1 \sin \alpha - \sqrt{(n_1^2 - \cos^2 \alpha)}}{m_1 \sin \alpha + \sqrt{(n_1^2 - \cos^2 \alpha)}}, \qquad (27.32)$$

where $m_1 = n_1^2$ in electrodynamics (the magnetic permeability is assumed to be unity), and $m_1 = \rho_1/\rho$ in acoustics.†

The expansion of $V_1(\alpha)$ in a power series in α, neglecting terms of the order α^3 and higher, has the form

$$V_1(\alpha) = -[1 - 2p_1 \alpha + 2p_1^2 \alpha^2 + \ldots], \qquad (27.33)$$

where
$$p_1 = \frac{m_1}{\sqrt{(n_1^2 - 1)}}. \qquad (27.34)$$

To within the same order of accuracy, we can contract Eq. 27.33 by writing $V_1(\alpha)$ in the form

$$V_1(\alpha) = -e^{-2p_1 \alpha}. \qquad (27.35)$$

Comparing the last expression with Eq. 27.24, we find $\phi_1 = 2ip_1 \alpha$. Straightforward calculations show that had we retained the term in α^3, we would have obtained

$$\phi_1 = 2ip_1 \alpha \left(1 - \frac{n_1^2 - 8 - 6m_1^2}{6(n_1^2 - 1)} \alpha^2 \right). \qquad (27.36)$$

However, we shall not use this higher approximation.

In the case of the reflection of sound waves, incident from a liquid onto its boundary with a solid, we use Eqs. 4.25 and 4.26; again transforming to the grazing angle α, we obtain Eq. 27.35, in which

$$p_1 = \frac{m_1[(2k^2 - \kappa_1^2)^2 - 4k^2 \sqrt{(k^2 - k_1^2)} \sqrt{(k^2 - \kappa_1^2)}]}{\kappa_1^4 \sqrt{(n_1^2 - 1)}}, \qquad (27.37)$$

where $\kappa = \omega/b_1$ is the wave number of the shear waves in the solid.

---

† In electrodynamics, the source is assumed to be a vertical dipole. If the waves are excited by a vertical "magnetic" dipole (current loop), all the formulas remain valid but we must set $m_1 = m_2 = 1$.

Replacing the subscript 1 by 2, we obtain a similar expression for the upper boundary.

Substituting $\phi = \phi_1 + \phi_2 \approx 2i(p_1 + p_2)\alpha$ into Eq. 27.26, we obtain

$$2kh \sin \alpha + 2i(p_1 + p_2)\alpha = 2\pi l.$$

Since we are interested in small α, we can set $\sin \alpha \approx \alpha$, after which we obtain

$$\alpha_l = \frac{\pi l}{kh + i(p_1 + p_2)}, \tag{27.38}$$

or, since kh is assumed to be large,

$$\alpha_l \approx \frac{\pi l}{kh} - i \frac{\pi l}{(kh)^2}(p_1 + p_2). \tag{27.39}$$

Comparing the last expression with Eq. 27.30, we see that the quantity Δ_l, which determines the attenuation, will be

$$\Delta_l = \frac{\pi l}{(kh)^2} \operatorname{Re}(p_1 + p_2). \tag{27.40}$$

In particular, if we assume that n_1 and n_2 are real, and that $n_1 < 1, n_2 < 1$, then Eq. 27.34 and the analogous one with the subscript 2 give $\operatorname{Re} p_1 = \operatorname{Re} p_2 = 0$, and, consequently, $\Delta_l = 0$, that is, there is no attenuation. This corresponds to the case of total internal reflection at the boundaries of the layer.

The method of calculating the poles which we have developed here has been applied by Ia. L. Al'pert[2] to the problem of the propagation over long distances of low frequency electromagnetic waves in the atmosphere.

We now find the field of the normal modes, given by the first term in Eq. 27.21. We assume that there are no accumulation points of poles in the finite region of the ϑ-plane. As regards the poles in the remote regions, it can be shown that they are equi-distantly situated on a line, asymptotically approaching the imaginary axis. Under these conditions, Cauchy's theorem of residues is completely valid[72], and the contribution of the infinitely remote poles is negligibly small. According to the usual rules for determining the residues, we must substitute $\vartheta = \vartheta_l = (\pi/2) - \alpha_l$ everywhere in Eq. 27.19 except in the denominator, and in the denominator, we must replace the function $f = 1 - V_1 V_2 e^{2bh}$ by $(df/d\vartheta)_{\vartheta=\vartheta_l}$. We then multiply the result of these operations by $2\pi i$, and sum over all l.

Writing the equation for the poles (Eq. 27.22) in the form $f(\vartheta_l) = 0$, it is not hard to show that

$$\left(\frac{df}{d\vartheta}\right)_{\vartheta_l} = -\left(\frac{1}{V_1}\frac{dV_1}{d\vartheta} + \frac{1}{V_2}\frac{dV_2}{d\vartheta} - 2ikh\sin\vartheta\right)_{\vartheta_l}. \qquad (27.41)$$

Since kh is assumed to be large, the first two terms can be neglected compared with the last two terms. Moreover, using the equation for the poles, we can prove the identity†

$$[\exp(-b_l z_0) + V_1(\vartheta_l)\exp(b_l z_0)]\{\exp[-b_l(h-z)] + V_2(\vartheta_l)\exp[b_l(h-z)]\}$$

$$= \frac{\exp(-b_l h)}{V_1(\vartheta_l)}[\exp(-b_l z) + V_1(\vartheta_l)\exp(b_l z)]$$

$$\times [\exp(-b_l z_0) + V_1(\vartheta_l)\exp(b_l z_0)],$$

where $b_l = ik\cos\vartheta_l$.

To simplify the formulas, the Hankel function $H_0^{(1)}(kr\sin\vartheta_l)$ in Eq. 27.21 can be replaced by its asymptotic representation. As a result, the field due to the normal waves is

$$\psi = \frac{1}{h}\sqrt{\left(\frac{\pi}{2kr}\right)}\exp[i(\pi/4)]\sum_{l=1}^{\infty}[\exp(-b_l z) + V_1(\vartheta_l)\exp(b_l z)]$$

$$\times [\exp(-b_l z_0) + V_1(\vartheta_l)\exp(b_l z_0)]\frac{\exp(ikr\sin\vartheta_l)}{V_1(\vartheta)\sin^{\frac{1}{2}}\vartheta_l}. \qquad (27.42)$$

We sum over all positive l, beginning with $l = 1$. The poles corresponding to negative l do not lie in the region involved in the deformation of the path of integration. On the other hand, Eq. 27.38 gives $\alpha_0 = 0$ when $l = 0$. However, $\alpha = 0$ is not a pole of the integrand because the numerator becomes zero as well as the denominator. Analyzing the indeterminacy, we find that the expression is equal to zero. Physically, the root $\alpha = 0$ would correspond to a plane wave. But in the present theory, a plane wave cannot exist since we have $V_1 \to -1$, $V_2 \to -1$ as $\alpha \to 0$, that is, the field is zero at the boundaries of the layer. Since the field of a plane wave does not vary over the layer, it would be zero everywhere.

A plane wave can be obtained only on the assumption $V_1 = V_2 = 1$. In this case, Eq. 27.24 gives $\phi_1 = \phi_2 = \pi$, $\phi = \phi_1 + \phi_2 = 2\pi$, and from Eq. 27.26, with $l = 1$, we have $\alpha_1 = 0$.

---

† The reader should not be disturbed by the fact that the expression we have obtained is nonsymmetrical in the subscripts 1 and 2. We could just as well have obtained a form in which V_2 would appear in place of V_1.

When the product $\Delta_l h$ is small, which signifies that the attenuation is small over a distance of the order of the layer thickness, Eq. 27.42 can be somewhat simplified. Under these conditions, we can set $\cos \vartheta_l \approx \pi l/kh$, $b_l \approx i\pi l/h$ everywhere in Eq. 27.42, except in the exponential $\exp (ikr \sin \vartheta_l)$, in which we must separate out the imaginary part of ϑ_l which causes the attenuation. Therefore, we will set

$$\vartheta_l \approx \arccos \frac{\pi l}{kh} + i\Delta_l.$$

Then
$$\sin \vartheta_l = \sqrt{\left[1 - \left(\frac{\pi l}{kh}\right)^2\right]} + i\,\frac{\pi l}{kh}\,\Delta_l.$$

Equation 27.42 can now be written in the form

$$\psi = \frac{1}{h} \sqrt{\left(\frac{\pi}{2kr}\right)} \exp\left[i(\pi/4)\right] \sum_{l=1}^{\infty} \left[\exp\left(-i\pi lz/h\right) + V_1(\vartheta_l^0) \exp\left(i\pi lz/h\right)\right]$$
$$\times \left[\exp\left(-i\pi lz_0/h\right) + V_1(\vartheta^0) \exp\left(i\pi lz_0/h\right)\right]$$
$$\times \frac{\exp\left(ikr \sin \vartheta_l^0\right)}{V_1(\vartheta_l) \sin^{\frac{1}{2}} \vartheta_l^0} \exp\left(-\pi l/h\right) \Delta_l r \qquad (27.43)$$

where $\vartheta_l^0 = \arccos \pi l/kh$. We see that the amplitude of each normal wave decreases as $[1/\sqrt(r)] \exp\left[-(\pi l/h)\Delta_l r\right]$, where the factor $1/\sqrt{r}$ is due to the cylindrical divergence of the wave, and the factor $\exp\left[-(\pi l/h)\Delta_l r\right]$ is due to attenuation at the boundaries of the layer. The expressions (27.42) and (27.43) for the field of the normal waves does not change when z and z_0 are interchanged. Consequently, these expressions are valid for all z within the limits $0 \leqslant z \leqslant h$.

4. Lateral waves

We will now examine the second and third (integral) terms in Eq. 27.21, which describe the lateral waves propagating in the media bounding the layer from below and from above. It is sufficient to consider any one of these terms, for example, the second. We will denote this term by W_1. The corresponding integral can be broken into two integrals, over each border of the cut,

$$W_1 = \int_{i\infty}^{A_1} \Phi(\vartheta) H_0^{(1)}(kr \sin \vartheta) \sin \vartheta \, d\vartheta$$
$$+ \int_{A_1}^{i\infty} \Phi^+(\vartheta) H_0^{(1)}(kr \sin \vartheta) \sin \vartheta \, d\vartheta, \qquad (27.44)$$

where $\Phi(\vartheta)$ and $\Phi^+(\vartheta)$ are the values of the function at neighboring points, lying respectively on the left and right borders of the cut. It is well known that as we go around the branch point, the sign of the corresponding radical, in the present case the square root $\sqrt{(n_1^2 - \sin^2 \vartheta)}$ (see above, Section 2), changes to its opposite. Consequently, Φ^+ is obtained from Φ by changing the sign of this square root.[†]

Reversing the direction of integration in the first of the two integrals, we can combine both integrals into one integral which will coincide with the last integral in Eq. 27.44, except that, instead of $\Phi^+(\vartheta)$, we will have the difference $\Phi^+(\vartheta) - \Phi(\vartheta)$. Using Eq. 27.19 for $\Phi(\vartheta)$, and letting

$$\delta_1 = \arcsin n_1, \tag{27.45}$$

we obtain, after some straightforward transformations,

$$W_1 = \frac{ik}{2} \int_{\delta_1}^{i\infty} \frac{(V_1^+ - V_1)\{V_2 \exp[b(h-z)] + \exp[-b(h-z)]}{[\exp(-bh) - V_1 V_2 \exp(bh)][\exp(-bh) - V_1^+ V_2 \exp(bh)]}$$
$$\times \{V_2 \exp[b(h-z_0)] + \exp[-b(h-z_0)]\}$$
$$\times H_0^{(1)}(kr \sin \vartheta) \sin \vartheta \, d\vartheta, \tag{27.46}$$

where V^+ is obtained from V_1 by changing the sign of the square root $\sqrt{(n_1^2 - \sin^2 \vartheta)}$. Equation 27.46 does not change when z and z_0 are interchanged, and is therefore valid for all z within the limits of the layer.

In the following calculations, we shall assume that

$$kr \, | \, 1 - n_1^2 \, | \gg 1,$$
$$\frac{h}{r} \ll \left| \frac{\sqrt{(1 - n_1^2)}}{n_1} \right| \tag{27.47}$$

(see below) and shall use the asymptotic expression

$$H_0^{(1)}(kr \sin \vartheta) \sim \frac{\sqrt{2}}{\sqrt{(\pi kr \sin \vartheta)}} \exp[i(kr \sin \vartheta - \pi/4)]. \tag{27.48}$$

Moreover, we evaluate integral (27.46) by the method of steepest descents. For this purpose, we deform the path of integration in such a way that it goes from the point $\vartheta = \delta_1$ along the contour on which the modulus of the exponential $\exp(ikr \sin \vartheta)$ decreases most rapidly. It is convenient to introduce a new variable s through the equation

$$\sin \vartheta = n_1 + is^2. \tag{27.49}$$

It is not hard to see that the path of most rapid decrease goes along the line corresponding to the variation of s from 0 to ∞. Taking the

---

[†] In $\Phi(\vartheta)$, that is, to the left of the cut, $\mathrm{Re}\sqrt{(n_1^2 - \sin^2 \vartheta)} > 0$. In fact, at the point $\vartheta = 0$, also lying to the left of the cut, we have $\sqrt{(n_1^2 - \sin^2 \vartheta)} = n_1$ but $\mathrm{Re}\, n_1 > 0$.

real parts of both sides of the last equation, we obtain the equation

$$\sin \vartheta' \cosh \vartheta'' = n_1, \tag{27.50}$$

for this path in the complex plane $\vartheta = \vartheta' + i\vartheta''$.

We have examined the case of real n_1. If n_1 has a significant imaginary part, the lateral waves will be strongly attenuated with distance, and, as a rule, will not be of interest.

The contour described by Eq. 27.50 is shown as a solid line in Fig. 123 for the case $n_1 < 1$ as well as for the case $n_1 > 1$. The dotted lines in these figures show the cuts.

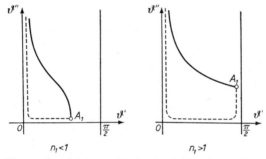

Fig. 123. The paths of integration in the derivation of the expression for the lateral waves.

Substituting Eqs. 27.49 and 27.48 into Eq. 27.46, we obtain the exponential $\exp(-krs^2)$ in the integrand.

We assume that the path of integration, which initially went along the line of the cut, has been transformed into the path of most rapid decrease.†

Since kr is assumed to be large, and the exponential $\exp(-krs^2)$ decreases rapidly, only small values of s, no greater than of the order

$$s_{\max} \sim \frac{1}{\sqrt{(kr)}} \tag{27.51}$$

will be important in the integrand.

Therefore, for an approximate calculation of the integral (27.46), the entire integrand, with the exception of the exponential $\exp(-krs^2)$ and the difference $V_1^+ - V_1$, can be taken outside the integral sign at the value $s = 0$. At $s = 0$, the difference $V_1^+ - V_1$ becomes zero, because

---

† The circuit of the poles which may be encountered as the path of integration is deformed from the initial path along the cut into the path of most rapid decrease, is examined below.

V_1 and V_1^+ differ from one another only by the sign of the square root $\sqrt{(n_1^2 - \sin^2 \vartheta)}$, which is equal to zero at $s = 0$.

Using Eq. 27.49, it is not difficult to obtain

$$\sqrt{(n_1^2 - \sin^2 \vartheta)} \approx \sqrt{(2n_1)} \exp\left[-i(\pi/4)\right] s.$$

We set

$$V_1 - V_1^+ = 2B_1 \sqrt{(n_1^2 - \sin^2 \vartheta)} = 2B_1 \sqrt{(2n_1)} \exp\left[-i(\pi/4)\right] s, \quad (27.52)$$

where B_1 is a constant. The last expression corresponds to the expansion of the difference $V_1 - V_1^+$ in a power series in s, and is limited to the first term of this expansion.

Using Eq. 22.1 for the reflection coefficient, we find

$$V_1 - V_1^+ = \frac{m_1 \cos \vartheta - \sqrt{(n_1^2 - \sin^2 \vartheta)}}{m_1 \cos \vartheta + \sqrt{(n_1^2 - \sin^2 \vartheta)}} - \frac{m_1 \cos \vartheta + \sqrt{(n_1^2 - \sin^2 \vartheta)}}{m_1 \cos \vartheta - \sqrt{(n_1^2 - \sin^2 \vartheta)}}$$

$$= \frac{4m_1 \cos \vartheta \sqrt{(n_1^2 - \sin^2 \vartheta)}}{n_1^2 - \sin^2 \vartheta - m_1^2 \cos^2 \vartheta}. \quad (27.53)$$

Setting $\sin \vartheta = n_1$, $\cos \vartheta = \sqrt{(1 - n_1^2)}$ everywhere except in the radical, we obtain

$$B_1 = -\frac{2}{m_1 \sqrt{(1 - n_1^2)}}. \quad (27.54)$$

In the transition from the variable ϑ to the variable s, we have, according to Eq. 27.49,

$$d\vartheta = \frac{2isds}{\cos \vartheta} = \frac{2isds}{\sqrt{[1 - (n_1 + is^2)^2]}} \approx \frac{2isds}{\cos \delta_1}. \quad (27.55)$$

Using this relation, and also using the second of integrals (19.15), we obtain from Eq. 27.46,

$$W_1 = \frac{iB_1 n_1}{kr^2 \cos \delta_1} \frac{\{V_2(\delta_1) \exp\left[i\gamma_1(h-z)\right] + \exp\left[-i\gamma_1(h-z)\right]\}}{\times \{V_2(\delta_1) \exp\left[i\gamma_1(h-z_0)\right] + \exp\left[-i\gamma_1(h-z_0)\right]\}} \exp(ikn_1 r),$$
$$\frac{}{[\exp(-i\gamma_1 h) - V_1(\delta_1) V_2(\delta_1) \exp(i\gamma_1 h)]^2}$$
$$(27.56)$$

where we have used the notation

$$\gamma_1 = k\sqrt{(1 - n_1^2)} = k \cos \delta_1. \quad (27.57)$$

Just as in the case of a single interface (see §21), the lateral wave, given by expression (27.56) propagates along the layer with a velocity equal to the propagation velocity in the lower medium, and its amplitude decreases with distance as $1/r^2$ (in the absence of absorption in the lower medium). The z-dependence of the amplitude will be oscillatory when $n < 1$, and will be a combination of exponential functions when $n > 1$. In the latter case, if n is large compared with unity, the second term in the square brackets can be neglected, and the expression for the lateral wave takes the form

$$W_1 = \frac{B_1 n_1}{kr^2 \sqrt{(n_1^2 - 1)}} \exp\left[-k\sqrt{(n_1^2 - 1)}(z + z_0)\right] \exp(ikn_1 r). \qquad (27.58)$$

In this case, the amplitude of the lateral wave decreases exponentially with distance from the lower boundary of the layer (for increasing z), i.e. the wave creeps, as it were, along the lower boundary.

Similarly, the integral over the borders of the second cut will yield the lateral wave W_2, which propagates along the layer with a velocity equal to the propagation velocity in the upper medium. Its analytic expression is obtained from Eq. 27.56 by replacing the subscript 1 by the subscript 2, $h - z$ by z, $h - z_0$ by z_0, and γ_1 by $\gamma_2 = k\sqrt{(1 - n_2^2)} = k\cos\delta_2$.

We note that if the media bounding the layer are nonabsorbing, so that the amplitude of the lateral waves decreases as $1/r^2$, practically the entire field in the layer (at sufficiently great distances from the source) will be determined by the field of the lateral waves. This occurs because the waves of the discrete spectrum, in propagating along the layer, are attenuated exponentially as a result of energy leakage through the boundaries of the layer. Therefore, at sufficiently great distances, their amplitude will be arbitrarily small compared with the amplitude of the lateral waves.

These considerations can be supplemented by a more picturesque description. The waves of the discrete spectrum, in propagating along the layer, gradually decrease in amplitude because of the energy leakage through the walls of the layer. But the lateral waves, at great distances, are continuously supplied with energy from the halfspaces bounding the layer.

For Eq. 27.56 to be valid, it is necessary that certain conditions be fulfilled. In Eq. 27.54, we neglected s^2 and s^4 compared with $1 - n^2$. Remembering that Eq. 27.51 gives the greatest significant values of s, we obtain the condition

$$kr\,|\,1 - n_1^2\,| \gg 1, \qquad (27.59)$$

which must be satisfied if the above-mentioned neglect of s^2 and s^4 is to be justified.

Furthermore, in the analysis of the integral (27.46), we took exponentials of the form $\exp[b(h - z)]$ outside the integral sign at the value $s = 0$. This can be done only if these exponentials are slowly varying functions compared with the exponential $\exp(-krs^2)$, which occurs when h is sufficiently small compared with r. A more exact criterion, which we shall not derive here, is

$$\frac{h}{r} \ll \left| \frac{\sqrt{(1 - n_1^2)}}{n_1} \right|. \tag{27.60}$$

It is of course understood that the same conditions, but with the subscript 1 replaced by 2, must be added to Eqs. 27.59 and 27.60.

We have not examined the problem of the poles which may be encountered as the path of integration along the cut (the dotted lines in Fig. 123) is deformed into the path of steepest descents (the solid lines). As an example, we investigate the case $n_1 > 1$.

We saw above that the path of integration Γ_3 (Fig. 122) can be deformed into the infinitely remote path Γ_4 which will hang over the cuts and the poles of the integrand. However, we can also consider an intermediate stage of the deformation process, in which the path Γ_3 is transformed into the path CA_1B (Fig. 124, solid line) plus the contours

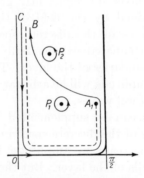

Fig. 124. Regarding the circuit of the poles in the analysis of the expression for the lateral wave.

enclosing the poles lying above the line A_1B (for example, P_2). Here, we can immediately choose the line A_1B such that it corresponds to the path of most rapid decrease, issuing from the branch point A_1. For the present, the poles lying below the line A_1B (for example, P_1) remain unaffected.

To calculate the integral over the contour CA_1, we transform this contour into the path of steepest descents A_1B by a continuous deformation. If we reverse the direction of integration, we obtain exactly the same integral as in Eq. 27.46, but taken over the contour of steepest descents. During such a deformation, it will be necessary to go around the poles lying below the line A_1B. As a result of all these operations, we obtain an integral over the path of steepest descents plus the residues at all the poles lying in the region $\vartheta = \vartheta' + i\vartheta''$, where $0 < \vartheta' < \pi/2$, $0 < \vartheta'' < \infty$, that is, we need add nothing to the solution studied previously.

However, it should be noted that in the above calculations, we assumed implicitly that none of the poles was too close to the branch point. Otherwise, our method of steepest descents is not applicable. In particular, if one of the poles in the complex plane coincides with the branch point $\vartheta = \delta_1$, the denominator in Eq. 27.56 for the lateral wave becomes zero, and the entire expression loses meaning. The case in which the position of the pole is arbitrary with respect to the branch point will be examined in § 29.

In conclusion, we investigate the lateral wave given by Eq. 27.56 for the case of sound propagating in a layer bounded from above by air, and from below by a liquid halfspace. The density of the liquid and the acoustic velocity in the layer will be denoted by ρ and c, respectively. The same quantities in the lower medium will be denoted by ρ_1 and c_1. For air, we can set $\rho_2 = 0$. According to Eq. 22.1, we have

$$V_1 = \frac{m_1 \cos \vartheta - \sqrt{(n_1^2 - \sin^2 \vartheta)}}{m_1 \cos \vartheta + \sqrt{(n_1^2 - \sin^2 \vartheta)}}, \qquad m_1 = \frac{\rho_1}{\rho}, \qquad n_1 = \frac{c}{c_1}. \qquad (27.61)$$

Assuming that $n_1 < 1$, and setting $\vartheta = \delta_1 = \arccos \sqrt{(1 - n_1^2)}$ in Eq. 27.61, we obtain $V_1(\delta_1) = 1$. As a result, Eq. 27.56 can be written in the comparatively simple form

$$W_1 = \frac{2ik_1}{m_1(kr)^2(1 - n_1^2)} \frac{\sin[\gamma_1(h - z)] \sin[\gamma_1(h - z_0)]}{\cos^2(\gamma_1 h)} \exp(ik_1 r). \qquad (27.62)$$

In the case $n_1 > 1$, the expression for the lateral wave will be written in the following form:

$$W = \frac{2ik_1}{m_1(kr)^2(n_1^2 - 1)} \frac{\sinh[\alpha_1(h - z)] \sinh[\alpha_1(h - z_0)]}{\cosh^2(\alpha_1 h)}, \qquad (27.63)$$

where $\alpha_1 = \sqrt{(n_1^2 - 1)}$.

§ 28. Application of the Theory to the Propagation of Electromagnetic Waves in Layers

1. *The field of the normal modes for the case in which one boundary of the layer reflects perfectly, and the other is arbitrary*

Many works have been devoted to the problem of the propagation of electromagnetic waves in layers (see, for example, Refs. 72, 97, 205). As an example, we shall apply the theory developed above to the propagation of electromagnetic waves in a layer, one of whose boundaries, let us say the upper, is perfectly reflecting. The semi-unbounded medium, bounding the layer from below, will be character-ized by the complex index of refraction n, where

$$n^2 = \epsilon + \frac{4\pi i \sigma}{\omega}, \tag{28.1}$$

and ϵ and σ are the dielectric permeability and the conductivity of the medium, respectively. We shall assume that the source is a vertical dipole. This problem was also studied in Ref. 72 by methods somewhat different from those applied below.

In the present case, we have

$$V_2 = 1, \quad V_1 = \frac{n^2 \cos \vartheta - \sqrt{(n^2 - \sin^2 \vartheta)}}{n^2 \cos \vartheta + \sqrt{(n^2 - \sin^2 \vartheta)}}. \tag{28.2}$$

We will study the field of the normal modes in the layer.†

According to Eqs. 27.19, 27.21 and 27.22, the field of the normal modes is

$$\psi = 2\pi i \sum_l \text{Res}\, \Phi(\vartheta_l)\, H_0^{(1)}(kr \sin \vartheta_l) \sin \vartheta_l, \tag{28.3}$$

where

$$\Phi(\vartheta_l) = ik \frac{\{\exp[-i(z_0/h)\,x] + V_1 \exp[i(z_0/h)\,x]\} \cos[(1 - z/h)\,x]}{\exp(-ix)[1 - V_1 \exp(2ix)]} \tag{28.4}$$

and for future convenience, we have introduced the notation

$$x \equiv kh \cos \vartheta. \tag{28.5}$$

As previously, h is the layer thickness, z_0 and z are the elevations of the source and the point of observation, respectively, above the lower boundary of the layer, and the ϑ_l are the roots of the transcendental equation

$$1 - V_1 e^{2ix} = 0. \tag{28.6}$$

We shall use the usual rules to calculate the residue of the function $\Phi(\vartheta)$ at the point $\vartheta = \vartheta_l$, denoted above by $\text{Res}\, \Phi(\vartheta_l)$. We set $\vartheta = \vartheta_l$, $x = x_l$ everywhere except in the parentheses in the denominator. The expression in the parentheses in the denominator, which becomes zero at $x = x_l$, must be replaced by its derivative with respect to ϑ, and then the substitution $x = x_l$ is made.

Taking Eq. 28.6 into account, we obtain

$$\exp[-i(z_0/h)\,x] + V_1 \exp[i(z_0/h)\,x] = 2 \exp(-ix) \cos\left[\left(1 - \frac{z_0}{h}\right)x_l\right]. \tag{28.7}$$

---

† If the layer is bounded from below by a dielectric in which absorption is very small, the lateral wave can be important. The appropriate expression can be obtained without difficulty from Eq. 27.56 by substitution of Eqs. 28.2 for V_1 and V_2.

Furthermore, we have

$$\frac{d}{d\vartheta}\left(1 - V_1\,e^{2ix}\right) = -\frac{dV_1}{d\vartheta}\,e^{2ix} - 2iV_1\,e^{2ix}\frac{dx}{d\vartheta}.$$

Performing the differentiation, and again using Eq. 28.6, we obtain

$$\left[\frac{d}{d\vartheta}\left(1 - V_1\,e^{2ix}\right)\right]_{x_l} = \frac{2n^2(n^2 - 1)\sin\vartheta_l}{\sqrt{(n^2 - \sin^2\vartheta_l)\,(n^4\cos^2\vartheta_l - n^2 + \sin^2\vartheta_l)}} + 2ikh\sin\vartheta_l.$$

Equation 28.3 is now written

$$\psi = 2\pi ik \sum_l \cos\left[\left(1 - \frac{z}{h}\right)x_l\right]\cos\left[\left(1 - \frac{z_0}{h}\right)x_l\right]$$

$$\times\left[kh + \frac{n^2(n^2 - 1)}{i\sqrt{(n^2 - \sin^2\vartheta_l)\,(n^4\cos^2\vartheta_l - n^2 + \sin^2\vartheta_l)}}\right]^{-1}H_0^{(1)}(kr\sin\vartheta_l),$$

$$(28.8)$$

where $x_l \equiv kh\cos\vartheta_l$ is found from Eq. 28.6, which, using Eq. 28.2, can be transformed into the form

$$in^2 x \tan x = \sqrt{[(kh)^2\,(n^2 - 1) + x^2]}. \qquad (28.9)$$

Using this equation, Eq. 28.8 can be written in a somewhat different form

$$\psi = 2\pi ik \sum_l \cos\left[\left(1 - \frac{z}{h}\right)x_l\right]\cos\left[\left(1 - \frac{z_0}{h}\right)x_l\right]$$

$$\times\left[kh + \frac{n^2(1 - n^2)\sin^2 x_l\cdot(kh)^3}{i[(n^2 - 1)\,(kh)^2 + x_l^2]^{\frac{3}{2}}}\right]^{-1}H_0^{(1)}(kr\sin\vartheta_l). \qquad (28.10)$$

References 82 and 94, in which the excitation of a surface wave in a system consisting of a thin dielectric layer lying on a perfectly conducting halfspace is investigated, are closely related to the case we are considering.

2. *The case in which the reflection coefficient of the second boundary is close to unity*

We now consider in more detail the case in which the reflection coefficient V_1 at the lower boundary is close to unity, which will occur when the modulus of n is sufficiently large (the complex dielectric constant is large). Then, we can neglect unity in Eq. 28.9 compared with n^2, and can neglect x^2 compared with $(kh)^2\,(n^2 - 1)$, so that the equation becomes

$$\tan x = \frac{kh}{inx}. \qquad (28.11)$$

We assume that the condition

$$\frac{kh}{|n|} \ll 1 \qquad (28.12)$$

is satisfied.

If on the contrary, the condition $kh/|n| \gtrsim 1$ were satisfied, then in view of our assumption that $|n|$ is large, this condition would mean that kh is large, and we would have the case that was studied in detail in the preceding Paragraph.

Condition (28.12) permits us to solve Eq. 28.11 by the method of successive approximations. Since the right hand side of the equation is small, we can set

$$\tan x = 0$$

as the zeroth approximation, when

$$x_l = l\pi, \quad l = 1, 2, \dots.$$

In the next approximation, setting $x_l = l\pi + \epsilon_l$, where ϵ_l is a small quantity, Eq. 28.11 can be written

$$\tan \epsilon_l \approx \epsilon_l = \frac{kh}{inl\pi}. \qquad (28.13)$$

Thus

$$x_l = l\pi + \frac{kh}{inl\pi}, \qquad (28.14)$$

where l takes on all integral values, beginning with unity. The value $l = 0$ is inadmissible because we would then obtain $x_l = l\pi = 0$ in the zeroth approximation, and the right hand side of Eq. 28.11 would not be small, as we have assumed, but on the contrary, would be infinite. We obtain the root corresponding to $l = 0$ by assuming that x is small, and $\tan x \approx x$. Then Eq. 28.11 yields

$$x_0^2 \approx \frac{kh}{in}, \quad x_0 = \sqrt{\left(\frac{kh}{n}\right)} \exp\left[-i(\pi/4)\right]. \qquad (28.15)$$

In calculating the field of the normal waves, the second term in the square brackets in Eq. 28.8 can be neglected compared with the first term for all l different from zero. At $l = 0$, we have

$$\frac{n^2(n^2-1)}{i\sqrt{(n^2-\sin^2\vartheta_l)}\,(n^4\cos^2\vartheta_l - n^2 + \sin^2\vartheta_l)} \approx \frac{1}{in\cos^2\vartheta_l} = \frac{(kh)^2}{inx_0^2} = kh.$$

Moreover, we can set $x_l \approx l\pi$ for all l in the arguments of the cosines in Eq. 28.8. As a result, we obtain

$$\psi = \frac{\pi i}{h}\left[H_0^{(1)}(kr\sin\vartheta_0) + 2\sum_{l=1}^{\infty} \cos\frac{l\pi z}{h}\cos\frac{l\pi z_0}{h} H_0^{(1)}(kr\sin\vartheta_l)\right]. \qquad (28.16)$$

We see that in the present approximation, the z-dependence of the field of each normal wave is the same as in the case of perfectly reflecting boundaries of the layer (see Eq. 26.16). In particular, the first term in the square brackets in the last expression corresponds to a plane wave, propagating in the layer. The z-dependences of the amplitudes of the first few normal waves are shown in Fig. 118. The essentially new feature, given by our theory, is the attenuation of the individual normal modes as they propagate along the layer.

At distances from the source which are large compared with the wavelength, the Hankel function in Eq. 28.16 can be replaced by its asymptotic representation. As a result, the r-dependence of the field of each normal wave will be given by the exponential $\exp{(ikr \sin \vartheta_l)}$. Consequently, the modes will be attenuated according to the law $\exp{(-\beta_l r)}$, where

$$\beta_l = k \operatorname{Im} \sin \vartheta_l. \tag{28.17}$$

We shall investigate this expression for the damping coefficient in more detail. From Eq. 28.5, we obtain

$$kh \sin \vartheta_l = \sqrt{[(kh)^2 - x_l^2]}. \tag{28.18}$$

We again set

$$x_l = l\pi + \epsilon_l, \tag{28.19}$$

where, according to Eqs. 28.14 and 28.15, we have

$$\epsilon_0 = \sqrt{\left(\frac{kh}{n}\right)} \exp{[-i(\pi/4)]}, \quad \epsilon_l = \frac{kh}{in l\pi}, \quad l = 1, 2, \dots. \tag{28.20}$$

We note that in all cases, $|\epsilon_l| \ll 1$.

Considering the case $l \neq 0$ for the present, we obtain

$$kh \sin \vartheta_l \approx \sqrt{[(kh)^2 - (l\pi)^2 - 2l\pi\epsilon_l]}.$$

Assuming that kh is not too close to $l\pi$ for any l, so that the inequality

$$2l\pi\epsilon_l \ll [(kh)^2 - (l\pi)^2],$$

is satisfied, we obtain

$$kh \sin \vartheta_l = \sqrt{[(kh)^2 - (l\pi)^2]} - \frac{l\pi\epsilon_l}{\sqrt{[(kh)^2 - (l\pi)^2]}}.$$

Substituting for ϵ_l from Eq. 28.20, we easily obtain the damping coefficient

$$\beta_l = \frac{k}{\sqrt{[(kh)^2 - (l\pi)^2]}} \operatorname{Re}\left(\frac{1}{n}\right), \quad l = 1, 2, 3, \dots. \tag{28.21}$$

This formula refers to those waves for which $l\pi < kh$. It is not hard to see that the wave attenuation increases with increasing order l. The normal modes for which $l > kh/\pi$ do not propagate in the layer (see § 26).

When $l = 0$, Eqs. 28.15 and 28.18 give

$$kh \sin \vartheta_0 = \sqrt{[(kh)^2 - \gamma_0^2]} = \sqrt{\left[(kh)^2 + \frac{ikh}{n}\right]}.$$

Here, the second term in the radical is small compared with the first, and we can write, approximately,

$$kh \sin \vartheta_0 \approx kh\left(1 + \frac{i}{2khn}\right).$$

Hence, we obtain $\qquad \beta_0 = k \operatorname{Im} \sin \vartheta_0 = \frac{1}{2h} \operatorname{Re}\left(\frac{1}{n}\right).$ $\qquad$ (28.22)

We note that had we extended Eq. 28.21 to the case $l = 0$ in spite of the fact that such an extension is not legitimate, we would have obtained the same result except for the coefficient $\frac{1}{2}$.

§ 29. THE PROPAGATION OF SOUND WAVES IN A LIQUID LAYER

The principal purpose of the present Paragraph is the presentation of the theory of the propagation of sound in a liquid layer. The results of the theory developed below can be applied practically to the propagation of sound in natural bodies of water (lakes, oceans, etc.) so long as their bottoms can be considered horizontal.

In order to attain a certain degree of generality, we shall begin by considering the case of a liquid layer, bounded on both sides by arbitrary, homogeneous, elastic halfspaces.

1. A liquid layer between two elastic halfspaces

Using the theory developed above, we shall study the propagation of sound waves in a liquid layer contained between two elastic halfspaces. The method of treatment will be different from that in the author's original work.[36]

As above, the elevation of the source above the lower interface will be denoted by z_0, and the elevation of the point of observation by z. The horizontal distance between these two points will be denoted by r. Omitting everywhere the factor $v_0/4\pi$, where v_0 is the volume velocity of the point source, the acoustic potential of the initial spherical wave radiated by the source will be written in the form

$$\psi_0 = \frac{e^{ikR}}{R}, \quad R = \sqrt{[(z - z_0)^2 + r^2]}.$$

As a result of the multiple reflections of this wave at the boundaries of the layer, an extremely complex interference pattern is established in the layer. The integral representation of the acoustic potential ψ at an arbitrary point in the layer will be given by Eq. 27.2, in which V_1 and V_2 will be the reflection coefficients of plane waves at the lower and upper boundaries of the layer, respectively.

The reflection coefficient at a liquid–solid boundary is given by Eq. 4.25. We shall apply this expression to the lower boundary of the layer and, using Eq. 4.26, shall write it in the form

$$V_1 = \frac{M_1 - 1}{M_1 + 1},$$ (29.1)

where

$$M_1 = \frac{\rho_1 c_1}{\rho c} \frac{\cos \vartheta}{\cos \vartheta_1} \left[\left(1 - \frac{b_1^2}{c^2} \sin^2 \vartheta \right)^2 + \frac{4 b_1^3}{c_1 . c^2} \cos \vartheta_1 \cos \gamma_1 \sin^2 \vartheta \right].$$ (29.2)

We recall that ρ and c are the density of the liquid and the propagation velocity in the liquid, ρ_1, c_1, and b_1 are the density of the lower medium and the propagation velocities, respectively, of the longitudinal and transverse waves in the lower medium, ϑ is the angle of incidence of the wave, and ϑ_1 and γ_1 are the angles of refraction of the longitudinal and the transverse waves, respectively, and are related to the angle ϑ by the law of refraction

$$\frac{\sin \vartheta}{c} = \frac{\sin \vartheta_1}{c_1} = \frac{\sin \gamma_1}{b_1}.$$

The quantity M_1 can be considered as the relative impedance of the lower medium with respect to the liquid. In particular, as $b_1 \to 0$, M_1 is transformed into the well-known expression for the relative impedance of two liquid media.

We obtain an expression similar to Eq. 29.1, with the subscript 1 replaced by 2, for the reflection coefficient from the medium bounding the layer from above.

Substituting the expression for V_1 and V_2 into Eqs. 27.2a and 27.2b, we obtain the integral expressions for the acoustic potential in the layer. Thus, for example, for the case $z > z_0$, we obtain

$$\psi = -ik \int_{\Gamma_1} [\sinh b(h-z) - M_2 \cosh b(h-z)]$$

$$\times [\sinh b z_0 - M_1 \cosh b z_0] H_0^{(1)}(kr \sin \vartheta) \frac{\sin \vartheta \, d\vartheta}{N},$$ (29.3)

where $N = -M_1 \cosh bh + \sinh bh + M_2(M_1 \sinh bh - \cosh bh),$

$$b = ik \cos \vartheta.$$ (29.4)

The path of integration Γ_1 is shown in Fig. 122. The case $z < z_0$ is obtained from Eq. 29.3 by interchanging z and z_0. The most interesting case in practice is a liquid layer bounded from above by air. Then, approximately, we can set $\rho_2 = 0$, $M_2 = 0$. In this case, Eq. 29.3 is written

$$\psi = -ik \int_{\Gamma_1} \sinh b(h-z)(\sinh bz_0 - M_1 \cosh bz_0)$$

$$\times H_0^{(1)}(kr\sin\vartheta)\frac{\sin\vartheta\,d\vartheta}{N}, \qquad (29.5)$$

where $$N = -M_1 \cosh bh + \sinh bh. \qquad (29.6)$$

This expression can be used as a basis for numerical calculations. The physical analysis of the solution, which was presented in § 27 for the general case, remains valid in the present case, except that instead of one lateral wave on the lower side of the layer, there will be two, one of which propagates along the layer with the velocity of longitudinal waves, and the other with the velocity of transverse waves in the bottom. In addition to the lateral waves, there will also be normal modes propagating in the layer. The propagation velocity and the damping of the normal modes will be determined by the roots of the equation

$$N = \sinh bh - M_1 \cosh bh = 0. \qquad (29.7)$$

2. The propagation of sound in shallow water

The theoretical analysis becomes most difficult when the thickness of the liquid layer is of the order of the wavelength of the sound wave or less ("shallow water").[†] We shall study this case in the present Section.

Let us suppose, to begin with, that the bottom is also liquid. Then we must set $b_1 = 0$ in Eqs. 29.2, 29.5 and 29.6. We then obtain

$$M_1 = m\frac{c_1\cos\vartheta}{c\cos\vartheta_1} = m\frac{\cos\vartheta}{\sqrt{(n^2-\sin^2\vartheta)}},$$

$$N = \frac{-m\cos\vartheta}{\sqrt{(n^2-\sin^2\vartheta)}}\cosh bh + \sinh bh, \qquad (29.8)$$

where, as previously, $m = \rho_1/\rho$, $n = c/c_1$. Introducing the notation

$$\nu^2 = 1-n^2, \quad bh = ix, \quad x = kh\cos\vartheta, \qquad (29.9)$$

Eq. 29.7 for the poles can be written in the form

$$\cot x = -\frac{1}{mx}\sqrt{(k^2h^2\nu^2 - x^2)}. \qquad (29.10)$$

---

† The case in which the thickness of the liquid layer is much greater than the wavelength was examined in sufficient detail in § 27.

The sign of the radical is chosen so as to satisfy the condition $\mathrm{Re}\,\sqrt{(k^2 h^2 \nu^2 - x^2)} > 0$. We examine the solution of this equation in detail as a function of the parameter $kh\nu$. At first, we consider this parameter as positive, which means that $n < 1$ or $c_1 > c$, that is, the velocity of sound in the bottom is assumed to be greater than in the water. Moreover, we assume that $m > 1$ $(\rho_1 > \rho)$, which is clearly satisfied in all cases of practical interest.

When $kh\nu = 0$, which occurs at zero frequency or at $n = 1$, the equation takes the form

$$\cot x = - i/m. \tag{29.11}$$

We seek a solution of this equation in the form

$$x = (2l + 1)\tfrac{1}{2}\pi + ix'', \quad l = 0, 1, 2, \ldots.$$

Then for x'', we obtain the equation

$$\tanh x'' = 1/m.$$

We thus see that Eq. 29.11 has an infinite number of roots situated on a straight line parallel to the real axis, with equal spacings of π between them.

Let us now increase the value of the parameter $kh\nu$. When $0 \leqslant kh\nu < \pi/2$, only complex values of x will satisfy Eq. 29.10. However, as $kh\nu$ increases, the roots move away from the above-mentioned straight line and approach the real axis. It can be shown that the first root, which is equal to $(\pi/2) + i \arctan 1/m$ when $kh\nu = 0$, is strongly displaced from the remaining ones, and approaches $x = \pi/2$ as $kh\nu$ approaches $\pi/2$. It is quite evident that Eq. 29.10 is satisfied by $x = \pi/2$ when $kh\nu = \pi/2$.

As $kh\nu$ is increased further, this root increases, remaining real, and approaches $x = \pi$ as $kh\nu \to \infty$. The value of the root as a function of the parameter $kh\nu$ is shown in Fig. 125 for $m = 2$.[†]

When $kh\nu = 3\pi/2$, Eq. 29.10 is also satisfied by the real root $x = 3\pi/2$. As kh increases, this second root remains real, and as $kh \to \infty$, the root approaches the value $x = 2\pi$. As $kh\nu$ is increased further, the number of real roots satisfying Eq. 29.10 becomes all the greater. We are assuming here that the liquid comprising the layer is not itself absorbing.

A separate normal mode corresponds to each root of the equation, and is obtained as the residue at the corresponding pole. Deforming the path of integration Γ_1 as was done in § 27, and remembering the notation

---

[†] The calculations were performed by Yu. L. Gazarian of the Acoustics Institute. The basic qualitative considerations concerning the roots of Eq. 29.10 also originated with him.

(29.9), we obtain

$$\psi = 2\pi i k \sum_l \sin\left[x_l\left(1 - \frac{z}{h}\right)\right]\left\{-M_1\cos\left(x_l\frac{z_0}{h}\right) + i\sin\left(x_l\frac{z_0}{h}\right)\right\}$$

$$\times H_0^{(1)}(kr\sin\vartheta_l)\frac{\sin\vartheta_l}{(dN/d\vartheta)_{\vartheta_l}} \qquad (29.12)$$

from Eq. 29.5.

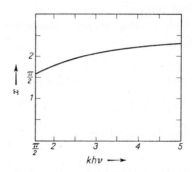

Fig. 125. The magnitude of the root which yields the first normal mode as a function of the layer thickness and the frequency.

Using the equation $N = -M_1\cos x_l + i\sin x_l = 0$, the curly brackets can without difficulty be transformed into the expression

$$-\frac{i}{\cos x_l}\sin\left[x_l\left(1 - \frac{z_0}{h}\right)\right].$$

Moreover, taking Eqs. 29.9 and 29.10 into account, Eq. 29.8 gives

$$\frac{1}{\sin\vartheta_l}\left(\frac{dN}{d\vartheta}\right)_{\vartheta_l} = -\frac{ikh}{\cos x_l}\left(1 - \frac{1}{x_l}\sin x_l\cos x_l - \frac{1}{x_l m^2}\sin^2 x_l\tan x_l\right).$$

As a result, Eq. 29.12 yields

$$\psi = \frac{2\pi i}{h}\sum_l x_l \frac{\sin\{x_l[1 - (z/h)]\}\sin\{x_l[1 - (z_0/h)]\}}{x_l - \sin x_l\cos x_l - (1/m^2)\sin^2 x_l\tan x_l}H_0^{(1)}\left(\frac{r}{h}\sqrt{[(kh)^2 - x_l^2]}\right)$$

$$(29.13)$$

for the field of the normal modes.

The last expression does not change when z and z_0 are interchanged, and is valid for any z within the region $0 \leqslant z \leqslant h$. The quantities x_l, $l = 1, 2, \ldots$, are the roots of Eq. 29.10. We note that in the limiting case of a completely rigid bottom ($m \to \infty$), all the terms in the denominator in Eq. 29.13, except the first, can be neglected. In fact, we see from Eq. 29.10 that as $m \to \infty$ we have $\cot x_l \approx \cos x_l \to 0$. However, while the quantity $m\cos x_l$ remains finite, $m^2\cos x_l \to \infty$.

Each normal mode in Eq. 29.13 is characterized by a definite amplitude distribution along the vertical (in the z-direction), and also by a definite phase velocity in the layer and attenuation. The amplitude distribution of the first normal mode along the vertical is shown in Fig. 126 for the case $c = 1500$ m/sec, $c_1 = 1501.5$ m/sec, $\rho_1 = 2\rho$, $h = 90$, and for various frequencies. The amplitude decreases exponentially with the depth in the lower medium, and the greater the frequency, the more rapid the decrease. At high frequencies, practically all the energy of the wave is contained in the layer.

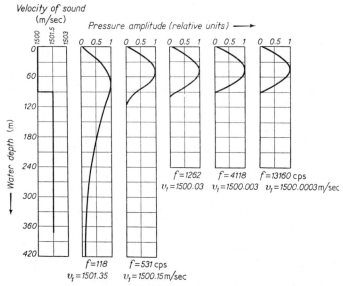

Fig. 126. The z-dependence of the amplitude of the first normal mode for various frequencies; v_1 is the phase velocity of the normal mode.

Remembering that the quantity $(1/h)\sqrt{[(kh)^2 - x_l^2]}$, which enters into the argument of the Hankel function, plays the rôle of a complex wave number, we obtain the following expressions for the phase velocity along the layer v_l and the absorption coefficient β_l:

$$v_l = \frac{khc}{\operatorname{Re}\sqrt{[(kh)^2 - x_l^2]}}, \qquad \beta_l = \frac{\operatorname{Im}\sqrt{[(kh)^2 - x_l^2]}}{h}. \tag{29.14}$$

The phase velocity v_1 of the first normal mode is shown in Fig. 126 for various frequencies. At the very lowest of these frequencies, v_1 is close to the acoustic velocity in the lower medium. As the frequency is increased, the velocity decreases, and approaches the acoustic velocity in the layer.

It is clear from Eq. 29.14 that the normal mode can be undamped only when x_l is real, and in addition, the condition $x_l < kh$ is fulfilled.

We saw above that the lowest kh for which a real root occurs is $kh = \pi/(2\nu)$. The thickness of the layer is then†

$$h = \frac{\pi}{2k\nu} = \frac{\lambda}{4\sqrt{(1-n^2)}}. \tag{29.15}$$

For a given h, the corresponding frequency, called the critical frequency, will be

$$f_k = \frac{\omega_k}{2\pi} = \frac{c}{4h\sqrt{(1-n^2)}}. \tag{29.16}$$

Waves with frequencies lower than f_k cannot propagate along the layer without damping. For the case shown in Fig. 126, the critical frequency is 93.3 cps.

The phase velocity of the wave of frequency f_k is obtained from Eq. 29.14 by substituting the values $x_l = \pi/2, kh = \pi/2\nu$ into the expression for v_l. As a result, we have $v_1 = c/n = c_1$. Thus, at the instant it is generated (at $f = f_k$), the phase velocity of the first normal mode will be equal to the propagation velocity of waves in the bottom.

Furthermore, from the relations $x_1 = kh \cos \vartheta = \pi/2, kh\nu = \pi/2$, which are valid for the first normal mode, we find

$$\cos \vartheta = \frac{\pi}{2kh} = \nu = \sqrt{(1-n^2)}, \quad \sin \vartheta = n.$$

It was shown above that ϑ is the angle of incidence at the boundary of each of two plane waves, the superposition of which gives the normal mode. Thus, the first undamped normal mode arises when the angle of incidence of the associated plane waves becomes equal to the angle of total internal reflection. This result has a simple physical significance. The absence of damping is due, evidently, to the presence of total internal reflection at the boundary.

As the frequency is increased further, the root x_1, as we saw above, remains real and approaches the value $x_1 = \pi$. Then, according to Eq. 29.14, the phase velocity v_1 approaches c, the acoustic velocity in the liquid comprising the layer.

When $kh\nu = 3\pi/2$, a second real root $x_2 = 3\pi/2$ appears, which signifies the appearance of a second undamped normal mode. It is easy to show that when $kh\nu = 3\pi/2$, the phase velocity of this wave

---

† For a perfectly rigid bottom, the equation for the poles has the form $\cos x_l = 0$, whence we obtain for the first root $x_1 = \pi/2$, for any $kh\nu$. From the condition $kh > x_1$, we immediately obtain $h = \lambda/4$ for the smallest depth for which at least one undamped normal mode exists. Thus, the depth of the water must not be less than a quarter of a wavelength.

is again equal to c_1, and the angle of incidence satisfies the equation $\sin \vartheta = n$. As the frequency is raised, the phase velocity approaches c, just as in the case of the first normal mode.

As khv is increased further, undamped normal modes of higher order will appear. For a given khv, the number of undamped normal modes is equal to the greatest odd number equal to or less than $(2/\pi) khv$.

We saw that when khv attains the value $\pi/2$, the first normal mode is transformed from a damped into an undamped wave. In this connection, its amplitude at some distance from the source increases sharply. However, this does not occur discontinuously, but gradually, as the root x_1 approaches the real axis.

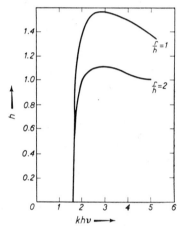

Fig. 127. The frequency dependence of the amplitude of the first normal mode for two different distances from the source.

The amplitude of the first normal mode is shown in Fig. 127 as a function of the frequency for two different distances from the source ($r/h = 1$ and $r/h = 2$). The abscissa represents the quantity khv, and the ordinate represents the product of the modulus of the acoustic potential and the layer thickness. The calculations were based on the first term of Eq. 29.13, with the assumptions $m = 2$, $z = 0$ and $z_0 = h/2$.

Equation 29.15 is a very important relation between the layer thickness h, the index of refraction n, and the limiting wavelength λ. Thus, knowing h, and having determined λ from experimental measurements of the limiting frequency, we can find n, and consequently, the acoustic velocity in the lower medium.

Relation (29.15) remains valid when the shear modulus in the lower medium is not zero.

In fact, according to Eq. 29.6 (and using the notation in Eq. 29.9), the equation for the poles in this case is written

$$\cot x = -i/M_1,$$

where M_1 is given by Eq. 29.2. It is clear from the latter that $1/M_1$ is proportional to $\cos \vartheta_1 \sim \sqrt{[x^2 - (kh\nu)^2]}$. Hence, when $kh\nu = x = \pi/2$, both sides of the last equation are zero, i.e. the equation is satisfied.

Now, however, when $kh\nu > \pi/2$, the normal modes will undergo some damping associated with the energy lost to the shear waves excited in the lower medium.

All that has been presented above referred to the case $n < 1$ in which the acoustic velocity in the lower medium is greater than the acoustic

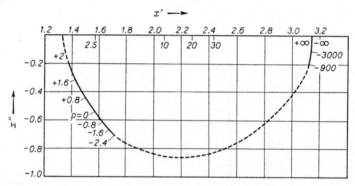

Fig. 128. The values of the root determining the first normal mode, shown in the complex plane as a function of the parameter $p = (kh)^2 (1 - n^2)$.

velocity in the layer. In the opposite case, in which $n > 1$, the physical picture of the propagation will be somewhat different. In particular, undamped normal modes will be absent because of the absence of total internal reflection. All the roots of Eq. 29.10 will have non-zero imaginary parts. It is true that in certain cases (for example, for very large n), the reflection coefficient at the lower boundary can be very close to unity, and the damping of the normal modes will be insignificant.

The values of the roots of Eq. 29.10 for the first normal mode are shown in Fig. 128 in the complex plane $x = x' + ix''$, with $m = 2$, and various values of the quantity $p = (kh\nu)^2 = (kh)^2 (1 - n^2)$, both positive $(n < 1)$ as well as negative $(n > 1)$. It is clear from the Figure that when p is positive, x will be real only when $p \geqslant (\pi/2)^2 = 2.46$. When p is negative, x is always complex. However, when $|p|$ is large, the imaginary part of x can be very small, and approaches zero as $p \to -\infty$. This case corresponds either to very high frequencies, when the absorbing action of the boundary has no effect (because of an extreme grazing angle of incidence at the boundary), or to very large n, when the boundary is almost perfectly reflecting.

It is worth noting that when n is finite but large, even a small increase of the frequency leads to a considerable increase of $|p| = kh(1-n^2)$, and consequently, to a decreased damping of the first normal mode. Therefore, at some distance from the source, the amplitude of the first normal mode will increase sharply with increasing frequency. As a result, we obtain an effect similar to the "generation" of a normal mode in the case $n < 0$ which is associated with total internal reflection.

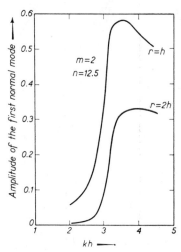

Fig. 129. The amplitude of the first normal wave as a function of kh for $m = 2$ and $n = 12.5$, and for two distances.

The amplitude of the first normal mode (in relative units) is shown in Fig. 129 as a function of kh, for $m = 2, n = 12.5$, and two distances. We see that the relationships are quite similar to those in Fig. 127.

3. Results of numerical calculations

It is clear from what has been presented that if the layer thickness is a small fraction of the wavelength (of the order of a quarter of a wavelength or less), all the normal modes will be very rapidly damped with distance. In this case, the sound field will be determined principally by the lateral wave, which at large distances is described by Eqs. 27.62 and 27.63. However, we were not able to obtain a simple expression for the lateral wave at distances comparable with the wavelength. Therefore, the only way to determine the field is, apparently, by

numerical integration in Eq. 29.5. Some results obtained in this way
will be presented in graphical form.

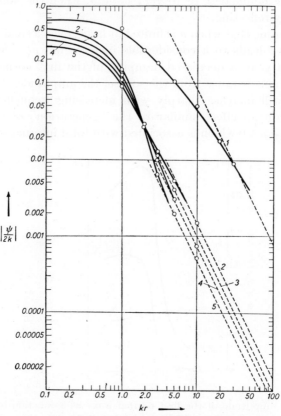

Fig. 130. The decrease of the sound field with distance for some
concrete cases.

TABLE 8

	c/c_1	ρ/ρ_1	kh	$\dfrac{z_0}{h}$	$\dfrac{z}{h}$
A	0.6	0.50	1.7952	0.70	0.4
B	2.0	0.50	0.84	0.75	0.0
C	2.0	0.75	0.84	0.75	0.0
D	2.0	1.00	0.84	0.75	0.0
E	2.0	1.50	0.84	0.75	0.0

The modulus of $\psi/2k$ as a function of the dimensionless distance kr
is shown for a number of cases in Fig. 130.[13] The values of the various
parameters used in the calculations are given in Table 8.

The solid lines in Fig. 130 represent the results obtained by numerical integration of the exact equation. The dotted straight lines correspond to the asymptotic values of the lateral wave as given by Eqs. 27.62 and 27.63.

The influence of the various parameters on the way in which the acoustic pressure decreases with distance can be seen in Figs. 131 and 132, also obtained by numerical integration of the exact equation.[†]

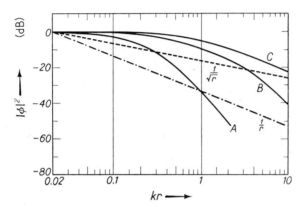

Fig. 131. The fall-off law of the modulus of the acoustic pressure for three different layer thicknesses.

TABLE 9

	kh	kz_0	kz
Curve A	0.5	0.25	0.15
Curve B	1.5	0.6	0.3
Curve C	2.2	1.0	0.5

Thus the fall-off law of the modulus of the acoustic pressure is shown in Fig. 131 for the case $m = \rho_1/\rho = 1.9$, $n = c/c_1 = 0.8$, and for three different layer thicknesses. In all cases, the source is situated either approximately or exactly in the center of the layer, and the receiver is located approximately one quarter of the layer thickness from the bottom. The values of the parameters in Fig. 131 are given in Table 9, where z and z_0 are measured from the bottom. The ordinate represents

---

† The calculations were performed at the Institute of Precision Mechanics and Calculational Technique of the Academy of Sciences, USSR.

the quantity

$$P = 10 \log \frac{|p|^2}{p_0^2}$$

where the initial level p_0 is arbitrary for each curve.

It is clear from Fig. 131 that as the thickness of the layer is decreased (or, equivalently as the frequency is lowered), the fall-off becomes more and more rapid. In all of these three cases, the value of h was sufficiently small so that there were no undamped normal modes. For comparison, dotted lines corresponding to the fall-off laws $|p| \sim 1/r$ and $|p| \sim 1/\sqrt{r}$ are also included in the Figure.

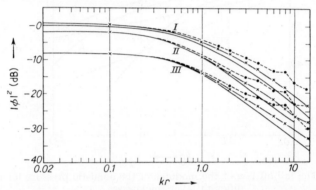

Fig. 132. Curves of the acoustic pressure fall-off for various parameters of the bottom and for various positions of the sources and the receivers.

The influence on the fall-off law of the acoustic pressure, exerted by the parameters of the bottom and the positions of the source and receiver, can be seen in Fig. 132. For all the curves, $kh = 2.2$. The curves are split into three groups with the following values of z and z_0:

$$\text{group \quad I}: kz_0 = 0.5, \quad kz = 1.0,$$

$$\text{group \quad II}: kz_0 = 1.0, \quad kz = 1.5,$$

$$\text{group III}: kz_0 = 1.0, \quad kz = 1.8.$$

In each group there are three kinds of curves, corresponding to various values of m and n, namely:

curves ——————— $m = 1.9, \quad n = 0.8,$

curves — × — × — × — $m = 2.7, \quad n = 0.8,$

curves — — — — $m = 2.7, \quad n = 0.5.$

We see from Fig. 132 that the values of z and z_0 affect the initial level of the acoustic pressure, but have practically no influence on the fall-off law (the three curves are at different levels, but are parallel to one another).

Changes in the values of m and n are also not too important, since there is very little divergence within each of the groups of curves. It is interesting to note that in the case $m = 2.7$, $n = 0.5$, there is one undamped normal wave (the limiting value of kh for this case is $(kh)_{\text{lim}} = (\pi/2)/\sqrt{[1 - (0.5)^2]} \approx 1.8$). This is apparently the cause of the somewhat peculiar, non-monotonic behavior of the curve.

We should mention that a partial study of the normal modes propagating in a liquid layer with an absorbing bottom is given in Ref. 134.

4. The lateral wave

The expression for the lateral wave in the present case is given by Eqs. 27.62 and 27.63. However, the first of these expressions, which pertains to the case of most interest to us, $n_1 < 1$ $(c_1 > c)$, becomes infinite and therefore loses meaning when $\gamma_1 h = (2l + 1)(\pi/2)$, where l is an integer. Using Eq. 27.57 for γ_1, the last condition can also be written in the form

$$khv = (2l + 1)(\pi/2), \quad v = \sqrt{(1 - n^2)}. \tag{29.17}$$

As we saw above, this relation is also the equation for the critical frequency of the normal mode of order l. If kh is increased, it will "give birth" to a new normal mode as it passes through the value given by Eq. 29.17.

It is not hard to see that the pole corresponding to this normal mode coincides with a branch point under these conditions.

In fact, we have $\vartheta = \delta_1$, $\cos \vartheta = \cos \delta_1 = \sqrt{(1 - n^2)} = v$ at the branch point. Then, according to Eqs. 29.9 and 29.17, we have

$$x = khv = (2l + 1)(\pi/2).$$

However, it is not hard to see that this value of x also satisfies Eq. 29.10 for the poles, which completes our proof.

When a pole coincides with a branch point, or is situated close to it, the method of steepest descents is inapplicable. We had a similar situation with regard to the *method of passage* in § 20. Using a method similar to that presented in § 20, Section 2, we will obtain an expression for the lateral wave, valid for any value of kh.†

We will start from Eq. 27.46. Setting $V_2 = -1$, and using Eq. 27.32 for V_1, where $\alpha = (\pi/2) - \vartheta$, we obtain, after some straightforward

---

† This problem was first studied by Yu. L. Gazarian.

transformations,

$$W_1 = 2ikm \int_{\delta_1}^{i\infty} \frac{\sqrt{(n^2 - \sin^2 \vartheta)} \sinh b(h-z)}{m^2 \cos^2 \vartheta \left[\dfrac{\cosh^2 bh - (n^2 - \sin^2 \vartheta) \sinh^2 bh}{m^2 \cos^2 \vartheta} \right]} \quad (29.18)$$

where the subscript 1 on m and n has been omitted for simplicity. The sign of the radical is chosen from the condition $\operatorname{Re} \sqrt{(n^2 - \sin^2 \vartheta)} > 0$.

We introduce the notation

$$\sqrt{(n^2 - \sin^2 \vartheta)} = x \quad (29.19)$$

and expand the square brackets in the denominator of the integrand in Eq. 29.18 in a power series in x up to x^2.

We then obtain

$$\nu^2 [\quad\quad] = \nu^2 \cos^2 kh\nu + x^2 \left(\frac{\sin^2 kh\nu}{m^2} - kh\nu \sin kh\nu \cos kh\nu \right),$$

where, as previously, $\nu = \sqrt{(1-n^2)}$.

As a result, we have

$$W_1 = \frac{2ik\nu}{m} \int_{\delta_1}^{i\infty} \frac{x \sinh b(h-z) \sinh b(h-z_0) H_0^{(1)}(kr \sin \vartheta) \sin \vartheta \, d\vartheta}{\nu^2 \cos^2 kh\nu + x^2 [(1/m^2) \sin^2 kh\nu - kh\nu \sin kh\nu \cos kh\nu]}. \quad (29.20)$$

As is clear from its definition in (29.19), the quantity x will take on real values from 0 to ∞ on the cut along which the path of integration passes.

It will be convenient for us in what follows to change the path of integration in such a way that it goes along the bisectrix of the fourth quadrant in the x-plane $\{x \sim \exp[-i(\pi/4)s]$, where s is real$\}$. There are no obstacles to such a change. The poles of the integrand in Eq. 29.20 lie approximately at the points $x = \pm im^2\nu^2 \cot kh\nu$ (since $\cos kh\nu$ is assumed small), and do not interfere with the indicated change of the path of integration. We shall use the asymptotic expansion of the Hankel function in the integrand in Eq. 29.20. Remembering that we obtain the exponential $\exp(ikr \sin \vartheta)$ in this case, we introduce a new variable of integration s, in place of ϑ, through the relation

$$\sin \vartheta = n + \frac{is^2}{kr}. \quad (29.21)$$

Then, to within s^2, $\qquad dn\vartheta = \dfrac{2is \, ds}{kr \cos \vartheta} \approx \dfrac{2is \, ds}{kr\nu},$

$$x = \sqrt{\left(\frac{2n}{kr} \right)} e^{-i(\pi/4)} s. \quad (29.22)$$

Now, we can set $\sin \vartheta = n$, $\cos \vartheta = \sqrt{(1-n^2)} = v$, $b = ik\cos\vartheta = ikv$ (which corresponds to $s = 0$) everywhere in the integrand in Eq. 29.18 except in the exponentials obtained in the asymptotic representation of the Hankel function, and in the square brackets in the denominator. As a result, the expression for the lateral wave takes the form

$$W_1 = \frac{2in}{km} \sin kv(h-z) \sin kv(h-z_0) \frac{e^{iknr}}{r^2} I, \qquad (29.23)$$

where

$$I = -\frac{4}{\sqrt{\pi}} \int_0^\infty \frac{e^{-s^2} . s^2 \, ds}{v^2 \cos^2 khv - \dfrac{2in}{rk} \left(\dfrac{\sin^2 khv}{m^2} - khv \sin khv \cos khv \right) s^2}. \qquad (29.24)$$

The integration with respect to s is taken over the real axis from 0 to ∞, which corresponds to the choice of the path of integration in the s-plane indicated above. Had we neglected the term s^2 in the denominator of the integrand in Eq. 29.24, we would have obtained the same expression for the lateral wave as in § 27.

Incidentally, this can be done in the case $n > 1$ (v imaginary), since in this case $\cos khv$ does not become zero for any value of kh.

We use the notation

$$w^2 = \frac{rk}{2n} \frac{v^2 m^2}{\tan^2 khv - khvm^2 \tan khv} \qquad (29.25)$$

in Eq. 29.24. The quantity w plays the rôle of the "numerical distance" introduced in § 20. Now,

$$I = \frac{-2iw^2}{v^2 \cos^2 khv} \frac{2}{\sqrt{\pi}} \int_0^\infty \frac{e^{-s^2} . s^2 \, ds}{s^2 + iw^2} \equiv \frac{-1}{v^2 \cos^2 khv} K(w^2). \qquad (29.26)$$

It can be shown that†

$$\int_0^\infty \frac{e^{-\alpha s^2} \, ds}{s^4 + 1} = \frac{\pi}{\sqrt{2}} \left\{ \cos \alpha \left[\frac{1}{2} - S(\alpha) \right] - \sin \alpha \left[\frac{1}{2} - C(\alpha) \right] \right\}, \qquad (29.27)$$

where

$$S(\alpha) = \int_0^\alpha \frac{\sin t}{\sqrt{(2\pi t)}} \, dt, \quad C(\alpha) = \int_0^\alpha \frac{\cos t}{\sqrt{(2\pi t)}} \, dt. \qquad (29.28)$$

---

† This identity can be proved by differentiating I twice with respect to α. Denoting the result by I'', it is not hard to obtain the equation

$$I'' + I = \int_0^\infty e^{-\alpha s^2} ds = \frac{1}{2} \sqrt{\left(\frac{\pi}{\alpha} \right)}.$$

Setting $I = c_1 \sin \alpha + c_2 \cos \alpha$, and using the method of variation of parameters, we obtain Eq. 29.27.

Now, remembering that $w^2 > 0$, we obtain

$$K(w^2) = 2iw^2 \llbracket 1 - \sqrt{(2\pi w^2)} \{\cos w^2[\tfrac{1}{2} - S(w^2)] - \sin w^2[\tfrac{1}{2} - C(w^2)]\}$$
$$- i\sqrt{(2\pi w^2)} \{\cos w^2[\tfrac{1}{2} - C(w^2)] + \sin w^2[\tfrac{1}{2} - S(w^2)]\} \rrbracket. \quad (29.29)$$

For large values of w^2, we have the asymptotic representation

$$K \approx 1 - \frac{15}{4w^4} + i\left(\frac{3}{2w^2} - \frac{105}{8w^6}\right), \quad (29.30)$$

and for small values of w^2, we have

$$K \approx \sqrt{(2\pi)}\, w^3(1 + w^2) - 4w^4 + i[2w^2 - \sqrt{(2\pi)}\, w^3(1 - w^2)]. \quad (29.31)$$

Thus, when $\cos khv$ is sufficiently small (i.e. when the frequency is sufficiently close to critical) so that w, as given by Eq. 29.25, is small and can be written

$$w^2 \approx \frac{rk}{2n} v^2 m^2 \cos^2 khv, \quad (29.32)$$

we obtain $K \approx 2iw^2$ from Eq. 29.31, or, remembering the meaning of w, Eq. 29.26 gives

$$I = -\frac{irk}{n} m^2. \quad (29.33)$$

We see from Eq. 29.23 that in this case the amplitude of the lateral wave decreases with distance as $1/r$.

Values of $K(w^2)$ are given in Table 10.

5. *Some remarks on sound propagation in shallow water, taking account of the shear elasticity of the bottom*

The dispersion equation (29.7) for the normal modes in a liquid layer lying on an elastic halfspace can also be written in the form

$$\tanh bh = -M_1, \quad b \equiv ik\cos\vartheta. \quad (29.34)$$

This expression coincides with the equation obtained by Sherman.[79]† The roots of this equation, particularly those corresponding to real values of the quantity $\xi = k\sin\vartheta$, were analyzed with great care by Sherman. These roots correspond to undamped normal modes, the amplitudes of which decrease only as $1/\sqrt{r}$ due to the cylindrical divergence of the waves. There are also interesting discussions regarding the normal modes in a liquid layer lying on an elastic base in Refs. 88 and 196. The same problem was studied by N. V. Zvolinskii.[51, 52]

Figure 133, taken from Ref. 88, shows the ratio v/c as a function of the ratio λ/h, where v is the phase velocity of the normal mode, c is the

---

† There are differences in notation, which must be taken into account.

acoustic velocity in the liquid, λ is the acoustic wavelength, and h is the thickness of the layer. The first three normal modes were considered, and are indicated by the numbers 1, 2 and 3. The following parameter values were used:

$$\rho/\rho_1 = 1, \quad b_1/c = 1.5, \quad \sigma = 0.5,$$

where b_1 is the transverse wave velocity in the bottom, and σ is the Poisson ratio. The cases $\sigma = 0.25$ and $\sigma = 0$ are also shown for the first normal mode.

TABLE 10

Values of $K(w^2)$

w^2	0.00	0.01	0.02	0.03	0.04	0.05	0.06	0.07
Re K	0.0000	0.00214	0.00525	0.0098	0.0143	0.0194	0.0245	0.0300
Im K	0.0000	0.0175	0.0330	0.0475	0.0609	0.0737	0.0860	0.0975

w^2	0.08	0.09	0.10	0.20	0.30	0.40	0.50	0.60
Re K	0.0357	0.0411	0.0470	0.11	0.16	0.22	0.27	0.32
Im K	0.109	0.120	0.131	0.20	0.26	0.30	0.32	0.34

w^2	0.70	0.80	0.90	1.00	1.50	2.00	2.50	3.00
Re K	0.36	0.39	0.43	0.46	0.60	0.68	0.76	0.79
Im K	0.35	0.36	0.37	0.38	0.38	0.36	0.34	0.31

w^2	4.00	5.00	6.00	7.00	8.00
Re K	0.84	0.89	0.92	0.94	0.95
Im K	0.27	0.24	0.21	0.19	0.17

Curves 2 and 3 contain nothing essentially new. They have the same form as the usual dispersion curves for the first two normal modes in a liquid layer lying on a liquid halfspace. In all cases, their velocity is greater than the velocity of sound in the liquid. The values $\lambda/h = 3.00$ and $\lambda/h = 1.46$ correspond to the critical frequencies for these waves.

The branches of curve 1 are of very great interest. There are no critical frequencies for these curves, and for all values of λ/h the corresponding wave propagates without damping. It is not hard to show that this wave is a surface wave propagating along the boundary between the liquid and the elastic halfspace, and is somewhat modified by the presence of the free boundary of the liquid. In fact, when λ is small, we have $v < c$, whence this wave must necessarily be attenuated exponentially in the z-direction. Moreover, we see that for small λ all three branches have a horizontal portion, which signifies that, for these values of λ, the velocity is independent of the layer thickness h. This is the case in which, because of the exponential attenuation of the wave in the vertical direction, its amplitude at the surface of the liquid will

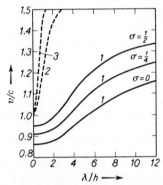

Fig. 133. Dispersion curves for waves in a layer, taking the shear elasticity of the bottom into account.

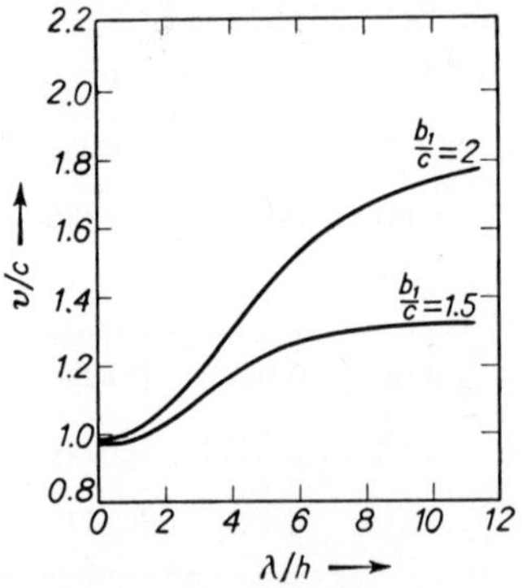

Fig. 134. The same as in Fig. 133, but with another density relation.

be so small that the surface will not play an essential rôle. Thus, we have here the case of surface waves on the boundary between a liquid and a solid halfspace. These waves are sometimes called Stonely waves.

As the ratio λ/h increases, the surface of the liquid begins to play a rôle, and the velocity v increases, passing through the value $v = c$ at some value of λ/h. As $\lambda/h \to \infty$, the velocity v approaches a limit, equal to the velocity of Rayleigh waves on the boundary of a free elastic halfspace (which for $\sigma = 0.5$ corresponds to $v = 1.432 = 0.955b_1$). Under these conditions, the liquid layer ceases to play any rôle whatsoever.

Thus, for any values of the frequency and the layer thickness, there will be a surface wave which for small values of λ will be a Stonely wave, and for large values of λ will be a Rayleigh wave.

Similar curves are shown in Fig. 134 for the cases $\rho/\rho_1 = 0.4$, $\sigma = 0.25$, $b_1/c = 1.5$ and 2. Reference 88 contains curves of the group velocities as well.

It should be noted that the influence of the shear elasticity of the bottom can also make itself felt as an increased attenuation of the acoustic field in the layer. This occurs because of the energy carried away from the layer by the shear waves.

§ 30. Sound Propagation in a Triple-layered Medium

1. *Integral expression for the field*

We shall now complicate the problem considered in § 29 by assuming that the liquid layer does not lie directly on an infinite homogeneous halfspace, but is separated from it by another layer. This is of practical importance in the study of sound propagation in a liquid layer lying on a layered bottom. We will assume that all three media are liquid. The notation for the parameters of these media is given in Fig. 135.

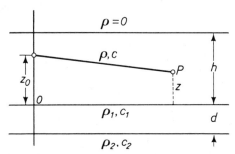

Fig. 135. Notation for the parameters of the media, and illustration of the positions of the source and the receiver in a three-layered medium.

The problem we are considering was investigated for the first time in Refs. 63 and 165, with the use of methods somewhat different from that employed below.

The source and the receiver will be situated in the upper layer at distances of z_0 and z, respectively, above its lower boundary. As previously, the origin of coordinates will be placed at the lower boundary of this layer. We shall also use the following notation

$$m_1 = \frac{\rho_1}{\rho}, \quad m_2 = \frac{\rho_2}{\rho}, \quad n_1 = \frac{c}{c_1}, \quad n_2 = \frac{c}{c_2},$$

$$k = \frac{\omega}{c}, \quad k_1 = \frac{\omega}{c_1}, \quad k_2 = \frac{\omega}{c_2}. \tag{30.1}$$

The field at any point P in the upper layer will be given by the general equations (27.2a) and (27.2b), in which we must set $V_2 = -1$ if the liquid is bounded by air at $z = h$. For V_1 we must use the value of the coefficient of reflection of a plane wave at a plane-parallel layer, obtained in §5. In using Eq. 5.10 for the reflection coefficient, we must remember that the notation used in §5 is somewhat different from that used in the present Paragraph because of the different numbering of the media. In particular, the parameters of the medium through which the incident wave travels were denoted in §5 by the subscript 3, whereas in the present Paragraph, these parameters will generally have no subscripts.

We shall use the notation

$$Z = \frac{\rho c}{\cos \vartheta}, \quad Z_1 = \frac{\rho_1 c_1}{\cos \vartheta_1}, \quad Z_2 = \frac{\rho_2 c_2}{\cos \vartheta_2},$$

$$\beta = k \cos \vartheta, \quad \beta_1 = k_1 \cos \vartheta_1 = k \sqrt{(n_1^2 - \sin^2 \vartheta)},$$

$$\beta_2 = k_2 \cos \vartheta_2 = k \sqrt{(n_2^2 - \sin^2 \vartheta)}, \tag{30.2}$$

where ϑ_1 and ϑ_2 are the angles of refraction, and are connected with the angle of incidence ϑ by the relations

$$k \sin \vartheta = k_1 \sin \vartheta_1 = k_2 \sin \vartheta_2. \tag{30.3}$$

The expression for V_1 can now be obtained from Eq. 5.10 with the following changes in notation:

$$Z_3 \to Z, \quad Z_2 \to Z_1, \quad Z_1 \to Z_2, \quad \alpha_2 \to \beta_1.$$

As a result, we obtain

$$V_1 = \frac{(Z_1 + Z_2)(Z_1 - Z)\exp(-i\beta_1 d) + (Z_2 - Z_1)(Z_1 + Z)\exp(i\beta_1 d)}{(Z_1 + Z_2)(Z_1 + Z)\exp(-i\beta_1 d) + (Z_2 - Z_1)(Z_1 - Z)\exp(i\beta_1 d)}. \tag{30.4}$$

Considering the case $z > z_0$ for definiteness, and substituting the values of V_1 and V_2 into Eq. 27.2a, we obtain

$$\psi = k \int_{\Gamma_1} \frac{1}{N} [m_1 \beta S \cos(\beta z_0) + \beta_1 \sin(\beta z_0)]$$
$$\times \sin[\beta(h - z)] H_0^{(1)}(kr \sin \vartheta) \sin \vartheta \, d\vartheta, \tag{30.5}$$

where

$$S = \frac{m_1 \beta_2 \tan(\beta_1 d) + i m_2 \beta_1}{m_1 \beta_2 - i m_2 \beta_1 \tan(\beta_1 d)}, \quad N = m_1 \beta S \cos x + \beta_1 \sin x, \quad x = \beta h. \tag{30.6}$$

The case $z < z_0$ is obtained from Eq. 30.5 by interchanging the rôles of z and z_0. The limiting case of a double-layered medium can be obtained from Eq. 30.5 in two ways:

(a) With $m_2 \to m_1$, $\beta_2 \to \beta_1$, which signifies physically that media 1 and 2 are given the same properties;

(b) With $d \to 0$, which signifies the vanishing of the intermediate layer. The carrying out of these limiting transitions is left to the reader.

The integral (30.5) will be analyzed in exactly the same way as in § 27. The path of integration Γ_1 is transformed into Γ_2 (Fig. 122), which in turn is pulled to infinity and passes into the path Γ_4. As a result, the integral is reduced to the sum of the residues of the integrand at the poles determined by the equation

$$N = m_1 \beta S \cos x + \beta_1 \sin x = 0, \qquad (30.7)$$

plus the integral over the borders of the cut, extending from the branch point of the integrand.

The integrand in Eq. 30.5 contains the two radicals β_1 and β_2. However, this expression is an even function of β_1 and therefore does not depend on the choice of the sign of the radical. As a result, we are left with the non-singlevaluedness of the function due to the two possible choices of the sign of the square root $\beta_2 = k \sqrt{(n_2^2 - \sin^2 \vartheta)}$. The branch point of the two-sheeted Riemann surface on which the integrand is singlevalued is the point

$$\vartheta_2 = \arcsin n_2. \qquad (30.8)$$

The integral over the borders of the cut extending from this point is calculated in exactly the same way as the corresponding integral in § 27, and yields a lateral wave, propagating along the layer with the velocity c_2, and with an amplitude decreasing as $1/r^2$. The amplitude of the wave also depends on the parameters of all three media.

The case $d \to \infty$ is interesting. As d increases, the amplitude of the lateral wave decreases, becoming zero in the limit. However, at the same time, the roots of Eq. 30.7 are split into two families. One family gives the roots considered in § 29, which correspond to the normal modes in a two-layered medium. As $d \to \infty$, the roots of the other family become infinitely close to one another, and the sum of the corresponding normal modes degenerates into an integral, yielding a lateral wave, propagating along the layer with the velocity c_1.

2. The normal modes in a three-layered medium

We now proceed to the consideration of the normal modes given by the residues of the integrand in Eq. 30.5.

We set $$\psi = \psi_N + \psi_L,$$

where ψ_N is the part of the field due to the normal modes, and ψ_L is the lateral wave. Using the rule for calculating residues, Eq. 30.5 gives

$$\psi_N = 2\pi i k \sum_l F(\vartheta_l) \frac{H_0^{(1)}(kr \sin \vartheta_l)}{(\partial N / \partial \vartheta)_{\vartheta_l}} \sin \vartheta_l, \qquad (30.9)$$

where ϑ_l $(l = 1, 2, \ldots)$ are the roots of Eq. 30.7, and

$$F(\vartheta_l) = [m_1 \beta S \cos(\beta z_0) + \beta_1 \sin(\beta z_0)] \sin[\beta(h - z)].$$

It can be shown that

$$\frac{1}{k \sin \vartheta_l}\left(\frac{\partial N}{\partial \vartheta}\right)_{\vartheta_l} = \left\{\frac{1}{\beta \beta_1 \cos x}\left[(\beta_1^2 - \beta^2)\sin x \cos x - \beta_1^2 x\right.\right.$$

$$- \frac{1}{m_1}\beta d \cos^2 x(m_1^2 \beta^2 + \beta_1^2 \tan^2 x)$$

$$\left.\left.- \frac{im_2 \beta(\beta_2^2 - \beta_1^2)\cos^2 x(m_1^2 \beta^2 + \beta_1^2 \tan^2 x)}{\beta_2(m_1^2 \beta_2^2 - m_2^2 \beta_1^2)}\right]\right\}_{\vartheta = \vartheta_l}, \quad (30.10)$$

$$[m_1 \beta S \cos(\beta z_0) + \beta_1 \sin(\beta z_0)]_{\vartheta_l} = -\frac{\beta_1 \sin[\beta_l(h - z_0)]}{\cos x}.$$

As a result, we obtain

$$\psi_N = -2\pi i \sum_l m_1 \left\{\frac{\beta_1^2 \beta}{[\ ]}\sin[\beta(h - z)]\sin[\beta(h - z_0)]\right\}_{\vartheta_l} H_0^{(1)}(kr \sin \vartheta_l),$$

$$(30.11)$$

where the symbol [   ] stands for the square brackets in Eq. 30.10.

As in the case of the two-layered medium, each term in this sum represents an individual normal mode, characterized by its own propagation velocity and z-dependence of its amplitude.

The variation with depth of the amplitude of the first normal mode is shown in Fig. 136 for various frequencies, beginning with $f = 157$ cps. The critical frequency for this case is approximately 103.5 cps. The functional form of the velocity of sound vs. depth, on which the calculations of Ref. 63 were based, is shown at the left in the Figure; the phase velocity v_1 is also shown in the Figure.

3. *The phase and group velocities of the normal modes*

We shall limit ourselves to the case $c_2 > c$, which is of the greatest interest for practical applications. In this case, total internal reflection occurs at angles exceeding some critical value. This means that undamped normal modes will exist if the frequency is greater than the critical frequency.

We assume that the condition $c < c_1 < c_2$ is also fulfilled. It can be shown that in this case the phase velocity v_l of each normal mode varies from the value c_2 at the corresponding critical frequency to c in the limit of very high frequencies.

The phase velocity is determined from Eq. 30.7, which can also be written

$$\tan x = -\frac{m_1 \beta S}{\beta_1}, \quad x = \beta h, \tag{30.12}$$

where S is given by Eq. 30.6. The quantities β, β_1 and β_2, entering into these relations, are connected with the phase velocity of propagation along the layer v by the equations

$$\beta = k \sqrt{\left[1 - \left(\frac{c}{v}\right)^2\right]}, \quad \beta_1 = k_1 \sqrt{\left[1 - \left(\frac{c_1}{v}\right)^2\right]} \quad \text{for} \quad v > c_1,$$

$$\beta_1 = ik_1 \sqrt{\left[\left(\frac{c_1}{v}\right)^2 - 1\right]} \equiv is_1 \quad \text{for} \quad v < c_1,$$

$$\beta_2 = ik_2 \sqrt{\left[\left(\frac{c_2}{v}\right)^2 - 1\right]} \equiv is_2. \tag{30.13}$$

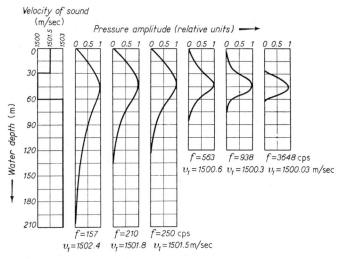

Fig. 136. The z-dependence of the amplitude of the first normal mode in a three-layered medium; v_1 is the phase velocity of the wave.

In the derivation of these equations, we took into account that

$$\frac{\omega}{v} = k \sin \vartheta,$$

where $k \sin \vartheta$ is the projection of the wave vector on the plane $z = \text{const}$. We also used Eqs. 30.2 and 30.3.

Equation 30.12 for the determination of the phase velocity v can now be written in the following two forms:

$$\tan x = -\frac{m_1\beta}{\beta_1}\frac{m_1 s_2 \tan(\beta_1 d) + m_2\beta_1}{m_1 s_2 - m_2\beta_1 \tan(\beta_1 d)} \quad \text{for} \quad v > c_1, \qquad (30.14)$$

$$\tan x = -\frac{m_1\beta}{s_1}\frac{m_1 s_2 \tanh(s_1 d) + m_2 s_1}{m_1 s_2 + m_2 s_1 \tanh(s_1 d)} \quad \text{for} \quad v < c_1. \qquad (30.15)$$

These equations can be used as a basis for numerical calculations.

We consider some special cases.

At the critical frequency, we have $v = c_2$ and $s_2 = 0$. Then Eq. 30.14 is reduced to the form

$$\tan x \tan y = \frac{m_1\beta}{\beta_1}, \qquad (30.16)$$

where

$$x = \beta h = kh\sqrt{\left[1-\left(\frac{c}{c_2}\right)^2\right]}, \quad y = \beta_1 d = k_1 d\sqrt{\left[1-\left(\frac{c_1}{c_2}\right)^2\right]}. \qquad (30.17)$$

Equation 30.16 allows us to find the critical frequency when the layer thicknesses and the characteristics of the media are known.

In particular, when $d \to 0$ (transition to the double-layered medium) Eq. 30.16 gives $\tan x \to \infty$, whence we obtain the relation between the critical frequencies ω_l $(l = 1, 2, \ldots)$ of the normal modes and the thickness of the layer

$$x_l = \left(l-\frac{1}{2}\right)\pi = \frac{\omega_l h}{c}\sqrt{\left[1-\left(\frac{c}{c_1}\right)^2\right]}, \qquad (30.18)$$

which was obtained in § 29.

§ 31. The Propagation of a Sound Pulse in a Liquid Layer

We saw in the preceding Section that the propagation of waves in a layer is always accompanied by dispersion. In connection with this, a pulse will change its form as it propagates. The task of the theory is to connect the characteristics of the layer with the way in which the pulse form changes. The solution of this problem has immediate applications, as, for example, the determination of the characteristics of the ocean bottom by the observation of the propagation of the sound of an explosion in the sea. Our presentation will be based on the work of Pekeris.[63] The bottom will be assumed to be liquid. See Ref. 181 for a generalization of the results to the case of a liquid layer lying on an elastic halfspace.

1. *Formal generalization of the solution to the case of a pulsed source*

An expression for the field of a normal mode from a point source of sound of frequency ω at an arbitrary point in a liquid layer was obtained in § 29 (see Eq. 29.13). We shall use the asymptotic representation of the Hankel function, and let k_l denote the wave number of the normal mode, where

$$k_l = k \sin \vartheta_l = k \sqrt{\left[1 - \left(\frac{x_l}{kh}\right)^2\right]}, \quad x_l = kh \cos \vartheta_l. \qquad (31.1)$$

Then the expression for the function ψ can be written in the form

$$\psi = \sum_{l=1}^{\infty} \exp\left[-i(\omega t - k_l r - \tfrac{1}{4}\pi)\right] Q_l(\omega), \qquad (31.2)$$

where

$$Q_l(\omega) = \frac{2\pi}{h} \sqrt{\left(\frac{2}{\pi k_l r}\right)} \left\{ \frac{x_l \sin\{x_l[1 - (z/h)]\} \sin\{x_l[1 - (z_0/h)]\}}{x_l - \sin x_l \cos x_l - [(\sin^2 x_l \tan x_l)/m^2]} \right\}. \qquad (31.3)$$

To obtain the complete solution of Eq. 31.2 for ψ, we must add on the lateral wave, given by the integral over the borders of the cut. However, we shall consider the field at great distances, where the lateral wave can be neglected. This approach is permissible because the amplitude of the lateral wave decreases with distance as $1/r^2$ or as $1/r$ when $khv \approx (2l + 1)(\pi/2)$, whereas the amplitude of the pulse decreases as $1/\sqrt{r}$ (we are considering the case $c_1 > c$, when total internal reflection occurs at the boundary of the layer).

Let the emitted pressure pulse have an arbitrary form, described by the function $f(t)$. We then expand this function in a Fourier integral†

$$f(t) = \frac{1}{2\pi} \int_{-\infty}^{\infty} e^{-i\omega t} g(\omega) \, d\omega, \qquad (31.4)$$

where the relation $g(-\omega) = g^*(\omega)$ holds because $f(t)$ is a real function.

The pressure at an arbitrary point will be given by the expression (the complex conjugate parts are omitted for the present)

$$p(r, z, t) = \frac{1}{2\pi} \int_{-\infty}^{\infty} g(\omega) \psi(\omega) \, d\omega = \sum_{l=1}^{\infty} p_l(r, z, t), \qquad (31.5)$$

where the expression

$$p_l(r, z, t) = \frac{1}{2\pi} \int_0^{\infty} \exp\{-i[\omega t - k_l(\omega) r - \tfrac{1}{4}\pi]\} g(\omega) Q_l(\omega, r, z) \, d\omega \qquad (31.6)$$

is the pressure in the lth normal mode.

---

† The function $f(t)$ prescribes the acoustic pressure, and not the acoustic potential in the initial pulse. Therefore, all the results obtained below will refer to the pressure and not to the potential.

We now investigate in more detail the propagation of an explosion pulse in a layer of water. In this case, we can set

$$f(t) = e^{-st} \quad \text{for} \quad t > 0 \atop f(t) = 0 \quad\quad \text{for} \quad t < 0 \Big\}, \quad \text{where} \quad s > 0. \tag{31.7}$$

Then

$$f(t) = \frac{1}{2\pi} \int_0^\infty \frac{e^{-i\omega t}}{s - i\omega} d\omega, \quad g(\omega) = \frac{1}{s - i\omega}, \tag{31.8}$$

$$p_l(r, z, t) = \frac{1}{2\pi} \int_0^\infty \frac{Q_l(\omega, r, z)}{s - i\omega} \exp\left\{-i[\omega t - k_l(\omega) r - \tfrac{1}{4}\pi]\right\} d\omega. \tag{31.9}$$

2. The group velocity of the normal modes

The integral in Eq. 31.9 is a superposition of sinusoidal waves, propagating with the various phase velocities $c_l = \omega/k_l(\omega)$. The amplitudes of these waves are

$$\frac{Q_l(\omega, r, z)}{|s - i\omega|} = \frac{Q_l(\omega, r, z)}{\sqrt{(s^2 + \omega^2)}},$$

and we can write the phase as

$$\phi(\omega, r, z) = \omega t - k_l(\omega) r - \tfrac{1}{4}\pi. \tag{31.10}$$

When ω, r and t are such that the phase varies rapidly with ω, the integral (31.9) will be small because the components of the integral with different ω will cancel. However, the cancellation will be minimum at values of ω for which $d\phi/d\omega = 0$ ("points of stationary phase"). At these frequencies, the mutual cancellation of the sinusoidal waves in the integrand will be least. Consequently, for given values of r and t, these frequencies will be predominant. The relation between t and r, corresponding to the point of stationary phase for a given ω, determines the propagation velocity of a group of waves composed of mutually interfering sinusoidal waves with frequencies close to ω, that is, the so-called group velocity U. The points of stationary phase are (the subscript l on k_l is omitted temporarily)

$$\frac{d\phi}{d\omega} = t - r \frac{dk(\omega)}{d\omega} = 0.$$

Hence, we obtain the well known expression for the group velocity:

$$U = \frac{r}{t} = \frac{d\omega}{dk}. \tag{31.11}$$

The last equation can also be written in the form

$$U = \frac{d\omega}{dk} = \frac{d(vk)}{dk} = v + k\frac{dv}{dk}, \qquad (31.12)$$

where v is the phase velocity. It is thus clear that the group velocity will always be smaller than the phase velocity, if the latter decreases with increasing frequency, as is the case here. The phase and group velocities of the first normal mode are shown in Fig. 137 as functions of

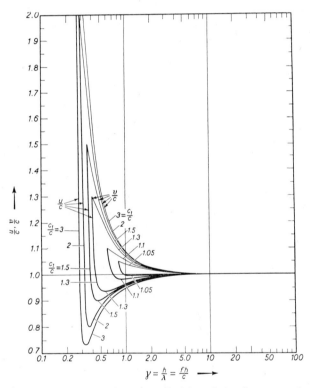

Fig. 137. The phase and group velocities of the first normal mode as functions of frequency for some concrete cases.

frequency, for several definite cases. The density of the bottom (assumed homogeneous) was chosen equal to 2. The velocity of sound in the bottom c_1 varies from $1.05c$ to $3c$, where c is the velocity of sound in the water. It is clear from the Figure that in all cases, the phase velocity passes through a minimum, and furthermore, that when $U < c$, two frequencies correspond to one value of the velocity. It will be shown below that these characteristics of the group velocity curves are of great significance in the study of the propagation of a pulse in a layer of water.

3. *Qualitative picture of pulse propagation*

Let us consider the qualitative picture of the propagation of a pulse in a layer, and the change of its form as it propagates.

For the present, we will take only the first normal mode into account. The group velocity as a function of frequency for this wave is shown schematically in Fig. 138. The curve in this Figure contains all the characteristic features of the actual group velocity curves, as, for example, those shown in Fig. 137.

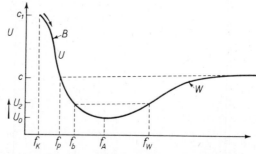

Fig. 138. Schematic illustration of the group velocity as a function of frequency.

The first arrival at the receiver is an almost sinusoidal wave of frequency f_k, where f_k is the critical frequency for the first normal mode (see Eq. 29.16)

$$f_k = \frac{c_1}{4h\sqrt{(1-n_1^2)}}, \qquad n_1 = \frac{c}{c_1}. \tag{31.13}$$

The arrival time of this wave is $t = r/c_1$ and corresponds to the propagation velocity in the bottom. In the course of time, the frequency and amplitude of this so-called ground wave must increase. The frequency increases because, as time goes on, portions of the wave arrive which have a smaller group velocity, which in Fig. 138 corresponds to moving along the left branch of the curve in the direction of higher frequencies (in the direction of the arrow). The reasons for the amplitude increase will be discussed later.

A very high frequency corresponding to the right branch (branch W) also arrives at some time $t = r/c$, in addition to the frequency f_k which corresponds to the left branch of the curve. We shall call this new wave, which arrives initially with very high frequencies, the *water wave*. Its initial phase propagates with the velocity of sound in the water.

The frequency of the water wave must decrease in the course of time, which corresponds to moving to the left along branch W. At an arbitrary instant of time $(r/c) < t < (r/U_0)$, where U_0 is the minimum group velocity, the first normal mode will be a superposition of two waves of different frequencies f_g and f_w, which we shall call the ground and water wave frequencies, respectively, as indicated by the subscripts. At $t = r/U_0$, the frequencies of these waves coincide, and we shall have a wave process with the single frequency f_A. This part of the arriving wave can be called the Airy wave. Its mathematical description is similar to Airy's calculation of the field near a caustic.

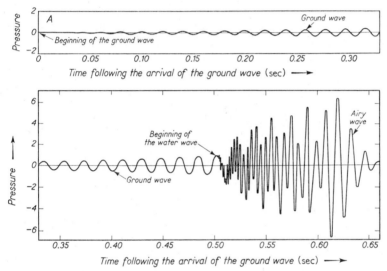

Fig. 139. The time dependence of the acoustic pressure for a concrete case.

The theoretically calculated time dependence of the sound pressure is shown in Fig. 139 for a definite case. In the Figure, we see the arrival of the ground wave (at $t = 0$), gradually increasing in amplitude and frequency, the arrival of the water wave (at $t \approx 0.5$ sec.), then the superposition of the two waves, with the ground wave continuing to increase in amplitude and frequency while the water wave increases in amplitude and decreases in frequency with time. The end of the record corresponds to the Airy wave, with a constant frequency and an exponentially decreasing amplitude.

4. *The complete theory of pulse propagation*

In the investigation of integral (31.9), we took into account that $Q_l(\omega)/(s - i\omega)$ is a comparatively slowly varying function of ω, whereas the exponential is a rapidly oscillating function. The significant part of the integral is obtained from the regions near points of stationary phase, determined by the equation $d\phi/d\omega \equiv \dot\phi = 0$, where the phase ϕ is given by Eq. 31.10.

In the neighborhood of a point of stationary phase ω_0, we can set $\omega = \omega_0 + u$, where u is a small quantity. Expanding $\phi(\omega)$ in a power series in u, we obtain

$$\phi(\omega) = \phi(\omega_0) - \frac{r\ddot{k}(\omega_0)}{2!} u^2 - \frac{r\dddot{k}(\omega_0)}{3!} u^3 - \dots. \qquad (31.14)$$

Only small values of u, both positive and negative, will be significant in the integral with respect to u, because the path of integration passes through the point ω_0, but does not begin there. Purely formally, the integration with respect to u can be taken between the limits $-\infty$ to $+\infty$, where, according to the above considerations, only small values of u will contribute significantly to the integral.

Thus,

$$\frac{1}{2\pi} \int_0^\infty \frac{Q_l}{s - i\omega} \exp[-i\phi(\omega)] \, d\omega = \frac{1}{2\pi} \frac{Q_l(\omega_0) \exp[-i\phi(\omega_0)]}{s - i\omega_0}$$

$$\times \int_{-\infty}^{+\infty} \exp\left\{ i\left[\frac{r\ddot{k}(\omega_0)}{2} u^2 + \frac{r\dddot{k}(\omega_0)}{6} u^3 + \dots \right] \right\} du,$$

where the dot denotes the derivative with respect to ω.

According to Pekeris,[63] we have†

$$\int_{-\infty}^{+\infty} \exp\left\{ i\left[\frac{r\ddot{k}}{2} u^2 + \frac{r\dddot{k}}{6} u^3 + \dots \right] \right\} du$$

$$\approx \sqrt{\left(\frac{2\pi}{r |\ddot{k}|} \right)} \exp[\pm i(\pi/4)] \left\{ 1 - \frac{i}{r} \left[-\frac{5(\dddot{k})^2}{24(\ddot{k})^3} + \frac{\ddddot{k}}{8(\ddot{k})^2} \right] + 0\left(\frac{1}{r^2} \right) \right\}, \qquad (31.15)$$

where in the exponent, we take the plus sign if $\dddot{k} > 0$, and the minus sign if $\dddot{k} < 0$.

Thus,

for $\dddot{k} > 0$:

$$p_l = \frac{1}{2\pi} \int_0^\infty \frac{Q_l(\omega)}{s - i\omega} \exp[-i\phi(\omega)] \, d\omega = \frac{Q_l(\omega_0) i}{2\pi(s - i\omega_0)} \exp\{i[k(\omega_0) r - \omega_0 t]\}$$

$$\times \sqrt{\left[\frac{2\pi}{r |\ddot{k}(\omega_0)|} \right]} \left\{ \ \right\}; \qquad (31.16)$$

----

† Let us note that in the exact evaluation of the integral, the terms of the order $1/r$ will include derivatives of $Q(\omega)$ and $g(\omega)$, which Pekeris neglects without qualification.

for $\ddot{k} < 0$:

$$p_l = \frac{Q_l(\omega_0)}{2\pi(s - i\omega_0)} \exp\{i[k(\omega_0)\,r - \omega_0 t]\}\sqrt{\left[\frac{2\pi}{r\,|\,\ddot{k}(\omega_0)\,|}\right]}\{\quad\}, \quad (31.17)$$

where the symbol $\{\}$ denotes the expression in curly brackets in Eq. 31.15.

We now recall that a complex conjugate term, in which the time dependence is given by the factor $e^{i\omega t}$, is also involved in Eq. 31.4. Following the same line of argument as above, we obtain expressions in this case which coincide with Eqs. 31.16 and 31.17, the only difference being that the sign in front of i will be reversed. Therefore, to obtain the complete expressions for the pressure, we must add the complex conjugate expressions to Eqs. 31.16 and 31.17. As a result (if, in addition, we set $\{\} \approx 1$, that is, we neglect the second and third terms in the parentheses) we obtain

$\ddot{k} > 0$:

$$p_l(r, z, t) = \frac{2Q_l(\omega_0)\cos[\omega_0 t - rk(\omega_0) - \arctan(\omega_0/s) - \tfrac{1}{2}\pi]}{[2\pi r\,|\,\ddot{k}(\omega_0)\,|\,(s^2 + \omega_0^2)]^{\frac{1}{2}}}; \quad (31.18)$$

$\ddot{k} < 0$:

$$p_l(r, z, t) = \frac{2Q_l(\omega_0)\cos[\omega_0 t - rk(\omega_0) - \arctan(\omega_0/s)]}{[2\pi r\,|\,\ddot{k}(\omega_0)\,|\,(s^2 + \omega_0^2)]^{\frac{1}{2}}}. \quad (31.19)$$

The case $\ddot{k} > 0$ is realized on the branch of the dispersion curve corresponding to the ground wave (to the left of the minimum of the curve in Fig. 138). In fact, on this branch, the group velocity U decreases with increasing frequency, i.e. the quantity $1/U = dk/d\omega$ increases. Consequently, $d^2 k/d\omega^2 > 0$. In the same way, it can be shown that the case $\ddot{k} < 0$ is realized on the branch of the dispersion curve corresponding to the water wave. Thus, Eq. 31.18 gives the ground wave, and Eq. 31.19 gives the water wave. Each of these equations represents a train of waves, modulated in frequency and amplitude. At each instant of time, the frequency ω_0 is found from the equation

$$Ut - r = 0, \quad \frac{1}{U} = \frac{dk}{d\omega}.$$

In the majority of cases, it is convenient to solve this equation graphically, after first having constructed the dispersion curve $U(\omega)$.

We now estimate the limits of applicability of Eqs. 31.18 and 31.19. In deriving these equations, we assumed that the condition

$$\frac{1}{r}\left[-\frac{5(\ddot{k})^2}{24(\ddot{k})^3} + \frac{\dddot{k}}{8(\ddot{k})^2}\right] \ll 1 \quad (31.20)$$

was fulfilled. This inequality is satisfied everywhere, except at small distances and in the region of the minimum of the group velocity curve, where $\ddot{k} = 0$. The neighborhood of this point requires special arguments, and will be treated in the next Section.

5. Airy waves

In the neighborhood of the group velocity minimum, where $\ddot{k}(\omega_0) = 0$, we set $\omega = \omega_0 + u$ in integral (31.9), and use the following power series expansion in u:

$$\phi(\omega) = \omega t - k(\omega)\,r - \tfrac{1}{4}\pi = \phi(\omega_0) + au + bu^3 + du^4 + \ldots, \quad (31.21)$$

where
$$a = t - r\dot{k}_0 = t - \frac{r}{U_0}, \quad b = -\frac{r}{6}\,\dddot{k}_0, \quad d = -\frac{r}{24}\,\ddddot{k}_0. \quad (31.22)$$

The subscript 0 on the derivatives of k signifies that the derivatives are to be taken at the fixed point $\omega = \omega_0$, at which $\ddot{k}(\omega) = 0$. Consequently, in distinction from the preceding Section, these derivatives are constants, depending neither on t nor on r.

Remembering that the complex conjugate expression must be added to Eq. 31.9, we obtain

$$p_l = \frac{1}{2\pi} \int_0^\infty Q_l(\omega) \left[\frac{e^{-i\phi(\omega)}}{s - i\omega} + \frac{e^{i\phi(\omega)}}{s + i\omega} \right] d\omega$$

$$= \frac{1}{\pi} \int_0^\infty \frac{Q_l(\omega)}{\sqrt{(s^2 + \omega^2)}} \cos\left[\phi(\omega) - \arctan\frac{\omega}{s} \right] d\omega.$$

We then have, approximately,

$$p_l \approx \frac{1}{\pi} \frac{Q_l(\omega_0)}{\sqrt{(s^2 + \omega_0^2)}} \int_{-\infty}^{+\infty} \cos\left[A + au + bu^3 + du^4 \right] du,$$

$$A \equiv \left[\omega_0 t - k(\omega_0)\,r - \arctan\frac{\omega_0}{s} \right] - \frac{\pi}{4}. \quad (31.23)$$

Furthermore:

$$\int_{-\infty}^{+\infty} \cos\left[A + au + bu^3 + du^4 \right] du$$

$$= 2\cos A \int_0^\infty \cos(au + bu^3)\cos(du^4)\,du$$

$$- 2\sin A \int_0^\infty \cos(au + bu^3)\sin(du^4)\,du$$

$$\approx 2\cos A \cdot T(a, b) + 2d\sin A \frac{\partial^2 T}{\partial a\,\partial b}, \quad (31.24)$$

where the integrals from $-\infty$ to $+\infty$ of the functions that are odd in u gave zero, and the integrals of the even functions are replaced by twice the integral from 0 to ∞. Furthermore, we have assumed that $\sin(du^4) \approx du^4$, and have used the notation

$$T(a,b) = \int_0^\infty \cos(au + bu^3)\, du = \frac{\pi}{3(2b)^{\frac{1}{3}}} E(v), \qquad (31.25)$$

where $E(v)$ denotes the Airy function, defined as

$$E(v) = v^{\frac{1}{2}}[I_{-\frac{1}{3}}(v) + I_{\frac{1}{3}}(v)], \quad t < \frac{r}{U_0}, \qquad (31.26)$$

$$E(v) = v^{\frac{1}{2}}[I_{-\frac{1}{3}}(v) - I_{\frac{1}{3}}(v)], \quad t > \frac{r}{U_0}, \qquad (31.27)$$

and
$$v = \frac{2}{3\sqrt{3}}\left|\frac{a^3}{b}\right|^{\frac{1}{2}} = \frac{4\sqrt{\pi}}{3\sqrt{(-\dot{Z}_0)}}\left(\frac{r}{h}\right)|\tau - \tau_m|^{\frac{3}{2}}. \qquad (31.28)$$

It was convenient to express v in terms of nondimensional parameters, for which we introduced the notation

$$Z = \frac{c^2}{h}\ddot{k}, \quad Z_0 = \frac{c^2}{h}\ddot{k}_0, \quad \tau = \frac{ct}{r} - 1, \quad \tau_m = \frac{c}{U_0} - 1,$$

$$\gamma = \frac{\omega h}{2\pi c} \quad \text{and} \quad \dot{Z}_0 = \frac{\partial Z_0}{\partial \gamma}. \qquad (31.29)$$

Furthermore, we find

$$\frac{\partial^2 T}{\partial a\, \partial b} = \frac{\pi}{b^{\frac{2}{3}}}\left(\frac{2}{3}\right)^{\frac{4}{3}} G(v), \qquad (31.30)$$

where
$$G(v) = \frac{3^{\frac{1}{3}}}{4}\{-\tfrac{2}{3}v^{\frac{2}{3}}[I_{-\frac{2}{3}}(v) - I_{\frac{2}{3}}(v)] + \tfrac{1}{2}v^{\frac{1}{3}}[I_{-\frac{1}{3}}(v) + I_{\frac{1}{3}}(v)]\},$$

$$t < \frac{r}{U_0}; \qquad (31.31)$$

$$G(v) = \frac{3^{\frac{1}{3}}}{4}\{-\tfrac{2}{3}v^{\frac{2}{3}}[I_{-\frac{2}{3}}(v) - I_{\frac{2}{3}}(v)] + \tfrac{1}{2}v^{\frac{1}{3}}[I_{-\frac{1}{3}}(v) - I_{\frac{1}{3}}(v)]\},$$

$$t > \frac{r}{U_0}. \qquad (31.32)$$

As a result,

$$\int_{-\infty}^{+\infty} \cos[A + au + bu^3 + du^4]\, du$$
$$\approx \frac{2\pi \cos A}{3(2b)^{\frac{1}{3}}} E(v) + \frac{2\pi d \sin A}{b^{\frac{2}{3}}(3/2)^{\frac{4}{3}}} G(v), \qquad (31.33)$$

where the second term can be treated as a correction term. It can be neglected under the condition that

$$\left|\frac{2^{\frac{2}{3}} d\, G(v)}{3^{\frac{1}{3}} b^{\frac{2}{3}} E(v)}\right| = \left|\frac{\dddot{k}G(v)}{r^{\frac{1}{3}}(-\ddot{k})^{\frac{4}{3}} E(v)}\right| = \frac{1}{(2\pi)^{\frac{2}{3}}}\left(\frac{h}{r}\right)^{\frac{4}{3}}\left|\frac{\dddot{Z}G(v)}{(-\ddot{Z})^{\frac{4}{3}} E(v)}\right| \ll 1. \qquad (31.34)$$

For this condition to be fulfilled, the distance must be sufficiently great, and the time t must not differ too much from $t_m = r/U_0$.

As a result, we find that in the neighborhood of t_m, in the case of an exponential pulse, the pressure p_l due to the lth normal mode, is expressed as follows:

$$p_l(r, z, t) = \frac{4 \cos\left[\omega_0 t - k_0 r - \arctan(\omega_0/s) - \frac{1}{4}\pi\right] E(v)}{3^{\frac{2}{3}} h \sqrt{[(s^2 + \omega_0^2)(k_0/2\pi)]}(-\ddot{k}_0)^{\frac{1}{3}} r^{\frac{5}{6}}}$$

$$\times \left\{ \frac{x_l \sin\{x_l[1 - (z/h)]\} \sin\{x_l[1 - (z_0/h)]\}}{x_l - \sin x_l \cos x_l - [(\sin^2 x_l \tan x_l)/m^2]} \right\}_{\omega_0}. \qquad (31.35)$$

We recall that ω_0 is found from the equation $\ddot{k}(\omega_0) = 0$, where $k(\omega)$ is the abbreviated notation for $k_l = \omega/v_l$, and v_l is the phase velocity of the normal mode of order l.

Equation 31.35 describes the final stage of the pulse, and represents a wave (it can be called the Airy wave) with a constant frequency, and an amplitude which decreases as the Airy function $E(v)$.

Using Eq. 31.35, we can calculate the pressure p_l for all significant $t > t_m$, and for a certain time interval preceding t_m. Equation 31.35 cannot be used for earlier instants of time because condition (31.34) will not be satisfied (v will be too large). However, Eqs. 31.18 and 31.19 are usually applicable for these values of t.

6. *The degree of excitation of the various normal modes*

Until now, our arguments have referred to one normal mode. Furthermore, it is frequently assumed in practice that the principal rôle is played by the first normal mode, and that the modes of higher order are only weakly excited. It can be shown, however, that this approach is not always valid.

Using the theory developed in the preceding Sections, we can determine exactly the relative strengths of the waves of various orders in the various stages of the pulse.

Substituting Eq. 31.3 for Q_l into Eqs. 31.18 and 31.19, and confining ourselves to the case in which the source and the receiver are situated on the bottom ($z = z_0 = 0$), we obtain the amplitude of the ground and water waves

$$|p_l| = \frac{4c}{hr \sqrt{(s^2 + \omega_0^2)}} G_l,$$

where
$$G_l = \frac{x_l \sin^2 x_l}{[x_l - \sin x_l \cos x_l - (\sin^2 x_l \tan x_l/m_1^2)] c \sqrt{(k |\ddot{k}|)}}. \qquad (31.36)$$

Thus, the amplitude of the normal mode at various frequencies is determined by the function G_l. We recall that, at frequencies for which

$\ddot{k} > 0$, we have the stage of the pulse called the ground wave, and for $\ddot{k} < 0$, we have the stage of the pulse called the water wave. It is convenient to introduce different symbols, G_l^g and G_l^w, for these waves, so that

$$\text{for } \ddot{k} > 0: \ G_l \equiv G_l^g,$$

$$\text{for } \ddot{k} < 0: \ G_l \equiv G_l^w. \qquad (31.37)$$

The amplitudes of the ground and water waves, calculated by Eq. 31.36 to the third order, are shown in Fig. 140 for various stages of the pulse. The abscissa represents the quantity $(T - T_0)/T_0$, which characterizes

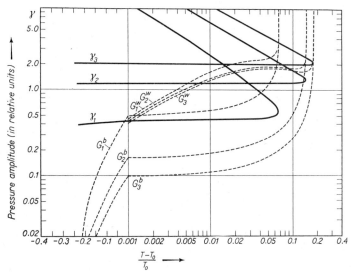

Fig. 140. The amplitudes of the bottom and water waves to the third order inclusively, at various stages of the received pulse.

the time elapsed after the arrival of the water wave. Here, T is the time reckoned from the instant of the explosion, and $T_0 = r/c$ is the time of arrival of the water wave. The following characteristics of the ground were used: $\rho_1/\rho = 2$, $c_1 = 1950$ m/sec.

The dotted curves in the Figure represent the curves G_1^g, G_2^g and G_3^g, which show the amplitudes of the ground waves of the first three orders, in arbitrary units. The analogous curves G_1^w, G_2^w and G_3^w for the water waves are also shown as dotted curves in the Figure.

It is clear from the Figure that in the ground wave, during almost the entire time of passage of the pulse, the amplitude of the first normal mode is approximately three times that of the second normal mode, and almost five times that of the third normal mode. On the other hand, the amplitudes of the waves of various orders in the water wave differ only slightly. It can therefore be expected that a receiver with a uniform frequency sensitivity will record a water wave consisting of several normal modes of almost the same amplitude, but different frequencies. Such a record will be quite complicated. However, by using a channel sensitive only to sufficiently low frequencies, we can separate out the first normal mode. The frequencies with which the first three normal modes arrive at the receiver are shown by the solid lines in Fig. 140, over the time interval covered by the Figure. The ordinate represents the quantity $\gamma_l = h/\lambda_l$, where λ_l is the wavelength in the water, and is connected with the associated frequency by the relation $\lambda_l = c/f_l$. We see that as the order of the normal mode increases, the quantity γ_l, and therefore the frequency, increases.

7. *The pulse characteristics from which conclusions can be drawn concerning the parameters of the bottom*

To conclude this Paragraph, we shall look into the question of how information regarding the parameters of the bottom can be obtained from the characteristics of the pulse, recorded at a sufficiently great distance from the explosion. We shall confine ourselves to the case in which the bottom can be treated as homogeneous, and shall refer the reader elsewhere[63] for the interesting details in the case of a two-layered bottom, for example.

Arrival time of the ground wave. As we saw above, the propagation velocity in the bottom, $c_1 = r/T_1$, can be determined by the arrival time T_1 of the ground wave. Experiment shows that the instant of arrival of the ground wave is determined most accurately by the low-frequency channel.

Dispersion in the water wave. Let us suppose that we are determining the arrival of the water wave with several channels, having different upper limits to their pass bands. Then, in accordance with the dispersion curves in Figs. 137 and 138, the arrival times will be different in the various channels. By constructing the experimental frequency dependence of the propagation velocity, and comparing it with the theoretical curves of the type shown in Fig. 137 for various parameters of the bottom, we can draw some conclusions regarding the characteristics of the bottom. In the case of an inhomogeneous bottom, the parameters obtained in this way will refer to the uppermost region of the bottom, the thickness of this region being of the order of a wavelength (corresponding to the upper boundary of the pass band) or less, since at high frequencies, the water waves do not penetrate very deeply into the bottom (see, for example, Fig. 126).

The frequency of the ground wave at the instant the water wave arrives.
It is not difficult experimentally to determine the frequency of the
ground wave at the instant immediately preceding the arrival of the
water wave ($t \approx 0.5$ sec in Fig. 139). This is the frequency f_R (see Fig.
138). It also has a direct connection with the characteristics of the
bottom. Figure 141 shows curves of $\gamma_R = h/\lambda_R = f_R h/c$ as a function
of the ratio c_1/c, for $m = \rho_1/\rho = 2$.

The critical frequency. The frequency of the ground wave, measured
at the very instant of its arrival will be the critical frequency f_k given
by Eq. 29.16, that is, the lowest frequency that can propagate without
severe attenuation in the layer. The dependence of $\gamma_k = h/\lambda_k$ on c_1/c
is shown graphically in Fig. 141.

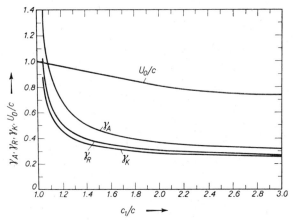

Fig. 141. Various characteristics of the received pulse as functions of the
ratio of the velocity of sound in the water and the bottom, for $\rho_1/\rho = 2$.

The frequency of the Airy wave. As has already been said, the frequency of the water wave decreases with time as the pulse passes,
whereas the frequency of the ground wave increases continuously,
until both frequencies become equal to one and the same frequency f_A,
called the Airy frequency, which also depends on the characteristics
of the bottom. At the instant that the wave with the Airy frequency
arrives, the amplitude of the water wave reaches a maximum, and as
time goes on, begins to decrease with its frequency remaining constant.
The Airy frequency f_A as well as the time of its arrival $t = r/U_0$,
associated with the minimum group velocity, can be determined
experimentally. Curves showing $\gamma_A = h/\lambda_A$ and U_0/c as functions of
c_1/c are given in Fig. 141.

Thus, by a careful analysis of the experimental records of pulses, we can obtain sufficiently complete information concerning the parameters of the bottom.

§ 32. Simplified Method for the Determination of the Characteristics of the Normal Modes[37]

It is clear from the two preceding Paragraphs that an important problem in the theory of wave propagation in layers is the determination of the characteristics of the various normal modes which can propagate in the layer. Among these characteristics are the phase and group velocities of the normal mode along the layer, its attenuation, and also the amplitude distribution of the normal mode over the layer thickness. We will show in this Paragraph that these characteristics can be obtained in a simpler and more elementary way than in the approach indicated above, which involved the investigation of integral expressions for the field in the complex plane. The method presented below permits us to draw some qualitative conclusions concerning the degree of excitation of the various normal modes, which in some cases is of interest. We shall consider the acoustic case. The method can be extended to the electromagnetic case without difficulty.

1. Sound waves in a liquid layer

It was shown in § 27 (see Eq. 27.22) that the characteristics of the normal modes in a layer are determined by the equation

$$V_1 V_2 \exp{(2ikh \cos \vartheta)} = 1, \tag{32.1}$$

where V_1 and V_2 are the reflection coefficients of a plane wave at the lower and upper boundaries of the layer, respectively, and are functions of the angle of incidence ϑ.

We obtain an infinite set of solutions ϑ_l, $l = 1, 2, \ldots$ of the last equation, where ϑ_l is the slope angle of the two plane waves which can represent the normal mode. The wave number of the normal mode, propagating along the layer, will be

$$k_l = k \sin \vartheta_l. \tag{32.2}$$

If k_l is real, i.e. the wave propagates along the layer without damping, its phase and group velocities will be

$$v = \frac{\omega}{k_l} = \frac{c}{\sin \vartheta_l}, \quad U = \frac{d\omega}{dk_l}, \tag{32.3}$$

where c is the wave velocity in the medium comprising the layer.

If k_l is complex, and if, furthermore, its imaginary part is considerably smaller than its real part, then the imaginary part will be the damping coefficient, and the real part, substituted into Eqs. 32.3 in place of k_l, will give v and U.

We shall now show that Eq. 32.1 can be obtained from simple physical considerations.

At comparatively large distances from the source, the waves can be considered plane. Remembering that as a result of wave reflection at the boundaries of the layer, there can be waves propagating in the positive and negative z-directions, the field in the layer, for a given frequency $\omega = kc$ can be written in the form

$$\phi = \phi_+ + \phi_-, \tag{32.4}$$

where

$$\phi_+ = A \exp[i(k_z z + k_x x)], \quad \phi_- = B \exp[i(-k_z z + k_x x)],$$

$$k_x = k \sin \vartheta, \quad k_z = k \cos \vartheta,$$

with ϑ an arbitrary angle, and A and B arbitrary constants. The acoustic potential is denoted by ϕ. The symbol ϕ_+ represents a wave propagating at an angle ϑ with respect to the positive z-axis. Similarly, ϕ_- is a wave propagating at the angle $\pi - \vartheta$ with respect to the z-axis.

Equation 32.4 satisfies the wave equation. We shall now demand that it also satisfy the boundary conditions.

The boundary conditions can be expressed in terms of the reflection coefficients V_1 and V_2. For this purpose, let us consider Fig. 142. At the

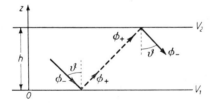

Fig. 142. On the approximate determination of the characteristics of the normal modes.

lower boundary of the layer, ϕ_- is the incident wave, and ϕ_+ is the reflected wave. Therefore, the condition

$$\left(\frac{\phi_+}{\phi_-}\right)_{z=0} = V_1,$$

must be satisfied, where V_1 is the reflection coefficient at the lower boundary. Substituting the expressions for ϕ_- and ϕ_+ given in Eqs. 32.4, we obtain

$$\frac{A}{B} = V_1. \tag{32.5}$$

At the upper boundary, ϕ_+ is the incident wave, and ϕ_- is the reflected wave. Therefore, the relation

$$\left(\frac{\phi_-}{\phi_+}\right)_{z=h} = V_2$$

must hold. Using Eq. 32.4, we obtain

$$\frac{B}{A}\exp\left(-2ik_z h\right) = V_2. \tag{32.5a}$$

Multiplying Eqs. 32.5 and 32.5a together, we obtain

$$V_1 V_2 \exp\left(2ik_z h\right) = 1, \tag{32.6}$$

which agrees with Eq. 32.1.

For any $\vartheta = \vartheta_l$, where ϑ_l $(l = 1, 2, \ldots)$ is a solution of the last equation, Eq. 32.4 for ϕ satisfies the wave equation and the boundary conditions. Therefore, the complete solution for the field will involve a sum of these expressions, with various values of l.

It is interesting to examine the limiting form of Eq. 32.6 as $h \to \infty$. We shall assume that the wave number k has some small imaginary part. We then have $\lim\limits_{h \to \infty} \exp\left(2ik_z h\right) = 0$ (in traveling from one boundary of the layer to the other, the wave is totally attenuated). Then Eq. 32.6 takes the form

$$\frac{1}{V_1 V_2} = 0$$

or, separately,

$$\frac{1}{V_1} = 0, \quad \frac{1}{V_2} = 0.$$

In some cases, this equation has no solutions (for example, when the liquid layer is bounded on both sides by liquid halfspaces), which signifies that, because of absorption in the layer, the normal modes, which are obtained as a result of multiple reflections from the boundaries of the layer, cannot form. However, if the layer is bounded by elastic halfspaces, the last equation will be satisfied for certain values, $\vartheta = \vartheta_1$ and $\vartheta = \vartheta_2$. Here, ϑ_1 and ϑ_2 are complex, and it can turn out that $V_1(\vartheta_1) \to \infty$ and $V_2(\vartheta_2) \to \infty$, separately. The values of ϑ_1 and ϑ_2 obtained in this way correspond to surface waves (similar to the Rayleigh wave), propagating along the lower and upper boundaries of the layer, respectively.

For any value of l, the constants A and B must be related by Eq. 32.5 or Eq. 32.5a. These two equations are equivalent at $\vartheta = \vartheta_l$. Therefore, ϕ can be written in either of the following two forms:

$$\phi = \phi_+ + \phi_- = \sum B_l[V_1(\vartheta_l)\exp(ikz\cos\vartheta_l) + \exp(-ikz\cos\vartheta_l)]$$
$$\times \exp(ikx\sin\vartheta_l), \quad (32.7)$$

or $\quad \phi = \phi_+ + \phi_- = \sum_l A_l\{\exp(ikz\cos\vartheta_l) + V_2(\vartheta_l)\exp[ik(2h-z)\cos\vartheta_l]\}$
$$\times \exp(ikx\sin\vartheta_l). \quad (32.8)$$

As a result, we have obtained the solution of the problem in the form of a sum of the normal modes. The expressions in the square brackets describe the z-dependence of the amplitude of each of the normal modes.

The coefficients B_l in Eq. 32.7, or A_l in Eq. 32.8, determine the degree of excitation of the corresponding normal mode. They are functions of the position and strength of the source, and cannot be found by the methods developed in this Paragraph.†

As an example, we consider a liquid layer bounded from below by a homogeneous elastic halfspace. The upper boundary of the layer is assumed to be free (for example, the boundary between water and air). The reflection coefficient at the upper boundary will be $V_2 = -1$. The reflection coefficient V_1 at the boundary between the liquid and the elastic halfspace is given by Eq. 4.25. Substituting the values of V_1 and V_2 into Eq. 32.6, we can write it in the form

$$i\tan k_z h = \frac{1}{Z}(Z_1\cos^2 2\gamma_1 + Z_t\sin^2 2\gamma_1).$$

We have thus obtained the dispersion equation for a liquid layer lying on an elastic halfspace. This equation was studied in detail by Sherman.[79]

In the present case, the equation $1/V_1 = 0$ is the dispersion equation for the surface waves on the liquid–solid boundary. It was studied in §4, Section 6. In particular, when $\rho/\rho_1 \to 0$, we obtain the well-known equation for the Rayleigh waves.

As another interesting example, we consider the dispersion equation for a triple-layered liquid halfspace (see §30). To apply Eq. 32.6 to this case, we set $V_2 = -1$, and for V_1 we use the expression, found in §5,

---

† The arguments used above, which led to Eqs. 32.7 and 32.8, can be used to obtain the field of the normal modes, but they are not applicable to the case of the lateral wave. In the latter case, it is essential to take the curvature of the wave front into account.

for the coefficient of reflection from a layer separating two different media. As a result, we obtain the dispersion equation (30.12).

2. Normal modes in a solid layer

The method developed above can also be applied to a solid layer. An interesting feature of this case is the presence in the layer of longitudinal and transverse waves, interacting with one another at the boundaries of the layer.[†]

Using the same assumptions as above, the potentials of the longitudinal and transverse waves in the layer can be written in the form

$\phi = \phi_+ + \phi_-$ the potential of the longitudinal waves (P-waves),

$\psi = \psi_+ + \psi_-$ the potential of the transverse waves (S-waves),

where
$$\phi_+ = A \exp(ik_z z), \quad \phi_- = B \exp(-ik_z z),$$
$$\psi_+ = C \exp(i\kappa_z z), \quad \psi_- = D \exp(-i\kappa_z z) \tag{32.9}$$

and
$$k_z = k \cos\vartheta, \qquad \kappa_z = \kappa \cos\gamma, \tag{32.10}$$

with the angle γ related to the angle ϑ by
$$k \sin\vartheta = \kappa \sin\gamma = k_x; \tag{32.11}$$

and k and κ are the wave numbers of the longitudinal and transverse waves, respectively. We omitted the common factor $\exp(ik_x x)$ in Eqs. 32.9.

We now consider the relations at the boundaries of the layer. As in § 24, we denote the various reflection coefficients at the lower boundary of the layer by $V_{11}^{(1)}$, $V_{21}^{(1)}$, $V_{22}^{(1)}$ and $V_{12}^{(1)}$, where $V_{11}^{(1)}$ characterizes the amplitude of the potential of the reflected P-wave with an incident P-wave, $V_{21}^{(1)}$ characterizes the amplitude of the P-wave with an incident S-wave, $V_{12}^{(1)}$ characterizes the amplitude of the S-wave with an incident P-wave, and $V_{22}^{(1)}$ characterizes the amplitude of the S-wave with an incident S-wave.

The coefficients $V_{11}^{(1)}$, ... are functions of the angles of incidence ϑ and γ, and are given by Eqs. 24.23–24.29.

At the lower boundary of the layer, ϕ_- and ψ_- are the incident waves, and ϕ_+ and ψ_+ are the reflected waves. The amplitude of the reflected wave ϕ_+ will be determined by the amplitudes of both incident waves ϕ_- and ψ_-, and by the reflection coefficients $V_{11}^{(1)}$ and $V_{21}^{(1)}$. The same holds for ψ_+.

[†] We consider the case in which the particle velocity in the medium is parallel to the xz-plane. If the particle velocity was perpendicular to this plane, the problem would become completely similar to that presented in the preceding Section.

Thus, at the lower boundary of the layer, we have the relations

$$z = 0 \begin{cases} \phi_+ = V^{(1)}_{11} \phi_- + V^{(1)}_{21} \psi_-, \\ \psi_+ = V^{(1)}_{12} \phi_- + V^{(1)}_{22} \psi_-. \end{cases} \tag{32.12}$$

At the upper boundary, on the other hand, ϕ_+ and ψ_+ will be the incident waves, and ϕ_- and ψ_- will be the reflected waves. We therefore have the analogous formulas

$$z = h \begin{cases} \phi_- = V^{(2)}_{11} \phi_+ + V^{(2)}_{21} \psi_+, \\ \psi_- = V^{(2)}_{12} \phi_+ + V^{(2)}_{22} \psi_+, \end{cases} \tag{32.13}$$

where $V^{(2)}_{11}, ..., V^{(2)}_{22}$ are the reflection coefficients at the upper boundary.

Substituting Eqs. 32.9 into Eqs. 32.12 and 32.13, we obtain a system of four homogeneous equations in A, B, C and D

$$A = V^{(1)}_{11} B + V^{(1)}_{21} D,$$

$$C = V^{(1)}_{12} B + V^{(1)}_{22} D,$$

$$B \exp(-ik_z h) = V^{(2)}_{11} A \exp(ik_z h) + V^{(2)}_{21} C \exp(i\kappa_z h),$$

$$D \exp(-i\kappa_z h) = V^{(2)}_{12} A \exp(ik_z h) + V^{(2)}_{22} C \exp(i\kappa_z h). \tag{32.14}$$

This system will have a nonzero solution only when the determinant of the system is equal to zero. As a result, after some fairly straightforward transformations, we obtain

$$\exp[i(k_z + \kappa_z) h] [V^{(1)}_{11} V^{(1)}_{22} - V^{(1)}_{21} V^{(1)}_{12}] [V^{(2)}_{11} V^{(2)}_{22} - V^{(2)}_{21} V^{(2)}_{12}]$$
$$- V^{(1)}_{22} V^{(2)}_{22} \exp[i(\kappa_z - k_z) h] - V^{(1)}_{11} V^{(2)}_{11} \exp[-i(\kappa_z - k_z) h]$$
$$- (V^{(1)}_{12} V^{(2)}_{21} + V^{(1)}_{21} V^{(2)}_{12}) + \exp[-i(\kappa_z + k_z) h] = 0. \tag{32.15}$$

This is the dispersion equation for the solid layer. Solving it, we find an infinite set of roots ϑ_l, $l = 1, 2, 3, ...$ As above, ϑ_l denotes the angle of incidence at the boundary of the layer of each of the two plane waves which are part of a specific normal mode and propagate in opposite directions with respect to the z-axis. The angle of incidence of the transverse waves which are part of the same normal mode are found from Eq. 32.11.

The phase velocity along the layer of the lth normal mode will be

$$v_l = \frac{\omega}{(k_x)_l} = \frac{c}{\sin \vartheta_l} = \frac{b}{\sin \gamma_l}, \tag{32.16}$$

where c and b are the propagation velocities of longitudinal and transverse waves, respectively, in the medium comprising the layer. In the general case, v_l will be complex, whence the damping can be determined.

As usual, the group velocity of the normal modes will be calculated by the formula

$$U = \frac{1}{[d(k_x)_l/d\omega]}.$$ (32.17)

If we set $V_{12}^{(1)} = V_{21}^{(1)} = V_{12}^{(2)} = V_{21}^{(2)} = 0$, which corresponds (often hypothetically) to the case in which interaction between the longitudinal and transverse waves is excluded, this equation takes the form

$$[\exp(ik_z h) \, V_{11}^{(1)} \, V_{11}^{(2)} - \exp(-ik_z h)]$$
$$\times [\exp(i\kappa_z h) \, V_{22}^{(1)} \, V_{22}^{(2)} - \exp(-i\kappa_z h)] = 0,$$

that is, it splits into two completely equivalent equations, one of which refers to the longitudinal waves, and the other to the transverse waves.

Finally, we note that, by solving Eq. 32.14, we can express three of the constants in terms of the fourth, for example, B, C and D in terms of A. Substituting the constants thus obtained into Eq. 32.9, we obtain functions characterizing the z-dependence of the field of any normal mode. Again, the one coefficient A remains undetermined for each normal mode. This coefficient is a function of the position and the strength of the source.

The reflection coefficients for the potentials ϕ and ψ enter into Eq. 32.15. Reflection coefficients, defined as the ratios of the vertical components of the particle velocities (or displacements) in the corresponding waves, are also frequently used in practice. We denote these coefficients by Q_{11}, Q_{12}, Q_{21} and Q_{22}, so that

$$Q_{11} = \frac{w_{pp}}{w_p}, \quad Q_{21} = \frac{w_{sp}}{w_s}, \quad Q_{12} = \frac{w_{ps}}{w_p}, \quad Q_{22} = \frac{w_{ss}}{w_s},$$ (32.18)

where w_p and w_s are the vertical displacements in the incident longitudinal and transverse waves,† and $w_{pp}, w_{ps}, \ldots$ are the vertical displacements, respectively, in the reflected PP, PS, etc., waves.

Let us connect the system of coefficients $V_{11}, V_{12}, \ldots$ with the coefficients $Q_{11}, Q_{12}, \ldots$. According to Eq. 4.2, the vertical component of the displacement is

$$w = \frac{\partial \phi}{\partial z} + \frac{\partial \psi}{\partial x},$$ (32.19)

where the first term gives the displacement in the longitudinal wave, and the second term the displacement in the transverse wave.

---

† Since we are considering a stationary oscillatory regime, the displacement refers to the amplitude of the particle oscillation.

We now examine the reflection at the lower boundary of the layer. The expressions for the potentials ϕ and ψ are given by Eqs. 32.9, where the minus sign indicates the waves incident on the boundary, and the plus sign indicates the waves reflected from the boundary. Remembering also that all the expressions must be multiplied by the factor $\exp[i(k_x x - \omega t)]$, we obtain, for $z = 0$,

$$w_p = \frac{\partial \phi_-}{\partial z} = -ik_z \phi_-, \quad w_s = \frac{\partial \psi_-}{\partial x} = ik_x \psi_-. \tag{32.20}$$

Taking Eqs. 32.12 into account, the vertical displacement in the reflected wave is

$$w = ik_z \phi_+ + ik_x \psi_+ = ik_z V_{11}^{(1)} \phi_- + ik_z V_{21}^{(1)} \psi_- + ik_x V_{12}^{(1)} \phi_-$$
$$+ ik_x V_{22}^{(1)} \psi_-. \tag{32.21}$$

Here, the four terms must be identified with w_{pp}, w_{sp}, w_{ps} and w_{ss}, respectively. As a result, using Eq. 32.18, we obtain

$$Q_{11}^{(1)} = -V_{11}^{(1)}, \quad Q_{21}^{(1)} = \frac{k_z}{k_x} V_{21}^{(1)},$$

$$Q_{12}^{(1)} = -\frac{k_x}{k_z} V_{12}^{(1)}, \quad Q_{22} = V_{22}^{(1)}. \tag{32.22}$$

In the same way, we obtain at the upper boundary

$$Q_{11}^{(2)} = -V_{11}^{(2)}, \quad Q_{21}^{(2)} = -\frac{k_z}{k_x} V_{21}^{(2)},$$

$$Q_{12}^{(2)} = \frac{k_x}{k_z} V_{12}^{(2)}, \quad Q_{22}^{(2)} = V_{22}^{(2)}. \tag{32.23}$$

The transition from Eq. 32.22 to Eq. 32.23 can be made formally by replacing k_z by $-k_z$.

It can now be shown without difficulty that Eq. 32.15 can be rewritten in such a way that the coefficients $Q_{11}^{(1)}, \ldots$ will appear instead of the coefficients $V_{11}^{(1)}, \ldots$, with the equation remaining otherwise unchanged.

In the case of greatest interest, that of undamped normal modes, Eq. 32.15 can be written in a somewhat different form, using Eqs. 24.23–24.29 for the reflection coefficients $V_{11}^{(1)}, \ldots$ and the analogous expressions for the coefficients $V_{11}^{(2)}, \ldots$. Since there will be only exponentially attenuating waves in the media on both sides of the layer, the quantities α' and β' in Eq. 24.20 will be real and positive.

We use the notation of § 24, and set

$$g = -\frac{i\mu'}{\mu}\alpha\beta', \quad h = \frac{-i\mu'}{\mu}\beta\alpha', \qquad (32.24)$$

$$j = \frac{1}{\mu^2\kappa^2\kappa'^2}[\sigma^2(\mu'\xi' - \mu\xi)^2 - \alpha'\beta'(\mu\xi - 2\mu'\sigma^2)^2],$$

$$k = -\frac{\alpha\beta}{\mu^2\kappa^2\kappa'^2}[(\mu'\xi' - 2\mu\sigma^2)^2 - 4\sigma^2\alpha'\beta'(\mu' - \mu)^2],$$

$$l = -\frac{4(\mu' - \mu)\,k\cos\vartheta}{\mu^2\kappa^2\kappa'^2}[\sigma^2(\mu'\xi' - 2\mu\sigma^2) + \alpha'\beta'(\mu\xi - 2\mu'\sigma^2)]. \qquad (32.25)$$

Then the reflection coefficients can be written

$$V_{11} = \frac{k - j + i(g - h)}{k + j + i(g + h)} = \Gamma\exp[-i(\bar{\epsilon} + \epsilon)],$$

$$V_{22} = \frac{k - j - i(g - h)}{k + j + i(g + h)} = \Gamma\exp[-i(\epsilon - \bar{\epsilon})], \qquad (32.26)$$

where
$$\Gamma = \sqrt{\left[\frac{(k - j)^2 + (g - h)^2}{(k + j)^2 + (g + h)^2}\right]},$$

$$\epsilon = \arctan\frac{g + h}{k + j}, \quad \bar{\epsilon} = \arctan\frac{h - g}{k - j}, \qquad (32.27)$$

and furthermore,

$$V_{12} = i\Lambda\,\mathrm{e}^{-i\epsilon}, \quad \Lambda = \frac{l}{\sqrt{[(k + j)^2 + (g + h)^2]}}, \qquad (32.28)$$

$$V_{21} = i\Delta\,\mathrm{e}^{-i\epsilon}, \quad \Delta = \frac{\beta\sigma^2}{\alpha}\Lambda, \qquad (32.29)$$

$$V_{11}V_{22} - V_{12}V_{21} = \mathrm{e}^{-2i\epsilon}. \qquad (32.30)$$

In what follows, we use a single prime to denote quantities referring to reflection at the lower boundary of the layer, and a double prime when these quantities refer to reflection at the upper boundary.

Then, substituting Eqs. 32.26–32.30 for the reflection coefficients into the dispersion equation (32.15), the latter equation can be written in the comparatively simple form†

$$\cos[(k_z + \kappa_z)h - \epsilon' - \epsilon''] = \Gamma'\Gamma''\cos[(k_z - \kappa_z)h - \bar{\epsilon}' - \bar{\epsilon}''] - \Lambda'\Delta''. \qquad (32.31)$$

Using this equation, it is not difficult to obtain the condition for the existence of surface waves. Just as in the case of the liquid layer, we carry out a limiting transition $(k_z + \kappa_z)h \to \infty$, $(\kappa_z - k_z)h \to \infty$. Assuming

---

† As a result, we have arrived at the equation obtained in Ref. 203.

that there is at least a small amount of attenuation in the layer, we obtain

$$\frac{\cos\left[(k_z + \kappa_z)h - \epsilon' - \epsilon''\right]}{\cos\left[(k_z - \kappa_z)h - \bar{\epsilon}' - \bar{\epsilon}''\right]} \to \exp\left(-2i\kappa_z h\right) \to \infty.$$

Under these conditions, Eq. 32.31 gives

$$\frac{1}{\Gamma'\Gamma''} \to 0. \tag{32.32}$$

The last equation can be split into two equations

$$\frac{1}{\Gamma'} \to 0 \quad \text{and} \quad \frac{1}{\Gamma''} \to 0, \tag{32.33}$$

representing the conditions for the excitation of surface waves at each of the boundaries of the layer.

We note that when this condition is fulfilled, the moduli of all the reflection coefficients V_{11}, etc., approach infinity, which indicates that the graphic representation of the surface wave, suggested in §4, is valid in this case.

3. *Approximate method for the determination of the characteristics of the normal modes*

Using a characteristic equation written in our form (32.1), Tolstoy[201] proposed a very convenient method for the approximate determination of the characteristics of the normal modes. We shall explain the concept of this method by considering the simplest cases of two and three layers.

Let a liquid layer of thickness h_1 (Fig. 143) have a free upper boundary, and be bounded from below by a homogeneous liquid half-space. The parameters of the media will be ρ_1, c_1 in the layer, and

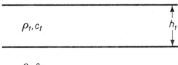

Fig. 143. On the calculation of wave propagation in a layer.

ρ_2, c_2 in the halfspace. We shall assume that $c_2 > c_1$. Let us now consider Eq. 32.1, in which the reflection coefficient at the upper boundary of the layer will be $V_2 = -1$. We write the reflection coefficient at the lower boundary as

$$V_1 = -\exp\left(2i\chi_{12}\right), \tag{32.34}$$

where χ_{12} is a quantity connected with the phase of the reflection at the boundary between media 1 and 2. Equation 32.1 can now be written in the form

$$\gamma_1 h_1 + \chi_{12} = \pi l, \tag{32.35}$$

where we have used the notation $\gamma_1 = k_1 \cos \vartheta$, $k_1 = \omega/c_1$ and l is the order of the normal mode.

We shall be interested in the undamped normal modes in the layer, which corresponds to the presence of total internal reflection at the boundary between media 1 and 2. From the known expression for the coefficient of reflection from an interface between two homogeneous media, we obtain

$$\chi_{12} = \arctan \left(p_{12} \frac{\rho_2}{\rho_1} \right), \tag{32.36}$$

where p_{12} is a function of the angle of incidence ϑ and the relation between the velocities in media 1 and 2:

$$p_{12} \equiv \frac{\cos \vartheta}{\sqrt{(\sin^2 \vartheta - n_{12}^2)}}, \quad n_{12} = \frac{c_1}{c_2}. \tag{32.37}$$

The angle ϑ is uniquely connected with the phase velocity along the layer $v = c_1/\sin \vartheta$. Then

$$p_{12} \equiv \sqrt{\left(\frac{v^2 - c_1^2}{c_1^2 - n_{12}^2 v^2} \right)}, \quad \gamma_1 = k_1 \sqrt{\left(1 - \frac{c_1^2}{v^2} \right)}. \tag{32.38}$$

Equation 32.35 is usually solved for v (or ϑ) with a given l, considering the frequency, that is, k_1 also as given. Tolstoy called attention to the fact that the equation is much more easily solved if v is considered as given, and the frequency is determined. In particular, since in the present case only γ_1 depends on the frequency in Eq. 32.35, we obtain

$$h_1 k_1 = \frac{\omega h_1}{c_1} = \frac{\pi l - \chi_{12}}{[1 - (c_1^2/v^2)]^{\frac{1}{2}}}.$$

We shall again have Eq. 32.35 in the case of a triple-layered medium (Fig. 144). However, the layer is bounded from below, not by a homogeneous halfspace, but by a system consisting of a layer and a halfspace. The coefficient of reflection from such a system is easily determined (see § 5), and it can be shown that

$$\chi_{12} = \arctan \left[p_{12} \frac{\rho_2}{\rho_1} \tan (\gamma_2 h_2 + \chi_{23}) \right],$$

$$\chi_{23} = \arctan \left(p_{23} \frac{\rho_3}{\rho_2} \right). \tag{32.39}$$

When there are n layers, we again obtain the equation

$$\gamma_1 h_1 + \chi_{12} = \pi l,$$

where
$$\chi_{12} = \arctan\left[p_{12}\frac{\rho_2}{\rho_1}\tan(\gamma_2 h_2 + \chi_{23})\right],$$

$$\chi_{23} = \arctan\left[p_{23}\frac{\rho_3}{\rho_2}\tan(\gamma_3 h_3 + \chi_{34})\right],$$

. . . .

$$\chi_{n-1,\,n} = \arctan\left[p_{n-1,\,n}\frac{\rho_n}{\rho_{n-1}}\tan(\gamma_n h_n + \gamma_{n,\,n+1})\right],$$

$$\chi_{n,\,n+1} = \arctan\left(p_{n,\,n+1}\frac{\rho_{n+1}}{\rho_n}\right),$$

and the subscript $n+1$ corresponds to the homogeneous halfspace lying under the system of layers.

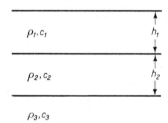

Fig. 144. On the calculation of wave propagation in a three-layered medium.

Tolstoy proposed an iterative method for the solution of this system of equations. With v given, this method yields γ_1 to within any accuracy, and can be programed for machine calculation. Tolstoy applied the method to the system shown in Fig. 144, and obtained interesting curves for the phase and group velocities as functions of the various parameters of the problem. Using our expression for the field of a point source situated in a layer bounded by arbitrary layered-inhomogeneous halfspaces (see § 34), Tolstoy indicated a way to calculate the excitation coefficients of the various normal modes.

§ 33. AVERAGED DECAY LAWS

We know from the work of the preceding Paragraphs that the field in a layer can be represented as a set of a number of normal modes and two lateral waves due to propagation in the media bounding the layer. In this Paragraph, we shall confine ourselves to the case of wave propagation in layers which are very thick compared with the wavelength. We shall also assume that the source is operating in the stationary regime. In this case, at not too great distances from the source, the field in the layer will be represented by a set of normal modes, the number of which is of the order of the ratio of the layer thickness to the half-wavelength.† The field in the layer will have a complex interference structure due to the superposition of all of these waves. Although, generally speaking, it is not difficult to calculate the amplitude and phase of each of the waves at an arbitrary point in the layer, the summation of all the waves and the analysis of the sum as a function of the coordinates is extremely difficult.

In this case, which is rather important practically, it is of most interest to find some averaged dependence on distance of the acoustic pressure or the intensities of the electromagnetic field. In the averaging process, the complex pattern of interference maxima and minima is smoothed out. The fact that in the majority of cases this "fine structure of the field" cannot be observed experimentally makes it all the more reasonable to eliminate it from the theoretical considerations. The difficulty in observing the fine structure experimentally stems either from lack of a perfectly monochromatic source or from non-stationary propagation conditions due to fluctuations of the properties of the medium (or of its boundaries). As a result, a measurement must be repeated many times, and can be used only as an average result. We shall be concerned with obtaining averaged laws in the present Paragraph.

1. *Averaged law of the decay of sound intensity with distance*

We shall obtain the averaged law of decay of the acoustic intensity in a liquid layer with a free upper boundary at distances from the source which are sufficiently large compared with the thickness of the layer (see below for the exact criterion).

We shall be interested in the mean square of the modulus of the acoustic potential $|\psi|^2$. The mean square of the acoustic pressure is then written

$$\overline{p^2} = \tfrac{1}{2}\rho^2\,\omega^2\,\overline{|\psi|^2}, \tag{33.1}$$

where ρ is the density of the medium, and ω is the angular frequency.

---

† In the present Paragraph, we are considering the field of the normal modes only. The amplitudes of the lateral waves decrease as $e^{-\gamma_i r}/r^2$ with distance, where the γ_i, $i = 1, 2$ are the absorption coefficients in the media bounding the layer.

The expression for $|\psi|^2$ will be a complex function of the coordinates r and z because of the interference structure of the sound field. However, this expression will be simplified considerably if we average in z over the layer thickness, and examine this averaged field as a function of the coordinate r. The averaging process reduces to applying the operation $\dfrac{1}{h}\displaystyle\int_0^h \ldots dz$ to $|\psi|^2$. Using Eq. 27.43 for ψ, we find

$$\overline{|\psi|^2} = \frac{\pi}{2krh^2}\Bigg\{\sum_l |\exp(-il\pi z_0/h) + V_1(\vartheta_l^0)\exp(i\pi l z_0/h)|^2$$

$$\times \frac{\exp(-2\pi l\Delta l r/h)}{|V_1(\vartheta_l^0)|^2\sin\vartheta_l^0}\cdot I_{ll} + \sum_{l,\,m\neq l}\{\exp(-il\pi z_0/h)+V_1(\vartheta_l^0)\exp[i(l\pi z_0/h)]\}$$

$$\times\{\exp(im\pi z_0/h)+V_1^*(\vartheta_m^0)\exp[-i(m\pi z_0/h)]\}$$

$$\times \frac{\exp[ikr(\sin\vartheta_l^0-\sin\vartheta_m^0)]}{V_1(\vartheta_l^0)\,V_1^*(\vartheta_m^0)\,\sqrt{(\sin\vartheta_l^0\sin\vartheta_m^0)}}\cdot\exp[-(\pi r/h)(l\Delta_l+m\Delta_m)]\,I_{lm}\Bigg\},$$

where $\qquad\qquad\qquad\qquad\qquad\qquad\qquad\qquad\qquad\qquad\qquad\qquad$ (33.2)

$$I_{lm} = \frac{1}{h}\int_0^h\{\exp[-i(l\pi z/h)]+V_1(\vartheta_l^0)\exp[i(l\pi z/h)]\}$$

$$\times\{\exp(im\pi z/h)+V_1^*(\vartheta_m^0)\exp[-i(m\pi z/h)]\}\,dz. \qquad (33.3)$$

We obtain I_{ll} from I_{lm} by replacing m by l.

Multiplying the curly brackets in Eq. 33.3 together, and integrating, we obtain

$$I_{l,\,m\neq l} = 0, \quad I_{ll} = 1 + |V_1(\vartheta_l^0)|^2. \qquad (33.4)$$

In what follows, we shall be interested in the comparatively weakly attenuated normal modes corresponding to small grazing angles. Under these conditions, the modulus of the reflection coefficient is close to unity. We therefore set

$$|V_1(\vartheta_l^0)| \approx 1 \qquad (33.5)$$

in the last equation.

It is useful to obtain the criterion for the validity of this assumption. At small grazing angles α, the reflection coefficient $V(\alpha)$ is given by Eq. 27.35. According to Eq. 27.39, the grazing angle corresponding to the normal mode of order l is, to a first approximation,

$$\tfrac{1}{2}\pi - \vartheta_l^0 \equiv \alpha_l \approx \pi l/kh. \qquad (33.6)$$

As a result, Eq. 27.35 gives

$$|V_1(\alpha_l)| = \exp[-2p_1'(\pi l/kh)], \qquad (33.7)$$

where $\qquad\qquad\qquad\qquad\qquad p_1' = \operatorname{Re} p_1. \qquad\qquad\qquad (33.8)$

We see that condition (33.5) will be satisfied if

$$p_1'\cdot(\pi l/kh)\ll 1. \qquad (33.9)$$

Taking Eq. 33.5 into account, Eq. 33.4 now gives $I_{II} \approx 2$.

The assumption (33.5) still says nothing about the phase of the reflection coefficient. Therefore, in Eq. 33.2, we set $V_1 = -\exp(i\phi_1)$, where ϕ_1 is an arbitrary phase, approaching zero as $\alpha \to 0$. Equation 33.2 then gives

$$\overline{|\psi|^2} = \frac{4\pi}{krh^2} \sum_l \sin^2\left(\frac{l\pi z_0}{h} + \frac{\phi_1}{2}\right) \exp\left(-2l\pi\Delta_l \frac{r}{h}\right). \tag{33.10}$$

If we now average over the source position, i.e. over z_0 within the same limits, we obtain

$$\overline{|\psi|^2} = \frac{2\pi}{krh^2} \sum_l \exp\left(-2l\pi\Delta_l \frac{r}{h}\right). \tag{33.11}$$

In the present case, in which one of the boundaries of the layer is perfectly reflecting, Eq. 27.40 for Δ_l is written

$$\Delta_l = [\pi l/(kh)^2]\, p_1'. \tag{33.12}$$

where p_1' is defined by Eq. 33.8. In particular, if we assume that the layer is bounded from below by a homogeneous medium, the quantity p_1 in Eq. 33.8 is given by Eq. 27.34.

Equation 33.11 is now written

$$\overline{|\psi|^2} = \frac{2\pi}{krh^2} \sum_l \exp\left[-2\left(\frac{\pi l}{kh}\right)^2 p'\left(\frac{r}{h}\right)\right]. \tag{33.13}$$

We shall investigate the field at sufficiently great distances, at which the condition

$$p'(r/h) \gg 1 \tag{33.14}$$

is fulfilled. In this case, only the terms for which

$$\pi l/kh \ll 1 \tag{33.15}$$

will be significant in the sum (33.13), which, according to Eq. 33.6, corresponds to small grazing angles. However, the result will not be changed if the summation is carried to $l = \infty$ because the contribution of the terms with large values of l (of the order of kh/π and above) will be negligible. We also introduce the nondimensional distance ρ, defined by the equation

$$\rho = 2p'\left(\frac{\pi}{kh}\right)^2 \cdot \frac{r}{h}. \tag{33.16}$$

The expression for the mean square of the acoustic potential is then written

$$\overline{|\psi|^2} = \frac{4\pi^3}{k^3 h^5} \frac{p'}{} f(\rho), \tag{33.17}$$

where
$$f(\rho) = \frac{1}{\rho} \sum_{l=1}^{\infty} \exp\left(-\rho l^2\right). \tag{33.18}$$

The case
$$\rho \ll 1 \tag{33.19}$$

can be realized within the limits of condition (33.14).

Under these conditions, a rather large number of terms will be significant in the sum (33.18), and furthermore, neighboring terms will not differ much from one another. The sum can therefore be replaced by an integral from 0 to ∞. As a result, we obtain†

$$f(\rho) = \frac{1}{2} \frac{\pi^{\frac{1}{2}}}{\rho^{\frac{3}{2}}} \tag{33.20}$$

and furthermore, according to Eq. 33.17,

$$\overline{|\psi|^2} = \frac{2\pi^3 \sqrt{(\pi)}\, p'}{k^3 h^5} \cdot \frac{1}{\rho^{\frac{3}{2}}}. \tag{33.21}$$

Now, expressing ρ in terms of r, we obtain

$$\overline{|\psi|^2} = \sqrt{\left(\frac{\pi}{2p'h}\right)} \cdot \frac{1}{r^{\frac{3}{2}}}. \tag{33.21a}$$

Thus, the acoustic intensity decreases with distance as $1/r^{\frac{3}{2}}$ within the interval of distances under consideration. This law of fall-off is intermediate between the cylindrical law $1/r$, associated with total internal reflection at the boundaries of the layer, and the spherical law $1/r^2$, associated with the absence of reflection at the boundaries. As we see, the "3/2" law takes into account the partial absorption of the waves as they undergo multiple reflections at the boundaries of the layer.

Conditions (33.14) and (33.19) for the applicability of the "3/2" law can be reduced to the one condition

$$1 \ll p' \frac{r}{h} \ll \left(\frac{kh}{\pi}\right)^2. \tag{33.22}$$

Hence, it is clear that for large values of kh/π, the distance interval over which this law is applicable can be very large (see the example below).

As the distance is increased further, we pass into the case

$$p \gtrsim 1. \tag{33.23}$$

---

† Using Euler's summation formula, it can be shown that replacing the sum by an integral leads to an error of order 1, which is small compared with the total value of the sum when condition (33.19) is fulfilled.

In this case, we need take into account only the one term for which $l = 1$, in the sum (33.18), and we obtain

$$\overline{|\psi|^2} = \frac{4\pi^3 p'}{k^3 h^5} \cdot \frac{1}{\rho} e^{-\rho}$$ (33.24)

which is a law of decay corresponding to cylindrical attenuation with additional exponential damping.

The function $\qquad f(\rho) = \frac{1}{\rho} \sum_{l=1}^{\infty} \exp(-\rho l^2),$ (33.25)

which, as we see, gives the mean acoustic intensity in the layer to within a constant factor, is shown in Fig. 145. It is clear from the

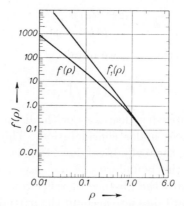

Fig. 145. Graph of the functions $f(\rho)$ and $f_1(\rho)$.

Figure that the $1/\rho^{\frac{3}{2}}$ law (a straight line on the scale of the Figure) is valid approximately up to $\rho = 0.5$, above which the $(1/\rho)e^{-\rho}$ law becomes valid. In preparing Fig. 145, numerical summation was used whenever the replacement of the sum by an integral was not sufficiently accurate. It is useful to note that under certain conditions, p' can be very small, which corresponds to small absorption at the boundaries of the layer. In particular, this occurs under conditions of total internal reflection at the boundaries of the layer. In fact, it is clear from Eq. 27.34 for p_1 that, if n_1 is real, and if furthermore, $n_1 < 1$ (which also corresponds to total internal reflection), then $p_1' = \operatorname{Re} p_1 = 0$. If the analogous condition is also fulfilled at the second boundary, then $p_2' = 0$, and, consequently, $p' = p_1' + p_2' = 0$. Then condition (33.14) will not be fulfilled at any distance, and the theory developed above will be inapplicable. In this case, thanks to the total reflection at the boundaries of the layer, the decay will follow the cylindrical law (if we neglect damping in the layer itself).

In practice, however, as a rule, n_1 or n_2 will have an imaginary part which must be taken into account. This corresponds to absorption in the media bounding the layer. In this case, p' will not be zero (although it may be very small), and the reflection will no longer be complete.

As an example, we consider acoustic propagation at a frequency of 4.5 kc in an ocean 30 meters deep. We make two different assumptions regarding the index of refraction and the ratio of the densities of the media at the water–bottom boundary:

$$(1) \qquad n_1 = 0.8(1+0.01i), \qquad m_1 = \rho_1/\rho = 2,$$

which corresponds to a wave velocity in the bottom of $c_1 = 1875$ m/sec, and damping such that the wave amplitude in the bottom is decreased by a factor of e over a distance of approximately fifteen wavelengths.

$$(2) \qquad n_1 = 1.1 \quad \text{or} \quad c_1 = 1360 \text{ m/sec}, \quad m_1 = 2.$$

In the first case, $p' = 0.06$. The "3/2" law zone will extend from $r = 30h$ to $r \approx 3 \times 10^5 h \approx 10^4$ km. As the frequency or the depth of the water is decreased, the upper limit of the region of applicability of the "3/2" law decreases. When $r \lesssim 30h$, the decay satisfies the cylindrical law.

In the second case, $p' = 4.4$. There is no zone of cylindrical decay, and the zone of the "3/2" law will extend from some depth to $r \approx 10^5 h$.

2. Averaged law of decay of an electromagnetic field in a layer

We now proceed to the averaged laws of decay for the electromagnetic field. As an example, we consider the field in a layer generated by a "vertical" radiating dipole.† Denoting the vertical component of the Hertz vector by $\psi(r, z)$, the components of the electromagnetic field are expressed in terms of it as follows:

$$E_r = \frac{\partial^2 \psi}{\partial r \, \partial z}, \quad E_z = -\frac{1}{r} \frac{\partial}{\partial r}\left(r \frac{\partial \psi}{\partial r}\right), \quad E_\phi = 0,$$

$$H_r = H_z = 0, \quad H_\phi = -ik \frac{\partial \psi}{\partial r}. \tag{33.26}$$

We shall use Eq. 27.43 for ψ. Carrying out the differentiation, we

---

† The transition to other forms of radiators presents no difficulties.

obtain

$$E_r = \frac{\partial^2 \psi}{\partial r\,\partial z} = \frac{k\pi^{\frac{3}{2}}\exp\left[i(\pi/4)\right]}{h^2\sqrt{(2kr)}} \sum_{l=1}^{\infty} l\left[\exp\left(-i\pi lz/h\right) - V_1(\vartheta_l^0)\exp\left(i\pi lz/h\right)\right]$$

$$\times\left[\exp\left(-i\pi lz_0/h\right) + V_1(\vartheta_l^0)\exp\left(i\pi lz_0/h\right)\right]$$

$$\times\frac{\exp\left(ikr\sin\vartheta_l^0\right)\sin^{\frac{1}{2}}\vartheta_l^0}{V_1(\vartheta_l)}\exp\left[-(\pi l/h)\,\Delta_l r\right]; \qquad (33.27)$$

$$E_z \approx -\frac{\partial^2 \psi}{\partial r^2} = \frac{k^2\sqrt{(\pi)}\exp\left[i(\pi/4)\right]}{h\sqrt{(2kr)}} \sum_{l=1}^{\infty}\left[\exp\left(-i\pi lz/h\right) + V_1(\vartheta_l^0)\exp\left(i\pi lz/h\right)\right]$$

$$\times\left[\exp\left(-i\pi lz_0/h\right) + V_1(\vartheta_l^0)\exp\left(i\pi lz_0/h\right)\right]$$

$$\times\frac{\exp\left(ikr\sin\vartheta_l^0\right)}{V_1(\vartheta_l^0)}\sin^{\frac{3}{2}}\vartheta_l^0\exp\left[-(\pi l/h)\,\Delta_l r\right]; \qquad (33.28)$$

and a similar expression for H_ϕ.

Our problem is now the determination of the r-dependence of the quantities $|E_r|^2$ and $|E_z|^2$, averaged over the position of the receiver and the position of the source. The calculations will follow the same course as in the preceding Section, in which the decay law of the averaged value of $|\psi|^2$ was determined.

As previously, it will be convenient to use the nondimensional distance

$$\rho = 2p'\left(\frac{\pi}{kh}\right)^2\cdot\frac{r}{h}, \qquad (33.29)$$

where

$$p' = p'_1 + p'_2,$$

and p'_1 and p'_2 are quantities characterizing the wave absorption at the lower and upper boundaries of the layer, respectively. They are determined by the following approximate representations of the reflection coefficients V_1 and V_2 at small grazing angles $\alpha = (\pi/2) - \vartheta$ (see § 27):

$$|V_1| \approx \exp\left(-2p'_1\alpha\right), \quad |V_2| \approx \exp\left(-2p'_2\alpha\right). \qquad (33.30)$$

The explicit expression for p'_1 in terms of the parameters of the media has the form

$$p'_1 = \mathrm{Re}\,\frac{n_1^2}{\sqrt{(n_1^2-1)}} = \mathrm{Re}\,\frac{\epsilon_1/\epsilon}{\sqrt{[(\epsilon_1/\epsilon)-1]}} \qquad (33.31)$$

and the expression for p'_2 is similar, with the subscript 1 replaced by 2. Here, ϵ is the dielectric constant of the medium comprising the layer, and ϵ_1 and ϵ_2 are the complex dielectric constants of the media bounding the layer from below and from above, respectively.

Carrying out the same calculations as in the first Section, we obtain

$$\overline{E_r^2} = \tfrac{1}{2} \overline{|E_r|^2} \approx \frac{2\pi^5 p'}{k h^7} f_1(\rho), \tag{33.32}$$

$$\overline{E_z^2} = \overline{H_\phi^2} = \tfrac{1}{2} \overline{|E_z|^2} \approx \frac{2k\pi^3 p'}{h^5} f(\rho), \tag{33.33}$$

where

$$f_1(\rho) \equiv \frac{1}{\rho} \sum_{l=1}^{\infty} l^2 \exp(-\rho l^2), \tag{33.34}$$

and $f(\rho)$ is given by Eq. 33.25. The functions $f(\rho)$ and $f_1(\rho)$ are shown in Fig. 145.

We see that the averaged laws of decay of $|E_z|^2$ and $|H_\phi|^2$ coincide with the analogous law for $|\psi|^2$. When $\rho \lesssim 0.5$, a large number of terms still play a significant role in the sums for $f(\rho)$ and $f_1(\rho)$ (the number of normal modes determining the field is still sufficiently large), and the sums can be replaced by integrals (see above). As a result, $f(\rho)$ is given by Eq. 33.20, and $f_1(\rho)$ by the expression

$$f_1(\rho) = \frac{\sqrt{\pi}}{4\rho^{\frac{5}{2}}}. \tag{33.35}$$

This signifies that the average values of the squares of E_z and H_ϕ decrease as $1/\rho^{\frac{3}{2}}$ with distance, while the averaged square of E_r falls off as $1/\rho^{\frac{5}{2}}$. When $\rho \gtrsim 1$, the sums can be limited to the first term, which corresponds to the first, least attenuated normal mode. We then obtain

$$f_1(\rho) \approx f(\rho) \approx (1/\rho) e^{-\rho}. \tag{33.36}$$

In this case, the fall-off law contains the factor $1/\rho$, which corresponds to cylindrical wave divergence, and the exponential factor $e^{-\rho}$, which takes into account the damping of the first normal wave due to absorption at the boundaries of the layer.

3. *Interpretation of the results from the point of view of the image source pattern*

The averaged laws obtained above can also be obtained from considerations based on geometrical acoustics (or optics).

In this approximation, we envisage a direct ray and a set of rays reflected various numbers of times from the boundaries of the layer, arriving at the receiver. Each such ray can be considered to arrive from some "image" source. All the image sources will be situated on a straight line perpendicular to the boundaries of the layer and passing through the source (see § 25 for a more detailed discussion of the image sources).

In this approach, the incoherent (energy) summation of the fields of the image sources corresponds to the averaging process which leads to a smoothing out of the interference structure of the field. We shall be interested in the field at distances from the source which are large compared with the thickness of the layer. The positions of the source and the receiver with respect to the boundaries of the layer are then unimportant. Therefore, for simplicity, we shall assume that the source and the receiver are situated in the middle of the layer. All the image sources will be situated at the same distance h from one another, where h is the thickness of the layer. The distance between the receiver and the lth image source will be

$$R_l = \sqrt{[r^2 + (lh)^2]}. \tag{33.37}$$

As an example, we shall consider an acoustic field in the layer. At the receiver, the square of the acoustic potential of each of the sources is written in the form†

$$\frac{M_1^p(\alpha_l)\, M_2^q(\alpha_l)}{R_l^2}, \tag{33.38}$$

where p and q are the number of reflections of the ray from the lower and the upper boundary of the layer, respectively, M_1 and M_2 are the reflection coefficients with respect to energy at these boundaries, i.e.

$$M_1(\alpha_l) = |V_1(\alpha_l)|^2, \quad M_2(\alpha_l) = |V_2(\alpha_l)|^2, \tag{33.39}$$

and finally, $$\alpha_l = \arctan(lh/r) \tag{33.40}$$

is the grazing angle between the ray and the boundaries of the layer.

Summing over all the image sources, we obtain

$$|\psi|^2 = \sum_{l=1}^{\infty} \left\{ \frac{M_1^l(\alpha_{2l})\, M_2^l(\alpha_{2l})}{R_{2l}^2} + \frac{[M_1(\alpha_{2l+1}) + M_2(\alpha_{2l+1})]\, M_1^l(\alpha_{2l+1})\, M_2^l(\alpha_{2l+1})}{R_{2l+1}^2} \right.$$
$$\left. + \frac{M_1^{l+1}(\alpha_{2l+2})\, M_2^{l+1}(\alpha_{2l+2})}{R_{2l+2}^2} \right\}. \tag{33.41}$$

At great (compared with the layer thickness) distances from the source, the terms of the last series will decrease very slowly with increasing l, and neighboring terms of the series will differ only slightly from one another. We can therefore replace the entire curly brackets by

$$\frac{4}{R_{2l}^2}\, M_1^l(\alpha_{2l})\, M_2^l(\alpha_{2l})$$

---

† We recall that the amplitude of the real source is chosen such that the acoustic potential is written as $\psi = e^{ikR}/R$ in free space.

and we can replace the sum by an integral. As a result, we obtain

$$|\psi|^2 = 4 \int_0^\infty \frac{M_1^l(\alpha_{2l}) M_2^l(\alpha_{2l})}{R_{2l}^2} \, dl. \tag{33.42}$$

At small grazing angles, which are the only ones of significance here, we obtain, using Eq. 27.35:

$$M_1(\alpha) M_2(\alpha) = |V_1(\alpha) V_2(\alpha)|^2 = \exp(-4p'\alpha). \tag{33.43}$$

Remembering furthermore that $\alpha_{2l} \approx 2lh/r$, Eq. 33.42 can be written

$$|\psi|^2 = 4 \int_0^\infty \frac{\exp[-8l^2 p'(h/r)]}{r^2 + 4l^2 h^2} \, dl. \tag{33.44}$$

Neglecting $4l^2 h^2$ compared with r^2, and substituting the known value of the integral (see Eq. 19.15), we obtain

$$|\psi|^2 \approx \sqrt{\left(\frac{\pi}{2p'h}\right)} \frac{1}{r^{\frac{3}{2}}}, \tag{33.45}$$

which agrees with Eq. 33.21a, obtained earlier.

In exactly the same way, we could have obtained the averaged laws of fall-off for the electromagnetic field, derived above.

We have neglected the term $4l^2 h^2$ compared with the term r^2 in Eq. 33.44. Thus, we are assuming that only those image sources whose distance from the real source is small compared with the distance to the receiver, are significant. This assumption can be made somewhat more precise. Because of the exponential in Eq. 33.44, the principal rôle will be played by those values of l whose order of magnitude is no greater than

$$l_{\max} \sim \frac{1}{2} \sqrt{\left(\frac{r}{p'h}\right)}. \tag{33.46}$$

The length of the radiating portion of the set of image sources is then

$$2hl_{\max} \sim \sqrt{\left(\frac{rh}{p'}\right)}. \tag{33.47}$$

Our assumption reduces to the condition

$$(2hl_{\max}/r)^2 \ll 1,$$

which, taking Eq. 33.47 into account, will agree with one of the conditions (33.22).

After all that has been presented, the averaged laws of fall-off can be given a very graphic and elementary treatment. We again confine ourselves to the "3/2" law for the acoustic intensity.

Since the length of the radiating portion of the image source network is small compared with the distance, we can express its field as $|\psi|^2 \sim W/r^2$ (omitting coefficients of proportionality), where W is the total energy radiated by the network. In view of our assumption regarding the incoherent summation of the fields of the image sources, W is proportional to the length of the radiating portion of the network that is, to $2hl_{max}$. According to Eq. 33.47, the latter increases as $\sqrt{r}$.

As a result, we obtain

$$|\psi|^2 \sim 1/r^{\frac{3}{2}},$$

i.e. the result obtained above by another method.

§ 34. WAVE PROPAGATION IN A LAYER BOUNDED BY INHOMOGENEOUS MEDIA

1. *General expressions for the field*

In the present Paragraph, we shall investigate wave propagation in a layer bounded from below and from above by layered-inhomogeneous media. The z-dependences of the parameters of these media (density, index of refraction) are assumed to be continuous. Moreover, we assume that as $z \to \pm\infty$, these parameters approach definite finite values.

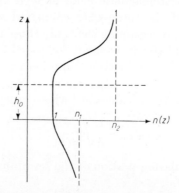

Fig. 146. Schematic illustration of the z-dependence of the index of refraction.

The assumed z-dependence of the index of refraction is shown schematically in Fig. 146. The medium is assumed to be homogeneous for $0 < z < h_0$. Without limiting the generality, the index of refraction can be taken equal to unity in this region. Furthermore, we shall use the notation $n_1 = \lim\limits_{z \to -\infty} n$, $n_2 = \lim\limits_{z \to +\infty} n$. The source is assumed to be situated in the homogeneous layer. In acoustics, the density $\rho(z)$ can also be a variable quantity.

We have obtained the expressions (27.2a) and (27.2b) for the field in a layer. These expressions will be valid in the present case, if we regard $V_1(\vartheta)$ and $V_2(\vartheta)$ as the reflection coefficients at the inhomogeneous media $z < 0$ and $z > h_0$, respectively.

Equation 27.5 will be valid for the field in the lower halfspace $z < 0$. The field in the upper halfspace $z > h_0$ will be written similarly. We write out these expressions here, remembering that in our present notation, the thickness of the homogeneous layer and the wave vector are h_0 and k_0, and not h and k as previously.

The acoustic potential or the vertical component of the Hertz vector is then

$z_0 < z < h_0$:
$$\psi = \frac{ik_0}{2} \int_{\Gamma_1} \frac{(e^{-bz_0} + V_1 e^{bz_0})(e^{-b(h_0-z)} + V_2 e^{b(h_0-z)})}{e^{-bh_0}(1 - V_1 V_2 e^{2bh_0})}$$
$$\times H_0^{(1)}(k_0 r \sin \vartheta) \sin \vartheta \, d\vartheta, \qquad (34.1)$$

$0 < z < z_0$:
$$\psi = \frac{ik_0}{2} \int_{\Gamma_1} \frac{(e^{-bz} + V_1 e^{bz})(e^{-b(h_0-z_0)} + V_2 e^{b(h_0-z_0)})}{e^{-bh_0}(1 - V_1 V_2 e^{2bh_0})}$$
$$\times H_0^{(1)}(k_0 r \sin \vartheta) \sin \vartheta \, d\vartheta, \qquad (34.2)$$

$z < 0$:
$$\psi_1 = \frac{ik_0}{2} \int_{\Gamma_1} \frac{e^{-b(h_0-z_0)} + V_2 e^{b(h_0-z_0)}}{e^{-bh_0}(1 - V_1 V_2 e^{2bh_0})}$$
$$\times f_1(z, \vartheta) H_0^{(1)}(k_0 r \sin \vartheta) \sin \vartheta \, d\vartheta, \qquad (34.3)$$

$z > h_0$:
$$\psi_2 = \frac{ik_0}{2} \int_{\Gamma_1} \frac{e^{-bz_0} + V_1 e^{bz_0}}{e^{-bh_0}(1 - V_1 V_2 e^{2bh_0})}$$
$$\times f_2(z, \vartheta) H_0^{(1)}(k_0 r \sin \vartheta) \sin \vartheta \, d\vartheta, \qquad (34.4)$$

where, as previously, z_0 is the coordinate of the source, and $b \equiv ik_0 \cos \vartheta$.

The function $f_1(z, \vartheta)$ is the expression for the field in the lower inhomogeneous medium when a plane wave of unit amplitude is incident at an angle ϑ on the plane $z = 0$ from the homogeneous layer. The reflection coefficient of a plane wave at the halfspace $z < 0$ can be expressed in terms of the functions f_1 and df_1/dz at the boundary $z = 0$.

To show this, we suppose for the time being that the lower inhomogeneous medium $z < 0$ is bounded not by a homogeneous layer, but by an infinite homogeneous halfspace. Let a plane wave be incident from the latter on the boundary $z = 0$. In the upper halfspace, the total field, consisting of the incident and the reflected waves, is written in the form [omitting the factor $\exp(-ik_0 x \sin \vartheta - i\omega t)$]

$$\psi = \exp(-ik_0 z \cos \alpha) + V_1 \exp(ik_0 z \cos \vartheta). \qquad (34.5)$$

The field for $z < 0$ is written in the form

$$\psi = f_1(z, \vartheta) = W_1 \phi(z, \vartheta),$$

where $W_1 = W_1(\vartheta)$ is independent of z, and $\phi(z, \vartheta)$ is the appropriate solution of the equation for the field in the lower halfspace, remaining bounded as $z \to -\infty$.

Both ψ as well as the derivative $d\psi/dz$ must be continuous at $z = 0$. Hence, we obtain two equations

$$1 + V_1 = f_1(0, \vartheta) \equiv W_1 \phi(0, \vartheta),$$

$$ik_0 \cos \vartheta (V_1 - 1) = (df/dz)_{z=0} \equiv W_1 (d\phi/dz)_{z=0}, \qquad (34.6)$$

which can be considered as the equations for the determination of $V_1(\vartheta)$ and $W_1(\vartheta)$. We could also write similar equations (with the subscript 1 replaced by 2, and $k_0 \cos \vartheta$ by $-k_0 \cos \vartheta$) for the upper boundary. The function $f_2(z, \vartheta)$ would then determine the field in the upper halfspace when a plane wave described by the expression $\exp\{ik_0[(z-h_0) \cos \vartheta + x \sin \vartheta]\}$ is incident on the boundary $z = h_0$.

Although there is no reason to doubt the derivation of Eqs. 34.1–34.4, we will show that these expressions satisfy all the necessary conditions, namely:

1. They satisfy the appropriate field equations for all z.

2. They satisfy the required continuity conditions at $z = 0$ and $z = h_0$.

3. The integrals in these expressions converge everywhere, except at the point $z = z_0$, $r = 0$, at which they have a singularity of the form $\psi \sim 1/\sqrt{[(z-z_0)^2 + r^2]}$.

The first condition is obviously satisfied because the integrands in Eqs. 34.1–34.4 satisfy the corresponding field equations in the layer and in the inhomogeneous halfspaces bounding the layer.

The continuity conditions, in expanded form, are written

$$(\psi)_{z=0} = (\psi_1)_{z=0}, \quad (\partial\psi/\partial z)_{z=0} = (\partial\psi_1/\partial z)_{z=0}, \qquad (34.7a)$$

$$(\psi)_{z=h_0} = (\psi_2)_{z=h_0}, \quad (\partial\psi/\partial z)_{z=h_0} = (\partial\psi_2/\partial z)_{z=h_0}. \qquad (34.7b)$$

We first examine the conditions (34.7a). Substituting Eqs. 34.2 and 34.3, differentiating under the integral sign (which is assumed permissible), and equating the integrands gives

$$1 + V_2 = f_1(0, \vartheta), \quad b(V_1 - 1) = (\partial f_1/\partial z)_{z=0}.$$

These equations agree with Eqs. 34.6, which were assumed to be satisfied. Thus, the validity of (34.7a) has been demonstrated.

The proof that conditions (34.7b) are satisfied is similar.

We now go on to condition 3. We consider the behavior of the integrand at infinitely remote points of the path of integration Γ_1 (see Fig. 122). These will be the points $\vartheta' = \pm \pi/2$, $\vartheta'' = \mp \infty$ in the plane $\vartheta = \vartheta' + i\vartheta''$.

We shall show in the next Section that the functions $V_1(\vartheta)$ and $V_2(\vartheta)$ tend toward finite values or zero as these points are approached. Moreover, at $\vartheta' = \pm \pi/2$, $\vartheta'' \to \mp \infty$, we have

$$b = ik \cos \vartheta = ik(\cos \vartheta' \cosh \vartheta'' - i \sin \vartheta' \sinh \vartheta'') \to -\infty.$$

As a result, the integrand in Eq. 34.1 is written in the form

$$e^{b(z-z_0)} H_2^{(1)}(k_0 r \sin \vartheta) \sin \vartheta \, d\vartheta \qquad (34.8)$$

on the remote portions of the path of integration. Introducing the new variable of integration $\xi = k_0 \sin \vartheta$ (the integration over which goes along the real axis from $-\infty$ to $+\infty$), and using the asymptotic representation of the Hankel function, the above expression is written, to within a constant factor, as follows:

$$\exp[i\xi r - (z-z_0)\sqrt{(\xi^2 - k_0^2)}] (d\xi/\sqrt{\xi}). \qquad (34.8a)$$

As a result, we obtain an integral which converges for all r and z, except at the point $r = 0$, $z = z_0$, which requires special treatment.

As $r \to 0$, $z \to z_0$, the principal rôle is played by the remote portions of the path of integration Γ_1, at which the moduli of $\sin \vartheta$ and $\cos \vartheta$ are large. The integrand in Eq. 34.1 can then be replaced by Eq. 34.8. As a result, we obtain Eq. 26.1 for a spherical wave. Thus, integral (34.1) converges everywhere, except at the point $r = 0$, $z = z_0$, in the neighborhood of which it behaves like a spherical wave, i.e. it has a singularity of the form $1/\sqrt{[(z-z_0)^2 + r^2]}$, which completes the proof.

Similar arguments can be used with Eq. 34.2.

To prove the convergence of the integrals (34.3) and (34.4), which give the fields in the inhomogeneous halfspaces, we must know the behavior of the functions $f_1(z, \vartheta)$ and $f_2(z, \vartheta)$ on the infinitely remote portions of the path of integration, i.e. when $\xi = k_0 \sin \vartheta \to \pm \infty$. For this purpose, we use Eqs. 17.5. For definiteness, we consider the upper halfspace. The function $f_2(z, \vartheta)$ is then the sum of the functions $P + R$ when the incident wave is given in the form

$$P = \exp(ik_0 z \cos \vartheta), \quad (z < h_0).$$

However, for any z, we have

$$f_2 = P + R = (1 + V)P. \qquad (34.9)$$

Since we know the behavior of V, we need only investigate P. In the first of Eqs. 17.5, the term containing γ can be neglected compared with the term containing m, since $|m| = |b| \to \infty$, while γ remains finite. The solution for P will then be

$$P = \exp\left(ik_0 h_0 \cos\vartheta\right)\exp\left(i\int_{h_0}^{z}\beta d\,z\right).$$

Since we have $ik_0 h_0 \cos\vartheta = b = -\infty$ for the values of ϑ of interest, it follows that $P = 0$ for all $z > h_0$.

From this, and also from Eq. 34.9, it follows that the function $f_2(z, \vartheta)$ approaches zero on the infinitely remote portions of the path of integration. Clearly, the function $f_1(z, \vartheta)$ will behave similarly. The proof that integrals (34.3) and (34.4) are convergent is now elementary.

We shall conclude our discussion of Eqs. 34.1–34.4 with the following useful remarks. It is easy to see from what has been presented above, that the choice of the positions of the boundaries of the layer is to a certain extent arbitrary. In particular, we can move the boundaries farther apart if the inhomogeneity of the medium is not too strong beyond the limits of the original positions of the boundaries. On the other hand, the thickness of the layer can always be decreased by moving the boundaries toward each other, and in the limit, the thickness can be made equal to zero. We shall show that, as the positions of the boundaries of the layer are changed in this way, Eqs. 34.1–34.4 do not change.

As an example, we consider the case $z_0 < z < h_0$, and suppose that the layer thickness is decreased from h_0 to $h_0' = h_0 - \Delta$ by choosing a new position of the upper boundary, without changing the properties of the medium under consideration. The phase of the reflection coefficient V_2 will then change by an amount equal to the change in phase of a wave as it travels across the thickness Δ and back. Consequently, if we let V_2' be the reflection coefficient for the new position of the boundaries, then $V_2' = V_2 \exp(2b\Delta)$. The coefficient V_1 does not change since, in the present case, the position of the lower boundary is not changed. We therefore set $V_1' = V_1$.

When h_0 is replaced by h_0', and V_2 by V_2', Eq. 34.1 remains unchanged, in view of the identity

$$\frac{e^{-b(h_0-z)} + V_2 e^{b(h_0-z)}}{e^{-bh_0}(1 - V_1 V_2 e^{2bh_0})} = \frac{e^{-b(h_0'-z)} + V_2' e^{b(h_0'-z)}}{e^{-bh_0'}(1 - V_1' V_2' e^{2bh_0'})},$$

which completes the proof.

The position of the upper boundary in the present example can also be chosen such that the point of observation (r, z) is outside the layer. As a result, Eq. 34.4 is appropriate, and it is easy to prove that, under these conditions, it is equivalent to Eq. 34.1. With the new choice of the boundary position, we must use f_2' instead of f_2, where f_2' is expressed in terms of the reflection coefficient V_2 at the "old boundary" by

$$f_2' = e^{b(z-h_0')}(1 + V_2 e^{2b(h_0-z)}).$$

Actually, as was said above, f_2' is the field at the point of observation due to a plane wave of unit amplitude incident on the upper boundary of the layer (in the present case $z = h_0'$). The factor $\exp[b(z - h_0')]$ in the last expression, multiplied by unity in the parentheses, gives a plane wave which travels from the boundary h_0' to the point of observation z. To this wave we must add the return wave which can be expressed in terms of the reflection coefficient V_2 at the old boundary, and the phase change as the wave traverses the distance between the point of observation and the boundary twice (the second term in the parentheses). We note, incidentally, that the last equation can also be written in the form

$$f_2' = e^{-b\Delta}(e^{-b(h_0-z)} + V_2 e^{b(h_0-z)}).$$

The presence of some freedom in the choice of the position of the boundaries of the layer enables us to carry out the transition to the limit $h_0 \to 0$. Then the two formulas (34.3) and (34.4), in which we must set $h_0 = 0$, will be sufficient for the description of the field. In this form, these formulas will be suitable for the description of the field in an arbitrary layered-inhomogeneous medium, since an infinitely thin homogeneous layer can be imagined as separated out in an arbitrary law of variation of the parameters of the medium. However, this problem will be the subject of a special investigation in the next chapter.

2. *Some properties of the reflection coefficient at an inhomogeneous layer*

We now investigate some of the properties of the functions $V_1(\vartheta)$ and $V_2(\vartheta)$, the reflection coefficients at the inhomogeneous media bounding the layer from below and from above, respectively. As above, we shall not confine ourselves to the investigation of a definite z-dependence of the parameters of the medium.

1. First of all, we shall prove the statement used in the preceding Section, namely, that when

$$\vartheta' = \pm\pi/2, \quad \vartheta'' \to \mp\infty, \tag{34.10}$$

V_1 and V_2 approach zero if the parameters of the medium are continuous functions of z, and approach finite values if the continuity is disrupted. For definiteness, we shall consider one of the reflection coefficients, say V_1, which is the ratio of the fields of the reflected and the incident waves at $z = 0$ (at the lower boundary of the layer). However, in a certain sense (see § 17) we can talk about this ratio at any point z in the inhomogeneous layer. We denote this quantity by $V = V(z)$. The function $V(z)$ is determined by Eq. 17.8.

In place of the function $V(z)$, we introduce a new function $u(z)$ through the relation

$$V(z) = \frac{q(z)\,u(z) - q_1}{q(z)\,u(z) + q_1},\qquad(34.11)$$

where

$$q(z) = \frac{k_0}{m}\sqrt{[n^2(z) - \sin^2\vartheta]},\quad q_1 \equiv (\lim_{z\to-\infty} q) = \frac{k_0}{m_1}\sqrt{(n_1^2 - \sin^2\vartheta)}.$$

Substituting Eq. 34.11 into Eq. 17.8, we obtain the following equation for $u(z)$:

$$\frac{du}{dz} = imq_1\left(1 - \frac{q^2}{q_1^2}u^2\right).\qquad(34.12)$$

Since $V(z)\to 0$ as $z\to-\infty$, the boundary condition required for the solution of Eq. 34.12 will be

$$z\to-\vartheta,\quad u\to 1.\qquad(34.13)$$

Remembering the definition of q (Eq. 17.18), Eq. 34.12 can also be written in the form

$$\frac{du}{dz} = ik_0\frac{m}{m_1}\sqrt{(n_1^2 - \sin^2\vartheta)}\left(1 - \frac{m_1^2}{m^2}\frac{n^2(z) - \sin^2\vartheta}{n_1^2 - \sin^2\vartheta}u^2\right),\qquad(34.14)$$

where the angle of incidence is denoted here by ϑ and not by ϑ_0 as in § 17.

We are interested in the value of the function $V(z)$ at the point $z = 0$ (at the upper boundary of the inhomogeneous layer). We therefore set $z = 0$ in Eq. 34.11. Here, $q(0) = k_0\cos\vartheta$. As a result, Eq. 34.11 gives

$$V_1 \equiv V(0) = \frac{m_1 u_0\cos\vartheta - \sqrt{(n_1^2 - \sin^2\vartheta)}}{m_1 u_0\cos\vartheta + \sqrt{(n_1^2 - \sin^2\vartheta)}},\quad u_0 \equiv u(0,\vartheta).\qquad(34.15)$$

We note that this expression differs from the usual "Fresnel" reflection coefficient at the boundary between two homogeneous media (see Eq. 19.7) only by the factor u_0 in the first term in the numerator and in the denominator.

When Eq. 34.10 is satisfied, we have

$$\sin \vartheta = \sin (\vartheta' + i\vartheta'') = \sin \vartheta' \cosh \vartheta'' + i \cos \vartheta' \sinh \vartheta''$$

$$= \pm \cosh \vartheta'' \rightarrow \pm \infty,$$

$$\cos \vartheta = \cos (\vartheta' + i\vartheta'') = \cos \vartheta' \cosh \vartheta'' - i \sin \vartheta' \sinh \vartheta''$$

$$= \pm i \sinh \vartheta'',$$

$$\sqrt{(n_1^2 - \sin^2 \vartheta)} \rightarrow i \sin \vartheta. \tag{34.16}$$

As a result, Eqs. 34.15 and 34.12 take the form

$$V_1 = \frac{m_1 u_0 - 1}{m_1 u_0 + 1}, \tag{34.17}$$

$$\frac{1}{\sin \vartheta} \frac{du}{dz} = k_0 \frac{m}{m_1} \left(\frac{m_1^2}{m^2} u^2 - 1 \right). \tag{34.18}$$

We now return to the proof of the statement made above regarding the behavior of V_1 as ϑ approaches infinity in accordance with Eq. 34.10. As $|\sin \vartheta| \rightarrow \infty$, Eq. 34.18 gives

$$u(z) = \frac{m(z)}{m_1}. \tag{34.19}$$

This solution also satisfies the boundary condition (34.13), since $m(-\infty) \equiv m_1$. From Eq. 34.19, we have

$$u_0 = 1/m_1,$$

after which Eq. 34.17 gives $V_1 = 0,$

which completes the proof.

In some cases, it is necessary to know the way in which V_1 approaches zero. In these cases, we use Eq. 34.19 for $u(z)$ as a first approximation; substituting it into the left hand side of Eq. 34.18, we find the second approximation for $u(z)$. Substituting the latter into Eq. 34.17, we find

$$V_1 \approx \frac{1}{4k_0 \sin \vartheta} \left(\frac{dm}{dz} \right)_{z=0}. \tag{34.20}$$

The results we have obtained refer to the case in which the function $m(z)$ is continuous. But if there are discontinuities in the z-dependence of m, then, carrying out the limiting transition (34.16) in Eq. 19.7, we obtain

$$V_1 \rightarrow \frac{m_1 - 1}{m_1 + 1}.$$

We note that in the case of horizontally polarized electromagnetic waves, when m is identically equal to unity, we obtain

$$V_1 \to 0$$

for both continuous as well as discontinuous variations of m.

2. We shall prove further that the reflection coefficient V_1 as a function of the grazing angle $\alpha = (\pi/2) - \vartheta$ has the property

$$V_1(-\alpha) = \frac{1}{V_1(\alpha)}. \tag{34.21}$$

Replacing ϑ by α in Eq. 34.15, we obtain the following equation for the reflection coefficient

$$V_1 = \frac{m_1 u_0 \sin \alpha - \sqrt{(n_1^2 - \cos^2 \alpha)}}{m_1 u_0 \sin \alpha + \sqrt{(n_1^2 - \cos^2 \alpha)}}. \tag{34.22}$$

Then Eq. 34.14 for the function $u = u(z, \alpha)$ can be written in the form

$$\frac{du}{dz} = ik_0 \frac{m}{m_1} \sqrt{(n_1^2 - \cos^2 \alpha)} \left(1 - \frac{m_1^2}{m^2} \frac{n^2(z) - \cos^2 \alpha}{n_1^2 - \cos^2 \alpha} u^2\right). \tag{34.23}$$

The limiting condition (34.13) remains unchanged.

The identity (34.21) now follows directly from Eq. 34.22, because, when the sign of α in this expression is changed, only the sign of $\sin \alpha$ changes; everything else, including (as is clear from Eq. 34.23) the function u, remains unchanged.

3. In § 27 we used a very simple formula for the reflection coefficient at an interface between two media, which was valid at small grazing angles (Eq. 27.35). We will now show that, in the case of reflection from an inhomogeneous layer, neglecting the small quantities of the order α^3 and higher, we obtain

for $\qquad n_1 \neq 1: \quad V_1 = -\exp(-2p_1 \alpha),$

for $\qquad n_1 = 1: \quad V_1 = -\exp(2ic_1 \alpha - 2d_1 \alpha^2), \tag{34.24}$

where p_1, c_1 and d_1 are constants. If the medium is not dissipative, then c_1 and d_1 are real.

When an analytic expression for V_1 as a function of α exists, these constants are found by expanding the right and left hand sides of Eq. 34.24 as power series in α, and comparing coefficients of the terms of the same degree. In the general case, however, p_1, c_1 and d_1 can be represented in the form of series, which we shall now derive.

It is clear from Eq. 34.22 that, to expand V_1 in a power series in α, neglecting the term in α^3, it is sufficient to know the series expansion of the function u up to the term in α. Since it follows from Eq. 34.23

that u depends only on even powers of α, we can set $\alpha = 0$ in Eq. 34.23. Then (to this approximation) we obtain the following equation for the function u

$$\frac{du}{dz} = iw_1 m(z) \left(1 - \frac{w^2(z)}{w_1^2} u^2\right), \tag{34.25}$$

where for brevity we have used the notation

$$w(z) = \frac{k_0}{m(z)} \sqrt{[n^2(z) - 1]}, \quad w_1 = \frac{k_0}{m_1} \sqrt{(n_1^2 - 1)}.$$

We shall solve this equation by Picard's method of successive approximations. Setting the right hand side of the equation equal to zero as the first approximation, we obtain $u = $ const, or, taking the limiting condition (34.13) into account, we obtain $u = 1$.

Substituting this value of u into the right hand side of the equation, and integrating, we obtain the second approximation, and so forth.

As a result, using the limiting condition for $u(0)$, we obtain the series

$$u = 1 - iw_1 I(0) + 2 \int_0^{-\infty} mw^2 I(z)\, dz - iw_1 \int_0^{\infty} mw^2 I^2(z)\, dz + ..., \tag{34.26}$$

where

$$I(z) = \int_z^{-\infty} m\left(1 - \frac{w^2}{w_1^2}\right) dz. \tag{34.27}$$

We could have continued this series as far as we pleased. It can be shown that the thinner the layer in which the inhomogeneity of the medium is mainly concentrated, the more rapidly will the series converge.

Now, substituting expression (34.26) for u into Eq. 34.22 and expanding V_1 as a power series in α, we obtain, up to the term in α^2 inclusively,

$$V_1 = -(1 - 2p_1\alpha + 2p_1^2\alpha^2) = -e^{-2p_1\alpha}, \tag{34.28}$$

where

$$p_1 = \frac{m_1 u_0}{\sqrt{(n_1^2 - 1)}}. \tag{34.29}$$

Thus, the first of Eqs. 34.24 can be considered proved.

It is convenient, in the case $n_1 = 1$, to use another equation instead of Eq. 34.12. This other equation is obtained from Eq. 34.12 by the substitution $u(z) = q_1/v(z)$, where $v(z)$ is a new unknown function.

Equation 34.11 for V is then written

$$V(z) = \frac{q(z) - v(z)}{q(z) + v(z)},$$

and at $z = 0$, we have

$$V_1 \equiv V_1(0) = \frac{k_0 \sin \alpha - v(0)}{k_0 \sin \alpha + v(0)}, \tag{34.30}$$

since, according to Eq. 34.11, $q(0) = k_0 \cos \vartheta = k_0 \sin \alpha$. The equation for the function $v(z)$ is found from Eq. 34.12 after the substitution $u = q_1/v$, and will be

$$\frac{dv}{dz} = im(q^2 - v^2) \tag{34.31}$$

with the limiting condition

$$z \to -\infty \quad v \to q_1. \tag{34.32}$$

Integrating Eq. 34.31 by the method of successive approximations, and using the limiting condition, we obtain

$$v(0) = q_1 + iM(0) - 2q_1 \int_0^{-\infty} mM(z)\,dz - i \int_0^{-\infty} mM^2(z)\,dz + \dots, \tag{34.33}$$

where

$$M(z) = \int_z^{-\infty} m(q_1^2 - q^2)\,dz.$$

According to Eq. 34.32, we have $q_1 = (k_0/m_1) \sin \alpha$ when $n_1 = 1$. Therefore, only the first and third of the terms written out in Eq. 34.33 will contain α to the first power. The expansions of the second and fourth terms will contain only even powers of α.

Now, substituting the value of $v(0)$ into Eq. 34.30, we can represent the reflection coefficient in the form

$$V_1 = -\frac{A_1 - ik_0(D_1 - 1)\,\alpha + B_1 \alpha^2}{A_1 - ik_0(D_1 + 1)\,\alpha + B_1 \alpha^2}; \tag{34.34}$$

where

$$A_1 = M_0 - \int_0^{-\infty} mM_0^2\,dz + \dots,$$

$$D_1 = \frac{1}{m_1}\left(1 - 2\int_0^{-\infty} mM_0\,dz + \dots\right), \tag{34.35}$$

and

$$M_0(z) = [M(z)]_{\alpha=0} = k_0^2 \int_z^{-\infty} \frac{1}{m}(1 - n^2)\,dz.$$

Expanding the right hand side of Eq. 34.34 in powers of α up to α^2 inclusively (the terms containing B_1 will drop out) and comparing with the expansion of the second of Eqs. 34.24 in powers of α, we obtain

$$c_1 = \frac{k_0}{A_1}, \quad d_1 = \frac{k_0^2 D_1}{A_1^2}. \tag{34.36}$$

We note that in practice, the integrals in Eq. 34.35 will be taken not from 0 to $-\infty$, but only between the limits of the layer in which the principal inhomogeneities of the medium are concentrated, let us say from 0 to some z_1. As the thickness of this inhomogeneous layer is decreased, the value of the integrals, and hence of A_1, decrease. The coefficients c_1 and d_1 increase under these conditions.

We note that at small grazing angles, we have, according to Eq. 34.24,

$$n_1 \neq 1: \quad |V_1|^2 = 1 - 4\,\mathrm{Re}\,p_1\,\alpha,$$

$$n_1 = 1: \quad |V_1|^2 = 1 - 4d_1\,\alpha^2,$$

that is, as $\alpha \to 0$, the modulus of the reflection coefficient approaches unity quite differently in the two cases. When $n_1 \neq 1$, the curve of $|V_1|^2$ as a function of α has a finite slope at $\alpha = 0$. When $n_1 = 1$, this slope is zero.

4. Since the square root $\sqrt{(n_1^2 - \cos^2\alpha)}$, the sign of which can be chosen in two ways, enters into Eqs. 34.22 and 34.23, it is clear that the reflection coefficient V_1 will be a double-valued function of the angle α. The branch point of this function will be the point $\alpha_1 = \mathrm{arc\,cos}\,n_1$. To determine the amplitude of the lateral wave (see § 27), we must have the expansion of the difference $V_1 - V_1^+$ as a series of powers of the radical $\sqrt{(n_1^2 - \cos^2\alpha)}$, where V_1^+ differs from V_1 by the sign of this radical.

Using Eqs. 34.30 and 34.33, and remembering the definition of q_1 (see Eq. 34.11) we obtain

$$V_1(\alpha) = \frac{k_0 \sin\alpha - E_1 - q_1 F_1}{k_0 \sin\alpha + E_1 + q_1 F_1}, \tag{34.37}$$

where

$$E_1 = iM_0 - i\int_0^{-\infty} mM^2(z)\,dz + \dots, \tag{34.38}$$

$$F_1 = 1 - 2\int_0^{-\infty} mM(z)\,dz + \dots. \tag{34.39}$$

By changing the sign in front of q_1 in Eq. 34.37, we obtain $V_1^+(\alpha)$. Then, forming the difference $V_1 - V_1^+$, and setting $\alpha = \alpha_1 = \mathrm{arc\,cos}\,n_1$ everywhere except in q_1, we obtain

$$V_1 - V_1^+ = 2B_1\sqrt{(n_1^2 - \cos^2\alpha)}, \tag{34.40}$$

where

$$B_1 = -\frac{2k_0^2 \sin\alpha_1 F_1}{m_1[(k_0 \sin\alpha_1 + E_1)^2 - q_1^2 F_1^2]}. \tag{34.41}$$

3. *Normal modes*

Having determined the reflection coefficients V_1 and V_2 at the lower and upper boundaries of the layer, the field of the normal modes is calculated in the same way as was done in § 27, Section 3 for a layer bounded by two homogeneous halfspaces. In particular, Eqs. 27.42 and 27.43, as well as Eq. 27.26 for the determination of the poles and its solutions (27.29) and (27.31), remain valid. We note, furthermore, that a very important statement regarding the absence of a point of accumulation of poles in a finite region of the α-plane can be proved.[41]

We now examine the solution of the equation for the poles (27.26) for small grazing angles. Remembering that the layer thickness is now denoted by h_0, and the wave number in the layer by k_0, this equation can be written in the form

$$2k_0 h_0 \sin \alpha + \phi = 2\pi l, \quad \text{where} \quad \phi = \phi_1 + \phi_2. \tag{34.42}$$

We first consider the case:

(a) $n_1 \neq 1, \quad n_2 \neq 1$.

Comparing the first of Eqs. 34.24 with the first of Eqs. 27.24, we find

$$\phi_1 = 2ip_1 \alpha. \tag{34.43}$$

We set $p_1 = p_1' - ip_1''$, where p_1' and p_1'' are real. We have a similar expression at the upper boundary of the layer (with the subscript 1 replaced by 2). As a result, Eq. 34.42 can be written

$$2k_0 h_0 \sin \alpha + 2i(p_1' + p_2') \alpha + 2(p_1'' + p_2'') \alpha = 2\pi l. \tag{34.44}$$

Neglecting quantities of the order of α^3, we set $\sin \alpha \approx \alpha$. Moreover, we use the notation

$$p_1'' \equiv k_0 h_1, \quad p_2'' \equiv k_0 h_2. \tag{34.45}$$

Then Eq. 34.44 is written

$$2k_0 h\alpha + 2i(p_1' + p_2') \alpha = 2\pi l, \tag{34.46}$$

where h denotes the "effective" thickness of the layer,

$$h \equiv h_0 + h_1 + h_2. \tag{34.47}$$

Letting

$$p' = p_1' + p_2' \tag{34.48}$$

and assuming that $k_0 h \gg p'$, Eq. 34.46 gives

$$\alpha_l \approx \frac{\pi l}{k_0 h} - i\Delta_l, \tag{34.49}$$

where

$$\Delta_l = \frac{\pi l}{(k_0 h)^2} p', \tag{34.50}$$

which agrees with Eq. 27.40, the only difference being that now the effective thickness of the layer appears. The quantity Δ_l determines the damping of the normal mode of order l (see Eq. 27.43).

We now consider the case:

(b) $n_1 = 1, n_2 \neq 1$. Here, we must use the second of Eqs. 34.24 for V_1. We then have

$$\phi_1 = 2c_1 \alpha + 2id_1 \alpha^2, \quad \phi_2 = 2ip_2 \alpha. $$

Substituting these expressions into Eq. 34.42, setting $\sin \alpha \approx \alpha$ therein and carrying out some straightforward transformations, we again obtain Eq. 34.49, in which

$$\Delta_l = \frac{\pi l}{(k_0 h)^2}\left(p_2' + \frac{\pi l}{k_0 h} d_1\right). \tag{34.51}$$

Here, the effective thickness h is determined by Eq. 34.47, in which $h_1 = c_1/k_0$, $h_2 = p_2''/k_0$.

It is interesting to note that Δ_l will depend only on p_2' when $(\pi l/k_0 h) d_1 \ll p_2'$, which is fulfilled when l is not too large, since we have assumed that $k_0 h$ is large. Consequently, in this case, the damping of the normal modes will be determined mainly by the absorption in the upper boundary of the layer.

Finally, we consider the case:

(c) $n_1 = n_2 = 1$.

In the same way as above, we obtain

$$\Delta_l = \frac{(\pi l)^2}{(k_0 h)^3} d, \tag{34.52}$$

where $d = d_1 + d_2$, $h = h_0 + h_1 + h_2$, $h_1 = c_1/k_0$, $h_2 = c_2/k_0$.

4. Lateral waves

In the analysis of Eqs. 34.1–34.4, we will transform the path of integration Γ_1 in the complex plane in the same way as was done in § 27, Section 3. If the conditions $n_1 \neq 1$, $n_2 \neq 1$ are fulfilled, then relations (27.20), which are essential for the transformation, as we showed above, will still be valid (see § 34, Section 2).

As a result, the integrals reduce to the sum of the residues at the poles of the integrands, plus the integrals over the borders of the cuts. The residues yield the lateral waves which we have already investigated. The lines along which the cuts are made will be the same as in the case considered in § 27, Section 4. The method of asymptotic evaluation of the integrals along these lines will also be the same. As a result, we obtain Eq. 27.56 for the lateral wave propagating in the lower halfspace, where now, B_1 is given by Eq. 34.41. The upper boundary of the layer can be treated similarly.

Now let $n_1 = 1$, but $n_2 \neq 1$. In this case $\sqrt{(n_1^2 - \cos^2 \alpha)} = \sin \alpha$, and $V_1(\alpha)$, in distinction from $V_2(\alpha)$ will be a single-valued function of α.† The corresponding branch point will be absent. However, a lateral wave will exist in this case. It will be given by the integral over the

---

† As above, we are here using the grazing angle $\alpha = (\pi/2) - \vartheta$ instead of the angle of incidence.

path Γ_2 (Fig. 122), which now will not be zero (it was equal to zero only under the conditions $n_1 \neq 1$, $n_2 \neq 1$).

We now obtain the analogous expression for this wave, and denote it by W_1†. For definiteness, we examine the case $z < z_0$. The integral expression for W_1 will then coincide with Eq. 34.2, the only difference being that the integral is taken over the path Γ_2. In terms of the variable α, it is written

$$W_1 = -\frac{ik_0}{2} \int_{-i\infty}^{i\infty} \Phi(\alpha) H_0^{(1)}(k_0 r \cos \alpha) \cos \alpha \, d\alpha, \qquad (34.53)$$

where

$$\Phi(\alpha) = \frac{(e^{-bz} + V_1 e^{bz})(e^{-b(h_0 - z_0)} + V_2 e^{b(h_0 - z_0)})}{e^{-bh_0}(1 - V_1 V_2 e^{2bh_0})}. \qquad (34.54)$$

Splitting the integral into two integrals, one from $-i\infty$ to 0 and the other from 0 to $i\infty$, and replacing the variable of integration α by $-\alpha$ in the second of these integrals, we obtain

$$W_1 = \frac{ik_0}{2} \int_0^{-i\infty} (\Phi + \Phi^+) H_0^{(1)}(k_0 r \cos \alpha) \cos \alpha \, d\alpha, \qquad (34.55)$$

where for brevity we have used the notation $\Phi \equiv \Phi(\alpha)$, $\Phi^+ \equiv \Phi(-\alpha)$. Substituting Eq. 34.54 for $\Phi(\alpha)$, letting $V_1(-\alpha) = V_1^+$, and remembering that since $n_2 \neq 1$ we have $V_2(-\alpha) = 1/V_2(\alpha)$ in accordance with the work of §34, Section 2, we obtain

$$W_1 = \frac{ik_0}{2} \int_0^{-i\infty} \frac{(1 - V_1 V_1^+)(V_2 e^{b(h_0 - z)} + e^{-b(h_0 - z)})(V_2 e^{b(h_0 - z_0)} + e^{-b(h_0 - z_0)})}{(e^{-bh_0} - V_1 V_2 e^{bh_0})(V_2 e^{bh_0} - V_1^+ e^{-bh_0})}$$
$$\times H_0^{(1)}(k_0 r \cos \alpha) \cos \alpha \, d\alpha. \qquad (34.55a)$$

When $k_0 r$ is large, the Hankel function can be replaced by its asymptotic representation. As a result, the exponential

$$\exp(ik_0 r \cos \alpha) \approx \exp(ik_0 r) \cdot \exp[-ik_0 r(\alpha^2/2)]$$

appears under the integral, whence it is clear that only small values of α will play a significant rôle. The integral can be evaluated by the method of steepest descents. For this purpose, it is expedient to make the change of variable $\alpha = \sqrt{(2)} \exp[-i(\pi/4)] \cdot s$, as a result of which the exponential is written in the form $\exp(-k_0 r s^2)$, and then to transform the path of integration in such a way that it corresponds to real values of s. In the α-plane, this will be the path going from the point $\alpha = 0$ to the bisector of the 4th quadrant. Only values of α not exceeding

---

† As previously, the lateral wave due to propagation in the medium bounding the layer from above will be given by the integral over the borders of the cut issuing from the branch point $\alpha = \arccos n_2$ (the point A_2 in Fig. 122).

$|\alpha_{max}| \sim 1/\sqrt{(k_0 r)}$ in order of magnitude will be significant under the integral. It is clear from the work in § 34, Section 3 that there are no poles of the integrand lying closer to the point $\alpha = 0$ than $\sim 1/k_0 h$.

We assume that the inequality

$$k_0 h \ll \sqrt{(k_0 r)} \tag{34.56}$$

is fulfilled. In this case, the presence of poles will have no effect on the value of the integral under consideration.

Taking the second of Eqs. 34.24 into account, the expression $1 - V_1 V_1^+$ in the integrand in Eq. 34.55a can be written (to within α^2) as

$$1 - V_1 V_1^+ \approx 1 - \exp(-4d_1 \alpha^2) \approx 4d_1 \alpha^2. \tag{34.57}$$

The other expressions in the integrand can be expanded similarly as series in α. As we have already indicated, we use the asymptotic value of the Hankel function and obtain the second of integrals (19.15), which as a result gives

$$W_1 = \frac{2id_1}{k_0 r^2} \frac{[ip_2' + k_0(h_0 + h_2 - z)][ip_2' + k_0(h_0 + h_2 - z_0)]e^{ik_0 r}}{(ip_2' + k_0 h)^2}, \tag{34.58}$$

where
$$h = h_0 + h_1 + h_2, \quad h_1 = \frac{c_1}{k_0}, \quad h_2 = \frac{p_2''}{k_0}. \tag{34.59}$$

Equation 34.58 remains unchanged when z and z_0 are interchanged, and is consequently valid not only for $z_0 < z < h_0$, but also for $0 < z < z_0$.

If not only $n_1 = 1$, but also $n_2 = 1$, there will not be a single branch point in the complex domain of α. However, the integral over the imaginary axis yields two lateral waves, one of which will be given by the expression

$$W_1 = \frac{2id_1}{k_0 r^2 h^2} (h_0 + h_2 - z)(h_0 + h_2 - z_0) e^{ik_0 r}, \tag{34.60}$$

where $h = h_0 + h_1 + h_2$, $h_1 = c_1/k_0$, $h_2 = c_2/k_0$, and the other is obtained from Eq. 34.60 by replacing the subscript 1 by 2, $h_0 - z$ by z and $h_0 - z_0$ by z_0. The case $n_1 = n_2 \neq 1$ is also interesting. Under these conditions, the integral over the imaginary axis vanishes, and both lateral waves are obtained from the integral over the borders of the one cut issuing from the branch point $\alpha_1 = \arccos n_1$. The analytic expression for one of these lateral waves coincides with Eq. 27.56, and the other is obtained from Eq. 27.56 by making the replacements indicated just above.

Thus, there are always two lateral waves. One of these waves is due to propagation in the lower halfspace, and the other to propagation in the upper halfspace.

5. Averaged decay laws

It was shown in § 33 that very simple laws of fall-off with distance can be obtained for the averaged field intensity. The situation is similar in the present case, in which the layer is bounded by inhomogeneous media.

Thus, for example, we obtain a formula completely identical to Eq. 33.11 for the mean value of the square of the acoustic potential, except that the quantity Δ_l will now be given by the formulas obtained in § 34, Section 3.†

The same calculations as in § 33 yield

$$\overline{|\psi|^2} = \frac{2\pi}{k_0 r h^2}\, \sigma(r), \tag{34.61}$$

where $\sigma(r)$ has different forms for the various cases, namely:

I. $n_1 \neq 1,\quad n_2 \neq 1$:

$$\sigma(r) \equiv \sigma_1(r) = \sum_{l=1}^{\infty} \exp\left[-2p'\left(\frac{l\pi}{k_0 h}\right)^2 \cdot \frac{r}{h}\right]. \tag{34.62}$$

II. $n_1 = 1,\quad n_2 \neq 1$:

$$\sigma(r) \equiv \sigma_2(r) = \sum_{l=1}^{\infty} \exp\left[-2\left(\frac{l\pi}{k_0 h}\right)^2 \cdot \frac{r}{h}\left(p_2' + \frac{l\pi}{k_0 h} d_1\right)\right]. \tag{34.63}$$

Assuming that

$$\frac{l\pi}{k_0 h} d_1 \ll p_2', \tag{34.64}$$

we have

$$\sigma_2(r) = \sum_{l=1}^{\infty} \exp\left[-2p_2'\left(\frac{l\pi}{k_0 h}\right)^2 \cdot \frac{r}{h}\right]. \tag{34.65}$$

III. $n_1 = n_2 = 1$:

$$\sigma(r) \equiv \sigma_3(r) = \sum_{l=1}^{\infty} \exp\left[-2d\left(\frac{l\pi}{k_0 h}\right)^3 \cdot \frac{r}{h}\right]. \tag{34.66}$$

Comparing Eqs. 34.62 and 34.65, we see that the decay law of the acoustic potential (and therefore, the acoustic pressure as well) with distance is the same in the first and second cases, except that the exponent contains p' in the first case, and p_2' in the second. In the third case, the fall-off law turns out to be different, and, as will be shown below, the fall-off is slower. This is connected with the smaller wave absorption upon reflection at small grazing angles in the case $n_1 = n_2 = 1$.

---

† Moreover, we must also assume that the inequality $(h_1 + h_2) \ll h_0$ is satisfied.

It is expedient in the analysis of Eqs. 34.62–34.66 to introduce dimensionless distances, defined as follows:

$$\text{I.} \quad \rho_1 = 2p'\left(\frac{\pi}{k_0 h}\right)^2 \cdot \frac{r}{h},$$

$$\text{II.} \quad \rho_2 = 2p_2'\left(\frac{\pi}{k_0 h}\right)^2 \cdot \frac{r}{h},$$

$$\text{III.} \quad \rho_3 = 2d\left(\frac{\pi}{k_0 h}\right)^3 \cdot \frac{r}{h}. \tag{34.67}$$

We then obtain†

$$\sigma_1 = \sum_1^\infty \exp\left(-\rho_1 l^2\right), \quad \sigma_2 = \sum_1^\infty \exp\left(-\rho_2 l^2\right), \quad \sigma_3 = \sum_1^\infty \exp\left(-\rho_3 l^3\right). \tag{34.68}$$

For small values of ρ_1, ρ_2 and ρ_3 for which the conditions

$$\sqrt{\rho_1} \ll 1, \quad \sqrt{\rho_2} \ll 1, \quad \sqrt[3]{\rho_3} \ll 1, \tag{34.69}$$

are fulfilled, the sums can be replaced by integrals, and we obtain approximately

$$\sigma_1 \approx \frac{1}{2}\sqrt{\left(\frac{\pi}{\rho_1}\right)}, \quad \sigma_2 \approx \frac{1}{2}\sqrt{\left(\frac{\pi}{\rho_2}\right)}, \quad \sigma_3 \approx \frac{0 \cdot 893}{\sqrt[3]{\rho_3}}. \tag{34.70}$$

Substituting these expressions in place of $\sigma(r)$ in Eq. 34.61, we obtain the following decay laws for the square of the acoustic potential:

$$\text{I.} \quad \overline{|\psi|^2} = \sqrt{\left(\frac{\pi}{2p' h}\right)} \cdot \frac{1}{r^{\frac{3}{2}}},$$

$$\text{III.} \quad \overline{|\psi|^2} = \frac{1.786}{\sqrt[3]{(2dh^2)}} \cdot \frac{1}{r^{\frac{4}{3}}}. \tag{34.71}$$

Case II differs from case I only in that p' is replaced by p_2'.

If, on the contrary, the numerical distances are large and the condition

$$\rho_1, \rho_2, \rho_3 \gg 1, \tag{34.72}$$

is fulfilled, we can limit the sums (34.68) to the respective first terms; as a result, Eq. 34.61 yields one and the same exponential fall-off law for all three cases:

$$|\psi|^2 = \frac{2\pi}{k_0 r h^2} e^{-\rho}, \tag{34.73}$$

where ρ takes the values ρ_1, ρ_2 or ρ_3 in the various cases. The functions $\sigma_1(\rho)$ and $\sigma_3(\rho)$ are shown graphically in Fig. 147 over an arbitrary range

---

† When $n_1 = n_2 = 1$, we confine ourselves to the case in which there is no wave absorption due to dissipation in the media bounding the layer [$n(z)$ is real]. Only in this case will ρ_3 be real.

of ρ. Numerical summation was used for intermediate values which did not satisfy conditions (34.69) or (34.72).

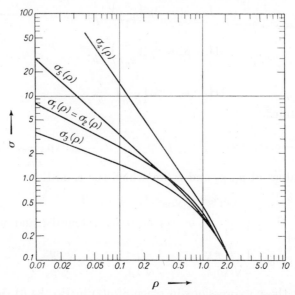

Fig. 147. Graphs of the functions determining the law of fall-off with distance of the acoustic pressure and the electromagnetic field intensity, for various cases.

We now proceed to the averaged laws of fall-off of the components of the electromagnetic field.

Exactly as in §33, we obtain in case I ($n_1 \neq 1$, $n_2 \neq 1$):

$$\overline{E_r^2} = \frac{k_0 \pi^3}{r h^4} \sigma_4(\rho_1), \quad \sigma_4(\rho_1) = \sum_{l=1}^{\infty} l^2 \exp(-\rho_1 l^2),$$

$$\overline{E_z^2} = \overline{H_\phi^2} = \frac{k_0^3 \pi}{r h^2} \sigma_1(\rho_1), \tag{34.74}$$

where ρ_1 and $\sigma_1(\rho_1)$ are determined by Eqs. 34.67 and (34.68), respectively. The function $\sigma_4(\rho_1)$ is shown graphically in Fig. 147. At small values of ρ_1, satisfying the condition $\rho_1^{\frac{3}{2}} \ll 1$, we replace the sum by an integral, and obtain

$$\sigma_4(\rho_1) = \frac{\sqrt{\pi}}{4\rho_1^{\frac{3}{2}}}. \tag{34.75}$$

According to Eqs. 34.74, this gives a decay law of the form $\sim 1/r^{\frac{5}{2}}$ for $\overline{E_r^2}$. The fall-off law will be $\sim 1/r^{\frac{3}{2}}$ for $\overline{E_z^2}$ and $\overline{H_\phi^2}$. When $\rho_1 \geqslant 2$, we will have

$$\sigma_4(\rho_1) \approx \sigma_1(\rho_1) \approx \exp(-\rho_1). \tag{34.76}$$

The fall-off law will be the same in case II ($n_1 = 1$, $n_2 \neq 1$), except that the numerical distance ρ_2 must be substituted in place of ρ.

In case III ($n_1 = n_2 = 1$), we obtain

$$\overline{E_r^2} = \frac{k_0\,\pi^3}{rh^4}\,\sigma_5(\rho_3), \quad \sigma_5(\rho_3) = \sum_{l=1}^{\infty} l^2 \exp\left(-\rho_3\,l^3\right), \tag{34.77}$$

$$\overline{E_z^2} = \overline{H_\phi^2} = \frac{k_0^3\,\pi}{rh^2}\,\sigma_3(\rho_3). \tag{34.78}$$

The function $\sigma_3(\rho_3)$ is defined by Eq. (34.66), and was investigated above. When $\sqrt[3]{\rho_3} \ll 1$, it is given by one of Eqs. 34.70, and we obtain the fall-off law $\sim 1/r^{\frac{4}{3}}$ for $\overline{E_z^2}$ and $\overline{H_\phi^2}$.

The function $\sigma_5(\rho_3)$ was not encountered previously. It is shown graphically in Fig. 147. When $\rho_3 \ll 1$, it is given by the expression

$$\sigma_5(\rho_3) \approx \frac{1}{3\rho_3}, \tag{34.79}$$

which gives the decay law $\sim 1/r^2$ for $\overline{E_r^2}$. When $\rho_3 \gtrsim 1$, we have

$$\sigma_5(\rho_3) \approx \sigma_3(\rho_3) \approx \exp\left(-\rho_3\right).$$

CHAPTER VI

THE FIELD OF A CONCENTRATED SOURCE IN A
LAYERED-INHOMOGENEOUS MEDIUM

In the present chapter, we shall investigate the field of a concentrated source situated in a layered-inhomogeneous medium. This is one of the foremost problems in modern radiophysics, acoustics and the physics of the earth's crust. Investigations of radiowave propagation, taking account of inhomogeneities in the atmosphere, of acoustic propagation in the atmosphere and in the sea, and of the propagation of elastic waves in the crust of the earth, all reduce to this problem. The problem we are presently considering has been studied by a great number of authors, who have suggested various methods for the treatment of the various questions. We do not intend to give a detailed review of these works, but rather, will elucidate the most important aspects of the problem using, in the main, methods and representations developed by the author. A number of very important theories will then remain untouched, among which we should mention first of all the theory of Fok.[78]

We shall not spend any time on the presentation of the theory of wave propagation in a spherically layered medium. If we limit ourselves to the investigation of wave propagation over distances not exceeding one quarter of the circumference of the earth, the problem with spherical symmetry reduces to the problem of propagation in a plane-layered medium (see § 40).

The author is well aware that the material presented below is not always completely rigorous mathematically. However, the consideration of purely mathematical questions would have made the presentation very unwieldy.

The main portion of the chapter (after the first two Paragraphs, which are of a general nature) will be devoted to the consideration of a number of concrete cases of wave propagation. These cases were selected with a view toward acquainting the reader with concrete results which could be useful in his practical work, and also, with the principal methods used in the theory. The material divides naturally into two main groups. The first of these groups is characterized by the presence of waveguide propagation, and the second by the presence of conditions which lead to "anti-waveguide" propagation. In the first of these cases, the inhomogeneities of the medium are such that a certain continuum of rays leaving the source return to the level at which the source is situated. This leads to the concentration of wave energy in a definite layer—the "waveguide". In the other case, all the rays leaving the source do not return to this level, but, as a rule, are bent away from this level by the action of the medium.

§ 35. REVIEW OF THE EXISTING SOLUTIONS

It is clear from the work of the preceding chapter, that to find the field of a point source in a layered-inhomogeneous medium, it is necessary, first of all, to know the plane wave reflection coefficients of the inhomogeneous halfspaces situated above and below the source.

Thus, at the first stage, our problem is reduced to a plane problem.

This is also clear from the fact that after the substitution $\psi = R(r)\,\Phi(z)$ and the separation of variables, the wave equation describing the field of a source in a cylindrical coordinate system†

$$\frac{1}{r}\frac{\partial}{\partial r}\left(r\frac{\partial \psi}{\partial r}\right) + \frac{\partial^2 \psi}{\partial z^2} + k^2(z)\,\psi = 0 \tag{35.1}$$

is written as the two equations

$$\frac{1}{r}\frac{d}{dr}\left(r\frac{dR}{dr}\right) + \xi^2\,R = 0, \tag{35.2}$$

$$\frac{d^2 \Phi}{dz^2} + [k^2(z) - \xi^2]\,\Phi = 0, \tag{35.3}$$

where ξ is the separation constant.

The first of these equations is solved in terms of cylindrical functions. The second coincides with the equation which arises in the problem of plane wave reflection from a layered-inhomogeneous medium (see, for example, Eq. 15.4).

However, having solved the plane problem, i.e. having found the solution of Eq. 35.3, it is by no means always possible to obtain useful results with regard to the problem of the field of a point source in a layered-inhomogeneous medium.

We shall give a short review of the known solutions of the plane problem and of the problem of the point source. We shall study the

---

† The conditions under which the equation of the field in an inhomogeneous medium reduces to the wave equation (35.1) are clear from the work in § 13.

following cases, which differ from one another by the form of the function $k(z)$:

1.
$$k^2(z) = k_0^2(1+az).\qquad(35.4)$$

In this case, the solution of Eq. 35.3 is expressed in terms of cylindrical functions of the $\frac{1}{3}$ order, and was apparently studied for the first time by Gans.[121] The reflection coefficient for a plane wave incident from a homogeneous medium $k(z) = k_0$ onto a medium described by Eq. 35.4, was found in § 15.

Hartree[130, 131] also found the coefficient of reflection from a layer of finite thickness, within which the index of refraction varied in accordance with Eq. 35.4, and outside of which it was constant. A solution of the problem of the reflection of an electromagnetic wave from a layer with a triangular form of $k^2(z)$ is given by Hartree, and also by Haddenhorst.[128]

The problem of the field of a point source in a medium described by Eq. 35.4 was considered by Pekeris.[174] The field of a vertical electric dipole in a medium of the form (35.4) was investigated by Grinberg.[6] By combining layers with different values of a (positive as well as negative), it is possible to construct a theory of the waveguide propagation of radiowaves under conditions close to those existing in the atmosphere.[14]

In the case of the standard atmosphere, the transition to the modified index of refraction (see § 40) is also given by Eq. 35.4.

An important case of wave propagation in a medium consisting of a homogeneous layer and a halfspace with the law (35.4) is investigated below in § 41.

2.
$$k^2(z) = k_0^2(1+az)^m.\qquad(35.5)$$

This case was investigated by Wallot[206] for normal incidence and arbitrary values of a and m. The reflection of plane waves for integral m and arbitrary angles of incidence was studied by Forsterling.[117] The latter author also made a detailed study of the field near $z = 0$ for the case in which the coefficient of Φ in Eq. 35.3 is represented by the series

$$k^2(z) - \xi^2 = a_m z^m + a_{m+1} z^{m+1} + \dots.\qquad(35.6)$$

If we go to the ray treatment of wave propagation, the point $z = 0$ corresponds to a turning point of the ray.

We should also mention the work of Forsterling and Wuster.[118]

3.
$$k(z) = \frac{k_0 a}{z+d}.\qquad(35.7)$$

This case was first studied by Rayleigh (Ref. 15, §148, b) for the case of normal incidence at the layer. Letting $z + d = \zeta$, the solution of Eq. 35.3 has the form

$$\Phi = A\zeta^{\frac{1}{2} \pm im}, \quad m^2 = (k_0 a)^2 - \tfrac{1}{4}.$$

The expression for the reflection coefficient of a wave incident normally on the boundary between a homogeneous medium and a medium in which $k(z)$ is expressed by Eq. 35.7 has the form

$$V = \frac{i}{2(k_0 a + m)}. \tag{35.8}$$

Rayleigh also found the coefficient of reflection from a layer within which the index of refraction varied in accordance with Eq. 35.7 and outside of which it was constant. The results for $|V|^2$—the intensity coefficient of reflection, are for real m:

$$|V|^2 = \frac{\sin^2(m \ln \mu)}{4m^2 + \sin^2(m \ln \mu)}, \tag{35.8a}$$

for imaginary m ($m = im'$):

$$|V|^2 = \frac{\sinh^2(m' \ln \mu)}{\sinh^2(m' \ln \mu) + 4m'^2}.$$

Here, $\mu = n_2/n_1$ is the ratio of the indices of refraction on both sides of the layer.

The more general case

4. $$k^2(z) = k_0^2[a^2 - (b^2/z^2)]$$

can be investigated. The two linearly independent solutions of Eq. 35.3 in this case are the functions

$$z^{\frac{1}{2}} H_p^{(1)}(\beta z) \quad \text{and} \quad z^{\frac{1}{2}} H_p^{(2)}(\beta z),$$

where $$p = (k_0^2 b^2 + \tfrac{1}{4})^{\frac{1}{2}}, \quad \beta^2 = k_0^2 a^2 - \xi^2.$$

The reflection of a plane wave from media of this type was investigated by Rytov and Iudkevich.[71]

The problem of the field of a point source was considered briefly in Ref. 176. In this connection, b^2 can be either positive or negative. When $a = 0$, we obtain the case in which the propagation velocity depends linearly on z: $c \sim c_0(z + d)$, that is, the Rayleigh case considered above. The field of a magnetic dipole in an atmosphere with a linearly varying wave velocity was studied by Eckart.[104] The analogous problem for elastic waves was considered by Alekseev.[31]

5.
$$k^2(z)/k_0^2 = 1 - P(z), \tag{35.9}$$

$$P(z) = c_2 z^2 + c_3 z^3 + c_4 z^4 + c_5 z^5 + c_6 z^6 + \dots.$$

The solution of Eq. 35.3 for this case was investigated by Mullin.[161] By letting

$$\Phi = P^{-\frac{1}{4}} \psi, \quad \zeta = \int_0^z P^{\frac{1}{2}} dz, \quad \epsilon = \frac{1 - \zeta^2}{k_0^2}, \tag{35.10}$$

the equation is transformed into the form

$$\frac{d^2 \psi}{dz^2} + \left(k_0^2 + \frac{k_0^2 \epsilon}{P} - Q \right) \psi = 0, \tag{35.11}$$

where
$$Q = - P^{-\frac{3}{4}} \frac{d^2 P^{-\frac{1}{4}}}{d\zeta^2} = P^{-\frac{1}{4}} \frac{d^2 P^{\frac{1}{4}}}{dz^2}. \tag{35.12}$$

We expand ζ and $(k_0^2 \epsilon / P) - Q$ in a power series in z and keep the first five terms, which corresponds to the representation of $P(z)$ in a power series in z up to the term in z^6 inclusively. Then

$$\frac{d^2 \psi}{d\zeta^2} + \left(\frac{3}{16\zeta^2} + \frac{\lambda_1}{\zeta} + \frac{\lambda_2}{\zeta^{\frac{1}{2}}} + \lambda_0 \right) \psi = 0, \tag{35.13}$$

where

$$\lambda_0 = k_0^2 + \frac{k_0^2 \epsilon}{2c_2} \left(-\frac{3c_4}{2c_2} + \frac{103c_3^2}{72c_2^2} \right)$$
$$+ \frac{3}{2c_2} \left(-\frac{5c_6}{3c_2} + \frac{5c_3 c_5}{2c_2^2} + \frac{109c_4^2}{96c_2^2} - \frac{1723c_3^2 c_4}{576c_2^3} + \frac{12061c_3^4}{13824c_2^4} \right), \tag{35.14}$$

$$\lambda_1 = \frac{1}{2c_2^{\frac{1}{2}}} \left[k_0^2 \epsilon - \frac{1}{8} \left(\frac{3c_4}{c_2} - \frac{7c_3^2}{4c_2^2} \right) \right], \tag{35.15}$$

$$\lambda_2 = \frac{\sqrt{2}}{c_2^{\frac{3}{4}}} \left[-\frac{k_0^2 \epsilon c_3}{3c_2} + \frac{3}{10} \left(-\frac{2c_5}{c_2} + \frac{8c_3 c_4}{3c_2^2} - \frac{28c_3^3}{27c_2^3} \right) \right]. \tag{35.16}$$

The solutions of Eq. 35.13 are the functions of a parabolic cylinder. The two linearly independent solutions can be chosen as

$$\psi_1 = \zeta^{\frac{1}{2}} D_n(y), \quad \psi_2 = \zeta^{\frac{1}{2}} D_{-n-1}(iy), \tag{35.17}$$

where
$$n = -\frac{1}{2} + \frac{i}{\lambda_0^{\frac{1}{2}}} \left(\lambda_1 - \frac{\lambda_2^2}{4\lambda_0} \right),$$

$$y = 2\lambda_0^{\frac{1}{4}} \left(\zeta^{\frac{1}{2}} + \frac{\lambda_2}{2\lambda_0} \right) \exp \left[-i(\pi/4) \right]. \tag{35.18}$$

Thus, the two linearly independent solutions of Eq. 35.3 will be

$$\Phi_1 = P^{-\frac{1}{4}}\zeta^{\frac{1}{4}}D_n(y), \quad \Phi_2 = P^{-\frac{1}{4}}\zeta^{\frac{1}{4}}D_{-n-1}(iy). \qquad (35.19)$$

We note, incidentally, that if we limit Eq. 35.9 to the term in z^2 and set $P(z) = c_2 z^2$, then when $c_2 > 0$ we obtain the Schrödinger equation for the harmonic oscillator.

The case $c_2 < 0$ was investigated in Ref. 141 in connection with the theory of the radio waveguide in the atmosphere adjacent to the earth.

6.
$$\frac{k^2(z)}{k_0^2} = 1 - N\frac{e^{mz}}{1+e^{mz}} - 4M\frac{e^{mz}}{(1+e^{mz})^2}. \qquad (35.20)$$

This case was investigated by Epstein[109] as a variation of the plane problem. In this case, the solutions of Eq. 35.3 are hypergeometric functions. The principal results were presented in § 14.

Rawer[184] showed that by using hypergeometric functions, solutions can be found for considerably more complicated functions $n(z)$ than Eq. 35.20.

Epstein's results are usually used in the study of the reflection of electromagnetic waves from ionospheric layers. However, Gazarian[46] recently investigated the field of a point source in an inhomogeneous medium corresponding to the law (35.20). On this basis, he constructed a very elegant theory of waveguide propagation in an inhomogeneous medium.

7.
$$k(z) = k_0 e^{\alpha z}. \qquad (35.21)$$

This case corresponds to an exponentially varying propagation velocity in the medium: $c = c_0 e^{-\alpha z}$. The reflection of a plane wave from a boundary between a homogeneous medium and a medium in which the wave number varies according to the law (35.21) was studied by Elias[108] and Heller[137]. If we let $v = (k_0/a)e^{\alpha z}$, the solutions of Eq. 35.3 will be the cylindrical functions $H_q^{(1)}(v)$ and $H_q^{(2)}(v)$, where

$$q = (k_0 \sin \vartheta)/\alpha.$$

At comparatively high frequencies and not too large angles of incidence ϑ, the energy coefficient of reflection has the form

$$R = \frac{1}{16(k_0/\alpha)^2 \cos^6 \vartheta + 1}. \qquad (35.22)$$

This expression can also be obtained by using the WKB approximation (see below). These references also contain a comparison of the results of approximate and the exact methods at various frequencies and angles of incidence. The reflection from a symmetrical layer, composed of two exponential laws, was considered in Ref. 128.

8. We will also mention some *approximate methods*. The best known of the approximation methods is applicable when the function $n(z)$ is a slowly varying function (the variation over a wavelength is small). In this case, the geometrical optics approximation is suitable at all points with the exception of those in some region surrounding the turning point of the ray. In the latter region, the solutions are cylindrical functions of the $\frac{1}{3}$ order. This case was examined in detail in §16. The method described in §16 is sometimes called the WKB method (Wentzel–Kramers–Brillouin).

A similar method can also be used in the problem of the field of a point source, in particular, in the determination of the characteristics of the normal modes.[14] In this case, the method is called the *phase integral* method. See Ref. 14 for a detailed presentation of this method. Briefly, the idea behind the method is the following. As is shown in §36 (see also §32), the angle of inclination ϑ_l of the lth normal wave in a layered-inhomogeneous medium can be found from the equation

$$V_1(\vartheta)\, V_2(\vartheta) = 1, \tag{35.23}$$

where $V_1(\vartheta)$ and $V_2(\vartheta)$ are respectively the plane wave reflection coefficients at the lower and upper halfspaces into which our inhomogeneous space is divided by some arbitrarily chosen plane $z = \text{const}$. Setting $V_1 = \exp(i\phi_1)$, $V_2 = \exp(i\phi_2)$, the last equation can be written in the form

$$\phi_1 + \phi_2 = 2\pi l, \quad l = 0, 1, 2, \dots. \tag{35.24}$$

We obtain approximate expressions for ϕ_1 and ϕ_2 from Eq. 16.65

$$\phi_2 = -\tfrac{1}{2}\pi + 2\int_z^{z_m''} k_z\, dz, \quad \phi_1 = -\tfrac{1}{2}\pi + 2\int_{z_m'}^{z} k_z\, dz,$$

where z_m' and z_m'' are the "turning" points of the rays in the lower and the upper halfspaces, respectively. Equation 35.24 can then be written

$$2\int_{z_m'}^{z_m''} k_z\, dz = \pi(2l+1), \quad l = 0, 1, 2, \dots, \tag{35.25}$$

or, since at an arbitrary point z we have $k_z = \sqrt{[n^2(z) - \sin^2\vartheta]}$ and $n(z) = c_0/c(z)$ is the index of refraction with respect to the selected level, then

$$2\int_{z_m'}^{z_m''} \sqrt{[n^2(z) - \sin^2\vartheta]}\, dz = \pi(2l+1). \tag{35.26}$$

The quantities z_m' and z_m'' are determined at a given ϑ by the conditions $n(z_m') = n(z_m'') = \sin\vartheta$. See §37 for a comparison of the results obtained by this method, with the results of exact calculations for one of the cases. A very consistent development of approximate considerations of this kind was given by Haskell.[133]

If we are considering propagation in a halfspace, on the lower boundary of which ($z'_m = 0$) we have $V_1 = -1$, that is, total reflection with phase reversal, then we must set $\phi_1 = \pi$ in Eq. 35.24. As a result, instead of Eq. 35.26, we now have

$$\int_0^{z''_m} \sqrt{[n^2(z) - \sin^2\vartheta]}\, dz = \pi(l + \tfrac{3}{4}).\qquad(35.26a)$$

The phase integral method gives fairly accurate results for the normal modes of high order (in this regard, see the end of § 37). The characteristics of the normal modes of the lowest orders can be determined in a number of cases by an approximate method suggested comparatively recently by Langer.[151]

Also of great importance are approximation methods whose applicability would not be limited to special forms of the function $k(z)$. A method of this kind was developed in § 17 for the approximate determination of the reflection coefficient of plane waves for plane-layered media with an arbitrary form of the function $k(z)$. The application of these results to the problem of the field of a point source was given in § 34.

The reflection of plane waves from layered-inhomogeneous media was also studied by Hines,[138] Schelkunoff [189] and Bailey.[84] We should also mention the work of Forsterling,[116] who investigated the reflection from a medium in which the index of refraction has an arbitrary z-dependence, but differs only slightly from unity.

A number of approximate methods (in particular, a variational method and a perturbation method) were developed in connection with the study of the propagation of electromagnetic waves of radar frequencies in an inhomogeneous atmosphere. Important results were obtained by using modern computing methods. The reader will find a partial review of these methods in Macfarlane[157] and Pekeris.[175] The case in which the inhomogeneity of the medium is described by the linear-exponential law

$$k^2(z) = k_0^2(a + bz + c\, e^{-dz})$$

is examined in detail in these references, and the results for the first normal mode, calculated by various methods, are compared. We shall touch upon these problems briefly in § 40.

§ 36. General Expressions for the Field

1. *Integral expressions for the field*

We shall obtain integral expressions for the field of a point source in a layered-inhomogeneous medium. For this purpose, we refer back to § 34 in which we investigated the field of a source situated in a homogeneous layer, bounded on both sides by layered-inhomogeneous media. The results obtained in § 34 are valid for a layer of any thickness. By letting the thickness approach zero, we obtain the case of a source situated in a layered-inhomogeneous medium.

Setting $h_0 = 0$ in Eqs. 34.3 and 34.4, and placing the origin at the point source (so that $z_0 = 0$), we obtain the following expressions for the field above and below the level at which the source is situated:

$z < 0$:

$$\psi_1 = \frac{ik_0}{2} \int_{\Gamma_1} \frac{1 + V_2}{1 - V_1 V_2} f_1(z, \vartheta) \, H_0^{(1)}(k_0 r \sin \vartheta) \sin \vartheta \, d\vartheta, \qquad (36.1)$$

$z > 0$:

$$\psi_2 = \frac{ik_0}{2} \int_{\Gamma_1} \frac{1 + V_1}{1 - V_1 V_2} f_2(z, \vartheta) \, H_0^{(1)}(k_0 r \sin \vartheta) \sin \vartheta \, d\vartheta, \qquad (36.2)$$

where Γ_1 is the path of integration in the complex plane shown in Fig. 93, k_0 is the wave number at the source level, $V_1 = V_1(\vartheta)$ is the plane wave coefficient of reflection from the lower halfspace, and $f_1(z, \vartheta)$ is a function describing the field in the lower halfspace when a plane wave of unit amplitude is incident on it at an angle ϑ. The symbols $V_2(\vartheta)$ and $f_2(z, \vartheta)$ have similar meanings.

Let us suppose that we have a medium in which the wave number $k(z)$ varies in accordance with the law shown in Fig. 148a. Now let us imagine that the entire space is divided into two halfspaces, and each halfspace is then made into a full space by the addition of a homogeneous medium (Figs. 148, b and c) in which $k(z) = k_0$. Then $V_1(\vartheta)$ is the reflection coefficient for a plane wave incident on the lower inhomogeneous halfspace from the homogeneous medium (Fig. 148b), and $f_1(z, \vartheta)$ is the field in this halfspace when the incident wave is of unit amplitude. Clearly, $f_1(z, \vartheta) = W_1(\vartheta) \phi_1(z, \vartheta)$, where $W_1(\vartheta)$ is some coefficient independent of z, and $\phi_1(z, \vartheta)$ is the linearly-independent solution of the equation for the field in the halfspace $z < 0$ which remains bounded as $z \to -\infty$. The quantities V_1 and f_1 are related by Eqs. 34.6. The symbols $V_2(\vartheta)$ and $f_2(z, \vartheta)$ have analogous meanings in the case of Fig. 148c.

Without carrying out a rigorous justification for the limiting transition ($h_0 \to 0$) which leads to Eqs. 36.1 and 36.2, we can show that:

1. Equations 36.1 and 36.2 satisfy the appropriate field equations for all z. It is clear that this condition is satisfied since we are assuming that $f_1(z, \vartheta)$ and $f_2(z, \vartheta)$ satisfy these equations.

2. Integrals 36.1 and 36.2 converge everywhere except at the point $z = 0$, $r = 0$, at which the field must have a singularity. The proof is exactly the same as that used in § 34, Section 1 for the case of finite h_0.

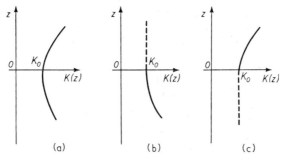

Fig. 148. On the determination of the coefficient of reflection from the lower and the upper inhomogeneous halfspaces.

3. The functions ψ_1 and $\partial\psi_1/\partial z$ transform continuously into ψ_2 and $\partial\psi_2/\partial z$ everywhere in the plane $z = 0$ except at the point $r = 0$. At the point $r = 0$, the quantity $(\partial\psi_2/\partial z) - (\partial\psi_1/\partial z)$ is discontinuous, which in acoustics characterizes the volume velocity of the source. We shall prove this. According to Eqs. 36.1 and 36.2,

$$(\psi_2 - \psi_1)_{z=0} = \frac{ik_0}{2}\int_{\Gamma_1} [(1+V_1)f_2(0, \vartheta) - (1+V_2)f_1(0, \vartheta)]$$
$$\times \frac{H_0^{(1)}(k_0 r \sin \vartheta)}{1 - V_1 V_2} \sin \vartheta \, d\vartheta.$$

Using the first of Eqs. 34.6 and the analogous relation with the subscript 2, we obtain $(\psi_2 - \psi_1)_{z=0} = 0$, that is, as asserted above, the field remains continuous at $z = 0$. Furthermore,

$$\left(\frac{\partial\psi_2}{\partial z} - \frac{\partial\psi_1}{\partial z}\right)_{z=0}$$
$$= \frac{ik_0}{2}\int_{\Gamma_1}\left[(1+V_1)\left(\frac{\partial f_2}{\partial z}\right) - (1+V_2)\left(\frac{\partial f_1}{\partial z}\right)\right]\frac{H_0^{(1)}(k_0 r \sin \vartheta)}{1 - V_1 V_2} \sin \vartheta \, d\vartheta.$$

Using the second of Eqs. 34.6 and the analogous relation with the subscript 2 (with $k_0 \cos \vartheta$ replaced by $-k_0 \cos \vartheta$), we obtain

$$\left(\frac{\partial\psi_2}{\partial z} - \frac{\partial\psi_1}{\partial z}\right)_{z=0} = -k_0^2 \int_{\Gamma_1} H_0^{(1)}(k_0 r \sin \vartheta) \cos \vartheta \sin \vartheta \, d\vartheta,$$

or, changing the variable of integration through the relation $k_0 \sin \vartheta = \xi$,

$$\left(\frac{\partial \psi_2}{\partial z} - \frac{\partial \psi_1}{\partial z}\right)_{z=0} = -\int_{-\infty}^{+\infty} H_0^{(1)}(\xi r)\, \xi\, d\xi = -2\int_0^{\infty} I_0(\xi r)\, \xi\, d\xi.$$

The last integral is equal to zero for all $r \neq 0$. It becomes infinite at $r = 0$, but it can be shown that the integral over the plane $z = 0$ is finite. Thus, Eqs. 36.1 and 36.2 satisfy all the basic requirements imposed on the solutions of our problem.

2. The presence of interfaces

Equations 36.1 and 36.2 are quite general. We apply them to the case of wave propagation in a halfspace bounded on one side by a perfectly reflecting boundary. Let the equation of the boundary be $z = z_1$ (where $z_1 < 0$; see Fig. 149). As previously, we assume that the source is situated at $z = 0$. The point of observation $P(r, z)$ can be

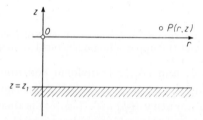

Fig. 149. On the calculation of wave propagation in an inhomogeneous halfspace.

located anywhere in the halfspace $z_1 < z < \infty$. We begin by setting the reflection coefficient $V = -1$ at the boundary, which means that the field at the boundary is assumed to be zero $[\psi(z_1) = 0]$. This case is realized in the propagation of sound in the ocean and, in certain cases, in the propagation of radiowaves over the surface of the earth.[14]

We assume that we know exact or approximate solutions of Eq. 35.3. We denote its two linearly-independent solutions by $\Phi_1(z)$ and $\Phi_2(z)$, where the latter solution corresponds to a wave traveling toward infinity as $z \to \infty$. The functions $V_2(\vartheta)$ and $f_2(z, \vartheta)$ in Eqs. 36.1 and 36.2 are found through the following considerations. We must imagine that the inhomogeneous halfspace $z > 0$ is supplemented by a homogeneous halfspace $z < 0$, from which a plane wave of unit amplitude is incident on the boundary $z = 0$.

The field in this fictitious homogeneous halfspace will then be (the factor $\exp(ik_0 x \sin \vartheta)$ is omitted)

$$\exp(ik_0 z \cos \vartheta) + V_2 \exp(-ik_0 z \cos \vartheta), \qquad (36.3)$$

and the field in the inhomogeneous halfspace will be

$$f_2(\vartheta, z) = W_2 \Phi_2(z). \tag{36.4}$$

For brevity, we temporarily omit ϑ in the argument of Φ_2.

The continuity conditions on the field and its z-derivative at $z = 0$ are written

$$1 + V_2 = W_2 \Phi_2(0),$$

$$b(1 - V_2) = W_2 \Phi_2'(0), \tag{36.5}$$

where
$$\Phi_2'(0) \equiv \left(\frac{d\Phi_2}{dz}\right)_{z=0}, \quad b \equiv ik_0 \cos\vartheta. \tag{36.6}$$

Equations 36.5 give

$$V_2 = \frac{M_2 - 1}{M_2 + 1}, \quad W_2 = \frac{2M_2}{\Phi_2(0)\,(M_2 + 1)},$$

$$M_2 \equiv \frac{bf_2(0)}{f_2'(0)} = \frac{b\Phi_2(0)}{\Phi_2'(0)}. \tag{36.7}$$

To determine $V_1(\vartheta)$ and $f_1(z, \vartheta)$ we must imagine that the inhomogeneous halfspace $z > 0$ is replaced by a homogeneous halfspace in which $k(z) = k(0) = k_0$, and from which a plane wave of unit amplitude is incident on the boundary $z = 0$. The field in this homogeneous halfspace, consisting of the incident and reflected waves, is written

$$\exp\left(-ik_0 z \cos\vartheta\right) + V_1 \exp\left(ik_0 z \cos\vartheta\right). \tag{36.8}$$

The field in the region $z_1 < z < 0$, which becomes zero at the boundary $z = z_1$, is written

$$f_1(z, \vartheta) = W_1[\Phi_1(z)\,\Phi_2(z_1) - \Phi_2(z)\,\Phi_1(z_1)]. \tag{36.9}$$

The continuity conditions at $z = 0$ again give

$$1 + V_1 = f_1(0),$$

$$-b(1 - V_1) = f_1'(0), \tag{36.10}$$

whence, using Eq. 36.9, we find

$$V_1 = \frac{M_1 - 1}{M_1 + 1}, \quad W_1 = \frac{2M_1}{M_1 + 1} \frac{1}{[\Phi_1(0)\,\Phi_2(z_1) - \Phi_2(0)\,\Phi_1(z_1)]}, \tag{36.11}$$

where
$$M_1 \equiv \frac{b[\Phi_1(0)\,\Phi_2(z_1) - \Phi_2(0)\,\Phi_1(z_1)]}{\Phi_2'(0)\,\Phi_1(z_1) - \Phi_1'(0)\,\Phi_2(z_1)}. \tag{36.12}$$

We now substitute the expressions for f_2, V_2, f_1 and V_1, given by Eqs. 36.4, 36.7, 36.9, 36.11 and 36.12, into the expressions for the field,

Eqs. 36.1 and 36.2. After some straightforward transformations, we obtain

$z < 0$:

$$\psi_1 = k_0^2 \int_{\Gamma_1} \frac{[\Phi_1(z)\,\Phi_2(z_1) - \Phi_2(z)\,\Phi_1(z_1)]\,\Phi_2(0)}{w\Phi_2(z_1)}$$
$$\times H_0^{(1)}(k_0 r \sin\vartheta)\cos\vartheta\sin\vartheta\,d\vartheta, \qquad (36.13)$$

$z > 0$:

$$\psi_2 = k_0^2 \int_{\Gamma_1} \frac{[\Phi_1(0)\,\Phi_2(z_1) - \Phi_2(0)\,\Phi_1(z_1)]\,\Phi_2(z)}{w\Phi_2(z_1)}$$
$$\times H_0^{(1)}(k_0 r \sin\vartheta)\cos\vartheta\sin\vartheta\,d\vartheta, \qquad (36.14)$$

where w denotes the Wronskian

$$w = \Phi_1'(0)\,\Phi_2(0) - \Phi_1(0)\,\Phi_2'(0). \qquad (36.15)$$

An integral representation of the solution, where the variable of integration is $\xi = k\sin\vartheta$, is often used. The path Γ_1 in the ϑ-plane corresponds to the real axis in the ξ-plane. Our solution is then written

$z < 0$:

$$\psi_1 = \int_{-\infty}^{+\infty} \frac{[\Phi_1(z)\,\Phi_2(z_1) - \Phi_2(z)\,\Phi_1(z_1)]\,\Phi_2(0)}{w\Phi_2(z_1)} H_0^{(1)}(\xi r)\,\xi\,d\xi,$$

$z > 0$:

$$\psi_2 = \int_{-\infty}^{+\infty} \frac{[\Phi_1(0)\,\Phi_2(z_1) - \Phi_2(0)\,\Phi_1(z_1)]\,\Phi_2(z)}{w\Phi_2(z_1)} H_0^{(1)}(\xi r)\,\xi\,d\xi. \qquad (36.16)$$

It can be shown in the same way that if at $z = z_1$ there is a perfectly reflecting boundary at which the condition $\partial\psi/\partial z = 0$ holds (a solid rigid wall in acoustics), the expressions for the field are written

$z < 0$:

$$\psi_1 = \int_{-\infty}^{+\infty} \frac{[\Phi_1(z)\,\Phi_2'(z_1) - \Phi_2(z)\,\Phi_1'(z_1)]\,\Phi_2(0)}{w\Phi_2'(z_1)} H_0^{(1)}(\xi r)\,\xi\,d\xi, \qquad (36.17)$$

$z > 0$:

$$\psi_2 = \int_{-\infty}^{+\infty} \frac{[\Phi_1(0)\,\Phi_2'(z_1) - \Phi_2(0)\,\Phi_1'(z_1)]\,\Phi_2(z)}{w\Phi_2'(z_1)} H_0^{(1)}(\xi r)\,\xi\,d\xi. \qquad (36.18)$$

Finally, if there are no boundaries, the corresponding expressions will be

$z < 0$:

$$\psi_1 = \int_{-\infty}^{+\infty} \frac{\Phi_1(z)\,\Phi_2(0)}{w} H_0^{(1)}(\xi r)\,\xi\,d\xi, \qquad (36.19)$$

$z > 0$:

$$\psi_2 = \int_{-\infty}^{+\infty} \frac{\Phi_1(0)\,\Phi_2(z)}{w} H_0^{(1)}(\xi r)\,\xi\,d\xi. \qquad (36.20)$$

The last equations can be obtained from Eqs. 36.16, for example, by letting $z_1 \to \infty$, and remembering that under these conditions, $\Phi_1(z_1) \to 0$.

3. *Analysis of the solution. Discrete and continuous spectra*

The analysis of the integral expressions for the field obtained above shows that they reduce to a sum of residues, representing the normal modes, and integrals over the borders of the cuts, representing lateral waves. We can also call the first part of the solution the discrete spectrum, and the second part, the continuous spectrum. An analysis of the general expressions for the field was given, for example, by V. A. Fok,[78] in a somewhat different representation. We will carry out a comparatively brief analysis of Eqs. 36.13 and 36.14, which refer to the case in which the inhomogeneous medium is bounded on one side by a perfectly reflecting boundary. In this connection, we assume that as $z \to \infty$, the propagation velocity in the medium approaches the constant value $c = c_1$. Accordingly, the wave number will be $k = k_1 = \omega/c_1$.

It can be shown, in a manner similar to that in § 34, Section 1, that the only singularities of the functions $\Phi_1(z)$ and $\Phi_2(z)$ in these formulas are branch points. The integral over the real axis can be transformed into an integral over the borders of the cut extending from the branch point $\xi = k_1$, plus the sum of the residues of the integrand at the poles lying in the upper halfplane. The positions of these poles are determined by the equation†

$$\Phi_2(z_1) = 0. \tag{36.21}$$

Methods for the asymptotic evaluation of the integral over the borders of the cut, which gives the lateral wave, were considered in sufficient detail in §§ 27 and 34. Let us consider the sum of the residues. We denote the roots of Eq. 36.21 by $\xi = \xi_l$, where $l = 0, 1, 2, \ldots$.‡ Then, setting $\xi = \xi_l$ everywhere in the integrand in Eqs. 36.16, replacing $\Phi_2(z_1)$ by $[\partial \Phi_2(z_1)/\partial \xi]_{\xi_l}$ in the denominator in accordance with the recipe for finding the residues, and multiplying the entire expression by $2\pi i$, we obtain the following expression for the sum of the residues.

$$\psi(z) = 2\pi i \sum_{l=0}^{\infty} \frac{\Phi_2(0, \xi_l)\, \Phi_2(z, \xi_l)\, H_0^{(1)}(\xi_l, r)\, \xi_l}{(\partial \Phi_2/\partial z)_{z_1,\, \xi_l} \cdot (\partial \Phi_2/\partial \xi)_{z_1,\, \xi_l}}, \tag{36.22}$$

---

† The points $w = 0$ are not poles of the integrands in Eqs. (36.13) and (36.14). The fact that the Wronskian becomes zero indicates that at the corresponding values of ϑ the solutions become linearly-dependent, and the numerators of the integrands also become zero.

‡ If Im $\xi_l = 0$, a pole will lie on the path of integration. This indeterminacy can be removed immediately, however, by assuming that the medium is to some (arbitrarily small) extent absorbing.

where for clarity we have again restored the arguments in the functions $\Phi_1(z, \xi)$ and $\Phi_2(z, \xi)$.

In the derivation of the last expression, we also used the fact that from the condition that the Wronskian be constant, and taking Eq. 36.21 into account, we have

$$w = \Phi_1'(0)\,\Phi_2(0) - \Phi_2'(0)\,\Phi_1(0) = \Phi_1'(z_1)\,\Phi_2(z_1) - \Phi_2'(z_1)\,\Phi_1(z_1)$$
$$= -\Phi_2'(z_1)\,\Phi_1(z_1). \tag{36.23}$$

Equation 36.22 is valid for all z within the limits $z_1 < z < \infty$. The coordinate of the source $z = 0$ and the coordinate of the receiver z have a completely equal status in this expression. The elevations of the source and the receiver above the boundary are frequently denoted by h_1 and h_2, respectively, and the functions $\Phi_2(h_1, \xi_l)$ and $\Phi_2(h_2, \xi_l)$ which characterize the amplitude variations of each normal mode, are called the *height factors*.

We note that Eq. 36.21, and consequently also its solutions, is independent of the source position. Consequently, the characteristics of the normal modes (damping, phase velocity, etc.) which propagate in a given inhomogeneous medium are also independent of the source position. Only the degree of excitation of the various normal modes depends on the position of the source.

If the condition $\partial \psi / \partial z = 0$ (the reflection coefficient $V = 1$) holds at the boundary $z = z_1$, the field of the normal modes is obtained similarly from Eqs. 36.17, and is given by

$$\psi(z) = -2\pi i \sum_{l=0}^{\infty} \frac{\Phi_2(0, \xi_l)\,\Phi_2(z, \xi_l)\,H_0^{(1)}(\xi_l, r)\,\xi_l}{\Phi_2(z_1, \xi_l)\,(\partial^2 \Phi_2 / \partial \xi \, \partial z)_{\xi_l,\, z_1}}. \tag{36.24}$$

§ 37. Waveguides in Inhomogeneous Media

1. *Theory of the simple waveguide*

By waveguide propagation in an inhomogeneous medium, we mean propagation under conditions in which the z-dependence of the index of refraction n and the character of the rays leaving the source have the general behavior illustrated schematically in Fig. 150. As a result of refraction, the rays which leave the source O at sufficiently small grazing angles return, at definite distances, to the level of the source. The case shown in Fig. 151 is also frequently encountered. In this case, one of the boundaries of the waveguide (in the present example, the lower boundary) is a completely or partially reflecting boundary between two media.

Unfortunately, ray constructions which are helpful in elucidating the nature of the wave propagation rather frequently turn out to be completely useless for a quantitative description. Therefore, a complete

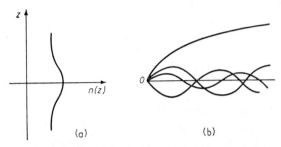

Fig. 150. The z-dependence of the index of refraction and the corresponding paths of the rays in the case of waveguide propagation.

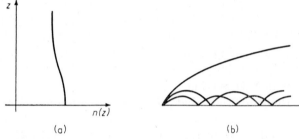

Fig. 151. The same as in Fig. 150, for the case in which the waveguide is bounded on one side by a perfectly reflecting boundary.

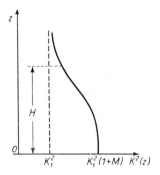

Fig. 152. The z-dependence of the wave number for the Epstein layer.

theory of waveguides in inhomogeneous media must inevitably be based on wave representations. Important work in this direction has been done by Fok[78] and by P. E. Krasnushkin.[54] Gazarian[46] was apparently the first one to investigate a simple case of waveguide propagation to the end. We shall begin with the results obtained by Gazarian since they contain all the fundamental features of this phenomenon.

Let us suppose that in the halfspace $z > 0$, the square of the wave number varies in accordance with the law (Epstein's symmetrical layer, see § 14)

$$k^2(z) = k_1^2 \left[1 + \frac{M}{\cosh^2 \frac{1}{2}(mz)} \right], \quad M > 0, \tag{37.1}$$

which is shown graphically in Fig. 152.

We shall assume that the plane $z = 0$ is a perfectly reflecting boundary on which the condition $\psi = 0$ (or equivalently, the reflection coefficient $V = -1$) holds. We describe the field in the halfspace $z > 0$ by the wave equation (35.1).†

As was shown in § 36, the vertical component of the Hertz vector or the acoustic potential is given by Eqs. 36.16, where $\Phi_1(z)$ and $\Phi_2(z)$ are linearly-independent solutions of Eq. 35.3. The analysis of Eqs. 36.16 yields a sum of normal modes (the discrete spectrum) and a lateral wave (the continuous spectrum) for ψ. We shall consider the normal modes in this Section. This part of the solution is given by Eq. 36.22. Remembering that now the equation of the reflecting boundary is $z_1 = 0$, and the elevations of the source and receiver are denoted by z_0 and z, respectively, Eq. 36.22 can be written

$$\psi = 2\pi i \sum_l \frac{\Phi(z_0, \xi_l) \, \Phi(z, \xi_l)}{[(\partial \Phi / \partial z) \cdot (\partial \Phi / \partial \xi)]_{z=0, \, \xi=\xi_l}} \, H_0^{(1)}(\xi_l r) \, \xi_l. \tag{37.2}$$

For brevity, we have omitted the subscript 2 in $\Phi_2(z, \xi)$. The quantities ξ_l are the roots of the equation

$$\Phi_{z=0} = 0. \tag{37.3}$$

Substituting Eq. 37.1 into Eq. 35.3, we obtain

$$\frac{d^2 \Phi}{dz^2} + \left\{ k_1^2 \left[1 + \frac{M}{\cosh^2 \frac{1}{2}(mz)} \right] - \xi^2 \right\} \Phi = 0. \tag{37.4}$$

---

† It was shown in § 13 that, strictly speaking, the wave equation (35.1) can only describe the field of a magnetic dipole in an inhomogeneous medium. In the case of a vertical electric dipole, and also in acoustics, the equation for the field can also be reduced to a wave equation in which the function $n^2(z)$ has additional terms. The more slowly the parameters of the medium vary with z, the smaller will be the additional terms.

It is expedient to introduce the dimensionless coordinate $\zeta = mz/2$, and to use the notation

$$\left(\frac{2k_1}{m}\right)^2 M = \left(\frac{2\omega}{mc_1}\right)^2 M = \nu(\nu - 1),$$

$$\nu = \tfrac{1}{2} + \sqrt{\left[\frac{1}{4} + \left(\frac{2k_1}{m}\right)^2 M\right]}, \tag{37.5}$$

while we take the arithmetical value of the radical. The quantity ν is analogous to the frequency, and increases from 1 to ∞ as ω varies from 0 to ∞.

Furthermore, we set

$$\mu^2 = (2/m)^2 \, (\xi^2 - k_1^2) \tag{37.6}$$

and agree, for real ξ, to consider

$$\operatorname{Re} \mu \geqslant 0, \quad \text{if} \quad \xi > k_1 \quad \text{and} \quad \operatorname{Im} \mu < 0, \quad \text{if} \quad \xi < k_1.$$

Equation 37.4 is then written

$$\frac{d^2 \Phi}{d\zeta^2} + \left[\frac{\nu(\nu - 1)}{\cosh^2 \zeta} - \mu^2\right] \Phi = 0. \tag{37.7}$$

The solution of this equation, which as $z \to \infty$ is a departing or an exponentially attenuating wave, will be[109]

$$\Phi = e^{-\mu\zeta}(1 + e^{-2\zeta})^\nu F(\nu + \mu, \nu, 1 + \mu, -e^{-2\zeta}). \tag{37.8}$$

As $\zeta \to \infty$, we have $\Phi \approx e^{-\mu\zeta}$. By expressing the spherical harmonics in terms of hypergeometric functions (see Ref. 16, No. 6, 704), Eq. 37.8 can be written

$$\Phi = \Gamma(1 + \mu) \, P_{\nu-1}^{-\mu}(\tanh \zeta). \tag{37.9}$$

Using the formula giving the value of the spherical harmonic of zero argument (see Ref. 16, No. 6, 756), Eq. 37.3 is now written

$$\frac{\Gamma(1 + \mu)}{\Gamma[\tfrac{1}{2}(1 + \mu + \nu)] \, \Gamma[1 + \tfrac{1}{2}(\mu - \nu)]} = 0. \tag{37.10}$$

We are interested only in those poles for which

$$\operatorname{Re} \mu > 0. \tag{37.11}$$

Only these poles correspond to normal modes which decrease (and do not increase) in amplitude as $z \to \infty$. Under these conditions, since we also have $\operatorname{Re} \nu > 0$, the arguments of all the gamma functions in Eq. 37.10, except for $\Gamma[1 + (\mu - \nu)/2]$, lie in the right halfspace, i.e. the values of these functions are neither zero nor infinity. Consequently, the roots of Eq. 37.10 of interest to us are found from the condition

and will be
$$\Gamma[1 + \tfrac{1}{2}(\mu - \nu)] = \infty \qquad (37.12)$$

$$1 + \tfrac{1}{2}(\mu_l - \nu) = -l, \quad \mu_l = \nu - s - 1, \quad s = 2l + 1, \quad l = 0, 1, 2, \dots \quad (37.13)$$

Substitution of these roots into Eq. 37.2 yields the field of the normal modes. In this connection,

$$\xi_l = \sqrt{[k_1^2 + (\tfrac{1}{2}m)^2 \mu_l^2]} \quad \mathrm{Im}\, \xi_l \geqslant 0. \qquad (37.14)$$

Let us calculate the derivatives in the denominator of Eq. 37.2. Using the expression for the spherical harmonic of zero argument (see Ryzhik and Gradshtein, No. 6, 756[16]) and expressing the sine in terms of a product of gamma functions according to the well-known formula

$$\frac{1}{\Gamma(p)} = \frac{\sin \pi p}{\pi} \Gamma(1 - p), \qquad (37.15)$$

and finally, using Eq. 37.18 (the recurrence relation), we obtain

$$\left[\frac{\partial \Phi(z, \xi_l)}{\partial z} \right]_{z=0} = -2^{\nu-1} m \frac{\Gamma(1 + \mu_l)\, \Gamma[\tfrac{1}{2}(\mu_l + \nu + 1)]}{\Gamma(\nu + \mu_l)\, \Gamma[\tfrac{1}{2}(\mu_l - \nu + 1)]}. \qquad (37.16)$$

Furthermore, if we use Eqs. 37.8 and 37.13, we obtain the following expression for one of the derivatives in the denominator of Eq. 37.2:

$$\left[\frac{\partial \Phi(0, \xi)}{\partial \xi} \right]_{\xi = \xi_l} = \left\{ \frac{2^{\nu-1}\,\Gamma(1 + \mu)\,\Gamma[\tfrac{1}{2}(\nu + \mu)]}{\Gamma(\nu + \mu)} \frac{\partial}{\partial \xi} \frac{1}{\Gamma[1 + \tfrac{1}{2}(\mu - \nu)]} \right\}_{\xi_l}.$$

We use the notation $1 + (\mu - \nu)/2 = p$; $p(\xi_l) = -l$. Using Eq. 37.15 we obtain

$$\left(\frac{\partial}{\partial \xi} \frac{1}{\Gamma(p)} \right)_{\xi_l} = \left\{ \frac{\partial}{\partial p} \left[\frac{\sin \pi p}{\pi} \Gamma(1 - p) \right] \right\}_{p=-l} \left(\frac{\partial p}{\partial \xi} \right)_{\xi_l} = (-1)^l \frac{2\xi_l}{m^2 \mu_l} \Gamma(1 + l).$$

As a result, we find

$$\left[\frac{\partial \Phi(0, \xi)}{\partial \xi} \right]_{\xi = \xi_l} = (-1)^l \frac{2^\nu \xi_l}{m^2 \mu_l} \frac{\Gamma(1 + \mu_l)\, \Gamma[\tfrac{1}{2}(\nu + \mu_l)]\, \Gamma(1 + l)}{\Gamma(\nu + \mu_l)}. \qquad (37.17)$$

We now substitute Eqs. 37.16 and 37.17 into Eq. 37.2 for the field of the normal modes. We use the gamma function relations (see Whittaker and Watson, pp. 8 and 11[22])

$$\Gamma(z + 1) = z\Gamma(z), \quad 2^{2z}\,\Gamma(z)\,\Gamma(z + \tfrac{1}{2}) = 2\sqrt{(\pi)}\,\Gamma(2z). \qquad (37.18)$$

We then obtain

$$\psi = i\pi m \sum_{l=0}^{\infty} \frac{\Gamma(\nu + \mu_l)\, \Phi_l(z_0)\, \Phi_l(z)}{\mu_l\, \Gamma^2(\mu_l)\, \Gamma(2l + 2)} H_0^{(1)}(\xi_l r), \qquad (37.19)$$

where μ_l and ξ_l are found from Eqs. 37.13 and 37.14, and for simplicity we have used the notation

$$\Phi(z_0, \xi_l) \equiv \Phi_l(z_0), \quad \Phi(z, \xi_l) \equiv \Phi_l(z). \tag{37.20}$$

The function Φ, given by Eq. 37.8, can also be written in the form†

$$\Phi = e^{-\mu\zeta} F\left(\nu, 1-\nu, 1+\mu, \frac{e^{-\zeta}}{e^{\zeta}+e^{-\zeta}}\right), \tag{37.21}$$

and, with the substitution $\xi = \xi_l$, also in the form[46]

$$\Phi_l = -2\sqrt{(\pi)}\,\Gamma(1+\mu_l)\,2^{2l+2-\nu}\,\frac{\Psi_2}{\Gamma(-\tfrac{1}{2}-l)\Gamma(\nu-l-1)}, \tag{37.22}$$

where Ψ_2 is another solution of the hypergeometric equation

$$\Psi_2 = \sinh\zeta\cosh^{-1-\mu}\zeta.F[1+\tfrac{1}{2}(\mu-\nu),\tfrac{1}{2}(1+\mu+\nu),\tfrac{3}{2},\tanh^2\zeta],$$

which can be expressed in terms of Jacobi polynomials. When $\xi = \xi_l$, we have

$$\Psi_2 = \frac{\tanh\zeta}{\cosh^{\nu-2-2l}\zeta}\,G_l(\nu-2l-\tfrac{1}{2},\tfrac{3}{2},\tanh^2\zeta), \tag{37.23}$$

where

$$\zeta = mz/2 = z/H. \tag{37.24}$$

The quantity H can be called the width of the waveguide (see Fig. 152). Here, in accordance with Eq. 37.5, ν is written

$$\nu = \tfrac{1}{2}+\sqrt{[\tfrac{1}{4}+(k_1 H)^2 M]}, \tag{37.25}$$

and μ_l in Eq. 37.24 will be

$$\mu_l = \nu - 2(l+1). \tag{37.26}$$

The quantity

$$N = (k_1 H)^2 M, \tag{37.27}$$

which we shall call the "wave parameter", is an important characteristic of the waveguide.

The function $\Phi_l(z)$ in Eq. 37.19 characterizes the z-dependence of the normal mode amplitude. Some of the first of the functions Φ_l are shown graphically by the solid lines in Fig. 153, for $N = 63.75$ ($\nu = 8.5$), when there are four undamped normal modes.

As is clear from Eq. 37.2, the phase velocity of each of the normal waves is $v_l = \omega/\xi_l$, or, using Eqs. 37.13, 37.14 and 37.24,

$$v_l = \frac{\omega}{k_1\sqrt{[1+(m\mu_l/2k_1)^2]}} = \frac{c_1}{\sqrt{\{1+1/(k_1 H)^2[\nu-2(l+1)]^2\}}}. \tag{37.28}$$

---

† See Rhyzik and Gradshtein,[16] Sections 7 and 231.

2. *The critical frequency of a waveguide*

As was mentioned above, we are interested only in those solutions of Eq. 37.12 for which the condition $\operatorname{Re} \mu_l > 0$ holds. We see from Eq. 37.26 that for each order l there is a minimum value of ν, and consequently, a minimum value of the frequency for which this condition is satisfied. In other words, there is a critical frequency for a normal mode of a given order. The lowest critical frequency, which corresponds to the normal mode of zero order ($l = 0$), can be called the critical frequency of the waveguide. As is clear from Eq. 37.26, this frequency can be found from the condition $\nu = 2$, or

$$\sqrt{[\tfrac{1}{4} + (k_1 H)^2 M]} = \tfrac{3}{2}.$$

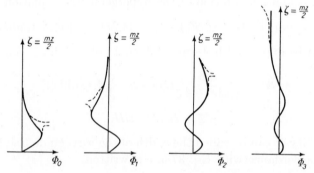

Fig. 153. The first four normal modes, for $\nu = 8.5$.

Hence, the critical wavelength (the maximum wavelength "restrained by the waveguide") is

$$(\lambda_1)_{\max}/H = \pi \sqrt{(2M)}. \tag{37.29}$$

Since M turns out to be very small in practice, $(\lambda_1)_{\max}$ will always be considerably less than the width of the waveguide H. Thus, for earth-adjacent atmospheric waveguides in the case

$$2M = 10^{-4} - 10^{-5}, \quad H = 100 \, \text{m}$$

we have $(\lambda_1)_{\max} \approx 1\text{–}3 \, \text{m}$.

3. *The continuous spectrum*

To obtain the complete expression for the field in the waveguide under consideration, we must supplement the sum of normal modes obtained above, by the addition of the integral over the borders of the cut extending from the branch point $\xi = k_1$. This integral represents a lateral wave, the amplitude of which, as a rule, decreases as $1/r^2$ with distance.

In the special case $1 - \nu = -l$ (l is an integer or zero), the continuous spectrum is represented by an integral, the absolute value of which decreases as $1/r$. In particular, as the inhomogeneity of the medium disappears, and $\nu \to 1$, $l \to 0$, this integral degenerates into the expression for the direct wave e^{ikR}/R.

4. A waveguide bounded on one side by a perfectly rigid wall

We again investigate wave propagation in a halfspace $z > 0$, in which the square of the wave number varies in accordance with the law (37.1). The condition at the boundary of the waveguide is now

$$\left(\frac{\partial \psi}{dz}\right)_{z=0} = 0; \tag{37.30}$$

which in acoustics corresponds to the case of a rigid wall.

In this case, the field of the normal modes will be given by Eq. 36.24, which in the notation of the present Paragraph is written

$$\psi(z) = -2\pi i \sum_{l=0}^{\infty} \frac{\Phi(z_0, \xi_l')\, \Phi(z, \xi_l')\, H_0^{(1)}(\xi_l', r)\, \xi_l'}{\Phi(0, \xi_l')\, (\partial^2 \Phi / \partial \xi\, \partial z)_{\xi = \xi_l',\, z = 0}}, \tag{37.31}$$

where ξ_l' are the roots of the equation

$$\left[\frac{\partial \Phi(\xi, z)}{\partial z}\right]_{z=0} = 0. \tag{37.32}$$

As above, after some manipulations, Eq. 37.8 yields

$$\left(\frac{\partial \Phi}{\partial z}\right)_0 = -\frac{2^{-\mu} m \sqrt{(\pi)}\, \Gamma(1+\mu)}{\Gamma[\frac{1}{2}(\mu+\nu)]\, \Gamma[\frac{1}{2}(1+\mu-\nu)]}. \tag{37.33}$$

Hence, it is clear that Eq. 37.32 can be satisfied when $\mu = \mu_l$, where μ_l is found from the equation

$$\tfrac{1}{2}(1 + \mu_l - \nu) = -l, \quad l = 0, 1, 2, \dots \tag{37.34}$$

or, in analogy with Eq. 37.13, the above equation can be written

$$\mu_l = \nu - s - 1, \quad s = 2l, \quad l = 0, 1, 2, \dots. \tag{37.35}$$

Through calculations similar to those used in the preceding Section Eq. 37.31 yields the following expression for the field of the normal modes

$$\psi = i\pi m \cdot \sum_{l=0}^{\infty} \frac{\Gamma(\nu + \mu_l)\, \Phi_l(z_0)\, \Phi_l(z)}{\mu_l\, \Gamma^2(\mu_l) \cdot \Gamma(2l+1)}\, H_0^{(1)}(\xi_l'\, r), \tag{37.36}$$

where $\Phi_l(z) \equiv \Phi_l(z, \xi_l')$. The first four of the functions $\Phi_l(z)$ are shown by the solid curves in Fig. 154, for $N = (k_1 H)^2 M = 63.75$. These curves are symmetrical with respect to $\zeta = 0$.

In the present case, the normal mode of zero order has no critical frequency (or more exactly, the critical frequency is equal to zero). In fact, when $s = l = 0$ we have $\mu_0 = \nu - 1$, which has a solution for any frequency, including arbitrarily low frequencies.

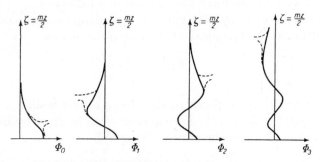

Fig. 154. The same as in Fig. 153, for another boundary condition at the boundary of the waveguide.

We also note that the results obtained in §37, Section 2 and §37, Section 4 allow us to obtain, without difficulty, the expression for the field of the normal modes when the waveguide has no boundary, and the law (37.1) extends to negative values of z. In this case, we are dealing with wave propagation in the symmetrical waveguide shown in Fig. 150. The field of the normal modes will then be expressed as a semi-sum of the "antisymmetric" part of Eq. 37.19 and the "symmetric" part of Eq. 37.36.

5. *Comparison with results obtained by the "phase integral" method*

The solution in the geometrical optics approximation was found in §16. The characteristics of the normal modes can be obtained from Eq. 35.26. If in addition we use the results of §36 and perform the necessary calculations, then, in place of Eq. 37.19, we obtain the approximate value for the field:

$$\psi = \frac{1}{H} \sum_{l=0}^{\infty} Z_l(z_0) Z_l(z) H_0 (\xi_l r). \tag{37.37}$$

We use the notation

$$\zeta = \frac{z}{H}, \quad n_1 = \frac{k_1}{k_0}, \quad \eta = \text{arc cosh}\left[1 - \frac{2l + \frac{3}{2}}{H k_0 \sqrt{(1 - n_1^2)}} \right]^{-1}. \tag{37.38}$$

We then have

$\zeta > \eta$:

$$Z_l(\zeta) = \frac{\exp(i\tfrac{1}{4}\pi)}{\sqrt{(2)}[1 - (\cosh^2\eta/\cosh^2\zeta)]^{\frac{1}{4}}} \exp\left\{Hk_0\sqrt{(1 - n_1^2)}\right.$$

$$\left. \times \left[\operatorname{arc\,cosh}\left(\frac{\tanh\zeta}{\tanh\eta}\right) - \frac{1}{\cosh\eta}\operatorname{arc\,cosh}\left(\frac{\sinh\zeta}{\sinh\eta}\right)\right]\right\}. \quad (37.39)$$

$\zeta < \eta$:

$$Z_l(\zeta) = -\frac{\sqrt{(2)}\exp[i(\tfrac{1}{4}\pi + l\pi)]}{[(\cosh^2\eta/\cosh^2\zeta) - 1]^{\frac{1}{4}}}\sin\left\{Hk_0\sqrt{(1 - n_1^2)}\right.$$

$$\left. \times \left[\frac{1}{\cosh\eta}\operatorname{arc\,sin}\left(\frac{\sinh\zeta}{\sinh\eta}\right) - \operatorname{arc\,sin}\left(\frac{\tanh\zeta}{\tanh\eta}\right)\right]\right\}. \quad (37.40)$$

For ξ_l we obtain

$$\xi_l \approx k_0\sqrt{\left\{n_1^2 + \left[\sqrt{(1 - n_1^2)} - \frac{2l + \tfrac{3}{2}}{k_0 H}\right]^2\right\}}. \quad (37.41)$$

The function Z_l coincides with Φ_l to within a constant factor. The approximate values of the latter are shown in Fig. 153 by the dotted lines.

When the waveguide is bounded on both sides by a "rigid wall", Z_l is again given by Eqs. 37.39 and 37.40, except that in the latter equation the symbol sin is replaced by $-\cos$. In Eqs. 37.38 and 37.41, $2l + 3/2$ is replaced by $2l + 1/2$. The results are otherwise unchanged. The values of Z_l are shown for this case by the dotted lines in Fig. 154.

It is important to note that the solution given by the phase integral method approaches the exact solution as l increases. However, with a normal mode of given order, and increasing frequency, the divergence between the exact and the approximate solutions does not approach zero.[46] Thus, for $l = 0$, it turns out in the case of the symmetrical solution that the principal term in the asymptotic expansion of Z_0 with respect to $1/k_0 H$ differs by 4% between the exact and the approximate solutions. This is not hard to understand when we remember that with increasing frequency we obtain a correct solution in the geometrical optics approximation only if we are considering the field of a plane wave which is incident on the inhomogeneous medium at a definite, *fixed* angle of incidence. Also, if we consider the order of the normal mode as given, and increase the frequency, the grazing angles of the corresponding normal modes will be decreased. Thus, the method under consideration (the WKB method) generally will not yield the asymptotic value of a normal mode of fixed order. However, the method will be asymptotic with respect to frequency and order simultaneously.

§ 38. RAY THEORY OF WAVEGUIDE PROPAGATION IN INHOMOGENEOUS MEDIA

When the width of the waveguide in the inhomogeneous medium is large compared with the wavelength, the number of normal modes participating in the propagation at great distances is very large, and the analysis of the field by the method used in the preceding Paragraph becomes difficult. In these cases, it is expedient to use ray representations, and we shall devote the present Paragraph to their development. It will be shown below how the ray picture is treated in a waveguide and how it is used to determine the intensity of the field. Furthermore, it will be shown how the ray representations follow from the exact theory, and how to calculate the field near a caustic and a point of tangency of caustics, where the ray representations in their usual form are inapplicable.

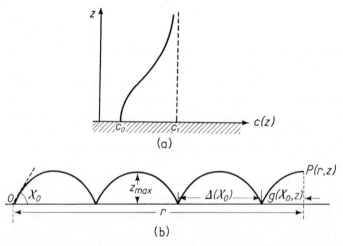

Fig. 155. Schematic illustration of the function $c(z)$ for a waveguide with a reflecting wall, and a sketch of one of the rays.

1. *The ray picture in a waveguide*

For simplicity, we will consider a waveguide in which the level of minimum velocity coincides with a perfectly reflecting boundary (Fig. 155a). We shall assume that the velocity increases monotonically with distance from the boundary. The function $c(z)$ is otherwise arbitrary. The source is assumed to be situated at the level $z = 0$. All of the features of the ray theory as applied to waveguides are brought out by this comparatively simple example. The extension of the theory to the general case presents no difficulty.

Various rays (one of them is shown in Fig. 155b) will arrive at an arbitrarily located observation point $P(r, z)$. Each of these rays can be characterized either by the angle at which it leaves the source, or by the number of times it is reflected from the boundary. We shall let χ_0 be the angle of inclination of the ray with respect to the boundary as it leaves the source, and $\Delta(\chi_0)$ be the distance along the horizontal traversed by the ray between two successive reflections from the boundary. Furthermore, we shall let $g(\chi_0, z)$ be the projection on the plane $z = 0$ of the path traversed by the ray as it travels from the boundary to the point P (Fig. 155b).

One of the following equations will hold:

$$r = N\Delta(\chi_0) + g(\chi_0, z),$$
$$r = (N+1)\Delta(\chi_0) - g(\chi_0, z), \quad N = 0, 1, 2, \dots \quad (38.1)$$

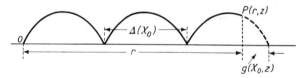

Fig. 156. The form of one of the rays.

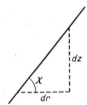

Fig. 157. A ray element.

The first of these two equations refers to the case in which P lies on a rising portion of the ray (Fig. 155b), the second refers to the case in which it lies on a descending portion (Fig. 156), while N is the number of reflections at the boundary $z = 0$. In Figs. 155b and 156, $N = 3$. Equations 38.1 can be considered as the equations for the angle χ_0 at which the ray leaves the source with given r, z and N.

We connect the functions $g(\chi_0, z)$ and $\Delta(\chi_0)$ with the law of variation of the velocity $c(z)$. An element of the ray is shown in Fig. 157. From Fig. 157, we have $dr = dz/\tan\chi$. Therefore

$$g(\chi_0, z) = \int_0^z \frac{dz}{\tan\chi}. \quad (38.2)$$

At an arbitrary point, the angle of inclination of the ray χ is connected by Snell's Law with the angle χ_0 at which the ray leaves the point O:

$$\frac{\cos\chi}{c(z)} = \frac{\cos\chi_0}{c_0}$$

or
$$n\cos\chi = \cos\chi_0, \tag{38.3}$$

where $n = c_0/c(z)$ is the index of refraction. Finding $\tan\chi$ from Eq. 38.3, we obtain

$$g(\chi_0, z) = \cos\chi_0 \int_0^z \frac{dz}{\sqrt{(n^2 - \cos^2\chi_0)}}. \tag{38.4}$$

Clearly,
$$\Delta(\chi_0) = 2g(\chi_0, z_{max}), \tag{38.5}$$

where z_{max} is the maximum distance from the boundary attained by the ray which left the source at the angle χ_0. Since $\chi = 0$ at a point of maximum distance from the boundary (a turning point of the ray), Eq. 38.3 yields an equation for the determination of z_{max}:

$$n(z_{max}) = \cos\chi_0. \tag{38.6}$$

Thus
$$\Delta(\chi_0) = 2\cos\chi_0 \int_0^{z_{max}} \frac{dz}{\sqrt{(n^2 - \cos^2\chi_0)}}. \tag{38.7}$$

Letting N take on the various values in Eq. 38.1, and solving each of the equations successively, we obtain the values χ_{ON} for the angles at which the ray must leave the source if it is to arrive, after a number of reflections, at the point of observation P. A concrete example of these calculations is given in §39.

Many rays, corresponding to various values of N and χ_{ON}, can arrive at each point. However, under certain conditions, there may be points at which neither of Eqs. 38.1 has a solution for any N. Not a single ray will arrive at such a point, and it is said that the point lies in the region of geometrical shadow.

If $c(z)$ increases without limit as $z \to \infty$, the angles χ_{ON} can lie in the entire interval $(0, \pi/2)$. But if $c(\infty)$ is equal to a finite value c_1, the waveguide will "capture" only those rays for which $\cos\chi_0 > n_1 = c_0/c_1$. In fact, if $\cos\chi_0 < n_1$, then, according to Eq. 38.3, as $z \to \infty$ we have a finite angle of inclination of the ray χ, determined by the relation $\cos\chi = \cos\chi_0/n_1$. This ray will travel to infinity without returning to the boundary. The limiting angle

$$\chi_0^{max} = \arccos n_1$$

is an important characteristic of the waveguide.

We now investigate the general case (no reflecting boundary). Let the source be situated at the point $r = 0$, $z = z_0$ (Fig. 158). We consider the two rays leaving the source at the grazing angles χ_0 and $-\chi_0$, respectively, and determine the values of r at which these rays pass through points of given z (the points $A_0, B_0, A_1, B_1, C_1, D_1, \ldots$; see Fig. 158). In Fig. 158, Δ is the length in the r-direction of a full cycle of the ray, $g(z_0)$ is the projection on the r-axis of the path traversed by the ray as it travels from the source to the vertex of the ray, and $g(z)$ is the similar projection as the ray travels from the level z to the vertex.

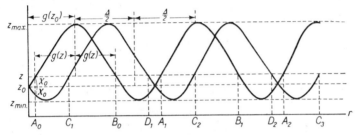

Fig. 158. On the calculation of the ray picture.

In analogy with Eq. 38.4, we have

$$g(z_0) = \cos\chi_0 \int_{z_0}^{z_{max}} \frac{dz}{\sqrt{(n^2 - \cos^2\chi_0)}}, \qquad g(z) = \cos\chi_0 \int_{z}^{z_{max}} \frac{dz}{\sqrt{(n^2 - \cos^2\chi_0)}}.$$

$$\Delta = 2\cos\chi_0 \int_{z_{min}}^{z_{max}} \frac{dz}{\sqrt{(n^2 - \cos^2\chi_0)}}, \tag{38.8}$$

where, as previously, $n(z) = c_0/c(z)$, and $c_0 = c(z_0)$ is the propagation velocity at the level of the source. We note that Δ, $g(z_0)$ and $g(z)$ are strongly dependent on χ_0.

It is not hard to see from Fig. 158 that the horizontal coordinates of the points at which the rays intersect the z-level are given by the relations

$$r_A = N\Delta + g(z_0) - g(z),$$

$$r_B = N\Delta + g(z_0) + g(z),$$

$$r_C = (N+1)\Delta - g(z_0) - g(z),$$

$$r_D = (N+1)\Delta - g(z_0) + g(z). \quad N = 0, 1, 2, \ldots. \tag{38.9}$$

It can turn out that not one of these equations will be satisfied at some arbitrary point of observation $P(r, z)$, with a given source position. This is the case in which P lies in the region of the geometrical shadow.

But if one or several rays arrive at the point P, then one or several equations of the system (38.8) will be satisfied, and their solutions will be the angles χ_0 at which the corresponding rays leave the source.

We now present some formulas which permit us to calculate the time required for a sound pulse to travel from the source to the receiver P along each of the rays. These formulas also permit us to determine the phase of the ray at any point. This information is necessary to calculate the result of the interference of a number of rays.

The time required to travel over an element of length ds of the ray (Fig. 157) is $dt = ds/c$, where $c = c(z)$ is the velocity at the given point, and $ds = dz/\sin\chi$. Using the law (38.3), we obtain

$$dt = \frac{1}{c_0} \frac{n^2\,dz}{\sqrt{(n^2 - \cos^2\chi_0)}}. \tag{38.10}$$

As a result, we obtain:

the time required to travel over a full cycle of the ray:

$$T = \frac{2}{c_0} \int_{z_{\min}}^{z_{\max}} \frac{n^2\,dz}{\sqrt{(n^2 - \cos^2\chi_0)}}; \tag{38.11}$$

the time required to travel over the segment of the ray from the level z_0 to the vertex:

$$t(z_0) = \frac{1}{c_0} \int_{z_0}^{z_{\max}} \frac{n^2\,dz}{\sqrt{(n^2 - \cos^2\chi_0)}}; \tag{38.12}$$

the time required to travel over the segment of the ray from the level z to the vertex:

$$t(z) = \frac{1}{c_0} \int_{z}^{z_{\max}} \frac{n^2\,dz}{\sqrt{(n^2 - \cos^2\chi_0)}}.$$

Equations 38.9 show how the rays arriving at various points are made up of the above ray segments. The total travel time is then easily calculated. Thus, for example, the travel time from the point O to the points $A_0, A_1, A_2, \ldots$ (Fig. 158) is

$$t_A = NT + t(z_0) - t(z), \quad N = 0, 1, 2, \ldots.$$

2. The focusing factor

For definiteness, we consider the sound field of a point, nondirectional acoustic source. In ray theory, the dependence on distance of the intensity of the wave corresponding to any one of the rays is determined by the law of expansion of the ray tube. We will investigate the "focusing factor" $f = I/I_0$, equal to the ratio of the acoustic intensity I at a given point in the inhomogeneous medium to the acoustic intensity I_0 in a homogeneous medium at the same distance. In general, f will differ from unity. As we shall see below, the case in which f is great compared with unity is of particular interest.†

We first find an expression for I at an arbitrary point P. The angle at which the pertinent ray leaves the source is denoted by χ_0. We also investigate the rays leaving the source at angles χ close to χ_0. The horizontal distance covered by an arbitrary ray in traveling to any point lying at the same level z as P is denoted by $r(\chi)$, while the grazing angle of the ray at the point is given by χ_P. Figure 159 shows two rays,

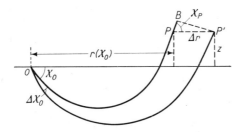

Fig. 159. On the calculation of the focusing factor.

which leave the source at the angles χ_0 and $\chi_0 + \Delta\chi$, and are the boundaries of the cross section of the ray tube in the plane of the Figure. The cross section of this tube is

$$BP' = PP' \sin\chi_P = \left(\frac{\partial r}{\partial\chi}\right)_{\chi_0} \Delta\chi \sin\chi_P.$$

Taking the cylindrical symmetry of the problem into account, the area of the wave front contained between these two rays is

$$ds = 2\pi r \sin\chi_P \left(\frac{\partial r}{\partial\chi}\right)_{\chi_0} \Delta\chi_0.$$

Let W be the total power emitted by the source. The energy radiated in the interval $d\chi_0$ is

$$dW = \tfrac{1}{2} W \cos\chi_0 \, \Delta\chi_0.$$

Consequently, the acoustic intensity at the point P is

$$I = \frac{dW}{ds} = \frac{W \cos\chi_0}{4\pi r} \cdot \frac{1}{\sin\chi_P (\partial r/\partial\chi)_{\chi_0}}. \tag{38.13}$$

---

† The case $f < 1$ is of course also possible. It would then be more natural to call f the divergence factor.

In the case of a homogeneous space, $I_0 = W/4\pi R^2$, where R is the distance OP from the source to the receiver. Therefore, the focusing factor will be

$$f = \frac{I}{I_0} = \frac{R^2 \cos \chi_0}{r(\partial r/\partial \chi)_{\chi_0} \sin \chi_P}.$$

(38.14)

Since we are mainly interested in the cases $r \gg z$, we can set $R \approx r$ and $\cos \chi_0 \approx 1$ in the last equation, and obtain

$$f = \frac{r}{(\partial r/\partial \chi)_{\chi_0} \sin \chi_P}.$$

(38.14a)

An example of the calculation of the focusing factor is given in the next Paragraph. We might mention that similar problems are examined in Ref. 164.

3. *The ray theory of waveguides in inhomogeneous media as a limiting case of the wave theory*

We shall show the conditions under which the exact theory of waveguide propagation in inhomogeneous media reduces to the ray picture. We shall also obtain the desired formulas for the calculation of the field near a caustic, which is the next step in the refinement of the ray theory.

In order not to complicate the picture by nonessential details, we again investigate the case of wave propagation in a halfspace, with the assumption that the reflection coefficient at the boundary $z = 0$ is equal to unity (Fig. 155a).

To simplify the arguments, we assume at first that a homogeneous layer of thickness h_0 is adjacent to the boundary. In the end, we can let h_0 approach zero.

By representing the initial field of the source (a spherical wave) as a superposition of plane waves, we can represent the field at any point in the halfspace as an infinite series of waves which are obtained by successive reflections from the boundary $z = 0$ and from the inhomogeneous halfspace.

The field in the lower halfspace was calculated in this way in § 27, Section 1, and as an example, four waves penetrating into the lower halfspace after a number of reflections are shown in Fig. 121. The discussion for the halfspace bounding the layer from above is, of course, the same, except that, as we have repeatedly mentioned, we must make the replacements $V_1 \gtrless V_2$, $z_0 \to h_0 - z_0$. As a result, after multiple reflections, each of the plane waves (with the wave vector components k_x, k_y, k_z) into which the spherical wave is expanded yields the following expression for the field at an arbitrary point $(r, z > h_0)$

$$\exp\left[i(k_x x + k_y y)\right] f(z, \vartheta) \sum_{N=0}^{\infty} \{\exp\left[ik_z(h_0 - z_0)\right]$$
$$+ V_1 \exp\left[ik_z(h_0 + z_0)\right]\} (V_1 V_2)^N \exp\left(2ik_z N h_0\right),$$

where the function $f(z, \vartheta)$ expresses the value of the acoustic potential at an arbitrary level z when a plane wave of unit amplitude is incident on the boundary $z = h_0$, from the homogeneous layer, at an angle of incidence ϑ.

Collecting all the plane waves which enter into the expansion of the spherical wave, i.e. integrating over ϕ and ϑ (see § 18), we obtain

$$\psi = -\frac{ik_0}{2} \sum_{N=0}^{\infty} \int_{\Gamma_1} \{\exp\left[b(h_0 - z_0)\right] + V_1 \exp\left[b(h_0 + z_0)\right]\}$$
$$\times (V_1 V_2)^N \exp\left(2bN h_0\right) f(z, \vartheta) H_0^{(1)}(k_0 r \sin \vartheta) \sin \vartheta \, d\vartheta, \quad (38.15)$$

where $b = ik_z = ik_0 \cos \vartheta$.

The path of integration Γ_1 is shown in Fig. 93. The last expression could also have been obtained from Eq. 34.4 by expanding $(1 - V_1 V_2 e^{2bh_0})^{-1}$ in the integrand in a power series in $V_1 V_2 e^{2bh_0}$.

As has already been mentioned, we assume that $V_1 = 1$, $h_0 = 0$ and $z_0 = 0$. Then Eq. 38.15 is written

$$\psi = ik_0 \sum_{N=0}^{\infty} \int_{\Gamma_1} V_2^N f(z, \vartheta) H_0^{(1)}(k_0 r \sin \vartheta) \sin \vartheta \, d\vartheta, \quad (38.16)$$

where $V_2 = V_2(\vartheta)$ is the plane wave reflection coefficient from the halfspace $z > 0$. In what follows, we omit the subscript 2 which is not needed, and in place of ϑ we use the new variable of integration $\xi = k_0 \sin \vartheta$. The path of integration in the ξ-plane lies on the real axis and extends from $-\infty$ to $+\infty$. The variable ξ represents the projection on the plane $z = 0$ of the wave vector of each of the plane waves. The expression for ψ can now be written

$$\psi = i \sum_{N=0}^{\infty} \int_{-\infty}^{+\infty} V^N f(z, \xi) H_0^{(1)}(\xi r) \left[\frac{\xi \, d\xi}{\sqrt{(k_0^2 - \xi^2)}}\right]. \quad (38.16a)$$

The sign of the radical is chosen such that

$$\left[\mathrm{Im} \sqrt{(k_0^2 - \xi^2)}\right]_{\xi \to \infty} > 0. \quad (38.17)$$

We shall now show that each term in the last expression gives the field of an individual ray, and that N has the same meaning (the number of vertices of the ray) as in the first Section.

16

The field in the halfspace $z > 0$, when a plane wave is incident on its boundary, was found in § 16, Section 4, and is given, far from a "turning point", by Eq. 16.52.

In this equation, $\vartheta = \vartheta(z)$ is a variable angle connected with ξ by Snell's law

$$k(z) \sin \vartheta(z) = k_0 \sin \vartheta = \xi, \tag{38.18}$$

whence

$$n(z) \cos \vartheta(z) = \frac{1}{k_0} \sqrt{[k^2(z) - \xi^2]}, \quad n(z) = \frac{k(z)}{k_0}.$$

According to Eq. 16.63, the constants A and B in Eq. 16.52 are related by $A = B e^{i\pi/3}$. To obtain the expression for $f(z, \xi)$, the constant B in $Z(z)$ must be chosen in such a way that the amplitude of the incident wave (the first term in the parentheses in Eq. 16.52 is equal to unity at $z = 0$. This means that we must set (in our notation)

$$B\left(\frac{a}{k_0}\right)^{\frac{1}{4}}\left\{\sqrt{\left(\frac{3}{\pi n \cos \vartheta}\right)} \exp\left[i\int_{z_m}^{z} \sqrt{[k^2(z) - \xi^2]} \, dz\right] \exp\left[i(5\pi/12)\right]\right\}_{z=0} = 1.$$

As a result, we find

$$(z, \xi) = \left(\frac{\beta_0}{\beta}\right)^{\frac{1}{4}} \left\{ \exp\left(i\int_0^z \beta \, dz\right) \right.$$
$$\left. + \exp\left[-\tfrac{1}{2}(i\pi) + 2i\int_0^{z_m} \beta \, dz\right] \exp\left(-i\int_0^z \beta \, dz\right)\right\}, \tag{38.19}$$

where for brevity, we have used the notation

$$\beta = \beta(z) = \sqrt{[k^2(z) - \xi^2]}, \quad \beta_0 = \beta(0) = \sqrt{(k_0^2 - \xi^2)}. \tag{38.20}$$

The signs of the radicals are chosen in accordance with Eq. 38.17. The first term in the square brackets in Eq. 38.19 represents the "direct" wave, propagating in the positive z-direction, and the second term represents the "return" wave. The ratio of these two waves at $z = 0$, which we call the reflection coefficient, will be

$$V = e^{i\phi}, \quad \phi = -\tfrac{1}{2}\pi + 2\int_0^{z_m} \beta \, dz, \tag{38.21}$$

which could also have been obtained from Eq. 16.65 by setting $z = 0$.

We recall that z_m is determined from the condition $\beta(z_m) = 0$. For brevity, we again use the notation

$$\gamma(z, \xi) = \int_0^z \beta \, dz. \tag{38.22}$$

Now, substituting Eq. 38.19 into Eq. 38.16a, and using the asymptotic representation of the Hankel function, we obtain

$$\psi = \sum_{N=0}^{\infty} (\psi_N^+ + \psi_N^-), \tag{38.23}$$

where

$$\psi_N^+ = \sqrt{\left(\frac{2}{\pi r}\right)} \exp\left(i\tfrac{1}{4}\pi\right) \int_{-\infty}^{+\infty} \exp\left[i(\gamma + N\phi + \xi r)\right] \sqrt{\left(\frac{\xi}{\beta_0 \beta}\right)} \, d\xi,$$

$$\psi_N^- = \sqrt{\left(\frac{2}{\pi r}\right)} \exp\left(i\tfrac{1}{4}\pi\right) \int_{-\infty}^{+\infty} \exp\left\{i[-\gamma + (N+1)\phi + \xi r]\right\} \sqrt{\left(\frac{\xi}{\beta_0 \beta}\right)} \, d\xi. \tag{38.24}$$

We shall show that in the limiting case of high frequencies, ψ_N^+ and ψ_N^-, $N = 0, 1, 2, \ldots$ give the field of each of the rays considered in the first Section. We will take ψ_N^+ as an example, and will evaluate the integral in the complex plane by the method of steepest descents. We will use the notation

$$w(r, z, \xi) = \gamma + N\phi + \xi r. \tag{38.25}$$

The passage point $\xi = \xi_0$ is found from the equation $(dw/d\xi)_{\xi_0} = 0$, or

$$r = -(N\phi' + \gamma')_{\xi_0}, \tag{38.26}$$

where the prime indicates the derivative with respect to ξ. Using the relation $\xi_0 = k_0 \sin \vartheta_0 = k_0 \cos \chi_0$; $(\chi_0 + \vartheta_0 = \pi/2)$, the passage point ξ_0 can be associated with an angle of incidence ϑ_0 or a grazing angle χ_0. It is not hard to show that

$$\Delta(\chi_0) = -(\phi')_{\xi_0}, \tag{38.27}$$

where $\Delta(\chi_0)$ is the horizontal displacement of the ray making an angle of χ_0 with respect to the boundary, between two successive reflections from the boundary. For this purpose, it is sufficient to differentiate Eq. 38.21 with respect to ξ, and to set $\xi = \xi_0 = k_0 \cos \chi_0$. The same result also follows from the general equation for the ray displacement, Eq. 8.25, in which p and α correspond to ξ and ξ_0.

In the same way, by differentiating Eq. 38.22 and comparing with Eq. 38.4, it is shown that

$$-(\gamma')_{\xi_0} = g(\chi_0, z).$$

As a result, Eq. 38.26 is written

$$r = N\Delta(\chi_0) + g(\chi_0, z),$$

which agrees with the first of Eqs. 38.1.

Had we considered the expression for ψ_N^-, we would have obtained a ray from the family given by the second of Eqs. 38.1. Thus, the solution of Eq. 38.26 for the determination of the saddle point yields the angle χ_0 at which the ray leaves the source.

We shall now evaluate the integral in the expression for ψ_N^+. The expression multiplying the exponential can be treated as a slowly varying function, and can be taken outside the integral sign at the value $\xi = \xi_0$. In this connection,

$$(\beta_0)_{\xi_0} = \sqrt{(k_0^2 - \xi_0^2)} = k_0 \sin \chi_0, \quad (\beta)_{\xi_0} = \sqrt{[k^2(z) - \xi_0^2]} = k_0 \sqrt{[n^2(z) - \cos^2 \chi_0]}$$

or, since $n(z) \cos \chi_P = \cos \chi_0$ according to the law of refraction, we have

$$(\beta)_{\xi_0} = k_0 n(z) \sin \chi_P,$$

where χ_P is the grazing angle of the ray at a point P at an arbitrary level z.

Furthermore, expanding the expression in the exponent in the integrand in a power series in $\xi - \xi_0$, and remembering that $w'_{\xi_0} = 0$, we obtain

$$w = w_0 + \tfrac{1}{2} w_0''(\xi - \xi_0)^2, \tag{38.28}$$

where for brevity, we have used the notation

$$w_0 \equiv w(r, z, \xi_0), \quad w_0'' = \left(\frac{\partial^2 w}{\partial \xi^2}\right)_{\xi_0}.$$

Now, Eq. 38.24 gives

$$\psi_N^+ = \left(\frac{2 \cos \chi_0}{\pi r \, k_0 \, n \sin \chi_0 \sin \chi_P}\right)^{\frac{1}{2}} \cdot \exp\left[i(w_0 + \tfrac{1}{4}\pi)\right] \int_L \exp\left[\tfrac{1}{2} i \, w_0''(\xi - \xi_0)^2\right] d\xi.$$

We let

$$\tfrac{1}{2} w_0''(\xi - \xi_0)^2 = is^2$$

in the integrand, and replace the integration over ξ from $-\infty$ to $+\infty$ by the integration over the "passage" path, on which s takes on all real values from $-\infty$ to $+\infty$.† Then, using the Poisson integral (19.15), we obtain

$$\psi_N^+ = 2 \sqrt{\left(\frac{\cos \chi_0}{n k_0 \, r w_0'' \sin \chi_P \sin \chi_0}\right)} \, e^{iw_0}. \tag{38.29}$$

Thus, the phase of the wave is

$$w_0 = (\gamma + N\phi + \xi r)_{\xi_0}. \tag{38.30}$$

Let us suppose, for the time being, that $z = 0$, and consequently, $\gamma = 0$. Then $w_0 = N\phi(\xi_0) + \xi_0 r$. Thus, the phase change along the ray can be determined by ascribing the wave number $\xi_0 = k_0 \cos \chi_0$

---

† It may turn out, in transforming from one path to the other, that we have to make a circuit around the cut extending from the branch point of the function $\phi(\xi)$. This gives the additional lateral wave, investigated in § 34.

(the velocity of $c_0/\cos\chi_0$) to it as it propagates in the r-direction, and taking into account that there is an addition ϕ to the phase at each reflection from the inhomogeneous medium.

However, it is not hard to show that the phase w_0 is sufficiently close to the value calculated from the travel time of a sound wave along the corresponding ray, using the formulas of the first Section. If the travel time is t, the phase will be $\Phi = \omega t = k_0 c_0 t$. As a result, taking Eq. 38.10 into account, we get for the phase

$$\Phi = 2k_0 N \int_0^{z_m} \frac{n^2\, dz}{\sqrt{(n^2 - \cos^2\chi_0)}} + k_0 \int_0^z \frac{n^2\, dz}{\sqrt{(n^2 - \cos^2\chi_0)}}. \tag{38.31}$$

The first integral gives the travel time over N complete cycles of the ray, while the second integral gives the travel time over the segment of the ray from the boundary $z = 0$ to the level z.

We subtract and add $\cos^2\chi_0$ to the numerator in each of the integrals. Then the first integral, for example, becomes

$$2k_0 N \int_0^{z_m} \sqrt{(n^2 - \cos^2\chi_0)}\, dz + 2k_0 N \cdot \cos^2\chi_0 \int_0^{z_m} \frac{dz}{\sqrt{(n^2 - \cos^2\chi_0)}},$$

or, using Eqs. 38.7 and 38.21,

$$N(\phi + \tfrac{1}{2}\pi)_{\xi_0} + N\Delta(\chi_0) \cdot k_0 \cos\chi_0.$$

Transforming the second term in Eq. 38.31, and remembering the meaning of r as given by the first of Eqs. 38.1, we finally obtain

$$\Phi = (\gamma + N\phi + \xi r)_{\xi_0} + \tfrac{1}{2}(N\pi).$$

Comparison with Eq. 38.30 shows that the phase calculated from the travel time along the ray differs only by $N(\pi/2)$ from the "exact" phase w_0. This difference is negligibly small under the conditions to which our theory is applicable (the properties of the medium vary slowly with z) and is obtained because in the ray theory we did not take into account the term $-\pi/2$ which is present in the exact expression (38.21).

We shall now show that the amplitude of the wave (38.29) can also be obtained with sufficient accuracy from the ray theory, by using the concept of the focusing factor, introduced above. Using Eq. 38.25, we obtain

$$w' = \gamma' + N\phi' + r.$$

We define the function $r(z, \xi)$ by the relation

$$r(z, \xi) = -(\gamma' + N\phi') \tag{38.32}$$

16*

and distinguish it from the prescribed quantity r which characterizes the position of the receiver.

According to Eq. 38.26, we have $r(z, z_0, \xi_0) = r$. In this notation, we have

$$w' = r - r(z, \xi), \qquad (38.32a)$$

and moreover, $(w')_{\xi_0} = 0$. Furthermore

$$w'' = -\frac{\partial r(z, \xi)}{\partial \xi}, \quad w_0'' = -\left(\frac{\partial r}{\partial \xi}\right)_{\xi_0}.$$

Remembering that $\xi = k_0 \cos \chi$,

$$\left(\frac{\partial}{\partial \xi}\right) = -\left(\frac{1}{k_0 \sin \chi}\right)\left(\frac{\partial}{\partial \chi}\right)$$

and $\xi_0 = k_0 \cos \chi_0$, we obtain

$$k_0 \sin \chi_0 \, w_0'' = \left(\frac{\partial r}{\partial \chi}\right)_{\chi_0}.$$

As a result, Eq. 38.29 gives

$$|\psi_N^+| = 2\left(\frac{\cos \chi_0}{nr \sin \chi_P \, (\partial r / \partial \chi)_{\chi_0}}\right)^{\frac{1}{2}}. \qquad (38.33)$$

Knowing the potential ψ_N^+, the acoustic intensity (energy flux) is obtained from the equation

$$I = |p\mathbf{v}| = \left|\rho \frac{\partial \psi_N^+}{\partial t} \operatorname{grad} \psi_N^+\right|.$$

But

$$\frac{\partial \psi_N^+}{\partial t} = -i\omega\psi_N^+, \quad \operatorname{grad} \psi_N^+ \approx |\psi_N^+| \, |\operatorname{grad} w_0| = |\psi_N^+| \, k_0 \, n(z),$$

where $k_0 \, n(z) = k(z)$ is the wave number at a given point.

Thus, at an arbitrary point P, the acoustic intensity will be

$$I = \rho\omega \, k_0 \, n(z) \, |\psi_N^+|^2,$$

or, using Eq. 38.33,

$$I = \frac{4\rho\omega \, k_0 \cos \chi_0}{r \sin \chi_P (\partial r / \partial \chi)_{\chi_0}}. \qquad (38.34)$$

The last expression differs from Eq. 38.13, which was obtained from the ray representation, only by constant factors which we have omitted. In particular, we omitted the factor characterizing the power of the source. These factors are of no interest to us in the present case because we shall immediately proceed to a relative quantity—the focusing factor.

In the case of a homogeneous halfspace, we obtain $|\psi| = 2/R$, because the source is situated at a rigid boundary, and consequently, the acoustic potential at any point is twice as great as it would be with the source situated in free space. Substituting this expression in place of ψ_N^+, and remembering that in this case $n(z) = 1$, we obtain the acoustic intensity I_0 in free space, to within a constant factor. Finally, for $f = I/I_0$ (the density is assumed constant), we obtain

$$f = \frac{R^2 \cos \chi_0}{r \sin \chi_P (\partial r/\partial \chi)_{\chi_0}}, \qquad (38.35)$$

which agrees with Eq. 38.14 obtained earlier.

4. The field near the caustic

In evaluating the integral (38.24) for ψ_N^+, we expanded the phase $w(r, z, \xi)$, given by Eq. 38.25, in a series about the saddle point, and limited ourselves to the quadratic term proportional to w_0''—the second derivative of the phase at the saddle point. This limitation is legitimate only when w_0'' is not too small. This is also clear from the fact that the expression we obtained for the potential, Eq. 38.29, increases without bound as $w_0'' \to 0$, which is devoid of physical meaning.

Let us now suppose that

$$w_0' = w'(r, z, \xi_0) = 0, \qquad w_0'' = w''(r, z, \xi_0) = 0. \qquad (38.36)$$

Treating ξ_0 as a parameter and eliminating it from these two equations, we obtain the equation of a curve

$$F(r, z) = 0. \qquad (38.37)$$

This curve is called the caustic. Ray methods cannot be used to calculate the field on and near this curve (the focusing factor increases without bound). In this Paragraph, we shall develop a method for calculating the field at these points.†

Using the notation (38.32a), Eqs. 38.36 can also be written in the form

$$r = r(z, \xi_0), \qquad \left[\frac{\partial r(z, \xi)}{\partial \xi} \right]_{\xi_0} = 0. \qquad (38.38)$$

Using Eqs. 38.38, a definite value of ξ_0 can be associated with each point on the caustic. In particular, the second of these equations does not contain the coordinate r, and can be used directly to express ξ_0 in terms of z or vice versa.

---

† From the geometrical point of view, the caustic is the envelope of the family of rays characterized by the parameter ξ_0 (see Fig. 162).

We now expand Eq. 38.25 for the phase in a power series in $\xi - \xi_0$, using arbitrary values of r and a constant value of z in w, which satisfy the second of Eqs. 38.38. Since this equation is obtained from $w''(\xi_0) = 0$, the quadratic term in the expansion will be absent. As a result,

$$w(r, z, \xi) = w(\xi_0) + w'(\xi_0)(\xi - \xi_0) + \tfrac{1}{6}w'''(\xi_0)(\xi - \xi_0)^3. \qquad (38.39)$$

Let us now return to the first of integrals (38.24). Using the above expansion for the phase, and taking the radical outside the integral sign at the value $\xi = \xi_0$ as a slowly varying function, we obtain

$$\psi_N^+ = \left(\frac{2\xi}{\pi r \beta_0 \beta}\right)_{\xi_0}^{\frac{1}{2}} \exp\{i[\tfrac{1}{4}\pi + w(\xi_0)]\}$$

$$\times \int_{-\infty}^{+\infty} \exp[iw'(\xi_0)(\xi - \xi_0) + \tfrac{1}{6}i\, w'''(\xi_0)(\xi - \xi_0)^3]\,d\xi, \qquad (38.40)$$

where
$$w'(\xi_0) = r - r(z, \xi_0), \quad w'''(\xi_0) = -\left[\frac{\partial^2 r(z, \xi)}{\partial \xi^2}\right]_{\xi_0}. \qquad (38.41)$$

If the point of observation is on the caustic, Eq. 38.38 gives $w'(\xi_0) = 0$, and only the cubic term remains in the exponent in the integrand in Eq. 38.40.

We now introduce a new variable of integration

$$s = 2^{-\frac{1}{3}}\,|\,w'''(\xi_0)\,|^{\frac{1}{3}}(\xi - \xi_0).$$

Moreover, we split the integral from $-\infty$ to $+\infty$ in Eq. 38.40 into two integrals, one from $-\infty$ to 0 and the other from 0 to $+\infty$. We replace s by $-s$ in the first of these integrals; as a result, the two integrals can be combined into one integral from 0 to ∞, and the integrand of the latter integral will contain a sum of exponents which can be combined into a cosine.

If, finally, we use the notation

$$t = \pm\, 2^{\frac{1}{3}} w'(\xi_0)\,|\,w'''(\xi_0)\,|^{-\frac{1}{3}} = \pm\, 2^{\frac{1}{3}}\{r - r(z, \xi_0)\}\,|\,w'''(\xi_0)\,|^{-\frac{1}{3}}, \qquad (38.42)$$

where we must choose the $+$ sign when $w''(\xi_0) > 0$, and the $-$ sign when $w''(\xi_0) < 0$, Eq. 38.40 is written

$$\psi_N^+ = 2^{11/6}\left(\frac{\xi}{r\beta_0\beta}\right)_{\xi_0}^{\frac{1}{2}}|\,w'''(\xi_0)\,|^{-\frac{1}{3}}\exp\{i[\tfrac{1}{4}\pi + w(\xi_0)]\}\cdot v(t), \qquad (38.43)$$

where $v(t)$ denotes the Airy integral

$$v(t) = \frac{1}{\sqrt{\pi}}\int_0^\infty \cos\left(\frac{s^3}{3} + st\right)ds. \qquad (38.44)$$

The function $v(t)$ has been thoroughly investigated.[23]

We have
$$v(0) = \frac{\sqrt{\pi}}{3^{\frac{2}{3}} \, \Gamma(\frac{2}{3})} = 0.62927. \tag{38.45}$$

The expansion

$$v(t) = v(0) \left[1 + \frac{t^3}{2 \cdot 3} + \frac{t^6}{(2 \cdot 5)(3 \cdot 6)} + \frac{t^9}{(2 \cdot 5 \cdot 8)(3 \cdot 6 \cdot 9)} + \cdots \right]$$

$$- 0.45874 \left[t + \frac{t^4}{3 \cdot 4} + \frac{t^7}{(3 \cdot 6)(4 \cdot 7)} + \frac{t^{10}}{(3 \cdot 6 \cdot 9)(4 \cdot 7 \cdot 10)} + \cdots \right]. \tag{38.46}$$

is valid for all values of t.

For large t, we can use the asymptotic expression

$$v(t) = \tfrac{1}{2} t^{-\frac{1}{4}} e^{-x} \left(1 - \frac{a_1}{x} + \frac{a_2}{x^2} - \frac{a_3}{x^3} + \cdots \right), \quad t > 0,$$

$$v(t) = v(-a) = a^{-\frac{1}{4}} \sin\left(x + \frac{\pi}{4} \right) \left[1 - \frac{a_2}{x^2} + \frac{a_4}{x^4} - \cdots \right]$$

$$- a^{-\frac{1}{4}} \cos\left(x + \frac{\pi}{4} \right) \left[\frac{a_1}{x} - \frac{a_3}{x^3} + \frac{a_5}{x^5} - \cdots \right], \quad t = -a < 0, \tag{38.47}$$

where

$$x = \tfrac{2}{3} t^{\frac{3}{2}} \quad \text{for} \quad t > 0 \quad \text{and} \quad x = \tfrac{2}{3} a^{\frac{3}{2}} \quad \text{for} \quad t < 0; \quad a_1 = \tfrac{5}{72};$$

$$a_2 = \frac{(5 \cdot 11) \cdot 7}{1 \cdot 2 \cdot (72)^2}; \quad a_3 = \frac{(5 \cdot 11 \cdot 17)(7 \cdot 13)}{1 \cdot 2 \cdot 3 (72)^3};$$

$$a_n = \frac{5 \cdot 11 \ldots (6n-1) \cdot 7 \cdot 13 \ldots (6n-5)}{1 \cdot 2 \ldots n (72)^n}.$$

We also give formulas relating the Airy function to the cylindrical functions:

$$v(t) = \tfrac{1}{3} \sqrt{(\pi t)} \left[I_{-\frac{1}{3}}(x) - I_{\frac{1}{3}}(x) \right], \quad t > 0,$$

$$v(t) = v(-a) = \tfrac{1}{3} \sqrt{(\pi a)} \left[I_{-\frac{1}{3}}(x) + I_{\frac{1}{3}}(x) \right], \quad t = -a < 0. \tag{38.48}$$

Finally, we include a table of the function $v(t)$ for $t \leqslant 9$ (Table 11).

Let us return to Eq. 38.43 for the field near the caustic. In this equation, $\xi_0 = k_0 \cos \chi_0$, and also, according to Eq. 38.20, $(\beta_0)_{\xi_0} = k_0 \sin \chi_0$ and $(\beta)_{\xi_0} = k_0 \sqrt{[n^2(z) - \cos^2 \chi_0]} = k_0 n(z) \sin \chi_P$. Furthermore, as previously, χ_P is the grazing angle of the ray tangent to the caustic at the given point, and χ_0 is the grazing angle of the same ray as it leaves the source.

As a result, according to Eq. 38.43, the field on the caustic itself ($t = 0$), at the point determined by the angle χ_0 at which the ray leaves the source, is

$$\psi_0 = 2^{11/6} [k_0 \, rn(z) \tan \chi_0 \sin \chi_P]^{-\frac{1}{2}} \times \left| \frac{\partial^2 r}{\partial \xi^2} \right|_{\xi_0}^{-\frac{1}{3}} \exp \{i [\tfrac{1}{4}\pi + w(\xi_0)]\} v(0), \tag{38.49}$$

TABLE 11
Airy Functions v(t)

t	v(t)	t	v(t)	t	v(t)
− 9	− 0.0392	− 3.8	− 0.3879	1.4	0.14541
− 8.9	− 0.2079	− 3.7	− 0.4999	1.5	0.12717
− 8.8	− 0.3581	− 3.6	− 0.5934	1.6	0.11084
− 8.7	− 0.4772	− 3.5	− 0.6656	1.7	0.09629
− 8.6	− 0.5550	− 3.4	− 0.7146	1.8	0.08337
− 8.5	− 0.5854	− 3.3	− 0.7394	1.9	0.07195
− 8.4	− 0.5665	− 3.2	− 0.7399	2.0	0.06190
− 8.3	− 0.5002	− 3.1	− 0.7167	2.1	0.05309
− 8.2	− 0.3928	− 3.0	− 0.6714	2.2	0.04539
− 8.1	− 0.2533	− 2.9	− 0.6060	2.3	0.03870
− 8.0	− 0.0934	− 2.8	− 0.5230	2.4	0.03289
− 7.9	0.0739	− 2.7	− 0.4255	2.5	0.02787
− 7.8	0.2355	− 2.6	− 0.3164	2.6	0.02355
− 7.7	0.3788	− 2.5	− 0.1991	2.7	0.019849
− 7.6	0.4932	− 2.4	− 0.0768	2.8	0.016680
− 7.5	0.5703	− 2.3	0.0473	2.9	0.013978
− 7.4	0.6050	− 2.2	0.1704	3.0	0.011683
− 7.3	0.5951	− 2.1	0.2898	3.1	0.009738
− 7.2	0.5421	− 2.0	0.4031	3.2	0.008096
− 7.1	0.4503	− 1.9	0.5083	3.3	0.006713
− 7.0	0.3266	− 1.8	0.6040	3.4	0.005552
− 6.9	0.1802	− 1.7	0.6888	3.5	0.004580
− 6.8	0.0215	− 1.6	0.7619	3.6	0.003769
− 6.7	− 0.1388	− 1.5	0.8229	3.5	0.003094
− 6.6	− 0.2898	− 1.4	0.8715	3.8	0.002534
− 6.5	− 0.4219	− 1.3	0.9080	3.9	0.002070
− 6.4	− 0.5267	− 1.2	0.9327		
− 6.3	− 0.5979	− 1.1	0.9462		
− 6.2	− 0.6317	− 1.0	0.9493		
− 6.1	− 0.6266	− 0.9	0.9429	t	1000 v(t)
− 6.0	− 0.5834	− 0.8	0.9280		
− 5.9	− 0.5054	− 0.7	0.9057	4.0	1.6866
− 5.8	− 0.3977	− 0.6	0.8771	4.1	1.3712
− 5.7	− 0.2670	− 0.5	0.8432	4.2	1.1122
− 5.6	− 0.1211	− 0.4	0.8051	4.3	0.9000
− 5.5	0.0315	− 0.3	0.7638	4.4	0.7267
− 5.4	0.1824	− 0.2	0.7201	4.5	0.5854
− 5.3	0.3236	− 0.1	0.6750	4.6	0.4705
− 5.2	0.4477	0	0.6293	4.7	0.3773
− 5.1	0.5486	0.1	0.5835	4.8	0.3019
− 5.0	0.6217	0.2	0.5383	4.9	0.2410
− 4.9	0.6638	0.3	0.4942	5.0	0.19204
− 4.8	0.6736	0.4	0.4515	5.1	0.15267
− 4.7	0.6511	0.5	0.4107	5.2	0.12111
− 4.6	0.5982	0.6	0.3719	5.3	0.09587
− 4.5	0.5178	0.7	0.3353	5.4	0.07574
− 4.4	0.4142	0.8	0.3010	5.5	0.05971
− 4.3	0.2925	0.9	0.2692	5.6	0.04697
− 4.2	0.1581	1.0	0.2398	5.7	0.03688
− 4.1	− 0.0172	1.1	0.2128	5.8	0.02889
− 4.0	− 0.1245	1.2	0.18810	5.9	0.02259
− 3.9	− 0.2613	1.3	0.16568	6.0	0.017632

TABLE 11 (*continued*)

t	$10^6 \, v(t)$	t	$10^6 \, v(t)$	t	$10^9 \, v(t)$
6.1	13.732	7.1	1.0148	8.1	62,43
6.2	10.675	7.2	0.7741	8.2	46.79
6.3	8.281	7.3	0.5894	8.3	35.00
6.4	6.412	7.4	0.4479	8.4	26.14
6.5	4.956	7.5	0.3398	8.5	19.492
6.6	3.822	7.6	0.2574	8.6	14.508
6.7	2.943	7.7	0.19455	8.7	10,781
6.8	2.261	7.8	0.14681	8.8	7.998
6.9	1.7345	7.9	0.11059	8.9	5.924
7.0	1.3279	8.0	0.08317	9.0	4.38

where for brevity, we have used the notation

$$(\psi_N^+)_{t=0} \equiv \psi_0.$$

In a homogeneous halfspace, with the source situated, as in the present case, at a perfectly rigid boundary, we would have

$$\psi_0 = 2[\exp(ik_0 R)/R].$$

Thus, the ratio of the field at the caustic to the field at the same point in a homogeneous medium (we denote this ratio by ζ) is

$$\zeta = 2^{\frac{5}{8}} R[k_0 \, rn(z) \tan\chi_0 \sin\chi_P]^{-\frac{1}{2}} \times \left|\frac{\partial^2 r}{\partial\xi^2}\right|_{\xi_0}^{-\frac{1}{4}} \exp\{i[\tfrac{1}{4}\pi + w(\xi_0) - k_0 R]\} v(0).$$

$$(38.50)$$

The last formula characterizes the amplification of the field at the caustic as compared with the field in a homogeneous medium. It can be shown that it is in no way necessary for the applicability of this expression that the source be situated at the boundary, as we have assumed.

At a point a distance Δr from the caustic, along the horizontal, we have, according to Eq. 38.43,

$$\psi(\Delta r) = \frac{\psi_0 v(t)}{v(0)}, \qquad (38.51)$$

where t, according to Eq. 38.42 and what has been said above, is

$$t = \pm 2^{\frac{1}{3}} \Delta r \left|\frac{\partial^2 r}{\partial\xi^2}\right|_{\xi_0}^{-\frac{1}{3}} \qquad (38.52)$$

with the same rule for the choice of the sign as in Eq. 38.42.

A graph of the function $v(t)$ is given in Fig. 160.

As we see, the function $v(t)$ decreases monotonically with increasing $|t|$ if $t > 0$, and decreases in an oscillatory manner if $t < 0$. The values $t < 0$ correspond to the convex side of the caustic. At each point near the caustic on this side, two rays intersect (see Fig. 162), and the oscillation is due to the interference between these two rays. On the other side of the caustic, there is shadow, in which the sound field decreases monotonically with depth.

The function $v(t)$ attains its maximum value at $t = -1.02$, that is, the field at the caustic itself $(t = 0)$ is not a maximum.

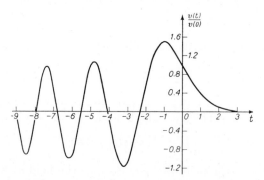

Fig. 160. Graph of the function $v(t)/v(0)$.

At sufficiently great distances from the caustic in the direction of negative t, where the condition $x = \frac{2}{3}|t|^{\frac{3}{2}} \gg 1$ holds, the sound field can be calculated as the field of the two rays, obtained for each of the rays by the usual method of steepest descents. It must be kept in mind here that the exponential $\exp[i(st + \frac{1}{3}s^3)]$ has two saddle points $s = \pm i\sqrt{t}$, determined, as usual, by the equation $(st + \frac{1}{3}s^3) = 0$. The path of integration over s from $-\infty$ to $+\infty$ can be made up of two contours, each of which will pass through its own saddle point, and will give the field of an individual ray (Ref. 28, p. 187).

5. An example of the calculation of the field in the neighborhood of the caustic

We illustrate the general results obtained above with a simple example in which the caustic is formed as a result of a single reflection from an inhomogeneous medium.

Let a homogeneous halfspace $z > 0$ adjoin an inhomogeneous halfspace $z < 0$ in which the propagation velocity is given by the law

$$n^2(z) = \left(\frac{c_0}{c(z)}\right)^2 = 1 + az, \quad z < 0. \tag{38.53}$$

The source of spherical waves will be situated at the point $z = z_0 > 0$. At an arbitrary point $P(r, z)$, the wave reflected from the inhomogeneous medium is given by Eq. 19.6, where $u \equiv kr \sin \vartheta$.

We set $V(\vartheta) = \exp[i\phi(\vartheta)]$ for the reflection coefficient and use the approximate expression (8.63) for ϕ, which is valid under the condition

$$\left(\frac{k_0}{a}\right)^{\frac{2}{3}} \cos^2 \vartheta \gg 1. \tag{38.54}$$

Furthermore, we use the asymptotic representation of the Hankel function in Eq. 19.6.

As a result,

$$\psi_{\text{refl}} = \sqrt{\left(\frac{k_0}{2\pi r}\right)} \exp(i\tfrac{1}{4}\pi) \int_{\Gamma_1} \exp\left[ik_0(z+z_0)\cos\vartheta + ik_0 r \sin\vartheta\right.$$
$$\left. + \left(\frac{4ik_0}{3a}\right)\cos^3\vartheta - \frac{i\pi}{2}\right] \sqrt{(\sin\vartheta)}\, d\vartheta. \tag{38.55}$$

We introduce the notation

$$\xi = k_0 \sin\vartheta, \quad \beta_0 = \sqrt{(k_0^2 - \xi^2)} = k_0 \cos\vartheta.$$

Then
$$\psi_{\text{refl}} = \frac{1}{\sqrt{(2\pi r)}} \exp[\tfrac{1}{4}(i\pi)] \int_{-\infty}^{+\infty} \exp(iw) \sqrt{\xi}\, \frac{d\xi}{\beta_0}, \tag{38.56}$$

where
$$w = (z+z_0)\beta_0 + \xi r + \frac{4}{3ak_0^2}\beta_0^3 - \frac{\pi}{2}. \tag{38.57}$$

The equation for the determination of the saddle point $\xi = \xi_0$ is

$$\left(\frac{\partial w}{\partial \xi}\right)_{\xi_0} = \left[r - \frac{(z+z_0)\xi}{\beta_0} - \frac{4\beta_0 \xi}{ak_0^2}\right]_{\xi_0} = 0; \tag{38.58}$$

and we set $\xi_0 = k_0 \cos\chi_0$, $\beta_0 = k_0 \sin\chi_0$, where χ_0 is the grazing angle of the ray. Then the last equation can be written in the form

$$r = (z+z_0)\cot\chi_0 + (2/a)\sin 2\chi_0, \tag{38.59}$$

which is nothing else than the equation of a ray incident on the boundary $z = 0$ at the grazing angle χ_0.

The term $(z+z_0)\cot\chi_0$ is the projection on the r-axis of the path traversed by the ray in the upper, homogeneous medium (Fig. 161), and the term $(2/a)\sin 2\chi_0$ is the similar projection of the path traversed in the inhomogeneous medium (see Eq. 8.71).

Figure 162, constructed on the basis of the last formula, shows a family of rays, incident on the boundary at various angles χ_0. We see that the rays form a caustic, indicated by the heavy line. To obtain the equation of the caustic, we set

$$\left(\frac{\partial^2 w}{\partial \xi^2}\right)_{\xi_0} = \left[\frac{4}{ak_0^2 \beta_0}(2\xi^2 - k_0^2) - \frac{k_0^2(z+z_0)}{\beta_0^3}\right]_{\xi_0} = 0. \qquad (38.60)$$

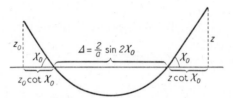

Fig. 161. On the calculation of the focusing factor upon reflection from an inhomogeneous halfspace.

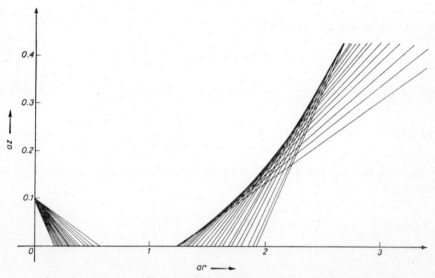

Fig. 162. The caustic formed upon reflection from an inhomogeneous halfspace.

The equation of the caustic is obtained by eliminating ξ_0 from the system of two equations (38.58) and (38.60). Having solved this system for $a(z+z_0)$ and ar, and transforming from ξ_0 to χ_0, we can write it in the form

$$ar = 4\sin 2\chi_0 \cos^2 \chi_0,$$

$$a(z+z_0) = 4\sin^2 \chi_0 \cos 2\chi_0. \qquad (38.61)$$

These expressions can be considered as the equation of the caustic in parametric form. In the case $\chi_0 \ll 1$, replacing the sines by their arguments, the cosines by unity, and eliminating χ_0, we obtain†

$$16a(z+z_0) = (ar)^2.$$

We will now proceed to the analysis of the field near the caustic. By using Eqs. 38.58 and 38.60, a definite value of ξ_0 can be associated with each point on the caustic. We expand the phase w [in the integral (38.56)] in a power series in $(\xi - \xi_0)$, assuming that in w, r is arbitrary, and $(z + z_0)$ has a constant value, determined for a given ξ_0 by Eq. 38.60. The quadratic term in the expansion will then be absent. Moreover, taking the slowly varying factor outside the integral at the value $\xi = \xi_0$ in Eq. 38.56, we obtain

$$\psi_{\text{refl}} = \left(\frac{\xi}{2\pi r \beta_0^2}\right)^{\frac{1}{2}}_{\xi_0} \cdot \exp\{i[\tfrac{1}{4}\pi + w(\xi_0)]\}$$
$$\times \int_{-\infty}^{+\infty} \exp\left[iw'(\xi_0)(\xi - \xi_0) + \tfrac{1}{6}i\,w'''(\xi_0)(\xi - \xi_0)^3\right]d\xi.$$

We have reduced the expression for ψ_{refl} to the form (38.40). In the same way as previously, we obtain the expression

$$\psi_{\text{refl}}^0 = 2^{\frac{5}{6}}(k_0 r \tan\chi_0 \sin\chi_0)^{-\frac{1}{2}}\left(\frac{\partial^2 r}{\partial \xi^2}\right)^{-\frac{1}{3}}_{\xi_0} \exp\{i[\tfrac{1}{4}\pi + w(\xi_0)]\}v(0) \quad (38.62)$$

for ψ_{refl} on the caustic. The last expression differs from Eq. 38.49 by a factor of 2, because previously, the source was situated on a perfectly rigid boundary, as a result of which, the field was doubled. Moreover, $\chi_P = \chi_0$ in the present case, since the upper medium is assumed to be homogeneous [$n(z) = 1$], and the rays in it are rectilinear.

In Eq. 38.62, the phase of the ray $w(\xi_0)$ is calculated by Eq. 38.57. Furthermore, differentiating Eq. 38.60, we obtain

$$\left(\frac{\partial^3 r}{\partial \xi^3}\right)_{\xi_0} = \left[\frac{4\xi}{ak_0^2\beta_0^3}(k_0^2 + 2\beta_0^2) - \frac{3k_0^2(z+z_0)\xi}{\beta_0^5}\right]_{\xi_0}.$$

Substituting the expression for $(z + z_0)$ from Eq. 38.61, and letting $\xi_0 = k_0 \cos\chi_0$, we obtain, after some straightforward manipulations,

$$\left(\frac{\partial^2 r}{\partial \xi^2}\right)_{\xi_0} \equiv -\left(\frac{\partial^3 w}{\partial \xi^3}\right)_{\xi_0} = \frac{16}{ak_0^2}\cot^2\chi_0 \cot 2\chi_0. \quad (38.63)$$

---

† The condition $\chi_0 \ll 1$ can be satisfied in spite of the condition (38.54), which can be written $(k_0/a)^{\frac{1}{3}} \cdot \chi_0^2 \gg 1$. For this to occur, the quantity a/k_0 must be sufficiently small.

Now, Eq. 38.62 gives the following expression for the amplitude of the potential on the caustic:

$$|\psi^0_{\text{refl}}| = (2k_0 r \tan\chi_0 \sin\chi_0)^{-\frac{1}{2}} \cdot (ak_0^2 \tan^2\chi_0 \tan 2\chi_0)^{\frac{1}{3}} v(0). \quad (38.64)$$

To obtain the potential $\psi_{\text{refl}}(\Delta r)$ at an arbitrary distance Δr from the caustic, we must multiply the last equation by $v(t)/v(0)$, where

$$t = -2^{\frac{1}{3}} \Delta r \left| \frac{\partial^2 r}{\partial \xi^2} \right|^{-\frac{1}{3}}_{\xi_0} = -\frac{\Delta r}{2} (ak_0^2 \tan^2\chi_0 \tan 2\chi_0)^{\frac{1}{3}}. \quad (38.65)$$

The rays and the caustic formed by them are constructed in Fig. 162. The axes represent the quantities az and ar.

In conclusion, we note that, to calculate the field near the caustic both in the present example as well as in similar cases, it is entirely unnecessary to return to the initial integral formulas of the type (38.55). We can start from Eqs. 38.49 or 38.50, obtained previously, and calculate the quantity $(\partial^2 r/\partial \xi_0^2)$ which enters into these equations by using the ray equation, which is different in the various cases. In particular, in the present example, this quantity can be obtained from Eq. 38.59, after having replaced χ_0 by $\xi_0 = k_0 \cos\chi_0$.

6. *The field near a point of contact of two caustics*

Equation 38.43 for the field near a caustic ceases to be valid near points at which $w'''(\xi_0) \to 0$. These are points at which two caustics are in contact, and are obtained from the three equations

$$w'(r, z, \xi_0) = 0, \quad w''(r, z, \xi_0) = 0, \quad w'''(r, z, \xi_0) = 0 \quad (38.66)$$

by eliminating ξ_0, and then solving for r and z.

If we assume that in any definite case, these equations are satisfied by one pair of values of r and z, then, letting N run through its various values in Eqs. 38.23 and 38.24, we obtain a series of points, as, for example, the points $A_1, A_2, \ldots$ in Fig. 172b.

To obtain the value of ψ_N^+ near a point at which caustics are in contact, Eq. 38.25 for the phase must be expanded in a power series in $(\xi - \xi_0)$, taking into account terms up to the fourth order. Then, at the very point of contact of the caustics, we obtain

$$\psi_N^+ = \left(\frac{2\xi}{\pi r \beta_0 \beta} \right)^{\frac{1}{2}}_{\xi_0} \exp\{i[\tfrac{1}{4}\pi + w(\xi_0)]\} \int_{-\infty}^{+\infty} \exp\left[(i/24) w_0^{\text{IV}} (\xi - \xi_0)^4 \right] d\xi, \quad (38.67)$$

which is similar to Eq. 38.40. We have used the notation

$$w_0^{\text{IV}} \equiv \left[\frac{\partial^4 w(r, z, \xi)}{\partial \xi^4} \right]_{r_0, z_0, \xi_0},$$

where r_0, z_0 and ξ_0 are the solutions of Eqs. 38.66.

We now introduce a new variable of integration in Eq. 38.67 through the relation

$$\xi - \xi_0 = \left[\frac{24}{w^{\mathrm{IV}}(\xi_0)}\right]^{\frac{1}{4}} t.$$

The field at the point of interest is then

$$\psi_N^+ = 2\left(\frac{\xi}{\pi r \beta_0 \beta}\right)_{\xi_0}^{\frac{1}{2}} \left(\frac{6}{w_0^{\mathrm{IV}}}\right)^{\frac{1}{4}} I_0, \tag{38.68}$$

where

$$I_0 = \int_{-\infty}^{+\infty} e^{it^4}\, dt = 1.813\, e^{i(\pi/8)}. \tag{38.69}$$

The value of ψ_N^+ can be obtained not only directly at the point of contact of the caustics, but also in its neighborhood, by replacing I_0 in Eq. 38.68 by the integral[171]

$$I(X, Y) = \int_{-\infty}^{+\infty} \exp\left[i(Yt + Xt^2 + t^4)\right] dt. \tag{38.70}$$

In order to clarify the meaning of the dimensionless quantities X and Y, we replace the coordinate system (r, z) by a local coordinate system (x, y) with its origin at the point of contact of the caustics, and with its x-axis in the direction of their common tangent. Then, near the point of contact, both caustics will be described by the equation[171]

$$x^3 = -\frac{9\sigma}{8} y^2, \tag{38.71}$$

where σ is a parameter determined by the form of the caustics in each definite case. The quantities X and Y are nondimensional coordinates, connected with x and y by the relations

$$X = \left(\frac{12\pi}{\sigma\lambda}\right)^{\frac{1}{2}} x, \quad Y = \left(\frac{192\pi^3}{\sigma\lambda^3}\right)^{\frac{1}{4}} y, \tag{38.72}$$

so that in these coordinates, the equation of the caustics is

$$X^3 = -(27/8)\, Y^2. \tag{38.73}$$

The caustics and the corresponding rays are shown in Fig. 163. Since only the relative curvature of the rays and the caustic is important, the rays in this Figure are shown as straight lines.

The modulus of the integral $I(X, Y)$ is shown in Fig. 164, and the phase in Fig. 165. Since the picture is symmetrical with respect to the x-axis, only its upper half is shown. The caustic is shown as a dotted line. We see from Fig. 164 that the maximum intensity, i.e. the focus, does not occur at the origin, but is displaced in the negative x-direction through a number of wavelengths, equal to

$$- 0.894 \, (\sigma/\lambda)^{\frac{1}{2}}.$$

In analyzing the phase change along the X-axis in Fig. 165, we can verify the well-known fact that the phase of the wave changes by 180° as the wave passes through the focus.

As we see in Fig. 163, there is a system of three rays between the caustics on the left hand side of the Figure. These rays interfere, and as a result, we obtain the maxima and minima seen in Fig. 164 to the left of the caustic. There is only one family of rays to the right of the caustic. Therefore, as we withdraw from the caustic, which leads to a weakening of the effect of the interfering rays, the decay law for the modulus I rapidly becomes monotonic.

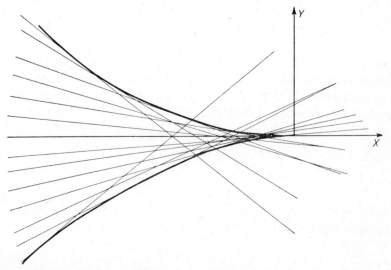

Fig. 163. The ray picture near a point of contact of caustics.

The picture we have drawn here coincides with the picture of the field of a focusing optical system in the neighborhood of its focus, when spherical aberration is present.

It is worth mentioning, in conclusion, that in the analysis of the field near a caustic and a point of contact of caustics, the use of Eqs. 38.21, 38.22 and 38.25 for the phase is possible only in a region sufficiently far from "turning" points of the rays, i.e. sufficiently far from points at which the rays forming the caustic are horizontal. In the opposite case, the theory based on geometrical optics must be refined, as was done in §16.

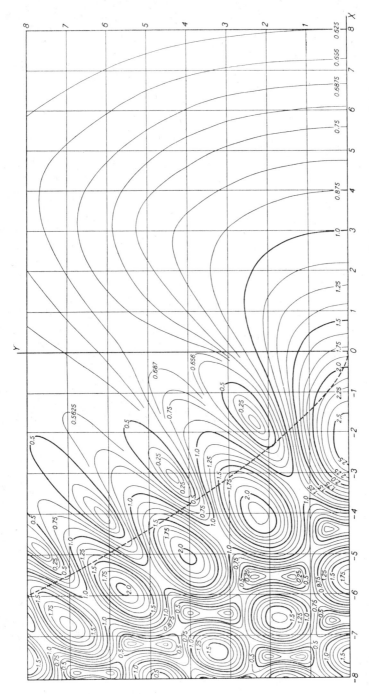

Fig. 164. The modulus of the integral $I(X, Y)$ in the neighborhood of a point of contact of caustics.

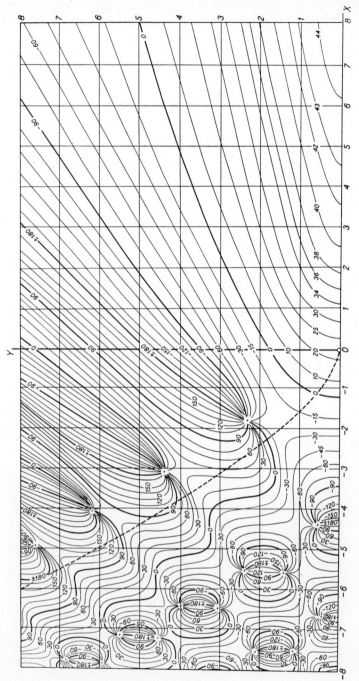

Fig. 165. The phase of the integral $I(X, Y)$.

§ 39. The Underwater Sound Channel

1. *The underwater sound channel and the cause of its formation*

The velocity of sound in sea water is a function of the temperature, salinity and hydrostatic pressure, which in turn vary with depth. In a sufficiently deep sea or ocean, this leads to the formation of an underwater acoustic waveguide, which is also called an underwater sound channel. A typical velocity profile is shown in Fig. 166, in which the velocity has a minimum at the level $z = z_0$. Above this level, the

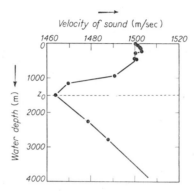

Fig. 166. Example of the velocity of sound as a function of depth in the ocean (Atlantic Ocean, 35° 57′ N. Lat., 69° 00′ W. Long., February 15, 1932).

velocity increases because of the temperature increase (due to the heating of the upper layers of the water by solar radiation), and below this level, the velocity of sound increases because of the increased hydrostatic pressure.† At the surface of the water, there is a layer in which the velocity of sound may remain constant on the average because of mixing, or may vary more or less irregularly with depth. As a rule, the variation of the salinity with depth does not change the character of the velocity curve. The curve in Fig. 166 was obtained in February in the Atlantic Ocean (35° 57′ N. Lat., 69° 00′ W. Long.).

In seas and oceans at more northerly latitudes, the level of the minimum sound velocity ("the axis of the sound channel") is closer to the surface (frequently 60–100 meters), and during the winter months may even reach the surface of the water. Moreover, the increase of the

---

† The velocity of sound as a function of the temperature, salinity and hydrostatic pressure is described by the following empirical formula (t is the temperature in °C, S is the salinity in °/$_{00}$, h is the depth in meters, c is the velocity of sound in meters/second):

$$c = 1410 + 4.21t - 0.037t^2 + 1.14S + 0.01h.$$

velocity of sound with depth at levels below the axis of the sound channel may be due in part to an increase of the water temperature with depth as a result of deep warm currents (as for example, in the Black and the Baltic Seas).

Let us suppose that a sound source is situated near the axis of the sound channel. As a result of the refraction of the sound rays under these conditions, a definite class of the rays leaving the source will be concentrated near the axis of the channel as they propagate, and will not reach the surface of the water. The behavior of the acoustic velocity with depth, and a picture of the sound rays, are shown schematically in Fig. 167.

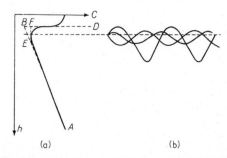

(a)   (b)

Fig. 167. Schematic illustration of the velocity of sound as a function of depth, and the corresponding picture of the sound rays.

The concentration of the sound rays in a layer of limited thickness and their propagation without scattering at the surface and absorption at the bottom leads to the possibility of the propagation of sound of comparatively low frequencies (for which the absorption in sea water is not great) to extremely great distances. Such long distance sound (SOFAR) propagation in the sea was discovered in the USSR in July of 1946 during the investigation of the propagation of sound from explosions. It is clear from previously classified wartime investigations, published in the USA, that this phenomenon was discovered there during the war years. It was shown that the explosion of several kilograms of TNT can be heard in the ocean at distances up to 6000 km. Long distance underwater sound propagation can have numerous peaceful and military applications. Thus, the SOFAR system, which consists of a group of several hydroacoustic receiving bases in the northeast Pacific Ocean, can receive signals from wrecks in the ocean and establish their location. This system can also be used to anticipate catastrophic tidal waves on the surface of the ocean, arising from the eruption of underwater volcanos and from underwater earthquakes.

The present Paragraph is devoted to the presentation of the results of a theoretical investigation of a number of fundamental characteristics of sofar propagation in an underwater sound channel.

2. *Elementary theory of sofar propagation*

In the vast majority of cases, the gradient of the acoustic velocity is considerably smaller below the axis of the sound channel than above it. This leads to "vigorous" refraction above the axis of the channel, and as a result, the rays do not enter very much into this region. Idealizing the picture somewhat, we assume that the variation of the velocity of sound with depth is described by the broken line *AEBFD* (Fig. 167). If we then suppose that the source is situated at the level *BD*, which will now act like a perfectly reflecting boundary, we obtain the ray picture shown in Fig. 168.†

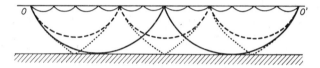

Fig. 168. The various types of sound rays.

The rays leaving the source O can be divided into two classes. The rays of the first class do not reach the bottom, and propagate by means of successive reflections at the upper boundary of the channel.

Some representatives of this class of rays are, for example, the ray shown as a dashed line in Fig. 168, and the ray shown as a solid line, which undergoes many reflections at the upper boundary. The rays of the second class undergo reflection at the bottom. One such ray is shown by the dotted line in Fig. 168. The ray tangent to the bottom, shown by a solid line in Fig. 168, lies on the boundary between the two classes of rays.

We now determine the form of each of the rays. Since the source is assumed to be located at the plane $z = 0$, the horizontal distance traversed by the ray as its vertical coordinate varies from 0 to z is (see § 38)

$$r = \int_0^z \frac{dz}{\tan \chi}. \tag{39.1}$$

---

† We note that sometimes the linear decrease of the velocity of sound with decreasing depth can extend up to the surface itself. The surface of the water will then serve as the upper boundary of the sound channel. In many cases, particularly at low frequencies, when the surface is not too agitated and the rays have small grazing angles, the scattering of sound at this boundary can be neglected, and the boundary can be considered to be perfectly reflecting.

At each point, the grazing angle of the ray χ is found from the equation

$$c(z) \cos \chi_0 = c_0 \cos \chi, \tag{39.2}$$

where $c(z)$ and χ are the variable values of the acoustic velocity and the grazing angle, and c_0 and χ_0 are their values at $z = 0$. We assume that the velocity variation obeys the linear law

$$c = c_0(1 + az). \tag{39.3}$$

If the velocity varies only because of the increase of the hydrostatic pressure, then $a = 0.012 \text{ km}^{-1}$, which corresponds to a velocity increase of approximately 1.8 m/sec per hundred meters of depth.

Using the relation between χ and z given by Eqs. 39.2 and 39.3, we can write Eq. 39.1 in the form

$$r = \int_0^z \frac{(1 + az) \cos \chi_0 \, dz}{\sqrt{[1 - (1 + az)^2 \cos^2 \chi_0]}},$$

which, after integration and elementary manipulations, gives

$$\left(r - \frac{1}{a} \tan \chi_0\right)^2 + \left(z + \frac{1}{a}\right)^2 = \frac{1}{a^2 \cos^2 \chi_0}. \tag{39.4}$$

We thus see that all the rays are arcs of circles.

We are interested, in the first place, in the horizontal distance traversed by the ray between two reflections from the boundary of the channel, and also in its maximum deviation from the boundary. We consider only a ray of the first class. Setting $z = 0$, Eq. 39.4 gives

$$\left(r - \frac{1}{a} \tan \chi_0\right)^2 = \frac{1}{a^2 \cos^2 \chi_0} - \frac{1}{a^2} = \frac{\tan^2 \chi_0}{a^2},$$

whence

$$r - \frac{1}{a} \tan \chi_0 = \pm \frac{\tan \chi_0}{a},$$

or

$$r = 0, \quad \frac{2 \tan \chi_0}{a}.$$

Consequently, the distance traversed between two reflections will be

$$\Delta r = \frac{2 \tan \chi_0}{a}. \tag{39.5}$$

The maximum deviation from the boundary $z = 0$ will occur at $r = \Delta r/2 = \tan \chi_0/a$, and, using Eq. 39.4, we obtain

$$z = z_{max} = \frac{1 - \cos \chi_0}{a \cos \chi_0}. \tag{39.6}$$

This result could also have been obtained from Eqs. 39.2 and 39.3 by remembering that z_{max} corresponds to a turning point of the ray, and at such a point, $\chi = 0$.

It is quite apparent that rays of the first class cause concentration of sound energy near the boundary of the channel. In fact, all the rays leaving the source at grazing angles between 0 and χ_0 will be concentrated in a layer of thickness $(1 - \cos \chi_0)/a \cos \chi_0$, or, for small χ_0, in a layer of thickness $\chi_0^2/2a$.

As χ_0 decreases, this thickness decreases as χ_0^2, whereas the angle, and consequently the energy radiated into the layer, decreases as χ_0, that is, considerably more slowly. This creates an increased concentration of energy in a thin layer near the boundary. At the boundary itself, the energy density becomes infinite in the ray approximation.

This effect is completely analogous to the whispering gallery effect explained by Rayleigh[189], which is caused by the increase of the sound energy density near curved surfaces (the striking effect of the "talking wall" in the Temple of Heaven in Peking is explained in the same way). The only difference between these effects and the phenomenon we are currently investigating is that in the case considered by Rayleigh the boundary is curved and the rays are rectilinear, whereas in our case, the boundary of the medium is plane, but the rays are curved.† As will be shown in §40, these two situations are completely equivalent, since only the relative curvature of the rays and the boundary is important.

We now investigate the propagation of a sound pulse (for example, an explosion) in a sound channel of the above idealized form. For the present, we shall consider only rays of the first class. We shall suppose that the receiver is also situated on the boundary, at the point O', at a distance r from O. Of all the rays leaving O at all possible angles, only those for which the distance r is an integral multiple N of a single "cycle" will arrive at the point O'. As a result, the angle χ_0 at which the ray leaves the source is given by

$$N \frac{2 \tan \chi_0}{a} = r, \tag{39.7}$$

whence
$$\tan \chi_0 = \frac{ar}{2N}. \tag{39.8}$$

---

† Strictly speaking, the upper boundary of the waveguide also has some curvature since the Earth is spherical. However, the curvature of the Earth is negligibly small compared with the curvature of the pertinent rays, and can therefore be neglected.

17

However, the grazing angles of the rays of the first class have a limited range. In particular, the maximum depth to which these rays penetrate, as given by Eq. 39.6, must be no greater than the depth of the sea h. This gives

$$\delta \cos\chi_0 \geqslant 1 - \cos\chi_0, \quad \text{where} \quad \delta = ah.$$

Setting $\cos\chi_0 = 1 - \chi_0^2/2$, we obtain

$$\chi_0 \leqslant \sqrt{(2\delta)}. \tag{39.9}$$

Thus, for $h = 1.9\,\text{km}$ and $a = 0.012\,\text{km}^{-1}$, we have $\delta = 0.023$ and $\chi_0 \leqslant 0.21$.

Setting $\tan\chi_0 = \chi_0$ in Eq. 39.7, and using the Eq. 39.9, we obtain $N \geqslant N_{\min}$, where

$$N_{\min} = \frac{ar}{2\sqrt{(2\delta)}}. \tag{39.10}$$

The smallest integer greater than $N_{\min}$ is the minimum number of cycles through which any ray connecting the points O and O' must pass. For $h = 1.9$ km, we have

$$N_{\min} = \frac{r}{35},$$

where r is in kilometers.

For a given r, each ray is characterized by its number N. This number can take on all integral values from $N_{\min}$ to ∞ for the various rays. As we see from Eq. 39.8, the greater N, the closer to grazing is the angle at which the ray leaves the point O, and the smaller is its deviation from the bounding plane as it propagates.[†]

We now consider the travel time of sound pulses from O to O' along each of the rays, and begin by calculating the travel time over one cycle between two successive reflections. The time required to traverse the element of length ds of the ray is equal to ds/c, where $ds = dz/\sin\chi$, and χ is the angle between the ray and the horizontal at a given point. Using Eqs. 39.2 and 39.3, and transforming from the differential dz to $d\chi$, we obtain

$$\Delta t = \frac{2}{ac_0} \int_0^{\chi_0} \frac{d\chi}{\cos\chi} = \frac{1}{ac_0} \ln \frac{1 + \sin\chi_0}{1 - \sin\chi_0} \tag{39.11}$$

for the travel time over one cycle.

---

† The ray theory ceases to be applicable[39] for rays leaving the source at grazing angles $\chi_0 \leqslant (a\chi/2\pi)^{\frac{1}{2}}$. However, we shall not go on to the extremely cumbersome wave picture here, because this difficulty will not be important in the refinement of the theory carried out in the next Section.

The total travel time from O to O' over a ray which goes through N cycles will be greater by a factor of N, that is,

$$t_N = \frac{N}{c_0 a} \ln \frac{1 + \sin \chi_0}{1 - \sin \chi_0}, \tag{39.12}$$

where χ_0 is determined by N through Eq. 39.7.

Using the fact that χ_0 is small, we expand the right hand side of Eq. 39.12 and the left hand side of Eq. 39.7 in series in χ_0; then, eliminating χ_0, we obtain

$$t_N = \frac{r}{c_0} \left(1 - \frac{a^2 r^2}{24 N^2} \right). \tag{39.13}$$

Thus, the pulse traveling along the bounding ray (the solid line in Fig. 168), for which $N = N_{\min}$, will have the least travel time. The greater the number of reflections of the ray from the boundary, the later will the pulse traveling along the ray arrive. The pulse propagating along the horizontal arrives last. The time interval between the arrival of successive pulses decreases continuously and approaches zero for pulses of infinitely high order. An increase of the sound intensity will be observed on a record of the process, because of the pulse concentration.

After the arrival of the horizontal ray, which is the last of the rays of the first class, the sound intensity decreases sharply, because the pulses arriving thereafter, which travel along rays of the second class, have undergone reflections at the bottom. Their intensity is small because of absorption at the bottom. Only at comparatively small ranges, at which a ray can arrive at the receiver O' with a small number of reflections from the bottom, will the record of the main signal, due to rays of the first class, be followed by a "tail", due to pulses arriving after reflections from the bottom. This effect is analogous to room reverberations due to reflections from the walls.

The duration of the main part of the signal is defined as the time interval between the arrivals of the first and last of the rays of the first class, i.e.

$$T_0 = t_\infty - t_{N_{\min}} = \frac{r}{24 c_0} \left(\frac{ar}{N_{\min}} \right)^2 = \frac{\delta r}{3 c_0}. \tag{39.14}$$

Thus, the duration of the signal is proportional to the distance.

The qualitative characteristics which we have obtained from the above simple theory are well confirmed by experiment. The record of a sound signal at a comparatively small distance (about 19 km) is shown in Fig. 169. Following the main signal phase, which lasts for about one second, we see a prolonged reverberation due to rays which arrive at the receiver after multiple reflections from the bottom. We note that at this distance, the arrival times of the various rays of the first class are practically all the same, and the duration of the first phase of the signal practically coincides with the duration of a separate radiated pulse (in the present case, an underwater explosion followed by pulsations of the gas bubble). Figure 170 is the record of the signal, taken at a distance

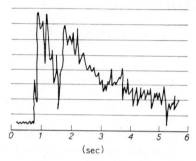

Fig. 169. The record of the acoustic signal from an explosion at a comparatively small distance.

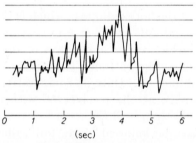

Fig. 170. The record of the sound of an explosion at a great distance (560 km).

of 560 km.[68] The slow increase of the sound intensity, and the comparatively rapid decrease at the end are quite evident. In this case, the duration of the signal is about 5 seconds, and there is no noticeable reverberation. The level recorded after $t > 5$ seconds is equal to the noise level. The vertical scale, in both Figures, is 5 db per division.

3. Refinement of the theory

We shall refine the theory somewhat in the present Section by assuming that the velocity of sound as a function of depth in the sea is described by the broken line $AEFD$ (Fig. 167). In comparison with the simplified theory of the preceding Section, we now have the homogeneous layer EF. The presence of this layer eliminates the problem of the infinite concentration of sound energy near the boundary BD in the ray approximation. Furthermore, we shall see below that it is possible to give a theoretical description of the interesting phenomenon of sound ray focusing at certain distances from the source.

Let H be the thickness of the homogeneous layer, h_0 the thickness of the layer between the upper boundary of the layer BD and the bottom of the sea; finally, let $h = h_0 - H$ (Fig. 171). We again assume that the velocity of sound is a linear function of depth below the homogeneous layer, and as previously, we use the notation $a = (1/c_0)(dc/dz)$, where c_0 is the acoustic velocity in the homogeneous layer.

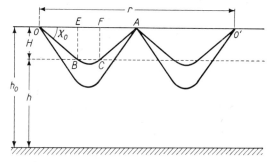

Fig. 171. On the calculation of the refined picture of the sound rays.

We investigate the sound ray picture when the source and the receiver are situated in the constant velocity layer. Their exact positions are not important, and we assume for definiteness that they are situated at the upper boundary of the layer.

We again consider only the rays of the first class, i.e. the rays which have not undergone reflection at the bottom.

A ray, which leaves O at the angle χ_0, will return to the boundary at some point A (Fig. 171). The distance OA is given by $OA = 2OE + EF$. The distance EF is given by Eq. 39.5, and $OE = H/\tan\chi_0$. Consequently,

$$OA = \frac{2H}{\tan\chi_0} + \frac{2\tan\chi_0}{a}. \qquad (39.15)$$

Only those rays for which the distance r is an integral multiple N of the distance OA will arrive at the point O'. Using this condition, and introducing the notation $p = aH$, we obtain the equation

$$\tan\chi_0 + \frac{p}{\tan\chi_0} = \frac{ar}{2N}, \qquad (39.16)$$

the solution of which is

$$\tan \chi_0 = \frac{ar}{4N} \pm \sqrt{\left[\left(\frac{ar}{4N}\right)^2 - p\right]}. \tag{39.17}$$

Thus, two rays with the same N will arrive at an arbitrary point O' (Fig. 171). As $p \to 0$, the value of $\tan \chi_0$ for one of these rays (with the + sign in the last equation) approaches the value given by Eq. 39.8. The other family of rays is new. The rays of this family (the ray $OBCA$ in Fig. 171) propagate principally in the homogeneous layer, and do not penetrate very deeply below the level BC.

It is clear from Eq. 39.17 that, in contrast to the case considered in the preceding Section, N cannot take on arbitrarily large values.

In fact, since $\tan \chi_0$ cannot be imaginary, we must have

$$N \leqslant N_{\max} = \frac{ar}{4\sqrt{p}}. \tag{39.18}$$

When $N = N_{\max}$, Eq. 39.17 gives

$$\chi_0 = \arctan \sqrt{p}. \tag{39.19}$$

It is not hard to see from Eq. 39.17 that the angle $\arctan \sqrt{p}$ lies on the boundary separating the two above-mentioned families of rays, corresponding to the two possible signs in front of the radical. The rays corresponding to the minus sign and propagating principally in the homogeneous layer have the grazing angles $\chi_0 < \arctan \sqrt{p}$, while the rays corresponding to the plus sign have the angles $\chi_0 > \arctan \sqrt{p}$. The grazing angles of the rays of the latter family have the upper bound $\chi_0 \leqslant \sqrt{(2\delta)}$, where $\delta = ah$. Combining this condition with Eq. 39.16, and using the fact that δ is small, we obtain the following expression for the minimum number of cycles through which a ray of this family must go:

$$N_{\min} = \frac{ar}{2\sqrt{(2\delta)}\,[1 + (p/2\delta)]}, \tag{39.20}$$

which is an insignificant refinement of Eq. 39.10, since $p/2\delta = H/2h$ is small.

On the other hand, N varies from $N = 1$ to $N = N_{\max}$ for the rays of the second family.†

Let us now consider the travel times of sound pulses along each of the rays. The travel time over the rectilinear portion OB in Fig. 171

---

† Our arguments are valid for distances which satisfy $r^2 \gg H/a$. In the opposite case, it may turn out that the rays of the classes we are considering will generally be absent.

will be $H/c_0 \sin \chi_0$. According to Eq. 39.11, the travel time over the portion BC will be

$$\frac{1}{c_0 a} \ln \frac{1 + \sin \chi_0}{1 - \sin \chi_0}.$$

Summing over the entire ray, the total travel time is

$$t_N = \frac{N}{ac_0} \left(\frac{2p}{\sin \chi_0} + \ln \frac{1 + \sin \chi_0}{1 - \sin \chi_0} \right), \tag{39.21}$$

where χ_0 is connected with N through Eq. 39.16. Using this equation, and also the fact that χ_0 is small, we expand t_N in a series in χ_0, and, introducing the notation

$$\tau_N = t_N - \frac{r}{c_0}, \tag{39.22}$$

we obtain the simpler formula

$$\tau_N = \frac{r}{c_0} \frac{3p - \chi_0^2}{6(p + \chi_0^2)} \chi_0^2. \tag{39.23}$$

Hence, it is clear that the ray for which $\chi_0 = \sqrt{(3p)}$ arrives simultaneously with the horizontal ray $\chi_0 = 0$ ($\tau_N = 0$ for these rays). Rays for which $\chi_0 < \sqrt{(3p)}$ arrive later than the horizontal ray; furthermore, investigation of Eq. 39.21 at its extremum shows that the most retarded ray is the one for which $\chi_0 = \sqrt{p}$. Rays for which $\sqrt{(3p)} < \chi_0 < \sqrt{(2\delta)}$ arrive earlier than the horizontal ray. The total duration of the signal will be

$$T_0 = (\tau_N)_{\chi_0 = \sqrt{p}} - (\tau_N)_{\chi_0 = \sqrt{(2\delta)}} = \frac{r}{c_0} \left(\frac{\delta}{3} - \frac{p}{2} \right). \tag{39.24}$$

This formula is also only an insignificant refinement of Eq. 39.14 since $p \ll \delta$.

Considering the possibility (for a crude analysis of the phenomenon) of incoherent (energy) addition of the sound fields of separate rays, the number of pulses, traveling along various paths, arriving at a given point per unit time (i.e. the quantity dN/dt), will be an important characteristic.

The time dependence of this quantity determines some averaged form of the received sound signal in time.

Using Eq. 39.16, Eq. 39.21 gives

$$\frac{dN}{dt} = ac_0 \left[\ln \frac{1 + \sin \chi_0}{1 - \sin \chi_0} - 2(1 - p) \sin \chi_0 \right]^{-1}. \tag{39.25}$$

We thus see that the pulse density is independent of the distance, and depends only on the grazing angle χ_0 of the ray as it leaves the point O.

Expanding the last expression into a power series in χ_0, and neglecting p and χ_0^2 compared with unity, we obtain the simpler formula

$$\frac{dN}{dt} = \frac{ac_0}{2\chi_0}\left(p + \frac{\chi_0^2}{3}\right)^{-1}.$$

(39.26)

The pulse density is therefore:
at the beginning of the signal

$$\left(\frac{dN}{dt}\right)_{\chi_0 = \sqrt{(2\delta)}} = \frac{3ac_0}{2(2\delta)^{\frac{3}{2}}},$$

(39.27)

at the end of the signal

$$\left(\frac{dN}{dt}\right)_{\chi_0 = \sqrt{p}} = \frac{3ac_0}{8p^{\frac{3}{2}}}.$$

(39.28)

Thus, compared with its value at the beginning, the sound intensity increases considerably at the end of the signal.

We note that Eq. 39.26 has no meaning when $\chi_0 \to 0$, at which the pulse density approaches zero. This results from the following: as N changes by one, the increment of χ_0 is not infinitesimal, as is assumed in the derivation of the equation, but is finite, and is comparable with χ_0, as a result of which our derivation is already invalid at not too great values of N.

4. *Focusing of sound rays*

So far, our development of the theory of sound propagation in an underwater sound channel has been based on ray acoustics. If the homogeneous layer is sufficiently thick, the ray theory will be applicable.

Without imposing very rigid demands on our theory, we can require that the ray representations be valid in the most important final phase of the signal, at which $\chi_0 \sim \sqrt{p}$.

In order that the geometrical theory describe the refraction of these rays, the condition (see § 8, Section 7)

$$\chi_0 \sim \sqrt{p} \gg \left(\frac{a\lambda}{2\pi}\right)^{\frac{1}{3}},$$

must be fulfilled, whence, remembering that $p = aH$, we obtain

$$H \gg \left(\frac{\lambda^2}{4\pi^2 a}\right)^{\frac{1}{3}}.$$

(39.29)

We shall consistently assume that this condition is satisfied. In this Section, we shall investigate other features of the ray picture, in particular, the emergence of a caustic or the focusing of sound rays.

The variation of the sound intensity along a ray is determined by the focusing factor f, defined as the ratio of the wave intensities in the case under consideration and, in the case of a homogeneous halfspace, at the same distance from the source.

If we know the function $r = r(z, z_0, \chi_0)$, which, for given values of z and z_0, is the distance traversed by a ray leaving the source at the angle χ_0, then, as shown in § 38, the focusing factor will be

$$f = \frac{r}{(dr/d\chi_0) \sin \chi(z)}. \tag{39.30}$$

As above, we assume for the present that $z = z_0 = 0$. Then r will be given as a function of χ_0 by Eq. 39.16. Calculating the derivative $dr/d\chi_0$, we obtain

$$f = \frac{\tan^2 \chi_0 + p}{\tan^2 \chi_0 - p} \cos \chi_0. \tag{39.31}$$

At $\chi_0 = \arctan \sqrt{p}$, we have $f = \infty$. This means that rays which leave the source at angles within an infinitesimal neighborhood of this angle, will again be brought to a point when they return to the boundary. In the ray approximation, the sound intensity will be infinite at these points. The locus of these points is simply the cross section of the caustic surface in the plane $z = 0$. It consists of concentric circles, with the source as the center. The radii of these circles are obtained by substituting $\chi_0 = \arctan \sqrt{p}$ into Eq. 39.16. As a result, we obtain $r = Nr_0$, where $r_0 = 4 \sqrt{(p)/a}$ is the radius of the first circle.

The acoustic potential at the caustic and in its vicinity is given by Eqs. 38.49 and 38.51. The ratio of the field at the caustic to the field in free space is given by Eq. 38.50. We have $R = r$, $n(z) = 1$ and $\chi_p = \chi_0$ in this formula, since we are considering the case in which the source is situated at the same level as the point of observation ($z = z_0 = 0$). Furthermore, r is given as a function of χ_0 by Eq. 39.16, from which, neglecting quantities of the order χ_0^2 compared with unity, we obtain

$$a \frac{\partial^2 r}{\partial \chi_0^2} = \frac{4N_p}{\chi_0^3}.$$

Moreover, we have

$$\left(\frac{\partial^2 r}{\partial \xi^2} \right)_{\xi_0} = \frac{\partial^2 r}{\partial \chi_0^2} \left(\frac{\partial \chi_0}{\partial \xi_0} \right)^2 + \frac{\partial r}{\partial \chi_0} \cdot \frac{\partial^2 \chi_0}{\partial \xi_0^2},$$

whence, remembering the relation $\xi_0 = k_0 \cos \chi_0$ and the fact that

$\partial r/\partial \chi_0 = 0$ at the caustic, we obtain

$$\left(\frac{\partial^2 r}{\partial \xi^2}\right)_{\xi_0} = \frac{4Np}{ak_0^2 \chi_0^5}. \tag{39.32}$$

At the caustic, we have $\chi_0 \approx \sqrt{p}$.

The focusing factor at the caustic can now be obtained from Eq. 38.50:

$$f = |\zeta|^2 = 2^{\frac{2}{3}} v^2(0) (k_0 rp)^{\frac{1}{3}} = 1.2574(k_0 rp)^{\frac{1}{3}}. \tag{39.33}$$

Thus, the focusing factor increases as $r^{\frac{1}{3}}$ with distance.

The amplitude distribution as a function of Δr, the distance from the caustic, will be given by the function $v(t)/v(0)$, where according to Eqs. 38.52 and 39.32

$$t = -2^{\frac{2}{3}} \pi \frac{p^{\frac{2}{3}}}{(r\lambda^2)^{\frac{1}{3}}} \Delta r. \tag{39.34}$$

The function $v(t)/v(0)$ is shown in Fig. 160. At negative t (to the right of the caustic), the field is oscillatory. The first zero of the function $v(t)$ occurs at $|t| = 2.34$. Thus, we can use the quantity

$$\Delta r \approx 0.5 \frac{(r\lambda^2)^{\frac{1}{3}}}{p^{\frac{2}{3}}} \tag{39.35}$$

as the width of the first maximum of the sound intensity. The ratio of the width of the maximum to the distance between neighboring caustics will be

$$\frac{\Delta r}{r_0} \approx \frac{a}{8p^{\frac{7}{6}}} (r\lambda^2)^{\frac{1}{3}}. \tag{39.36}$$

We should also note that a definite number of other "non-focusing" rays can arrive at a point on the caustic. The sound intensity for these rays is calculated by the usual methods.

We have considered only those points of the caustic which lie in the horizontal plane passing through the source. The complete picture of the configuration of the caustic surfaces is very complicated, and we shall look into it only qualitatively.

From the fact that $dr/d\xi_0 = 0$ on the caustic, it follows that the ray which leaves the source at the grazing angle $\chi_0 = \alpha = \arctan \sqrt{p}$ traverses the shortest distance in returning to the plane $z = 0$. It turns out that the rays which leave the source at the angles $\chi_0 > \alpha$ and $\chi_0 < \alpha$, form separate caustics. The rays corresponding to $\chi_0 > \alpha$, and the caustics formed by them, which we will call the external caustics, are shown in Fig. 172a.

The rays for which $\chi_0 < \alpha$, and the "internal" caustics corresponding to them, are shown in Fig. 172b. Finally, the complete system of caustics for the case under consideration is shown in Fig. 172c. The caustics are tangent to one another at the points A_1, A_2, ...

In the present paragraph, we have considered the family of caustics, for the case in which the velocity of sound c was a simple function of the vertical coordinate z. It may turn out, in more complicated cases, that not all of the caustics are easily delineated by means of a graphical ray construction.

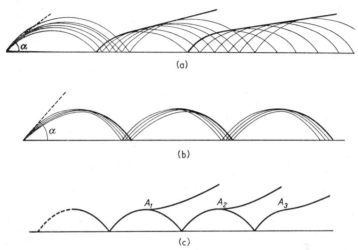

Fig. 172. (a), the rays forming the exterior caustic; (b), the rays forming the interior caustic; (c), the complete system of caustics in the underwater sound channel of the type under consideration.

In these cases, we can use the equation of the ray family, which in a general form can be written as $f(r, z, \chi_0) = 0$. As above, the angle χ_0 at which the ray leaves the source is a parameter. From this equation we obtain $r = r(z, \chi_0)$ and $(\partial r/\partial \chi_0)_z$. We then plot $(\partial r/\partial \chi_0)_z$ as a function of χ_0, and at a given z, we find the value of χ_0 for which $(\partial r/\partial \chi_0)_z = 0$. Using this value of χ_0 and the equation of the ray, we find the point (r, z) lying on the caustic. In the same way, we can find the angle χ_0 for which $(\partial z/\partial \chi_0)_r = 0$, and then we find the points lying on the caustic.

§ 40. Long Distance Radiowave Propagation in the Atmosphere

1. *Conditions for the formation of an atmospheric waveguide*

The variation of the dielectric constant of air with height is an extremely important factor in the propagation of radiowaves in the meter and the centimeter range. The index of refraction as a function of the main factors—the temperature, pressure and humidity—is given by the empirical formula

$$n = 1 + \frac{79}{T}\left(p_{mb} + \frac{4800 l_{mb}}{T}\right) 10^{-6}, \tag{40.1}$$

where T is the absolute temperature, p_{mb} is the pressure of the air in millibars, and l_{mb} is the pressure of the water vapor in the same units.

As a rule, the index of refraction decreases with height. By international agreement, the "standard atmosphere" is defined by $T = 273° + 15°$, $dT/dz = 0.0065$ deg/m, $dl/dz = -0.0033$ mb/m and $l = 10$ mb at sea level. Using these values, and the barometric formula, it can be shown that in this case, the gradient of the index of refraction will be constant with height, and will be equal to

$$dn/dz = -3.9 \times 10^{-8}\, m^{-1}.$$

Under such conditions, which actually correspond to some average state of the atmosphere, rays will be deflected towards the surface of the earth, and will have the form of a circular arc of some constant radius. The ray picture in a homogeneous atmosphere is shown in Fig. 173a. The analogous ray picture in the standard atmosphere is shown in Fig. 173b.

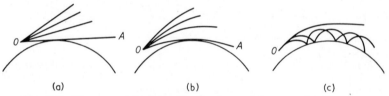

(a)   (b)   (c)

Fig. 173. The ray pictures in the fundamental cases of radiowave propagation in the atmosphere.

The theory of wave propagation under the conditions of the standard atmosphere indicates[1] that only the relative curvature of the ray and the surface of the earth, and not the curvature of either one separately, is important. Therefore, the concept of the so-called "equivalent" radius of the surface of the earth has become rather widely used in the theory. From this point of view, the atmosphere is considered homogeneous, the rays rectilinear, and the curvature of the surface of the earth is not given by its true value, but by an effective value equal to the relative curvature of the ray and the surface.

In this representation, cases (a) and (b) in Fig. 173 will appear as shown in Fig. 174. The rays are now rectilinear in both cases. In Fig. 174a, the equivalent radius of the earth is equal to its true value, and in Fig. 174b it is greater than the true value. For the standard atmosphere, we have $a_{equiv} \approx \frac{4}{3} a_{true} \approx 8500$ km.

However, as a rule, the concept of the equivalent radius of the earth cannot be used in the case of greatest interest, shown in Fig. 173c, in which we have a so-called earth-attached waveguide. In this case, the gradient of the index of refraction, and, consequently, the curvature of the rays, usually varies with height.† Therefore, in what follows, we shall use another representation, namely, the modified index of refraction representation. In this approach, the surface of the earth is considered to be plane, and instead of the true index of refraction $n(z)$, we use a modified index of refraction, given by

$$n_{\text{mod}} = n(z) + (z/a). \tag{40.2}$$

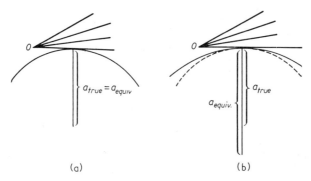

Fig. 174. On the representation of the equivalent radius.

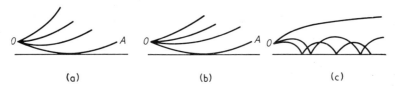

Fig. 175. The ray representations with the introduction of the concept of the modified index of refraction.

Thus, the index of refraction now has the additional term z/a, which leads to an additional term in the gradient dn/dz. The latter causes additional curvature of the rays. Thus, in this case, the earth is flattened; to conserve the relative curvature of the earth and the rays, the latter are given some additional curvature. Cases (a), (b) and (c) in Fig. 173 will now appear as shown in Fig. 175. Cases (a) and (b) are

---

† If, however, the propagation picture shown in Fig. 173c is realized with a constant value of dn/dz, independent of height, the equivalent radius representation can be used, but the equivalent radius turns out to be negative.

now qualitatively similar, the only difference being the magnitude of the curvature of the rays.

The modified index of refraction representation in the theory of atmospheric waveguides has been generally accepted. It is especially convenient for us because all the representations of wave propagation in plane-layered inhomogeneous media, developed above, remain valid.

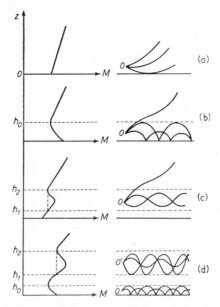

Fig. 176. A number of M-profiles and the corresponding ray pictures.

In the next Section we pay special attention to the basis of this representation and to the determination of its region of applicability. Now, however, we consider several examples which will illustrate how the dependence of n_{mod} on z immediately determines the character of the radiowave propagation.

We note, incidentally, that in practice we usually use the quantity

$$M = (n_{\text{mod}} - 1) \times 10^6,$$

and the dependence of M on z is called the M-profile. Several M-profiles and the corresponding ray pictures are shown in Fig. 176. Case (a) corresponds to a modified index of refraction which increases linearly with height, and is realized, in particular, in the case of the standard atmosphere. Case (b) corresponds to an earth-attached waveguide extending from the surface of the earth to the height h_0. A certain portion of the rays which leave the source O, situated within the waveguide, are "captured" by the waveguide and propagate in it. If the source is situated outside the waveguide, none of the rays will be captured by it. However, the more exact wave theory shows that even in this case, a certain portion of the radiated energy is captured by the waveguide and propagates in it.

Case (c) corresponds to an elevated waveguide, extending from the height h_1 to h_2. Finally, in case (d), an earth-attached waveguide and an elevated waveguide exist simultaneously. These waveguides capture the greatest amount of the radiated energy when the radiating antenna is situated within the waveguide, and sufficiently far from its boundaries (for example, O or O'). However, as has already been mentioned, a certain portion of the energy will propagate in the waveguides, even when the antennas are situated outside the waveguides, Another very interesting phenomenon is the transfer of energy from one waveguide into another as the waves propagate in the horizontal direction.[54]

The ray representations which we have developed here are frequently inapplicable for an exact quantitative description of the electromagnetic fields of waves in atmospheric waveguides. However, they always give a correct qualitative representation of the character of the propagation. That they are at all applicable, even to a limited extent, is due to the fact that the thickness of the atmospheric waveguides, which varies from tens to hundreds of meters, is many times greater than the wavelengths of the radiation (in the centimeter and meter range).

2. The basis of the modified index of refraction method

We shall now look into the theoretical basis of a method which enables us, with sufficient accuracy, to reduce our problem to the investigation of wave propagation in a plane-layered medium (instead of a spherically-layered medium) with a modified index of refraction which varies with height.

As was shown in Ref. 78, the field of any simple source in a spherical-layered medium can be expressed in terms of two scalar functions.

Following Ref. 177, we shall limit ourselves to the case in which the relative change of the index of refraction over a change of height of the order of a wavelength is sufficiently small. In this case, each of the two scalar functions will satisfy the wave equation

$$\nabla^2 \psi + k^2 n^2(r) \psi = 0, \qquad (40.3)$$

where $k = 2\pi/\lambda$ is the wave number in vacuum, and $n = n(r)$ is the index of refraction which is a function only of the radial coordinate r, reckoned from the center of the earth. In the present case, it is sufficient to investigate the solution for ψ which is symmetrical with respect to the axis passing through the source. Then ψ can be built of elementary solutions of the form

$$\psi(r, \vartheta) = p(\vartheta) . U(r). \tag{40.4}$$

Substituting Eq. 40.4 into Eq. 40.3 and using the expression for the Laplacian in spherical coordinates, we obtain two equations for the functions $P(\vartheta)$ and $U(r)$:

$$\frac{d^2 P}{d\vartheta^2} + \frac{\cos \vartheta}{\sin \vartheta} \frac{dP}{d\vartheta} + a^2 k_m^2 P = 0, \tag{40.5}$$

$$\frac{d^2 U}{dr^2} + \frac{2}{r} \frac{dU}{dr} + \left[k^2 n^2(r) - \frac{a^2 k_m^2}{r^2} \right] U = 0, \tag{40.6}$$

where the separation constant has been denoted by $a^2 k_m^2$ [instead of the usual notation $n(n+1)$], where a is the radius of the earth, and the quantity k_m has the dimensions of reciprocal length.

In order that the function ψ be the solution of our problem, it must satisfy not only Eq. 40.3, but also the following conditions:

(a) at small distances R from the source, it must have the form e^{ikR}/R;

(b) at large distances from the source, it must represent a diverging wave;

(c) the tangential components of the electric and magnetic fields must be continuous at the surface of the earth.

We now replace the spherical coordinates r and ϑ by the horizontal coordinate $\rho = a\vartheta$ measured along the surface of the earth, and the height coordinate $z = r - a$, measured from the surface of the earth. Equations 40.5 and 40.6 then become

$$\frac{d^2 P}{d\rho^2} + \frac{1}{\rho} \frac{dP}{d\rho} + k_m^2 P = \frac{1}{\rho} \frac{dP}{d\rho} \left(\frac{1}{3} \frac{\rho^2}{a^2} + \frac{1}{45} \frac{\rho^4}{a^4} + \ldots \right), \tag{40.7}$$

$$\frac{d^2 U}{dz^2} + \left[k^2 n^2(z) - k_m^2 + 2k_m^2 \frac{z}{a} \right] U = -\frac{2}{a} \frac{dU}{dz} + 3k_m^2 \frac{z^2}{a^2} U + \ldots, \tag{40.8}$$

where we have expanded $\cos \vartheta / \sin \vartheta$ and $1/r$ in powers of ρ/a and z/a, respectively. If we set the right hand sides of these equations equal to zero, we obtain the same equations as in the case of the plane earth, except that, in place of $k^2 n^2(z)$, the quantity

$$k^2 n^2(z) + 2k_m^2 (z/a) = k^2 n_{\text{mod}}^2 (z)$$

appears in Eq. 40.8.

Since we can set $k_m \approx k$ under actual conditions, and since z/a is small, the last expression yields Eq. 40.2 for n_{mod}.

Solving Eqs. 40.7 and 40.8 by successive approximations, Pekeris found that:

(a) as the horizontal distance ρ is increased, the plane earth with a modified index of refraction approximation can be used up to distances equal to one half of the radius of the earth, and the error in the value of the field will not exceed two per cent, at any frequency;

(b) as the height of the receiver is increased, this approximation can be used as long as the factor

$$1 + i\frac{\sqrt{6}}{5}\frac{kz^{\frac{3}{2}}}{a^{\frac{1}{2}}},$$

which enters into the more exact expression for the field, can be taken as unity.

Thus, for example, when the wavelength is 10 cm, the error in the function $U(z)$ is of the order of 17 per cent at a height of 1.5 km.

It is useful to note that if, instead of the distance $\rho = a\vartheta$ measured along the spherical surface of the earth, we use the distance $\rho = a \sin \vartheta$ measured from an axis passing through the source and the center of the earth, the approximation is considerably worse.

3. *The bilinear profile*

The propagation of centimeter and decimeter waves in atmospheric waveguides received considerable attention during the Second World War. During this period and several years thereafter, basic results, which we shall present below briefly, were obtained in this region. The work of large groups of scientists and engineers was concentrated on theoretical investigations and calculations of a number of concrete cases, which were characterized by a definite dependence of the modified index of refraction on height, i.e. by a definite M-profile. The methods used and the results obtained will be described briefly below.

The bilinear profile was studied in detail in the USA by W. Furry, J. Freehafer and others.[14] In this case, the quantity n^2_{mod} in the equation

$$\frac{d^2 U_m}{dz^2} + (k_0^2\, n^2_{\text{mod}} - k_m^2)\, U_m = 0 \qquad (40.9)$$

(to which our problem now reduces) is given by

$$n^2_{\text{mod}} = 1 + p(z - d), \quad z < d,$$
$$n^2_{\text{mod}} = 1 + q(z - d), \quad z > d. \qquad (40.10)$$

Thus, n_{mod} is taken as unity at $z = d$. In the present case, the solutions of Eq. 40.9 can be expressed in terms of Airy functions. We are interested in the case of an earth-attached atmospheric waveguide, for which p will be negative.

The mathematical calculations will be simplified somewhat if we transform to the nondimensional coordinate

$$Z = z/H, \quad \text{where} \quad H \equiv (qk_0^2)^{-\frac{1}{3}}, \tag{40.11}$$

and also introduce the notation

$$Y(Z) = (k_0/q)^{\frac{2}{3}}(n_{\text{mod}}^2 - 1) = (1/qH)(n_{\text{mod}}^2 - 1),$$
$$k_m^2 = k_0^2(1 - qHA_m). \tag{40.12}$$

Equation 40.9 is then written

$$\frac{d^2 U_m}{dZ^2} + (Y + A_m) . U_m = 0, \tag{40.13}$$

where the U_m are the eigenfunctions, and the A_m are the eigenvalues.

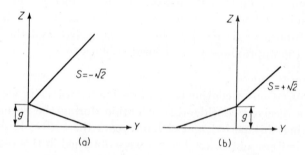

Fig. 177. Examples of the bilinear profile.

The function $Y = Y(Z)$ is written:

$$Y = Z - g, \quad Z > g,$$
$$Y = s^3(Z - g), \quad Z < g, \tag{40.14}$$

where $g = d/H$ is the nondimensional height of the waveguide, and

$$s^3 \equiv p/q. \tag{40.15}$$

We shall assume in what follows that s is real. In the present case, s is negative. The function $Y(Z)$, with $s = -\sqrt{2}$, is shown in Fig. 177a.

The case $s = \sqrt{2}$, for which the waveguide is absent, is shown in Fig. 177b.

The solution of the Airy equation

$$\frac{d^2 y}{d\zeta^2} + \zeta y = 0, \tag{40.16}$$

will be denoted by $y(\zeta)$.

The solutions of Eq. 40.13 can then be written

$$U_m(Z) = y_1(Z - g + A_m), \quad Z \geqslant g,$$
$$U_m(Z) = y_2(sZ - sg + s^{-2}A_m), \quad Z < g, \qquad (40.17)$$

where y_1 and y_2 are the appropriately chosen solutions of Eq. 40.16. These solutions, and also the eigenvalues, must be determined such that they satisfy the conditions:

(a) as $Z \to \infty$, $U_m(Z)$ must represent an outgoing wave;

(b) at $Z = g$, $U_m(Z)$ must be continuous;

(c) at $Z = 0$, the appropriate boundary conditions must be satisfied; in the simplest case we have $U_m(0) = 0$.

We can obtain approximate formulas, valid for large and small g, to use in the calculation of $U_m(Z)$ and A_m. However, at moderate values of g, we must use tedious numerical methods. We shall not discuss the details of the calculations, but shall only present some of the results.

The eigenvalue A_1 for the first normal mode was calculated for a wide range of values of g, and for the following values of s: -3, -2, $-\sqrt{2}$, -1, 0, $\sqrt{\frac{1}{2}}$, $\sqrt{\frac{2}{3}}$, $\sqrt{\frac{4}{5}}$, $\sqrt{\frac{9}{10}}$, $\sqrt{2}$, 2.

The case $s \geqslant 0$, corresponding to the absence of a waveguide, is also of great practical interest.

At great distances from the source, the dependence of each of the normal modes on the horizontal distance r is given by $(1/r)e^{ik_m r}$. If we use the second of Eqs. 40.12 connecting k_m with A_m, introduce the "natural" unit of length in the horizontal direction

$$L = 2(k_0 qr)^{-\frac{1}{3}}, \qquad (40.18)$$

and remember that k_m and k_0 are close to one another, we obtain

$$k_m = k_0 \sqrt{(1 - qHA_m)} \approx k_0 - (A_m/L). \qquad (40.19)$$

Consequently, the wave will be damped with distance according to the law

$$(1/\sqrt{r}) \exp[-C_m(r/L)],$$

where

$$C_m = -\operatorname{Im} A_m.$$

The quantity C_m can be called the damping constant. A plot of this constant for the first normal mode and various values of s and g is shown in Fig. 178. We see that when s is negative, the attenuation decreases sharply when the thickness of the waveguide g is increased above some definite value. When g is small, all values of C_1 approach a value somewhat greater than 2. This case corresponds to a strongly damped normal mode in the absence of a waveguide (propagation in the standard atmosphere). If $s > 0$, there is no waveguide for any g. Curves of the damping constant for this case are also shown in Fig. 178.

The amplitude of the first normal mode is shown in Fig. 179 as a function of the dimensionless height Z, that is, the so-called "height function". This function is assumed to be normalized by the condition

$$\int_0^\infty U_1^2(Z)\,dZ = 1.$$

It was convenient to let the ordinate represent $20 \log |U_1(Z)|$ rather than $U_1(Z)$ itself.

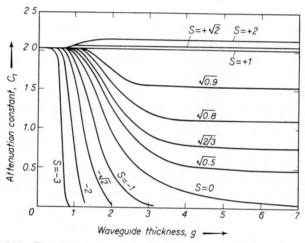

Fig. 178. The damping constant of the first normal mode as a function of the waveguide thickness g (in dimensionless units) for the bilinear profile.

All four curves in the Figure refer to the case $s = -1$. The values 1.37, 1.93, 2.68 and 4.34 were chosen for g. As is clear from the curve corresponding to $s = -1$ in Fig. 178, the damping constant is comparatively large in the first two of these four cases. These cases do not correspond to waveguide propagation. In the third case, C_1 is considerably smaller than in the first two cases, but is still noticeable. This is the case of the not too strongly defined waveguide. Finally, the fourth case, in which C_1 is very small, is typical of a waveguide.

These considerations are fully confirmed by the dependence of $|U_1|$ on the height Z, shown for all four cases in Fig. 179. As we see, the amplitude of the first normal mode increases continuously with height in the first two cases, and the field is not concentrated in any definite region.

In the third case, there is an amplitude maximum near $Z = 1.5$, which shows the presence of a waveguide effect, although not too strongly pronounced. Following the maximum is a broad minimum.

In the fourth case, the field is mainly concentrated in the region of a well-defined waveguide, the upper boundary of which lies at $Z = 4.34$. When the height is increased further, the amplitude of the wave decreases continuously.

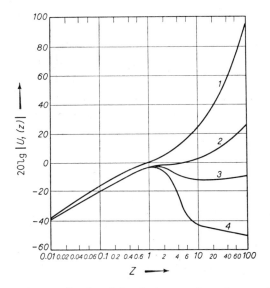

Fig. 179. The amplitude of the first normal mode as a function of the dimensionless vertical coordinate Z, for various values of the waveguide thickness: *1*, $g = 1.37$; *2*, $g = 1.93$; *3*, $g = 2.68$; *4*, $g = 4.34$. In all cases, s was set equal to -1.

As we see, the cases of the presence and the absence of a waveguide are not sharply divided from one another; there are also transitional cases.

Using the phase integral method, we can obtain an approximate criterion for the presence of a waveguide. This criterion can be useful in discussions of a qualitative nature.

According to this criterion, a normal mode of order m, which is a solution of Eq. 40.9, will be restrained by the waveguide if the following inequality is satisfied:

$$\int_0^{z^{(m)}} \sqrt{(k_0^2 \, n_{\text{mod}}^2 - \text{Re} \, k_m^2)} \geqslant (m - \tfrac{1}{4}) \, \pi, \quad m = 1, 2, \ldots, \qquad (40.20)$$

where $z^{(m)}$ is the smallest root of the equation

$$k_0^2\, n_{\mathrm{mod}}^2 - \operatorname{Re} k_m^2 = 0.$$

With the present profile, this criterion gives

$$\int_0^g Y^{\frac{1}{2}}\, dZ = \tfrac{2}{3}(-sg)^{\frac{3}{2}} > \tfrac{3}{4}\pi = 2.356 \qquad (40.21)$$

for the first normal mode ($m = 1$) (see Ref. 14, § 2.19).

In the four cases considered above, the quantity $\tfrac{2}{3}(-sg)^{\frac{3}{2}}$ was equal to 1.07, 1.69, 2.93 and 6.03, respectively. Thus, by using this criterion, we could have predicted that the first two cases would correspond to the absence of a waveguide, the third to a weak waveguide, and the fourth case to a rather well-defined waveguide.

A shortcoming of the bilinear profile is the fact that it has a break at some height. Although the modified index of refraction is continuous, its derivative with respect to height will be discontinuous. The question as to whether this break will have an effect on the laws of waveguide propagation is important. To clarify this question, calculations were made with a profile consisting of two linear laws joined by a segment of a parabola, which yields a continuous derivative. The width of this segment 2τ was made small so that the calculations could be performed by expanding the solution in a power series in τ. The calculations showed that rounding the break which exists in the bilinear profile does not change the results significantly.

4. The power profile

The efforts of a large group of English scientists were concentrated on the study of the laws of propagation of radiowaves in the case of the power profile

$$n_{\mathrm{mod}}^2 = 1 + q\left[z - \frac{d}{n}\left(\frac{z}{d}\right)^n \right]. \qquad (40.22)$$

The parameter q can be chosen to correspond to the standard atmosphere. The quantity d, assumed positive, is the waveguide thickness, and n is between 0 and 1.

A shortcoming of the power profile is the fact that n_{mod}^2 does not approach the standard value $1 + qz$ asymptotically as $z \to \infty$. The second term in the square brackets increases without bound, even though it becomes negligibly small compared with the first term. A second peculiarity of this profile is the fact that the derivative of n_{mod} with respect to z becomes infinite at the surface of the earth (at $z = 0$). The dependence of n_{mod} on z is shown in Fig. 180 for $n = \tfrac{1}{2}, \tfrac{1}{5}$, and also for the standard atmosphere (the dotted straight line).

A power profile with $n = \frac{1}{2}$ was proposed by Booker.[89] This case was investigated extensively by the phase integral method; the eigenvalues and eigenfunctions were determined, and the field was calculated at various distances, taking one or several normal modes into account. In this case, the integrals involved in the phase integral method are evaluated in closed form.

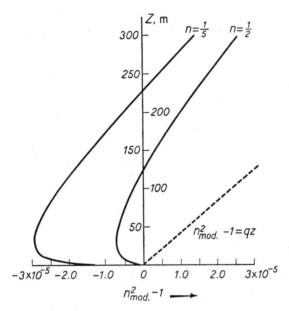

Fig. 180. The modified index of refraction as a function of altitude for three cases.

The case $n = \frac{1}{5}$ was studied by a group of scientists directed by D. Hartree. Their work was based principally on numerical integration of the pertinent wave equations or of the Riccati equations obtained from them and was carried out in Manchester and Cambridge on differential analyzers. A variational method of the Rayleigh–Ritz type was also used in a number of cases. We shall not go into the details of the calculations, which the reader can find elsewhere.[132]

As an illustration, the calculated radar range is shown in Fig. 181 for the standard atmosphere and also for the power profiles with $n = \frac{1}{2}$ and $n = \frac{1}{5}$. In this Figure, the abscissa represents the horizontal range, and the ordinate represents the height of the target. The waveguide is 100 feet (30.5 meters) thick. The modified index of refraction as a function of height is shown for all three cases in Fig. 180.

The wavelength is $\lambda = 10$ cm. Under these conditions, only the first normal mode is important. The radar antenna is at an elevation of 50 feet. The characteristics of the radar and the target were chosen such that the range in a free homogeneous space would be 80.5 km (50 miles). It is clear from the Figure that the range in the presence of the waveguide is considerably greater than in the standard atmosphere, even when the height of the target is several times greater than the waveguide thickness. However, at great heights, the waveguide ceases to play a role, and the range becomes approximately the same in all three cases.

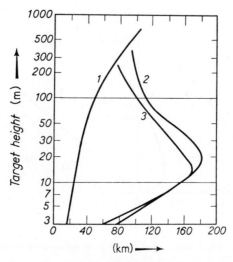

Fig. 181. Calculated radar ranges for the standard atmosphere (*1*), and also for the power profiles with $n = \frac{1}{2}$ (*2*) and $n = \frac{1}{5}$ (*3*).

5. *The linear-exponential profile. Approximate methods for the treatment of the problem of waveguide propagation*

We shall now investigate the profile given by the equation

$$n^2_{\text{mod}} = 1 + qz + \alpha' \exp(-\gamma' z),$$

where α' and γ' are constants. It is again natural (although not obligatory) to choose q to correspond to the standard atmosphere. If we transform to the dimensionless coordinate Z in accordance with Eq. 40.11, the equation for the eigenfunctions is written in the form (40.13), with

$$Y(Z) = Z + \alpha \exp(-\gamma Z). \tag{40.23}$$

This expression yields an atmospheric earth-attached waveguide when $\alpha > 0$. The quantity γ is related to the thickness of the waveguide.

In the present case, Eq. 40.13 does not have an exact solution in terms of any known functions. Using this example, we shall examine various approximate methods for treating the problem of the waveguide propagation of radiowaves in the atmosphere (even though we have discussed these methods previously).

a. *The perturbation method* was developed by Pekeris.[174] In this method, the solution of Eq. 40.13 is expressed as a superposition of the normal waves U_m^0 corresponding to the standard atmosphere ($\alpha = 0$). The eigenvalues A_m are then the solutions of the equation obtained by setting the "secular" determinant equal to zero. In his calculations, Pekeris limited himself to a determinant of the sixth order. The root A_1, corresponding to the first normal mode, was obtained by an iterative process. The accuracy of the results is evident from the convergence of the values for A_1 obtained successively from determinants of the fourth, fifth and sixth order. Of course, the accuracy of the method can also be estimated by comparing the results with those obtained by other methods.

It is quite natural that this method is most fruitful and most accurate when α is comparatively small, and the profile does not differ too greatly from the standard profile. In the case $\alpha > 0$, that is, when a waveguide is present, the method is sufficiently accurate as long as the value of the damping constant $C_1 = - \operatorname{Im} A_1$ does not fall to $\frac{1}{3}$ or $\frac{1}{2}$ of standard. On the other hand, when $\alpha < 0$, the method yields sufficiently accurate results up to $-\alpha = 25$. Calculations were not carried out for greater values of $-\alpha$.

b. *The variational method.* A variational method was first applied by D. Hartree and his co-workers[157] to the problem of radiowave propagation in an inhomogeneous atmosphere. This method was also used by Pekeris[175] to explain some experimental results. The method yields sufficient accuracy (at least 1 per cent) for any value of α and can therefore be used for calculations both in the case of waveguide propagation as well as in the case of "anti-waveguide" propagation. In Ref. 172, a variational method was used to evaluate the accuracy of other approximate methods. We will follow this reference in our presentation below.

As is well known, the problem of solving the differential equation

$$\frac{d^2 U_m}{dZ^2} + (Z + A_m + \alpha \, e^{-\gamma Z}) U_m = 0 \tag{40.24}$$

is equivalent to the problem of minimizing the integral I, where

$$I = \int_0^\infty \left[\left(\frac{dU_m}{dZ} \right)^2 - (Z + A_m + \alpha e^{-\gamma Z}) U_m^2 \right] dZ. \qquad (40.25)$$

As a trial solution, we will use some function $U_m(Z) = F(a_1, a_2, ..., Z)$ which satisfies the boundary conditions. The arbitrary parameters are then determined from the system of equations $\partial I / \partial a_k = 0$, after which the eigenvalues A_m are found from the equation $I = 0$.

Two types of trial functions were used in Ref. 172:

$$\text{I. } U(Z) = AZ e^{-aZ},$$

$$\text{II. } U(Z) = A(Z + bZ^2) e^{-aZ}.$$

TABLE 12

α	γ	A_{I}	A_{II}	A_{exact}	$A_{\text{I}}/A_{\text{exact}}$	$A_{\text{II}}/A_{\text{exact}}$
0	—	$-1.238 + 2.145i$	$-1.177 + 2.038i$	$-1.169 + 2.025i$	1.059	1.0066
10	1.271319	$-2.847 + 0.211i$	—	$-2.72 + 0.187i$	1.047	—
10	0.593776	-5.116	—	$-5.123 + 0.015i$	0.9986	—
15	1.584000	$-3.106 + 0.0025i$	$3.080 + 0.010i$	$-3.10 + 0.05i$	1.0018	0.8935
20	2.645805	$-1.994 + 0.176i$	—	$-1.85 + 0.27i$	1.071	—
20	1.788576	-3.668	—	$-3.74 + 0.008i$	0.981	—
10	2.4	$-1.439 + 0.879i$	$-1.246 + 0.840i$	$-1.237 + 0.887i$	1.114	0.987

We shall not enter into the details of the calculations here. The eigenvalues A_{I} and A_{II} (corresponding to trial functions I and II, respectively) for the first normal mode are given in Table 12 for various values of α and γ. Also tabulated is A_{exact}, the exact value of A obtained by D. Hartree and others by use of the differential analyzer.

After the eigenvalues have been determined, the eigenfunctions are obtained from Eq. 40.24 by numerical integration. For this purpose, it is only necessary to know $(\partial U / \partial Z)_{Z=0}$, which can be determined by a method given in Ref. 172.

c. *The phase integral method.* The fundamental concept of the phase integral method was presented above in §35. In the present case, the characteristics of the normal modes can be obtained approximately by finding Z_1, satisfying the equation

$$\int_0^{Z_1} \sqrt{(Z + \alpha e^{-\gamma Z} - Z_1 - \alpha e^{-\gamma Z_1})} . dZ = \pi(m - \tfrac{1}{4}), \quad m = 1, 2, ..., \qquad (40.26)$$

and then setting $\qquad\qquad A_m = Z_1 + \alpha e^{-\gamma Z_1}.$

After the eigenvalues have been thus determined, the eigenfunctions are obtained as the asymptotic solutions of Eq. 40.24. Near values of Z for which the expression in the parentheses on the left hand side of Eq. 40.24 becomes zero or is small, the solution is expressed in terms of cylindrical functions of the 1/3 order (see § 16).

Pekeris[173] developed the phase integral method substantially, and also found the successive terms of the expansion for certain cases. The phase integral method also allows us to obtain the damping coefficient of the normal mode if it is not too great (when the imaginary part of A_m in Eq. 40.24 is considerably smaller than the real part).

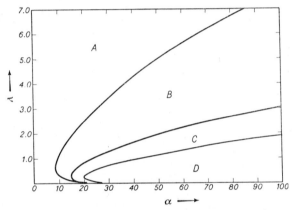

Fig. 182. Lines of transition of normal modes of various orders from non-attenuating to strongly attenuating, for the linear-exponential profile.

d. Calculational method for the transitional case. The phase integral method cannot be used in principle to calculate the characteristics of the normal mode for the frequencies at which the waveguide ceases to restrain the normal mode and it is transformed into a strongly damped wave. This case is characterized by the quantity Z_1 in Eq. 40.26, which represents the upper boundary of the layer in which the energy of the corresponding normal mode is mainly concentrated (or the turning level of the ray corresponding to the given normal wave), becoming equal to Z_0, the height of the upper boundary of the waveguide. The latter is defined as the height at which n^2_{mod} is a minimum.

The lines of transition, in the $\alpha\gamma$-plane, from slightly attenuating to strongly attenuating, for the first three normal modes, are shown in Fig. 182, taken from Ref. 172. In region A, the normal modes of all orders are substantially damped. The waveguide is too narrow and cannot support even a single normal mode. In region B, the waveguide can support only the first normal mode, and the remaining ones are rapidly attenuated. Two and three normal modes are supported in regions C and D, respectively. We see that for a given α, the smaller λ, that is, the greater the width of the waveguide, the greater the number of normal modes supported by the waveguide.

Pekeris[173] also developed an approximate theory for the cases in which the values of α and λ correspond to points lying near transition lines in the $\alpha\lambda$-plane. The idea of the method is to expand the solutions of Eq. 40.24 in a power series in $Z_0 - Z_1$. We shall not dwell on the details of the calculations. The reader is referred to Ref. 172 for a detailed comparison of the results obtained for the linear-exponential profile with the various approximate methods.

§ 41. WAVE PROPAGATION UNDER CONDITIONS OF SHADOW ZONE FORMATION

1. Statement of the problem and review of the cases encountered

It is convenient, in the study of wave propagation in layered-inhomogeneous media, to distinguish two main cases. In the first case, which was studied in the preceding three Paragraphs, waveguide propagation occurs. Here the medium plays the role of a concentrator of the radiated energy; moreover, the latter is concentrated in some bounded layer. In the other case, which could appropriately be called "anti-waveguide", the medium plays the opposite rôle, creating conditions under which the radiated energy is drawn away from the propagation path, and beginning at certain distances, shadow zones, in which the field intensity is comparatively small, are formed along the path.

A well-known example of anti-waveguide propagation is radiowave propagation over the spherical earth under the conditions of the standard or homogeneous atmosphere. Under these conditions, shown schematically in Fig. 173a, b, there is a limiting ray OA, and none of the rays emitted by the source enter the region to the right of and below this ray. As a result, a zone of so-called geometrical shadow is formed. Wave energy penetrates into this zone only because of diffraction effects. The same behavior is evident in Fig. 175a, b, in the representation in which the earth is considered flat.

A completely analogous case occurs in underwater sound under the conditions of so-called negative ray refraction in the surface layers of water (Fig. 183). In this case, the acoustic velocity decreases with depth (almost always because the temperature decreases with depth), and the ray picture has the form shown in Fig. 183b. It has exactly the same form as in the case of radiowave propagation (Fig. 175a, b), except that it is inverted.

The region of geometrical shadow in Fig. 183b is shaded.

However, under practical conditions, the ideal geometrical shadow may not occur. Indeed, let us suppose that as we approach the boundary (in the case of Fig. 183, the boundary is the surface of the water), the gradient of the acoustic velocity becomes zero, as may occur in the sea, for example, because of the mixing of the upper layers of

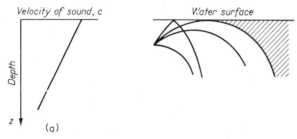

Fig. 183. Sound propagation in the sea in the presence of sound shadow.

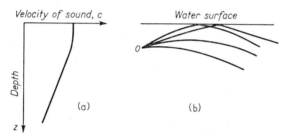

Fig. 184. Sound propagation in the presence of a homogeneous layer adjoining the water surface.

water. In this case (Fig. 184), rays striking the upper "homogeneous" layer at small grazing angles can in theory propagate to arbitrarily great distances within the layer. It would then not be possible to find any point in the entire halfspace which could not be connected by a ray with the source O. However, the acoustic intensity at these points will be very small at great distances from the source because of the very strong expansion of the ray tubes. In this case, the zone of weakened wave intensity can be called the "effective" shadow zone.

We shall study the theory of the geometric and effective shadows below. The ray methods, which were very useful in the description of the qualitative aspects of the phenomena, cannot be used for quantitative calculations, and we shall have to develop an exact wave theory. It will be convenient to consider the case in which the wave velocity is constant in some layer of thickness h adjacent to the boundary; outside the layer it varies in such a way that the square of the index of refraction is a linear function of z:

$$0 \leqslant z \leqslant h \quad n = 1, \qquad\qquad c = c_0,$$

$$z \geqslant h \qquad n^2 = 1 + 2a(z - h), \quad c = c_0/\sqrt{[1 + 2a(z - h)]}, \qquad (41.1)$$

where a characterizes the gradient of the propagation velocity, and has the dimensions of inverse length. In what follows, we call z the "depth", even though in the case of radiowave propagation, for example, it will be the height. Since we shall be interested in cases for which $a(z - h) \ll 1$, the law (41.1) can also be written in the form

$$0 \leqslant z \leqslant h \quad c = c_0,$$

$$z \geqslant h \qquad c = c_0[1 - a(z - h)]. \qquad (41.1a)$$

The velocity c as a function of depth is shown graphically in Fig. 185a.

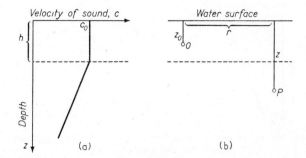

Fig. 185. On the sound field in the presence of a homogeneous layer.

It is interesting to note that, if the source is situated within the homogeneous layer, the rays which propagate along this layer in a homogeneous medium must, in the ray form of the theory, decay in the same way as in an infinite homogeneous space. However, in reality, the ray is gradually diffused, and the inhomogeneous medium lying below the layer begins to play an important rôle, drawing wave energy out of the homogeneous layer as a result of "negative refraction". This drawing off of energy leads to additional attenuation of the waves propagating in the homogeneous layer. This process has a specifically wave nature, and will also be examined below.

The case we are now considering is a special case of the bilinear profile considered by Furry and others.[14]

In the case $h = 0$, the theory developed below gives the field of radiowaves in the region of geometric shadow (beyond the horizon) under the conditions of the standard atmosphere, and also gives the acoustic field, investigated by Pekeris,[176] in the region of geometrical shadow, under water.

Our presentation will be based principally on Refs. 38 and 39.

2. The solution in a general form

We assume, at first, that the source is situated in the homogeneous layer at a depth z_0 (Fig. 185b). We must determine the value of the acoustic potential or the appropriate component of the Hertz vector at an arbitrary point P with the coordinates r and z. As everywhere above, the source power will be assumed normalized such that the potential near the source will be given by e^{ikR}/R.

The problem, formulated in this way, can be solved by the general methods developed in §§ 27 and 34. In particular, we can use Eqs. 34.1 and 34.2. Since we are now assuming that the z-axis is directed downward, V_1 in these formulas will be the reflection coefficient of plane waves at the upper boundary of the layer ($z = 0$).

We set $V_1 = -1$, which will be satisfied unconditionally if we are considering the propagation of sound waves in water and the upper boundary of the layer is the free surface of the water, and will be satisfied approximately for radiowave propagation, in the centimeter and meter range, over the surface of the earth.[14]

To simplify the notation, we let $V_2 = V(\vartheta)$ be the coefficient of reflection from the inhomogeneous halfspace $z > h$, and let $f_2(z, \vartheta) = f(z, \vartheta)$ be the acoustic potential in the inhomogeneous halfspace when a plane wave of unit amplitude is incident on its boundary from the homogeneous layer. We then obtain the following expressions for the acoustic potential in the various regions of space:

$0 \leqslant z \leqslant z_0$:

$$\psi = -ik_0 \int_{\Gamma_1} \frac{[V e^{b(h-z_0)} + e^{-b(h-z_0)}]\sinh bz}{(e^{-bh} + V e^{bh})} H_0^{(1)}(k_0 r \sin \vartheta) \sin \vartheta \, d\vartheta,$$

$z_0 \leqslant z \leqslant h$:

$$\text{(41.2)}$$

$$\psi = -ik_0 \int_{\Gamma_1} \frac{[V e^{b(h-z)} + e^{-b(h-z)}]\sinh bz_0}{e^{-bh} + V e^{bh}} H_0^{(1)}(k_0 r \sin \vartheta) \sin \vartheta \, d\vartheta,$$

$z \geqslant h$:

$$\text{(41.3)}$$

$$\psi = -ik_0 \int_{\Gamma_1} \frac{\sinh bz_0 f(z, \vartheta)}{e^{-bh} + V e^{bh}} H_0^{(1)}(k_0 r \sin \vartheta) \sin \vartheta \, d\vartheta. \qquad \text{(41.4)}$$

The symbol Γ_1 denotes the path of integration from $-(\pi/2)+i\infty$ to $(\pi/2)-i\infty$ in the complex plane of ϑ (see Fig. 122).

The functions $V(\vartheta)$ and $f(z,\vartheta)$ were obtained in §15. In particular, according to Eq. 15.19, we have

$$V = \frac{iH_{\frac{1}{3}}^{(1)}(w_0) - H_{-\frac{1}{3}}^{(1)}(w_0)}{iH_{\frac{1}{3}}^{(1)}(w_0) + H_{-\frac{1}{3}}^{(1)}(w_0)} = -e^{i\eta},$$

where

$$\eta = -2\arctan\frac{H_{\frac{1}{3}}^{(1)}(w_0)}{H_{-\frac{1}{3}}^{(1)}(w_0)}, \tag{41.5}$$

with

$$w_0 = \frac{k_0}{3a}\cos^3\vartheta. \tag{41.6}$$

Furthermore, Eqs. 15.16, 15.17 and 41.5 give

$$f(z,\vartheta) = 2\left(\frac{w}{w_0}\right)^{\frac{1}{3}}\frac{H_{\frac{1}{3}}^{(1)}(w)}{H_{\frac{1}{3}}^{(1)}(w_0) - iH_{-\frac{1}{3}}^{(1)}(w_0)}, \tag{41.7}$$

where

$$w = (k_0/3a)[\cos^2\vartheta + 2a(z-h)]^{\frac{3}{2}}, \tag{41.8}$$

so that

$$w_0 \equiv w_{z=h}.$$

We now show that Eqs. 41.2–41.4 for the acoustic potential can be transformed into the sum of the residues at the poles of the integrands, i.e. the field of the source can be represented as a sum of normal modes. As an example, we consider Eq. 41.4.

The integral over the path Γ_1 can be represented as a sum of integrals over the paths Γ_2 and Γ_3 (Fig. 122). Just as in §27, Section 3, it can be shown that the integral over the path Γ_2 is identically zero. In fact, in terms of the variable $\alpha = (\pi/2)-\vartheta$, this integral will extend over the imaginary axis from $i\infty$ to $-i\infty$, and will have the form

$$ik_0\int_{i\infty}^{-i\infty}\frac{\sinh bz_0 f(z,\vartheta)}{e^{-bh} + Ve^{bh}}H_0^{(1)}(k_0 r\cos\alpha)\cos\alpha\, d\alpha, \tag{41.9}$$

where

$$b = ik_0\sin\alpha.$$

We now prove that the integrand is an odd function with respect to α. From Eq. 41.6, we have $w_0 = (k_0/3a)\sin^3\alpha$. Consequently, when α is changed to $-\alpha$, the argument w_0 in the Hankel functions in Eqs. 41.5 and 41.7 is changed to $w_0 e^{3\pi i}$.

We use the recurrence relation (Ref. 3, p. 90),

$$H_\nu^{(1)}(u\,e^{m\pi i}) = \frac{\sin(1-m)\nu\pi}{\sin\nu\pi}H_\nu^{(1)}(u) - e^{-\nu\pi i}\frac{\sin m\nu\pi}{\sin\nu\pi}H_\nu^{(2)}(u). \tag{41.10}$$

ed in §34, Eqs. 41.2–41.4 give

h:

$$\psi = 4\pi k_0 \sum_l \left[\frac{\sinh bz \sinh bz_0}{(\partial/\partial\vartheta)\,(1 + V\,e^{2bh})}\right]_{\vartheta_l} H_0^{(1)}(k_0 r \sin \vartheta_l) \sin \vartheta_l, \qquad (41.14)$$

$$= 2\pi k_0 \sum_l \left[\frac{e^{bz} \sinh bz_0}{(\partial/\partial\vartheta)\,(1 + V\,e^{2bh})}\right]_{\vartheta_l} f(z, \vartheta_l)\, H_0^{(1)}(k_0 r \sin \vartheta_l) \sin \vartheta_l. \qquad (41.15)$$

ook into account only the poles which lie in the region $< \vartheta' < \pi/2$, $\vartheta'' > 0$, involved in the deformation of the path of ation.†

membering our notation $V = -e^{i\eta}$, and using Eq. 41.13, we obtain

$$\partial\vartheta)\,[1 + V\exp(2bh)]\}_{\vartheta_l} = \{(\partial/\partial\vartheta)\,[1 - \exp(i\eta + 2bh)]\}_{\vartheta_l}$$
$$= -[\exp(i\eta + 2bh)\,(i\eta' - 2ik_0 h \sin \vartheta)]_{\vartheta_l} = -i(\eta' - 2k_0 h \sin \vartheta)_{\vartheta_l}.$$

Eqs. 41.14 and 41.15 can be written in the form

$$\psi = \sum u_l(z_0)\, u_l(z)\, H_0^{(1)}(k_0 r \sin \vartheta_l), \qquad (41.16)$$

$$u_l(z) = \exp\,\left[\tfrac{1}{4}(i\pi)\right] \sqrt{\left[\frac{4\pi k_0 \sin \vartheta_l}{\eta'(\vartheta_l) - 2k_0 h \sin \vartheta_l}\right]} \cdot \sinh b_l z$$
$$\text{for}\quad z \leqslant h,$$

$$u_l(z) = \exp\,\left[i(\tfrac{1}{4}\pi)\right] \sqrt{\left\{\frac{\pi k_0 \sin \vartheta_l}{[\eta'(\vartheta_l) - 2k_0 h \sin \vartheta_l]}\right\}} \cdot e^{b_l h} \cdot f(z, \vartheta_l)$$
$$\text{for}\quad z \geqslant h. \qquad (41.17)$$

far, we have examined the case in which the source is situated in the homogeneous layer $(0 \leqslant z_0 \leqslant h)$. In this case, the first of 41.17 is valid for $u_l(z_0)$. However, it can be proved that Eq. 41.16 he field is valid for any positions of the source and receiver, and ermore, Eqs. 41.17 with z replaced by z_0 are valid for the function).‡

Actually, the poles lie in the more restricted region $0 < \vartheta' < \pi/2$, $\vartheta'' > 0$. is also useful to note that the points given by the solutions of the equation $w_0) - iH_{-\frac{1}{4}}^{(1)}(w_0) = 0$, at which the function $f(z, \vartheta)$ becomes infinite, are not of the integrand in Eq. (41.4), because V also becomes infinite at these s, and the integrand as a whole remains finite.

The proof is carried out without difficulty by using Eqs. 36.1 and 36.2. ever, to save space, we shall not go into the details.

Setting $u = w_0$ and $m = 3$, we obtain

$$H_{\frac{1}{3}}^{(1)}(w_0 \, e^{3\pi i}) = -H_{\frac{1}{3}}^{(1)}(w_0), \quad H_{-\frac{1}{3}}^{(1)}(u$$

Now, using Eqs. 41.5 and 41.7 for V and $f($
w does not change when the sign of α is chan
that the integrand in Eq. 41.9 is odd, and c
zero.

We will now pull the path Γ_3 toward pos
shown in Fig. 122. It can be shown that th
vanishes, and as a result, we are left with t
of the cuts extending from the possible bran
at the poles. Let us analyze the possible bra

It is clear from Eq. 41.7 that the fun
quantity w, which, according to Eq. 41.8, is
of ϑ. As a result, it may seem that all th
singlevalued if we do not forbid circuits arou
by the relations $w = 0$ or $\cos \vartheta = \pm i \sqrt{[2a(z-}$
this is not so, and the function $f(z, \vartheta)$ as wel
the integrand in Eq. 41.4 returns to its ori
around these branch points. First of all, it is e
and $H_{-\frac{1}{3}}^{(1)}(w_0)$ in Eq. 41.7 do not change under
prove that the quantity $w^{\frac{1}{3}} H_{\frac{1}{3}}^{(1)}(w)$ also do
purpose, we use the notation

$$w^{\frac{1}{3}} = \left(\frac{k_0}{3a}\right)^{\frac{1}{3}} [\cos^2 \vartheta + 2a(z-h)$$

After one circuit around the branch point, the
Therefore, after the circuit, instead of

$$w^{\frac{1}{3}} H_{\frac{1}{3}}^{(1)}(w) = p H_{\frac{1}{3}}^{(1)}(p^3).$$

we have $-p H_{\frac{1}{3}}^{(1)}(p^3 e^{3\pi i})$, or, according to Eq.
i.e. the same value as before the circuit. T
essentially not a branch point.

It can be shown that the same will hold for t

As a result, the integrands in Eqs. 41.2–41.4
and reduce to the sum of the residues at the po
given by the equation

$$1 + V e^{2bh} = 0.$$

We denote the roots of this equation by $\vartheta =$
using the rules for the determination of residu

18

3. Analysis of the poles

We will carry out an approximate analysis of the equation for the poles (41.13). Using Eq. 41.5, and remembering our notation $b = ik_0 \cos \vartheta$, this equation can also be written in the form

$$\eta + 2k_0 h \cos \vartheta = 2\pi l, \quad l = 0, \pm 1, \pm 2, \ldots \tag{41.18}$$

$$\eta = -2 \arctan W, \quad W = \frac{H_{\frac{1}{3}}^{(1)}(w_0)}{H_{-\frac{1}{3}}^{(1)}(w_0)}. \tag{41.19}$$

We shall find it necessary, in what follows, to use asymptotic expansions of the Hankel functions for large values of the argument, and series expansions in the neighborhood of zero. It will therefore be expedient to write out these expansions initially.

We first consider the expansion for small values of the argument. We use the relation between the Hankel and Bessel functions

$$H_{\nu}^{(1)}(w) = \frac{e^{-i\nu\pi} J_{\nu}(w) - J_{-\nu}(w)}{-i \sin \nu\pi}, \tag{41.20}$$

and will also use the series expansion of the Bessel function

$$J_{\nu}(w) = \left(\frac{w}{2}\right)^{\nu} \sum_{m=0}^{\infty} \frac{(-1)^m}{m! \, \Gamma(m+\nu+1)} \left(\frac{w}{2}\right)^{2m}. \tag{41.21}$$

Then, using Eqs. 41.5 and 41.7, we obtain, with no special difficulty,

$$f(z, \vartheta) = \frac{2 \exp\left[-\frac{1}{6}(i\pi)\right] \gamma w_0^{\frac{1}{3}}}{1 + \exp\left[-\frac{1}{6}(i\pi)\right] \gamma w_0^{\frac{1}{3}}} \{1 - (3/2\gamma) \exp\left[-\frac{1}{3}(i\pi)\right] w^{\frac{2}{3}} + O(w^{\frac{4}{3}})\},$$

$$W = \exp\left[-\frac{1}{3}(2\pi i)\right] \gamma w_0^{\frac{1}{3}} \{1 - (3/2\gamma) \exp\left[-\frac{1}{3}(\pi i)\right] w_0^{\frac{2}{3}} + O(w_0^{\frac{4}{3}})\},$$

$$\eta = -2\gamma \exp\left[-\frac{1}{3}(2\pi i)\right] w_0^{\frac{1}{3}} - 2\gamma_1 w_0,$$

$$\gamma = \frac{1}{2^{\frac{1}{3}}} \frac{\Gamma(\frac{1}{3})}{\Gamma(\frac{2}{3})} = 1.5703, \quad \gamma_1 = \frac{3}{2} - \frac{1}{3}\gamma^3 = 0.2093. \tag{41.22}$$

The asymptotic expansions of the Hankel functions for large values of the argument have the form

$$H_{\nu}^{(1)}(u) = \sqrt{\left(\frac{2}{\pi u}\right)} \exp\left\{i[u - \tfrac{1}{2}(\nu\pi) - \tfrac{1}{4}\pi]\right\} \left(1 + \frac{4\nu^2 - 1}{-8iu} + \ldots\right),$$

$$-\pi < \arg u < 2\pi, \quad (41.23)$$

$$H_{\nu}^{(2)}(u) = \sqrt{\left(\frac{2}{\pi u}\right)} \exp\left\{-i[u - \tfrac{1}{2}(\nu\pi) - \tfrac{1}{4}\pi]\right\} \left(1 + \frac{4\nu^2 - 1}{8iu} + \ldots\right),$$

$$-2\pi < \arg u < \pi. \quad (41.24)$$

These expressions are valid only for certain values of the argument. It is therefore necessary to ascertain the limits within which $\arg w$ and $\arg w_0$ vary in the present case.

The quantity w is related to ϑ by Eq. 41.8.

As will become clear below, we are interested in the values of $\vartheta = \vartheta' + i\vartheta''$ which lie in the region $0 < \vartheta' < \pi/2$, $\vartheta'' > 0$. Then, as is easily verified, $-\pi/2 < \arg \cos \vartheta < 0$, that is, $\cos \vartheta$ lies in the fourth quadrant. The quantity $\cos^2 \vartheta + 2a(z-h)$ lies in the lower halfplane, whence

$$-\tfrac{1}{2}(3\pi) \leqslant \arg w \leqslant 0. \tag{41.25}$$

Using Eq. 41.23, we can obtain an asymptotic expansion of $H_{\frac{1}{3}}^{(1)}$, not over this entire sector, but only over the region within the limits $-\pi < \arg w \leqslant 0$. To obtain the expansion in the remaining part of the sector, in which $-3\pi/2 < \arg w \leqslant -\pi$, we use the recurrence relation (41.10).

Setting $\qquad m = -1, \quad u = w e^{i\pi} = \tilde{w}, \quad \nu = \tfrac{1}{3},$

in the recurrence relation, we obtain

$$H_{\frac{1}{3}}^{(1)}(w) = H_{\frac{1}{3}}^{(1)}(\tilde{w}) + \exp\left[-\tfrac{1}{3}(i\pi)\right] H_{\frac{1}{3}}^{(2)}(\tilde{w}). \tag{41.26}$$

Assuming that $-\pi < \arg \tilde{w} < \pi$, and using the usual asymptotic representations (41.23) and (41.24) for the functions on the right hand side, we obtain an expansion for the function $H_{\frac{1}{3}}^{(1)}(w)$ in the sector $-2\pi < \arg w < 0$, which includes the sector of interest to us.

As a result, using Eq. 41.23, we obtain

$$H_{\frac{1}{3}}^{(1)} \approx \sqrt{\left(\frac{2}{\pi w}\right)} \exp\left[i(w - \tfrac{5}{12}\pi)\right], \quad -\pi \leqslant \arg w \leqslant 0, \tag{41.27}$$

and using Eq. 41.26,

$$H_{\frac{1}{3}}^{(1)} \approx \sqrt{\left(\frac{2}{\pi w}\right)} \exp\left[-i(w + \tfrac{11}{12}\pi)\right] + \sqrt{\left(\frac{2}{\pi w}\right)} \exp\left[i(w - \tfrac{5}{12}\pi)\right],$$
$$-2\pi < \arg w < 0. \tag{41.28}$$

We can neglect the first term in Eq. 41.28 in the sector $-\pi < \arg w < 0$ (since $|w| \gg 1$), and Eq. 41.28 is transformed into Eq. 41.27.

Similarly, we obtain

$$H_{-\frac{1}{3}}^{(1)}(w) = \sqrt{\left(\frac{2}{\pi w}\right)} \exp\left[-i(w + \tfrac{5}{12}\pi)\right] + \sqrt{\left(\frac{2}{\pi w}\right)} \exp\left[i(w + \tfrac{1}{12}\pi)\right],$$
$$-2\pi < \arg w < 0. \tag{41.29}$$

We can replace w by w_0 in the above formulas. Using these formulas, Eqs. 41.7 and 41.19 give

$$f(z, \vartheta) \approx \left(\frac{w_0}{w}\right)^{\frac{1}{6}} \{\exp(iw) + \exp[-iw - \tfrac{1}{2}(i\pi)]\} \exp(-iw_0),$$
$$-2\pi < \arg w < 0, \quad (41.30)$$

$$\eta = -2w_0 + \tfrac{1}{2}\pi. \quad (41.31)$$

It should be noted that the last expressions are invalid if $\arg w_0$ is equal to zero (which for real k_0 corresponds to a real angle of incidence) or is small. In this case, we must use the asymptotic expansion (41.27) which is valid for $\arg w = 0$, and the analogous expansion for $H_{-\frac{1}{3}}^{(1)}$. It is then necessary to take into account the next term (not written out in Eq. 41.27) of the expansion.† As a result, we obtain

$$V = \frac{1}{12w_0} \exp(-\tfrac{1}{2}i\pi),$$

$$\eta = i \ln(12w_0) + \tfrac{1}{2}\pi. \quad (41.32)$$

We note, finally, that when w_0 is small but w is large, it is possible that at small grazing angles and at large distances from the homogeneous layer [$a(z-h)$ sufficiently large] the Hankel functions of the argument w_0 will have to be expanded in a series in the neighborhood of zero, and the functions of w will have to be represented in an asymptotic form. As a result, we obtain

$$f(z, \vartheta) = 2^{-\frac{1}{6}} \sqrt{\left(\frac{3}{\pi}\right)} \Gamma\left(\frac{1}{3}\right) \frac{(w_0^2/w)^{\frac{1}{6}}}{\{1 + \gamma w_0^{\frac{1}{3}} \exp[-\tfrac{1}{6}(i\pi)]\}} \exp[i(w - \tfrac{5}{12}\pi)]. \quad (41.33)$$

The function W, which depends only on w_0, will, as previously, be given by Eq. 41.22.

We now go on to the analysis of Eq. 41.18 for the determination of the poles, and use the approximate values of the function η, found above. We start by investigating this equation at large values of w_0. Substituting Eq. 41.31 into Eq. 41.18, the equation for the poles can be written as the cubic equation

$$x^3 - 3sx + (l - \tfrac{1}{4})\pi = 0, \quad (41.34)$$

where we have used the notation

$$w_0^{\frac{1}{3}} = x = \left(\frac{k_0}{3a}\right)^{\frac{1}{6}} \cos\vartheta, \quad s = \tfrac{1}{3}k_0 h\left(\frac{3a}{k_0}\right)^{\frac{1}{6}}. \quad (41.35)$$

---

† If we do not take this term into account, we obtain $V = 0$.

As we have already pointed out, we are interested in the solutions of the equation x, which lie in the fourth quadrant $(-\pi/2 \leqslant \arg \cos \vartheta \leqslant 0)$.†
An elementary analysis of the roots of Eq. 41.34 indicates that we must then take $l > 0$. Moreover, the smallest value of l must be chosen such that the pole corresponding to this value of l does not lie too close to the real axis, because Eq. 41.31 is invalid for real w_0.

To find the poles situated near the real axis, we must use the approximate expression (41.32). Substituting Eq. 41.32 into Eq. 41.18, we obtain

$$\tfrac{1}{2}(3i)\ln x + 3sx + i\,1.2425 - (l - \tfrac{1}{4})\,\pi = 0. \tag{41.36}$$

If $s \geqslant 1$, the last equation can be solved by the method of successive approximations, remembering that according to our assumptions $|x| \gg 1$, and neglecting the first term in the first approximation. Then, in the second approximation, we obtain

$$x = \left[\frac{(l - \tfrac{1}{4})\,\pi}{3s} - \frac{1}{2s}\arctan\frac{\ln 12}{2(l - \tfrac{1}{4})\,\pi}\right]$$
$$- i\left\{\frac{\ln 12}{6s'} + \frac{1}{4s}\ln\left[\left(l - \frac{1}{4}\right)^2\left(\frac{\pi}{3s}\right)^2 + \left(\frac{\ln 12}{6s}\right)^2\right]\right\}. \tag{41.37}$$

Let us now go on to the case of small w_0. Substituting Eq. 41.22 into Eq. 41.18, we obtain

$$\gamma_1 x^3 - \{3s + \tfrac{1}{2}[1 + i\sqrt(3)]\gamma\}x + \pi l = 0. \tag{41.38}$$

Since x is assumed to be small in the present case, this equation can be solved by successive approximations, neglecting the first term in the first approximation. Using the notation

$$p = (3s + \tfrac{1}{2}\gamma) + i\frac{\sqrt{3}}{2}\gamma, \tag{41.39}$$

the first approximation gives $x = \pi l/p$, and the second approximation gives

$$x = \frac{\pi l}{p} + \gamma_1\frac{(\pi l)^3}{p^4}. \tag{41.40}$$

In the above equations, l must be taken as positive, because only for positive l will the poles lie in the fourth quadrant.

---

† Formally, we would have to take into account the poles lying in the broader region $-\pi/2 < \vartheta' < \pi/2$, $\vartheta'' > 0$, which is involved in the deformation of the path of integration in the plane of $\vartheta = \vartheta' + i\vartheta''$. However, there are no poles in the region $-\pi/2 < \vartheta' < 0$. This follows from our approximate equations. Moreover, we can evidently prove this in the more general case, because poles lying in this region would correspond to normal waves running from infinity to the source, which would have no physical meaning.

As we see, the parameter s, which involves the layer thickness h, plays a fundamental rôle in our theory. When $h = 0$, we have $s = 0$. This case has already been investigated by Pekeris.[176] Our equations permit an exact solution (Table 13).

But if we use the approximate formulas, we must start from Eq. 41.34, which was obtained by using asymptotic expansions of the Hankel functions, because here, $|x| > 1$. Setting $s = 0$ in Eq. 41.34, the roots lying in the fourth quadrant are given by

$$x_l = \exp\left[-\tfrac{1}{3}(i\pi)\right]\left[(l - \tfrac{1}{4})\,\pi\right]^{\frac{2}{3}}.$$

The complex values of x for various values of s are summarized in Table 13. The real and imaginary parts of x as functions of the parameter s, for various values of l, are shown in Figs. 186 and 187.

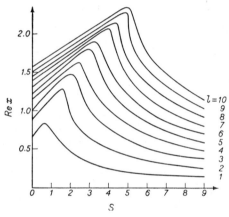

Fig. 186. The real part of x_l as a function of s for various l.

4. Analysis of the expressions for the field

After finding the values of the roots x_l, we determine the quantities ϑ_l through Eq. 41.35. The acoustic potential is then given as a set of normal modes by Eqs. 41.16 and 41.17.

The damping coefficients of the various normal modes will be

$$\beta_l = k_0 \operatorname{Im} \sin \vartheta_l = k_0 \operatorname{Im} \sqrt{\left[1 - \left(\frac{3a}{k_0}\right)^{\frac{2}{3}} x_l^2\right]}.$$

In all practically important cases, the quantity $(3a/k_0)^{\frac{1}{3}} x_l$ is much less than unity. The damping coefficients can therefore be written in the form

$$\beta_l = k_0 \operatorname{Im}\left[1 - \frac{1}{2}\left(\frac{3a}{k_0}\right)^{\frac{2}{3}} x_l^2\right] = (9a^2 k_0)^{\frac{1}{3}} x_l' x_l'',$$

TABLE 13

l	$s = 0$		$s = 0.1$		$s = 0.2$		$s = 0.4$		$s = 0.6$		$s = 0.8$		$s = 1$		$s = 1.5$	
	x'_l	x''_l	x'_l	x''_l	x'_l	x''_l	x'_l	x''_l	x'_l	x''_l	x'_l	x''_l	x'_l	x''_l	x'_l	x''_l
1	0.6653	−1.1525	0.703	−1.0874	0.736	−1.024	0.813	−0.8975	0.8845	−0.740	0.8322	−0.4013	0.7425	−0.2911	0.5608	−0.1499
2	0.8825	−1.5285	0.9107	−1.4794	0.937	−1.430	0.9964	−1.3335	1.050	−1.229	1.1053	−1.1248	1.1589	−1.0144	1.2879	−0.6898
3	1.0259	−1.777	1.051	−1.733	1.072	−1.689	1.123	−1.607	1.168	−1.519	1.2183	−1.4344	1.2661	−1.3449	1.3808	−1.1049
4	1.1377	−1.9706	1.156	−1.924	1.176	−1.885	1.225	−1.659	1.268	−1.739	1.3125	−1.6637	1.3552	−1.5840	1.4606	−1.3782
5	1.231	−2.1322	1.264	−2.177	1.277	−2.046	1.312	−1.817	1.356	−1.923	1.3925	−1.849	1.4327	−1.7771	1.5308	−1.5905

l	$s = 2$		$s = 3$		$s = 4$		$s = 5$		$s = 6$		$s = 7$		$s = 8$		$s = 9$	
	x'_l	x''_l	x'_l	x''_l	x'_l	x''_l	x'_l	x''_l	x'_l	x''_l	x'_l	x''_l	x'_l	x''_l	x'_l	x''_l
1	0.4461	−0.0912	0.3156	−0.0443	0.2431	−0.0260	0.1976	−0.0171	0.1664	−0.0121	0.1439	−0.0090	0.1266	−0.0070	0.1129	−0.0056
2	0.9021	−0.1805	0.6346	−0.0907	0.4876	−0.0526	0.3958	−0.0344	0.3331	−0.0243	0.2881	−0.0181	0.2532	−0.0140	0.2250	−0.0111
3	1.4913	−0.8198	0.9607	−0.1415	0.7940	−0.0804	0.5952	−0.0521	0.5004	−0.0366	0.4325	−0.0272	0.3801	−0.0210	0.3391	−0.0167
4	1.5627	−1.1520	1.2972	−0.1987	0.9868	−0.1057	0.7963	−0.0705	0.6686	−0.0492	0.5774	−0.0365	0.5073	−0.0281	0.4524	−0.0223
5	1.6266	−1.3921	1.8111	−0.9164	1.2332	−0.1312	0.9998	−0.0897	0.8376	−0.0622	0.7230	−0.0459	0.6349	−0.0353	0.5659	−0.0280

where $x_l = x'_l - ix''_l$. In particular, when $s = 0$ (the Pekeris case), we have $x'_1 = 0.6653$, $x''_1 = 1.1525$ for the least damped wave $l = 1$, whence $\beta_1 = 0.767(9a^2 k_0)^{\frac{1}{3}}$. The amplitude of each normal mode as a function

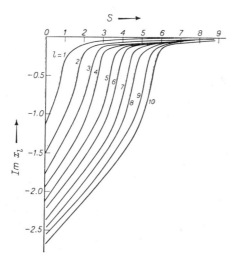

Fig. 187. The imaginary part of x_l as a function of s for various l.

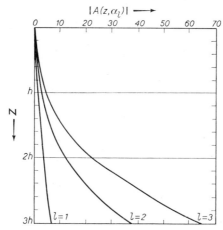

Fig. 188. The amplitudes of the first three normal modes as functions of depth.

of the depths of the source and the receiver is determined by the form of the function $u_l(z)$, given by Eq. 41.17. In the homogeneous layer, this function is proportional to $\sinh b_l z$, where $b_l = ik_0 \cos \vartheta_l = i(3ak_0^2)^{\frac{1}{3}} x_l$, and in the inhomogeneous halfspace, it is proportional to the function

$f(z, \vartheta_l)$, given by Eq. 41.7 or the approximate expressions (41.22), (41.30) and (41.33).

The amplitudes of the first three normal modes ($l = 1, 2, 3$) are shown as functions of depth in Fig. 188, with $s = 0.4$ and a conventional scale.

Every normal mode is attenuated as the horizontal distance increases, and the higher the order of the mode, the more rapid the attenuation. At sufficiently great distances, the field is determined by one or several normal modes and is readily calculated. This is the zone of "effective" shadow, in which the amplitude is considerably less than it would be if the propagation took place in a homogeneous unbounded space.

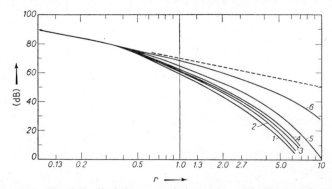

Fig. 189. The fall-off of the acoustic intensity with distance for various s: *1*, $n = 0.2$, $m = 0.2$, $s = 0.2$; *2*, $n = 1$, $m = 0.2$, $s = 0.2$; *3*, $n = 0.2$, $m = 0.4$, $s = 0.4$; *4*, $n = 1$, $m = 0.4$, $s = 0.5$; *5*, $n = 0.4$, $m = 0.2$, $s = 0.8$; *6*, $n = 0.4$, $m = 0.2$, $s = 1$. The dashed line indicates the quadratic law of fall-off.

On the other hand, at sufficiently small distances, the attenuation of the various normal modes is still small. A large number of normal modes, differing only slightly in amplitude, will then play a rôle. In this case, the summation over all the normal waves can be replaced by integration. Then using the method of steepest descents, we obtain expressions for the field which could have been obtained at the very beginning by using the methods of ray theory. It can be shown that the distances r at which such an approach is valid satisfy the condition

$$r \ll (h^2/\lambda)\, s.$$

It is clear from Eq. 41.16 for the field that if we use the asymptotic representation for the Hankel function, the r-dependence of the field of each of the normal modes is determined by the factor

$$\exp\left(ik_0\, r \sin\vartheta_l\right) = \exp\left\{ik_0\, r\,\sqrt{\left[1 - \left(\frac{3a}{k_0}\right)^{\frac{2}{3}} x_l^2\right]}\right\}$$

$$\approx \exp\left(ik_0\, r\right)\exp\left[-\tfrac{1}{2}(ir)\,(9a^2\, k_0)^{\frac{1}{3}} x_l^2\right] = \exp\left(ik_0\, r\right)\exp\left[-\tfrac{1}{2}(iqx_l^2)\, r\right].$$

The parameter $q = (9a^2 k_0)^{\frac{1}{3}}$, with the dimensions of inverse length, is the second essential parameter of the theory.

The decay law of the sound intensity with distance in the homogeneous layer is shown in Fig. 189 for the various layer thicknesses, characterized by the quantity s taking on the values 0.2, 0.4, 0.5, 0.8 and 1.0. The ordinate represents the quantity $10 \log |\psi|^2$, that is, the sound intensity in decibels. The zero decibels reference was chosen arbitrarily, which corresponds to the determination of ψ to within a constant factor.

The parameters m and n determine the depth of the source and the receiver, respectively, and are equal to $m = z_0/h$, $n = z/h$.

It is clear from Fig. 189 that the position of the receiver within the limits of the homogeneous layer has no significant effect on the fall-off law, and that the sound intensity decreases with distance more rapidly, the smaller s. At small distances, the decay law is close to $1/r^2$, shown by the dotted line in the Figure, which corresponds to the limits of applicability of the ray theory. The greater the value of s, the greater the distance to which the square law of fall-off extends.

REFERENCES

I. Monographs and Textbooks

1. AL′PERT, YA. L., GINZBURG, V. L., and FEINBERG, E. L., "Rasprostranenie radiovoln" (Radiowave Propagation) Gostekhizdat (1953).
2. AL′PERT, YA. L., Rasprostranenie élektromagnitnykh voln nizkoi chastoty (The propagation of low frequency electromagnetic waves) Izd-vo Akad. Nauk S.S.S.R. (1955).
3. WATSON, G. N., "Theory of Bessel Functions" (Russ. transl.), Part 1 IL (1949).
4. GINZBURG, V. L., "Teoriya rasprostraneniya radiovoln v ionosfere" (The Theory of Radiowave Propagation in the Ionosphere), § 14. Gostekhizdat (1949).
5. GREBENSHCHIKOV, N. V., VLASOV, A. G., NEPORENT, B. S., and SUIKOVSKAYA, N. V., "Prosvetlenie optiki" (Reflection Reduction of Optical Systems). Gostekhizdat (1956).
6. GRINBERG, G. A., Izbrannye voprosy matematicheskoi teorii elektricheskikh i magnitnykh yavlenii (Selected problems of the theory of electric and magnetic phenomena). Izd-vo Akad. Nauk S.S.S.R. (1948).
7. COURANT, R., and HILBERT, D., "Methoden der Mathematischen Physik" (Russ. transl.), Vol. 1, p. 501. GTTI (1933).
8. COURANT, R., "Geometrical Theory of Functions of a Complex Variable (Russ. transl.), ONTI (1934).
9. LANDAU, L., and LIFSHITZ, E., "Mekhanika sploshnykh sred" (Mechanics of Continuous Media), § 56. Gostekhizdat (1944).
10. LOVE, A., "A Treatise on the Mathematical Theory of Elasticity" (Russ. transl.), p. 171. ONTI (1953).
11. MORSE, P., "Vibration and Sound" (Russ. transl.). Gostekhizdat (1949).
12. PETROVSKII, I. G., "Lektsii po teorii obyknovennykh differentsial'nykh uravnenii", §§ 15, 16, 31, 32. Gostekhizdat (1947).
13. Collected papers on "Sound Propagation in the Ocean" (Russ. transl.), p. 7. IL (1951).
14. "Propagation of Short Radio Waves". Edited by D. Kerr (Russian transl. edited by B. A. Shillerov). "Sovetskoe radio" (1954).
15. RAYLEIGH, Lord, "Theory of Sound" (Russ. transl.) I, II. Gostekhizdat (1940).
16. RYZHIK, I. M., and GRADSHTEIN, I. S., "Tablitsy integralov, summ, ryadov i proizvedenii" (Tables of Integrals, Sums, Series and Products). Gostekhizdat (1956).
17. Sávarenskii, E. F., and Kirnos, D. P., "Elementy seismologii i seismometrii" (Elements of Seismology and Seismometry). Gostekhizdat (1949).
18. SMIRNOV, V. I., "Kurs vysshei matematiki dlya fizikov i tekhnikov" (Advanced Mathematics for Physicists and Engineers), Vol. III. ONTI.

19. STRATTON, J. A., "Electromagnetic Theory" (Russ. transl.), p. 453. Gostekhizdat (1948).
20. TAMM, I. E., "Osnovy teorii elektrichestva" (Fundamentals of the Theory of Electricity). Gostekhizdat (1946).
21. TITCHMARSH, E., "Introduction to the Theory of Fourier Integrals" (Russ. transl.), §§ 2, 8. Gostekhizdat (1948).
22. WHITTAKER, E. T., and WATSON, G. N., "A Course of Modern Analysis" (Russ. transl.). ONTI (1934).
23. FOK, V. A., "Tablitsy funktsii Airi" (Tables of Airy Functions). Moscow (1946).
24. FRANK, F., and VON MISES, R., "Die Differential- und Integral-gleichungen der Mechanik und Physik" (Russ. transl.). ONTI (1937).
25. FURDUEV, V. V., "Elektroakustika" (Electroacoustics) (1948).
26. KHARKEVICH, A. A., "Spektry i analiz" (Spectra and Analysis), § 9. Gostekhizdat (1952).
27. ZWIKKER, C., and KOSTEN, C. W., "Sound Absorbing Materials" (Russ. transl.). IL (1952).
28. BREMMER, H., "Terrestrial Radio Waves. Theory of Propagation". Elsevier Publ. Co. (1949).
29. BRILLOUIN, L., "Wave Propagation in Periodic Structures". McGraw-Hill (1946).
30. MORSE, P., and FESHBACH, H., "Methods of Theoretical Physics". McGraw-Hill (1953).

II. Original Works

31. ALEKSEEV, A. S., Some laws of wave propagation in an inhomogeneous medium. *Doklady Akad. Nauk S.S.S.R.* **103**, 989 (1955).
32. BREKHOVSKIKH, L. M., The reflection of spherical waves at a plane interface between two media. *J. Tech. Phys.* (U.S.S.R.) **18**, 455 (1948).
33. BREKHOVSKIKH, L. M., The reflection and refraction of spherical waves. *Uspekhi Fiz. Nauk* **38**, 1 (1949).
34. BREKHOVSKIKH, L. M., The limits of applicability of some approximate methods used in architectural acoustics. *Uspekhi Fiz. Nauk* **32**, 464 (1947).
35. BREKHOVSKIKH, L. M., The field of the refracted electromagnetic waves in the problem of a point source. *Izvest. Akad. Nauk S.S.S.R., Ser. Fiz.* **12**, 322 (1948).
36. BREKHOVSKIKH, L. M., The propagation of sound waves in a layer of liquid between two absorbing halfspaces. *Doklady Akad. Nauk S.S.S.R.* **47**, 422 (1945).
37. BREKHOVSKIKH, L. M., Concerning the dispersion equation for normal modes in layered media. *Akust. Zhur.* **2**, 341 (1956); *Soviet Phys. Acoustics* **2**, 362 (1956).
38. BREKHOVSKIKH, L. M., A type of sound propagation in an inhomogeneous medium. *Doklady Akad. Nauk S.S.S.R.* **27**, 715 (1952).
39. BREKHOVSKIKH, L. M., and IVANOV, I. D., Broadening the region of application of ray theory in the investigation of wave propagation in layered media. *Doklady Akad. Nauk S.S.S.R.* **83**, 545 (1952).
40. BREKHOVSKIKH, L. M., Reflection of plane waves from layered-inhomogeneous media. *J. Tech. Phys.* (*U.S.S.R.*) **19**, 1126 (1949).

41. BREKHOVSKIKH, L. M., The field of a point source in a layered-inhomogeneous medium. *Izvest. Akad. Nauk S.S.S.R., Ser. Fiz.* **31**, 505 (1949).

42. BREKHOVSKIKH, L. M., Sound propagation in an underwater sound channel. *Doklady Akad. Nauk S.S.S.R.* **69**, 157 (1949).

43. BREKHOVSKIKH, L. M., The reflection of spherical waves at "weak" interfaces. *J. Tech. Phys. (U.S.S.R.)* **18**, 473 (1948).

44. BREKHOVSKIKH, L. M., The reflection of wave beams and pulses. *Uspekhi Fiz. Nauk* **50**, 539 (1953).

45. VOIT, S. S., The reflection and refraction of spherical sound waves upon transition from a stationary to a moving medium. *Priklad. Matem. i Mekh.* **17**, 157 (1953).

46. GAZARYAN, YU. L., On the problem of waveguide sound propagation in inhomogeneous media. *Akust. Zhur.* **2**, 133 (1956) ; *Soviet Phys. Acoustics* **2**, 134 (1956).

47. GOGOLADZE, V. G., Rayleigh waves on a boundary between two solid elastic media. *Doklady Akad. Nauk S.S.S.R.* **33**, 16 (1941).

48. GUBANOV, A., Rayleigh waves on a boundary between a solid and a liquid. *J. Exptl. Theoret. Phys. (U.S.S.R.)* **15**, 497 (1945).

49. ZAITSEV, L. P., and ZVOLINSKII, N. V., Investigation of the head wave arising at an interface between two liquids. *Izvest. Akad. Nauk S.S.S.R., Ser. Geofiz. i Geogr.* **15**, 20 (1951).

50. ZVOLINSKII, N. V., Multiple reflections of elastic waves in a layer. *Trudy Geofiz. Inst.*, No. 22 (149).

51. ZVOLINSKII, N. V., Plane waves in an elastic halfspace and a liquid layer covering the halfspace. *Doklady Akad. Nauk S.S.S.R.* **56**, 21 (1947).

52. ZVOLINSKII, N. V., Plane surface waves in an elastic halfspace and a liquid layer covering the halfspace. *Doklady Akad. Nauk S.S.S.R.* **56**, 363 (1947).

53. KEILIS-BOROK, V. I., Surface waves in a layer lying on an elastic halfspace. *Izvest. Akad. Nauk S.S.S.R., Ser. Geofiz.*, No. 2, 17 (1951).

54. KRASNUSHKIN, P. E., The method of normal modes applied to the problem of long-range radio communications. *Izd.* MGU (Published by the Moscow State University, 1947).

55. KUPRADZE, V. D., and SOBOLEV, S. L., The propagation of elastic waves on an interface between two media with different elastic properties. *Trudy Seismol. Inst. Akad. Nauk S.S.S.R.*, No. 10 (1930).

56. LEVIN, M. L., The propagation of a plane electromagnetic wave in a periodical layered medium. *J. Tech. Phys. (U.S.S.R.)* **81**, 1399 (1948).

57. LEONTOVICH, M. A., A method for the solution of the problem of the propagation of electromagnetic waves along the surface of the earth. *Izvest. Akad. Nauk S.S.S.R., Ser. Fiz.* **8**, 16 (1944).

58. LEONTOVICH, M. A., and FOK, V. A., A solution of the problem of the propagation of electromagnetic waves along the surface of the earth, using the method of parabolic equations. *J. Exptl. Theoret. Phys. (U.S.S.R.)* **16**, 557 (1946)

59. MARKHASEV, G. S., Head waves with a plane boundary. *Priklad. Matem. i Mekh.* **19**, 165 (1955).

60. NAIMARK, M. A., Concerning the roots of the frequency equation of an elastic lying on an elastic halfspace. *Trudy Geofiz. Inst. Akad. Nauk S.S.S.R.*, No. 1 (1949).

61. NAIMARK, M. A., The oscillations of a thin elastic layer lying on an elastic halfspace, excited by a concentrated harmonic force applied to the free surface of the layer. *Trudy Seismol. Inst. Akad. Nauk S.S.S.R.*, No. 119 (1947).

62. NAIMARK, M. A., An estimate of the lower limit of the absolute value of the real roots of the frequency equation of an elastic layer lying on an elastic halfspace. *Trudy Seismol. Inst. Akad. Nauk S.S.S.R.*, No. 127 (1948).

63. PEKERIS, C. L., Theory of propagation of explosive sound in shallow water. Report of Columbia University Division of War Research, No. 6.1— sr 1131—1891 (Jan. 1945).

64. PETRASHEN', G. I., The propagation of elastic waves in layered-isotropic media, separated by parallel planes. *Uch. zap. LGU* (scientific notes, Leningrad State University) **26**, No. 162 (1952).

65. RIZNICHENKO, YU. V., The propagation of seismic waves in discrete and heterogeneous media. *Izvest. Akad. Nauk S.S.S.R., Ser. Geofiz. i Geogr.* **13**, 115–128 (1949).

66. RIZNICHENKO, YU. V., Seismic quasi-anisotropy. *Izvest. Akad. Nauk S.S.S.R., Ser. Geofiz. i Geogr.* **13**, 518–544 (1949).

67. ROZENBERG, G. V., Multiray interferometry and optical interference filters. I *Uspekhi Fiz. Nauk* **47**, 3 (1952) ; II *Uspekhi Fiz. Nauk* **47**, 173 (1952).

68. ROZENBERG, L. D., A new phenomenon in hydroacoustics. *Doklady Akad. Nauk. S.S.S.R.* **69**, 175 (1949).

69. RYTOV, S. M., The electromagnetic properties of a finely layered medium. *J. Exptl. Theoret. Phys. (U.S.S.R.)* **29**, 605 (1955) ; *Soviet Phys. JETP* **2**, 466 (1956).

70. RYTOV, S. M., The acoustical properties of a finely layered medium. *Akust. Zhur.* **2**, 71 (1956) ; *Soviet Phys. Acoustics* **2**, 67 (1956).

71. RYTOV, S. M., and YUDKEVICH, F. S., Electromagnetic wave reflection from a layer with a negative dielectric constant. *J. Exptl. Theoret. Phys. (U.S.S.R.)* **10**, 285 (1946).

72. RYAZIN, P. A., and BREKHOVSKIKH, L. M., The field of radiowaves between two semiconducting media. *Izvest. Akad. Nauk S.S.S.R., Ser. Fiz.* **10**, 285 (1946).

73. TARTAKOVSKII, B. D., The passage of sound waves across boundaries between solid and liquid media. *J. Exptl. Theoret. Phys. (U.S.S.R.)* **21**, 1194 (1951).

74. TARTAKOVSKII, B. D., On the theory of plane wave propagation across homogeneous layers. *Doklady Akad. Nauk S.S.S.R.* **71**, 465 (1950).

75. TARTAKOVSKII, B. D., Acoustical transitional layers. *Doklady Akad. Nauk S.S.S.R.* **75**, 29 (1950).

76. TARKHOV, A. G., On the anisotropy of the elastic properties of rock layers. *Mater. Vses. n.-i. geol. inst.* (Reports of the All-Union Scientific Research Geol. Inst.), Obshchaya Seriya, sb. (collection) **5**, 209–222 (1940).

77. TIKHONOV, A. N., and MUKHINA, G. V., The determination of the variable electric field in a layered medium. *Izvest. Akad. Nauk S.S.S.R., Ser. Geofiz. i Geogr.* **14**, 99 (1950).

78. FOK, V. A., The theory of radiowave propagation in an inhomogeneous atmosphere for an elevated dipole. *Izvest. Akad. Nauk S.S.S.R., Ser. Fiz.* **14**, 70 (1950).

79. SHERMAN, D. I., Wave propagation in a liquid layer lying on an elastic base. *Trudy Seismol. Inst. Akad. Nauk S.S.S.R.*, No. 115 (1945).

80. EIKHENVAL'D, On energy motion in the total internal reflection of light. *Zhur. Russ. Fiz.-Khim. Obshchestva, Fizich. chast'* (Journal of the Russian Physiochemical Society, Physical Part) **41**, 131 (1909).

81. ARENBERG, D., Ultrasonic solid delay lines. *J. Acoust. Soc. Am.* **20**, 1 (1948).

82. ATTWOOD, S. A., Surface wave propagation over a coated plane conductor. *J. Appl. Phys.* **22**, 504–509 (1951).

83. BACKHAUS, H., Das Schallfeld der kreisförmigen Kolbenmembran. *Ann. Phys.* **5**, 1 (1930).

84. BAILEY, V. A., Reflection of waves by an inhomogeneous medium. *Phys. Rev.* **96**, 865–868 (1954).

85. BERANEK, L. L., Acoustical properties of homogeneous, isotropic rigid tiles and flexible blankets. *J. Acoust. Soc. Am.* **19**, 556 (1947).

86. BERANEK, L. L., and WORK, G. A., Sound transmission multiple structures containing flexible blankets. *J. Acoust. Soc. Am.* **21**, 419 (1949).

87. BERGMANN, P. G., The wave equation in a medium with a variable index of refraction. *J. Acoust. Soc. Am.* **17**, 329 (1946).

88. BIOT, M. A., The interaction of Rayleigh and Stonely waves in the ocean bottom. *Bull. Seismol. Soc. Am.* **42**, 81 (1952).

89. BOOKER, H., and WALKINSHAW, W., The mode theory of tropospheric refraction and its relation to wave-guides and diffraction. Meteorological factors in radiowave propagation. Report of a Conference held on 8 April 1946 at the Royal Institute, London.

90. BREMMER, H., On the theory of spherically symmetric inhomogeneous wave-guides in connection with tropospheric radio propagation and underwater acoustic propagation. *Philips Research Repts.* **3**, 102–120 (April 1948).

91. BREMMER, H., The propagation of electromagnetic waves through a stratified medium and its W. K. B. approximation for oblique incidence. *Physica* **15**, 593 (1949).

92. BREMMER, H., The troposphere as a medium for the propagation of radio waves. *Philips Tech. Rev.* **15**, No. 5, 148–159 (1953).

93. BREMMER, H., The W. K. B. approximation as the first term of a geometric-optical series. *Commun. Pure Appl. Math.* **4**, 105–115 (June 1951).

94. BRICK, D. B., The excitation of surface waves by a vertical antenna. *Proc. IRE* **43**, 721–727 (1955).

95. BRILLOUIN, L., Sur une méthode de calcul approchée de certaines intégrales, dite méthode de col. *Ann. de l'école norm. supér.* **33**, 17 (1916).

96. BRUGGEMAN, D. A. G., Berechnung verschiedener physikalischer Konstanten von heterogenen Substanzen. *Ann. Phys.* **24**, 636–679 (1935).

97. BUDDEN, K. G., The propagation of radio atmospheric. I. *Phil. Mag.* **42**, 1 (1951); II. *Phil. Mag.* **43**, 1179 (1952).

98. CAFFERATA, H., The calculation of input or sending end impedance of feeders and cables terminated by complex loads. *Marconi Rev.* **6**, No. 64 (1937).

99. CONSTABLE, J. E. R., Acoustical insulation afforded by double partitions constructed from similar components. *Phil. Mag.* **18**, 321 (1934).

100. CONSTABLE, J. E. R., Acoustical insulation afforded by double partitions constructed from dissimilar components. *Phil. Mag.* **26**, 253 (1938).

101. CREMER, L., Über die Analogie zwischen Einfallswinkel und Frequenz-problemen. *Arch. electr. Übertragung* **1**, 28 (1947).

102. CROOK, A. W., The reflection and transmission of light by any system of parallel isotropic films. *J. Opt. Soc. Am.* **38**, 954 (1948).

103. DOAK, P. E., The reflexion of a spherical acoustic pulse by an absorbent infinite plane and related problems. *Proc. Roy. Soc. (London)* **A215**, 233 (1952).

104. ECKART, G., Le dipôle magnétique dans une atmosphère stratifiée. *L'Onde Électrique* **29**, 378 (1949).

105. ECKART, G., Étude des échos des ondes acoustiques dans le milieu stratifié de la troposphère. *Acustica* **2**, 256 (1952).

106. ECKART, G., and LIÉNARD, P., Analogie incomplète des impédances caractéristiques électrique et acoustique et conséquences relatives à l'écho dans les milieux stratifiés continus. *Acustica* **2**, 157 (1952).

107. ECKERSLEY, T. L., Radio transmission problems treated by phase integral methods. *Proc. Roy. Soc. (London)* **A136**, 499 (1932).

108. ELIAS, G. I., Das Verhalten elektromagnetischen Wellen bei räumlich veränderlichen electrischen Eigenschaften. *Elektr. Nach. Tecknik.* **8**, 4 (1931).

109. EPSTEIN, P., Reflection of waves in an inhomogeneous absorbing medium. *Proc. Nat. Acad. Sci. U.S.A.* **16**, 627 (1930).

110. FATOU, P., *Bull. Soc. math. France* **57**, 98 (1928).

111. FAY, R. D., and FORTIER, O. V., Transmission of sound through steel plates immersed in water. *J. Acoust. Soc. Am.* **23**, 339 (1951).

112. FILDMAN, B., Propagation in a non-homogeneous atmosphere. *Commun. Pure Appl. Math.* **4**, 317 (1951).

113. FISCHER, F. A., Über die Totalreflexion von ebenen Impulswellen. *Ann. Physik* **2**, 113 (1948).

114. FLOQUET, G., *Ann. école norm.* **12**, 47 (1883).

115. FOCKE, I., Asymptotische Entwicklungen mittels der Methode der Stationären Phase. *Berichte über die Verhande Sächsischen Akad. der Wiss. Math.-naturwiss. Klasse* **101**, No. 3 (1954).

116. FÖRSTERLING, K., Lichtfortpflanzung in inhomogenen Medien (Theorie der Lippmannschen Farbenphotographie). I. *Phys. Z* **14**, 265 (1913); II. *Phys. Z.* **15**, 225 (1914); III. *Phys. Z.* **15**, 940 (1914).

117. FÖRSTERLING, K., Über die Ausbreitung des Lichtes in inhomogenen Medien. *Ann Physik.* **11**, 1 (1931).

118. FÖRSTERLING, K., and WÜSTER, H. O., On reflection in inhomogeneous media. *Ann. Physik.* **8**, 129 (1950). In German.

119. FRAGSTEIN, C. V., Zur Seitenversetzung des totalreflektierten Lichtstrahles. *Ann. Physik.* **4**, 271 (1949).

120. FRIEDLANDER, F. G., On the total reflection of plane waves. *Quarterly J. Mech. Appl. Math.* **1**, 376 (1948).

121. GANS, R., Fortpflanzung des Lichtes durch ein inhomogenes Medium. *Ann. Physik.* **47**, 709 (1915).

122. GERJUOY, E., Total reflection of waves from a point source. *Commun. Pure Appl. Math.* **6**, 73 (Feb. 1953).

123. GERJUOY, E., Refraction of waves from a point source into a medium of a higher velocity. *Phys. Rev.* **73**, 1442 (1948).

124. GOOS, F., and HÄNCHEN, H., Ein neuer und fundamentaler Versuch zur Total-reflexion. *Ann. Physik.* **1**, 333 (1947).

125. GOOS, F., and LINDBERG-HÄNCHEN, H., Neumessung des Strahlversetzungseffektes bei Totalreflexion. *Ann. Physik.* **5**, 251 (1949).

126. GÖTZ, J. Über den Schalldurchgang durch Metallplatten in Flüssigkeiten bei schrägem Einfall einer ebenen Welle. *Akust. Z.* **8**, 145 (1943).

550 REFERENCES

127. GREENLAND, K. M., Interference filters in optics. *Endeavour* **11**, 143 (1952).
128. HADDENHORST, H. G. Durchgang von elektromagnetischen Wellen durch inhomogene Schichten. *Z. angew. Phys.* **8**, 487 (1955).
129. HADLEY, L. W., and DENNISON, D. M., Reflection and transmission interference filters. *J. Opt. Soc. Am.* **37**, 451 (1947).
130. HARTREE, D. R., The propagation of electromagnetic waves in a stratified medium. *Proc. Cambridge Phil. Soc.* **25**, 97 (1929).
131. HARTREE, D. R., Optical and equivalent paths in a stratified medium treated from a wave standpoint. *Proc. Roy.Soc.*(*London*) **A131**, 428 (1931).
132. HARTREE, D., MICHEL, J., and NICHOLSON, P., Practical methods for the solution of the equations of tropospheric refraction. Meteorological factors in radiowave propagation. Report of a Conference held on 8 April 1946, at the Royal Institute, London.
133. HASKELL, N. A., Asymptotic approximation for the normal modes in sound channel wave propagation. *J. Appl. Phys.* **22**, 157 (1951).
134. HATKIN, L., Analysis of propagating modes in dielectric sheets. *Proc. IRE* **42**, 1565 (October, 1954).
135. HEELAN, P. A., On the theory of head waves. *Geophysics* **18**, 871 (1953).
136. HEELAN, P. A., Radiation from a cylindrical source of finite length. *Geophysics* **18**, 685 (1953).
137. HELLER, G. S., Reflection of acoustic waves from an inhomogeneous fluid medium. *J. Acoust. Soc. Am.* **25**, 1104 (1953).
138. HINES, C. O., Reflection of waves from varying media. *Quart. Appl. Math.* **11**, 9–31 (1953).
139. HUFFORD, G. A., A note on the wave propagation through an inhomogeneous medium. *J. Appl. Phys.* **24**, 268 (1953).
140. HURST, D., The transmission of sound by a series of equidistant partitions. *Can. J. Research.* **12**, 398 (1935).
141. IAMADA, R., On the radio wave propagation in a stratified atmosphere. *J. Phys. Soc. Japan* **10**, 71 (1955).
142. INGARD, U., On the reflection of a spherical sound wave from an infinite plane. *J. Acoust. Soc. Am.* **23**, 329 (1951).
143. JEFFREYS, H., The reflection and refraction of elastic waves. *Monthly Notices Roy. Astron. Soc. Geophys. Suppl.* **1**, 321 (1926).
144. KAHAN, T., and ECKART, G., A general account of asymptotic expansions in wave propagation. *Rev. Sci., Paris* **87**, 3 (Jan.–Feb. 1949).
145. KING, P., and LOCKHART, L. B., Two-layered reflection-reducing coatings. *J. Opt. Soc. Am.* **36**, 513 (1946).
146. KOFINK, W., Reflexion elektromagnetischer Wellen an einer inhomogenen Schicht. *Ann. Phys.* **1**, 119 (1947).
147. KOPPE, H., Über Rayleigh-Wellen an der Grenzfläche zweier Medium. *Z. angew. Math. u Mech.* **28**, 355 (1948).
148. KORNHAUSER, E. T., and RANEY, W. P., Attenuation in shallow-water propagation due to an absorbing bottom. *J. Acoust. Soc. Am.* **27**, 689 (1955).
149. KRONIG, P., BLAISSE, B. S., and SANDE, J. J., Optical impedance and surface coating. *Appl. Sci. Res.* **B1**, 63 (1947).
150. LAMB, H., On waves in an elastic plate, *Proc. Roy. Soc.* (*London*) **A93**, 114 (1917).
151. LANGER, R., On the wave equation with small quantum numbers. *Phys. Rev.* **75**, 1573 (1949).

152. LAUE, M., Die Spiegelung und Brechung des Lichtes an der Grenze zweier isotroper Körper. *Handb. exp. Phys.* **18**, 149.

153. LAWHEAD, R. B., and RUDNICK, I., Acoustic wave propagation along a constant normal impedance boundary. *J. Acoust. Soc. Am.* **53**, 546 (1951).

154. LOCKHART, L. B., and KING, P., Three-layered reflection-reducing coatings. *J. Opt. Soc. Am.* **37**, 689 (1947).

155. LONDON, A., Transmission of reverberant sound through double walls. *J. Acoust. Soc. Am.* **22**, 270 (1950).

156. LONDON, A., Transmission of reverberant sound through single walls. *J. Research Nat. Bur. Standards* **42**, 605 (1949).

157. MACFARLANE, G. G., A variational method for determining eigenvalues of the wave equation applied to tropospheric refraction. *Proc. Cambridge Phil. Soc.* **43**, 11, 213 (1947).

158. MAECKER, H., Die Grenze der Totalreflexion strahlenoptischen Näherung mit der Wolterschen Strahldefinition. *Ann. Physik* **10**, 115 (1952).

159. MAKINSON, K. R., Transmission of ultrasonic waves through a thin solid plate at the critical angle for the dilational wave. *J. Acoust. Soc. Am.* **24**, 202 (1952).

160. MUCHMORE, R., Optimum band width for two layer anti-reflection films. *J. Opt. Soc. Am.* **38**, 20 (1948).

161. MULLIN, C. J., Solution of the wave equation near an extremum of the potential. *Phys. Rev.* **92**, 1323 (1953).

162. MUSKAT, M., and MERES, M. W., Reflection and transmission coefficients for plane waves in elastic media. *Geophysics* **5**, 115 (1940).

163. NIESSEN, K. F., Über die entfernten Raumwellen eines vertikalen Dipolsenders. *Ann. Physik* **18**, 893 (1933).

164. NOBLE, W. J., On the focusing effect of reflection and refraction in a velocity gradient. *J. Acoust. Soc. Am.* **27**, 888 (1955).

165. OFFICER, C. B., Normal mode propagation in three layered liquid half-space by ray theory. *Geophysics* **16**, 207 (1951).

166. OSBORNE, M. F. M., and HART, S. D., Transmission, reflection and guiding of an exponential pulse by a steel plate in water. I. Theory. *J. Acoust. Soc. Am.* **17**, 1 (1945). II. Experiment. *J. Acoust. Soc. Am.* **18**, 170 (1946).

167. OTT, H., Reflexion und Bruchung von Kugelwellen. 1. Effekte. 2. Ordnung. *Ann. Physik* **41**, 443 (1942).

168. OTT, H., Die Sattelpunktsmethode in der Umgebung eines Pols. *Ann. Physik* **43**, 393 (1943).

169. OTT, H., Die Bodenwelle eines Senders. *Z. angew. Phys.* **3**, 123 (1951).

170. OTT, H., Bemerkung zum Beweis von W. H. Wise über die Nichtexistenz der Zenneckschen Oberflächenwelle im Antennenfeld. *Z. Naturforsch.* **8a**, 100 (1953).

171. PEARCEY, T., The structure of an electromagnetic field in the neighbourhood of a cusp of a caustic. *Phil. Mag.* **37**, 311 (1946).

172. PEKERIS, C. L., and AMENT, W. C., Characteristic values of the first normal mode in the problem of propagation of microwaves through an atmosphere with a linear exponential modified index of refraction. *Phil. Mag.* **38**, 801 (1947).

173. PEKERIS, C. L., Asymptotic solutions for the normal modes in the theory of microwave propagation. *J. Appl. Phys.* **17**, 1108 (1946).

174. PEKERIS, C. L., Perturbation theory of the normal modes for an exponential M-curve in non-standard propagation of microwaves. *J. Appl. Phys.* **17**, 678 (1946).

175. PEKERIS, C. L., Wave theoretical interpretation of propagation of 10-cm. and 3-cm. waves in low-level ocean ducts. *Proc. IRE* **35**, 453 (1947).

176. PEKERIS, C. L., Theory of propagation of sound in a half-space of variable sound velocity under conditions of formation of a shadow zone. *J. Acoust. Soc. Am.* **18**, 295 (1946).

177. PEKERIS, C. L., Accuracy of the earthflattening approximation in the theory of microwave propagation. *Phys. Rev.* **70**, 518 (1946).

178. PEKERIS, C. L., Ray theory vs. normal mode theory in wave propagation problems. *Proc. Symposium Appl. Math.* **2**.

179. POLSTER, H. D., Reflection from a multilayer filter. *J. Opt. Soc. Am.* **39**, 1038 (1949).

180. PORITSKY, H., Extension of Weyl's integral for harmonic spherical waves to arbitrary wave shapes. *Commun. Pure Appl. Math.* **4**, 33 (June 1951).

181. PRESS, F., and EWING, M., Propagation of explosive sound in a liquid overlying a semi-infinite elastic solid. *Geophysics* **15**, 426 (1950).

182. PRESS, F., and EWING, M., Propagation of elastic waves in a floating ice sheet. *Trans. Am. Geophys. Union* **32**, 673 (Oct. 1951).

183. RAYLEIGH, LORD (J. W. STRUTT), On waves propagated along the plane surface of an elastic solid. *Sci. Pap.*, II, 441 (1886).

184. RAWER, K., Elektrische Wellen in einem geschichteten Medium. *Ann. Physik.* **35**, 385 (1939).

185. RUDNICK, I., The propagation of an acoustic wave along a boundary. *J. Acoust. Soc. Am.* **19**, 348 (1947).

186. RYDBECK, O. E. H., "On the propagation of radio waves". Göteborg (1944).

187. SALZBERG, B., Propagation of electromagnetic waves through a stratified medium. *J. Opt. Soc. Am.* **40**, 465 (1950).

188. SATO, I., Study of surface waves. Velocity of surface waves propagated upon elastic plates. *Bull. Earthquake Research Inst. Tokyo Univ.* **29**, 223 (June 1951). Love waves with double superficial layer. *Bull. Earthquake Research Inst. Tokyo Univ.* **29**, 435 (Sept. 1951). Equivalent single layer to double superficial layer. *Bull. Earthquake Research Inst. Tokyo Univ.* **29**, 519 (Dec. 1951). Love waves propagated upon heterogeneous medium. *Bull. Earthquake Research Inst. Tokyo Univ.* **30**, 1 (March 1952).

189. SCHELKUNOFF, S. A., Remarks concerning wave propagation in stratified media. *Commun. Pure Appl. Math.* **4**, 117 (1951).

190. SCHOCH, A., Schallreflexion, Schallbrechung und Schallbeugung. *Ergeb. exakt. Naturw.* **23**, 127 (1950).

191. SCHOCH, A., Der Schalldurchgang durch Platten. *Acustica* **2**, 1 (1952).

192. SCHOCH, A., Seitliche Versetzung eines total reflektierten Strahles bei Ultraschallwellen. *Acustica* **2**, 17 (1952).

193. SCHOLTE, J. G., The range of existence of Rayleigh and Stonely waves. *Monthly Notes Roy. Astron. Soc. Geophys. Suppl.* **5**, 120 (1947).

194. SCHUSTER, K., Anwendung der Vierpoltheorie auf der Probleme der optische Reflexionsminderung, Reflexions, Verstärkung und der Interferenzfilter. *Ann. Physik* **4**, 352 (1949).

195. SOMMERFELD, A., Über die Ausbreitung der Wellen in der drahtlosen Telegraphie. *Ann. Physik* **28**, 665 (1909) ; **81**, 1135 (1926).

REFERENCES

196. STONELY, R., The effect of the ocean on Rayleigh waves. *Monthly Notes Roy. Astron. Soc. Geophys. Suppl.* **1**, 349 (1926).

197. STONELY, R., Elastic waves at the surface of separation of two solids. *Proc. Roy. Soc.* (*London*) **A106**, 416 (1924).

198. THOMSON, W. T., Transmission of elastic waves through a stratified solid material. *J. Appl. Phys.* **21**, 89 (1950).

199. THOMSON, W. T., The equivalent circuit for the transmission of plane elastic waves through a plate at oblique incidence. *J. Appl. Phys.* **21**, 1215 (1950).

200. TIMOSHENKO, S. P., On the transverse vibrations of bars of uniform cross-section. *Phil. Mag.* **43**, 125 (1922).

201. TOLSTOY, I., Dispersion and simple harmonic point sources in wave ducts. *J. Acoust. Soc. Am.* **27**, 897 (1955).

202. TOLSTOY, I., Note on the propagation of normal modes in inhomogeneous media. *J. Acoust. Soc. Am.* **27**, 274 (1955).

203. TOLSTOY, I., and USOLIN, E., Dispersive properties of stratified elastic and liquid media. A ray theory. *Geophysics* **18**, 844 (1953).

204. ULLER, K., Elastische Oberflächen Planwellen. *Ann. Physik* **56**, 463 (1918).

205. WAIT, I. R., Radiation from a vertical electric dipole over a stratified ground. *Trans. IRE* IAP-1, No. 1, 9 (1953) ; II. AP-2, No. 4, 144 (1954).

206. WALLOT, J., Der senkrechte Durchgang elektromagnetischer Wellen durch eine Schicht räumlich veränderlicher Dielektrizitätskonstante. *Ann. Physik* **60**, 734 (1919).

207. WEIL, H., Ausbreitung elektromagnetischer Wellen über einem ebenen Leiter. *Ann. Physik* **60**, 481 (1919).

208. WEINSTEIN, W., The reflectivity and transmissivity of multiple thin coatings. *J. Opt. Soc. Am.* **37**, 546 (1947).

INDEX